INTERNATIONAL SERIES OF MONOGRAPHS IN
NATURAL PHILOSOPHY
GENERAL EDITOR: D. TER HAAR

VOLUME 55

A SURVEY OF HIDDEN-VARIABLES THEORIES

A SURVEY OF HIDDEN-VARIABLES THEORIES

BY

F. J. BELINFANTE

Department of Physics,
Purdue University, Lafayette,
Indiana, U.S.A.

PERGAMON PRESS
OXFORD · NEW YORK · TORONTO
SYDNEY · BRAUNSCHWEIG

Pergamon Press Ltd., Headington Hill Hall, Oxford
Pergamon Press Inc., Maxwell House, Fairview Park, Elmsford, New York 10523
Pergamon of Canada Ltd., 207 Queen's Quay West, Toronto 1
Pergamon Press (Aust.) Pty. Ltd., 19a Boundary Street, Rushcutters Bay, N.S.W. 2011, Australia
Vieweg & Sohn GmbH, Burgplatz 1, Braunschweig

First edition 1973

Library of Congress Cataloging in Publication Data

Belinfante, Frederik Jozef.
A survey of hidden-variable theories.

(International series of monographs on natural philosophy, v. 55)
Includes bibliographical references.
1. Mathematical physics. 2. Variables (Mathematics) 3. Quantum theory. I. Title.
QC20.B42 1973 530.1'5 72-13519
ISBN 0-08-017032-3

Printed in Hungary

CONTENTS

PREFACE

THIS work was started in 1970 with the idea of writing three papers for publication in a journal of the type of *American Journal of Physics*, explaining the ideas, merits, and failures of hidden-variables theories in a way understandable to those who have had in wave mechanics an introductory course only and who are not particularly mathematically minded. As the work outgrew the size of journal articles, it is here presented as a book in three parts.

The author's interest was triggered when in the Fall of 1969 Clauser, Horne, Shimony, and Holt published their plans of performing an experiment of which they claimed that it would decide unambiguously whether "local" hidden-variables theory or whether quantum theory is wrong, depending upon whether the measured value of a certain quantity Δ would be found to be positive or negative. The author then found that the proof of $\Delta \leq 0$ for this kind of hidden-variables theories could be much simplified. (See Section 4.7 of Part III of this book.)

While lecturing on this, he was made aware of the existence of a different kind of hidden-variables theories (called in this book "theories of the first kind") which cannot be disproved by the experiments proposed by Clauser *et al.* without simultaneously disproving quantum theory, as these theories and quantum theory would make identical predictions. On the other hand, the literature was full of claims that hidden-variables theories would not be possible at all. The kind of hidden-variables theories that were shown to be impossible was, of course, somewhat different from the "first" kind mentioned above, as well as from the "second" ("local") kind that Clauser *et al.* planned to prove or disprove experimentally. The "impossible" kind of hidden-variables theories is called here the "zeroth" kind.

While the impossibility of theories of the zeroth kind can be shown without performing new experiments, the question whether theories of the first or of the second kind agree or disagree with the behavior of nature is a *physical* question, and can be anwered only by experiments hunting for the effects in disagreement with quantum theory which these theories predict. That also theories of the first kind may predict such effects was first considered seriously by Papaliolios (1967).

While most of this book is a *review* (partially a summary, partially an elaboration) of the work of others, it contains some novel points. Some of the so-called proofs of impossibility of hidden-variables theories are based on results of Kochen and Specker and of Gleason, which here, by sacrificing some mathematical rigor, are presented in Chapter 3 of Part I in a more understandable form, with proofs even simpler than those proposed a few years ago by Bell.

In Chapter 2 of Part II, Bohm's 1952 hidden-variables theory (in which the particles' positions are the hidden variables) is applied to a minimum-uncertainty wave packet for clarifying the difference in interpretation of free-particle motion in this theory and in ordinary wave mechanics. The application of this theory of Bohm to photons starts from a reformulation of conventional quantum electrodynamics in which one avoids the use of creation and annihilation operators, and one describes the photon field by applying the principles of familiar wave mechanics to certain harmonic oscillators describing the field. This formulation of pure quantum-theoretical electrodynamics has the advantage of giving a clearer description of finite-size photon wave packets in x, y, z-space than the more usual treatment of quantum electrodynamics gives. An account of this treatment is given in Appendix H of Part II, and in Section 2.4–2.6 of Part III its connection is shown with Akhiezer–Berestetskii's treatment of photon wavefunctions.

For those acquainted with Dirac's relativistic wave mechanics, it is shown in Section 2.15 of Part II why it is not well possible to generalize Bohm's 1952 theory to a relativistic theory.

In Chapters 3 and 4 of Part II the intimate relation between the theories of the first kind constructed by Wiener and Siegel on the one hand, and by Bohm and Bub on the other hand, is shown. Therefore both theories predict the same deviations from quantum theory that could in principle be observed experimentally if these theories would be correct. Not only are the experiments of Papaliolios discussed, which failed to find evidence for the existence of these effects, but in Chapter 5 of Part II some additional experiments of a similar nature are proposed, and it is predicted what outcome such experiments should have if these theories would be any good. This outcome would depend here upon the type of stochastic chaos that would exist for the polarizations of light emitted by the light source.

The work of Gleason and of Kochen and Specker discussed in Part I is usually regarded as proving that ambiguity enters into the application of the theories of Wiener–Siegel and Bohm–Bub upon measurements of observables with degenerate eigenvalues. In Section 3.7 and Appendix S of Part II it is shown how a suggestion made in 1969 by Tutsch (see Section 3.6 of Part II) can remove these ambiguities and then lead to a formula suggested already in 1953 by Wiener and Siegel. However, in Appendix K of Part II it is shown how application of this rule of Tutsch may sometimes lead to paradoxical results.

In Sections 4.3–4.4 of Part III the experiments performed by Freedman and Clauser and by Holt are discussed. Their analysis in Sections 2.7–2.9 of Part III using photon wavefunctions is kept so simple that it will be understandable also to those for whom the fuller discussion of the meaning of these photon wavefunctions in Sections 2.4–2.6 of Part III is too mathematical. The latter discussion was inserted primarily for justifying the use of two-photon waves with quantum number $j = 1$ for Holt's experiments by qualifying the claim found in the literature that two-photon wavefunctions with $j = 1$ would not exist. This claim is found not to be valid for the experiments of Holt, although it is valid for experiments using photons from annihilation of electron pairs, as in the experiments of Bleuler and Bradt, Wu and Shaknov, Langhoff, and of Kasday, Ullman, and Wu, discussed in Chapter 5 of Part III.

In Chapter 3 of Part III explicit examples of theories of the second kind are given, which may serve as models for which the validity may be verified of the Clauser–Horne–Shimony inequality $\Delta \leqslant 0$. The derivation of $-1 \leqslant \Delta \leqslant 0$ for these theories given in Section 4.7 of Part III is more direct and therefore more easily understandable than the derivation given by Clauser *et al.* themselves, who started from an inequality proved by Bell (see Section 4.9 of Part III).

In conclusion, Chapter 6 of Part III ends with an expression of opinion about the present status of hidden-variables theory.

It is hoped that this book may acquaint many, in the first place, with the ingenious attempts made by many trying to construct a (crypto)deterministic theory "explaining" quantum theory or "correcting its shortcomings"; in the second place, with the possibility of experimental verification or falsification of such theories; and, finally, with the refusal of nature, in experiments thus far performed, to give any indication of submitting itself to the regulations which such theories would impose upon nature's otherwise in the small unpredictable behavior. (See, however, footnote 25b on page 290.)

ACKNOWLEDGEMENTS

I GRATEFULLY acknowledge conversations and correspondences with many workers in the hidden-variables field which have helped me to clarify my understanding and often to remove misunderstandings on various aspects of the theory. Among those who have been especially helpful are: J. Bub, J. F. Clauser, R. A. Holt, M. A. Horne, J. M. Jauch, L. R. Kasday, C. Papaliolios, P. Pearle, A. Shimony, A. Siegel, J. H. Tutsch, and E. P. Wigner. I also acknowledge financial support of this work by the National Science Foundation by its grants GP-9381 and GP-29786.

PART I

THREE KINDS OF HIDDEN-VARIABLES THEORIES AND SO-CALLED PROOFS OF IMPOSSIBILITY

FOREWORD TO PART I

THE primary aim of this book in three parts is an attempt to make the literature on hidden variables understandable to those who are confused by the original papers with their controversies and often excessive use of mathematical methods unknown to the average reader of physics papers. It is aimed at those who have an open mind for trying out novel ideas but are also willing to discard any results that do not stand up against criticism based upon physical rather than philosophical argument. Physicists call a theory satisfactory if (1) it agrees with the experimental facts, (2) it is logically consistent, and (3) it is simple as compared to other explanations. Our stress on these physical considerations about the theory means that (in Parts II and III) we will pay attention to the question how these theories may be tested experimentally. In fact, the author's interest in hidden-variables theories was kindled only when recently he became aware of the possibility of such experimental tests.

On the other hand, we do not want to ignore the metaphysical implications of the theory.

In this Part I we review the motives which have led different people to developing different types of hidden-variables theories. The quest for determinism led to theories of the first kind; the quest for theories that look like causal theories when applied to spatially separated systems that interacted in the past led to theories of the second kind. The latter contradict quantum theory, and experiments are underway for verifying who is right: the quantum theorist, or the hidden-variables theorist of the second kind.

Theories of the first kind are not that easily distinguishable experimentally from quantum theory. The hidden variables, according to theories of the first kind, have a distribution of values which tends irreversibly and rapidly to an equilibrium distribution, much like a velocity distribution of molecules in a gas tends to a Maxwell distribution. Quantum theory then follows when the equilibrium distribution of hidden variables is reached. It is, however, possible to willfully and predictably perturb such an equilibrium distribution. When this is done, deviations from quantum theory should occur according to theories of the first kind. Preliminary experiments trying to detect such deviations have had negative results. One might say that this seems to contradict the validity of hidden-variables theory of the first kind. However, another possible explanation is that the perturbed hidden-variable distribution relaxed to its equilibrium distribution before the experiment was able to measure the predicted deviations from quantum theory.

Various authors have tried to prove that hidden-variables theories would be impossible. These attempts seem strange in view of the fact that several hidden-variables theories do exist. We investigate in this Part I the question what causes the apparent discrepancy. It turns out that each of these authors defined hidden-variables theory by some set of postulates which then later they proved to be self-contradictory. Such "theories" defined by a self-contradictory set of postulates, we call theories of the zeroth kind. Their impossibility is obvious.

It is then interesting to study *which* properties postulated by these authors made these theories impossible. We then should be careful, in developing more realistic hidden-variables theories (of the nonzeroth kind), to avoid the postulates that caused the troubles. In particular, we shall discuss in Part II how hidden-variables theories of the first kind circumvent these postulates. Some methods for achieving this are indicated already in this Part I.

One of the major pitfalls to be avoided in constructing a hidden-variables theory lies in what the hidden variables can or cannot predict about future measurements. The work of Gleason shows that in general it is impossible to predict uniquely on the basis of any given hidden variables, whether or not the measurement of a quantity A will lead to the result $A = A_n$, where A_n is an eigenvalue of A, or to predict in case of a reproducible measurement whether or not the wavefunction ψ by the measurement will be reduced to the eigenfunction ϕ_n of A_{op}. What hidden-variables theories *do* predict is, from a given complete orthonormal set of eigenfunctions $\{\phi_i\}$ describing the possible outcomes of a reproducible measurement, to which eigenfunction ϕ_n will ψ be reduced when that measurement is made. The difference seems subtle, but lies in the fact that the choice of ϕ_n from the set $\{\phi_i\}$ may be changed into the choice of a different final state of the measurements when the orthonormal set $\{\phi_i\}$ in Hilbert space is rotated around ϕ_n. Such a rotation is understood to correspond to a change of the experimental setup of the measurement, but this change may be unperceivably small when the rotation in Hilbert space is in first approximation merely a change of a basic set of eigenfunctions for some degenerate eigenvalue of the observable measured. Moreover, this degenerate eigenvalue may be different from the eigenvalue A_n which the measurement would have found before the change was made.

As Gleason's proof of the result mentioned above is rather abstract, we derive his result from the more understandable work of Kochen and Specker, which discusses a special case of Gleason's ideas and which is very revealing. We give a much simplified proof of Kochen and Specker's result, simpler yet than a proof of Gleason's result given by Bell. Also at various other places in our survey, we clarify points by novel approaches that simplify the reasoning. Our main aim here is making the conclusions more easily understandable or easier to conceive visually. We do not aim at utmost mathematical rigor, and would prefer a plausibility argument to a lengthy rigorous proof.

CHAPTER 1

INTRODUCTION

THIS book of hidden-variables theories consists of three parts. In Part I we shall roughly outline what a hidden-variables theory is, what it might achieve, and what it cannot achieve. The latter question is related to so-called proofs of the impossibility of hidden-variables theories, and we shall look into these proofs.

In discussing what a hidden-variables theory is, we shall find that there exist different theories trying to achieve different things. Therefore we shall be led to consider separately what we shall call "theories of the first kind" and "theories of the second kind." Part II of this book will discuss in some detail the most important theories of the first kind thus far published and known to this author. Also, the question of experimental verification of such theories will be discussed.

In Part III we shall consider theories of the second kind and experiments which serve to verify or disprove theories of this kind.

1.1. Some fundamentals of quantum theory

Hidden-variables theories of the first and of the second kind have been proposed by people who were not satisfied about certain aspects of quantum theory. So let us first very briefly review some of the methods of quantum theory and mention the points that arose the displeasure of these people.

When a quantum theorist wants to make some predictions about a physical system, he will admit at once that in most cases he can make at most some probability predictions. After he has ascertained in what way the system was prepared, he will choose an initial state vector in Hilbert space,[1] which in simple cases often is called the "wavefunction" of the initial state. This ψ will describe all systems that underwent a similar preparation. These systems taken together form an ensemble (E_ψ), and ψ will describe this entire ensemble.

When now an observable[2] A is measured which is represented by a hermitian operator

[1] Or something similar to a Hilbert space, like Gupta's space with indefinite metric with auxiliary conditions imposed to prevent negative probabilities.

[2] An *observable*, in this book, is a quantity which we imagine to be measurable. In quantum theory, observables A are represented by hermitian operators $A_{op} = A_{op}^\dagger$. The opposite need not be true. One can construct hermitian operators for which it is hard to imagine that they could be measured. In some axiomatic quantum theories created by mathematicians, like the one of von Neumann (1932, 1955), the idea of an "observable"

A_{op} with a complete orthonormal set of eigenfunctions ϕ_i and eigenvalues A_i, quantum theory postulates that (but for experimental errors) each measurement on a single physical system that is a member of the ensemble E_ψ will find for A one of the eigenvalues A_i, but different members of E_ψ may yield different eigenvalues A_i. When the measurement is of the kind that is reproducible,[3] we may describe the state after the measurement by[4]

$$\psi \to \phi_i, \tag{1}$$

where the arrow indicates a change between initial and final state vector which we make personally by choice ("reduction of the wavefunction"), since for the purpose of describing future measurements on this system we want to replace the ensemble E_ψ by the new ensemble E_{ϕ_i} of systems in which the preparation now includes the finding of the measurement result ϕ_i.[5]

is generalized to "any" quantity represented by a hermitian operator in Hilbert space for which certain mathematically convenient conditions are met. Against this usage of the word "observable," Bohm has made objections [see Bohm (1953b), on page 277].

[3] Measurements are not always reproducible. For instance, a nicol may measure whether an incident light wave is polarized horizontally or vertically, and might absorb the former kind of wave. Then a measurement resulting in finding the wave polarized vertically may be repeated by nicols behind the first one or by other means. When no light comes through the first nicol, we may pronounce it to *have been* horizontally polarized, but this can no more be verified by placing a second nicol behind the first one. In that case the measurement was not reproducible and, if it found that the polarization was one described by an eigenfunction ϕ_i, the state after the measurement will *not* have become describable by eq. (1).

A different example of a nonreproducible measurement is the following. Of an atom with a given set of energy levels, we know that it is excited and want to measure to which energy level it was excited. We look for a photon to be emitted by the atom. The photon reaches a photographic plate via a grating determining its wavelength. We recognize the black spot on the plate as being created by a photon emitted as the atom jumped down from state i to state j. We thus have made a nonreproducible measurement of the state i.

However, in this case we also have made a reproducible measurement of the state j, which is the final state obtained in this emission of a photon and which may serve as an initial state for further developments preceding a next measurement.

The examples given here show that *reproducible measurements may also serve as preparations of initial states.* When in this paper we talk about measurements, we mainly refer to these reproducible measurements which prepare states. Ballentine (1970) has correctly remarked (on page 366 of his paper) that this is a poor usage of language. The name "measurement" ought to be confined to measurements of the past, which includes nonreproducible measurements. Reproducible measurements should be called "state preparations." Nevertheless, we continue in this book this poor usage of language, which is common in what is known as the "theory of *measurement* in quantum theory."

The Ballentine paper quoted gives on pages 364–367 a very clear discussion of the uncertainty principle, what it really means, and how it is often misinterpreted.

[4] In case of a measurement of a degenerate eigenvalue A_i, all we know for sure is that ϕ_i is one of the eigenfunctions belonging to A_i, but it is generally postulated that ϕ_i in eq. (1) is to be the renormalized orthogonal projection of the initial ψ onto the Hilbert subspace corresponding to A_i, and therefore it is said that the final state after a reproducible measurement is determined by the "*projection postulate.*" [More explicitly, we would construct in the subspace of Hilbert space corresponding to the eigenvalue A_i a complete orthonormal set of eigenfunctions of A_{op} such that all of these eigenfunctions except one are orthogonal to ψ. Then ϕ_i in eq. (1) is this one exceptional eigenfunction.] Compare, Section 3.6, eq. (119), of Part II.

[5] Due to misunderstandings about the meaning of ψ, there have been all kinds of quarrels about the meaning of the reduction (1) of the wavefunction in quantum theory. The fictitiousness of much of this problem is illuminated by some of the conclusions of Pearle (1967), who shows that the choice between reduction and nonreduction of ψ is largely a question of taste and convenience, as long as in case of nonreduction the formalism is properly complicated otherwise.

Even without such complications, it is clear that reduction or nonreduction is a question of choice depending upon whether ψ for future measurements is to provide probabilities of future results in the en-

While according to quantum theory we cannot predict for a given single system which eigenvalue A_i will be the result of measurement of A, quantum theory does predict that performance of the same measurement upon all members of the ensemble E_ψ will yield for the various A_i a frequency distribution described by a probability w_i for any final[6] state ϕ_i, where

$$w_i = |a_i|^2, \quad \psi = \sum_i a_i \phi_i, \quad a_i = \int \phi_i^* \psi. \tag{2}$$

The above assumed that the quantum theorist knew unambiguously how to describe the initial state by means of a unique state vector ψ. If he is in doubt, he will consider all possible state vectors $\psi^{(\alpha)}$ that might describe the system prepared as it was, and for each $\psi^{(\alpha)}$ he will calculate by (2) sets $\{a_i^{(\alpha)}\}$ and $\{w_i^{(\alpha)}\}$. He will assign proper weights p_α to all of his choices, which add up (by summing or by integration) to 1, and then average his results $w_i^{(\alpha)}$ with these weights:

$$w_i = \int_{(\alpha)} p_\alpha w_i^{(\alpha)}, \quad \int_{(\alpha)} p_\alpha = 1. \tag{3}$$

The expected average of the measuring results then will be

$$\langle A \rangle_{\text{mix}} = \int_{(\alpha)} p_\alpha \langle A \rangle_{\psi(\alpha)} = \sum_i w_i A_i = \int_{(\alpha)} p_\alpha \int \psi^{(\alpha)*} A_{\text{op}} \psi^{(\alpha)}, \tag{4}$$

which is not a prediction for a single measurement, but is a prediction for the weighted mean of the averages of results of measurements on all members of each of the ensembles $E_{\psi(\alpha)}$.

The averaging over α which takes place in these cases of "mixed states" is due to a lack of knowledge by the theorist which is *not* typically quantum mechanical in nature. A similar lack of knowledge about the initial state might exist in classical mechanics. The "classical" nature of the probabilities p_α is illustrated in eqs. (3)–(4) by simple summations of the probabilities p_α. The nature of the probabilities p_α is therefore quite different from the nature of the probabilities $w_i^{(\alpha)} = |a_i^{(\alpha)}|^2$. If the expansions of the $\psi^{(\alpha)}$ are inserted in the expression for the expected average of a second observable B, we find

$$\langle B \rangle_{\psi(\alpha)} = \sum_i \sum_j a_i^{(\alpha)*} a_j^{(\alpha)} \int \phi_i^* B_{\text{op}} \phi_j = \sum_i \sum_j B_{ij} a_i^{(\alpha)*} a_j^{(\alpha)}. \tag{5}$$

This expression features "interference terms" with $i \neq j$, which are characteristic for probability distributions that are of a quantum-theoretical origin.

Because the averaging over α, where necessary, can be performed as an afterthought, we shall in the following ignore this complication and deal with states as if they are quantum-mechanically "*pure.*" That is, we shall do as if ψ is uniquely given.

semble E_{ϕ_i}, or in the ensemble $E_{\psi(t)}$ as altered from E_ψ during the time t by the external interaction with the measuring apparatus, which will, of course, affect ψ even if we do not reduce the final $\psi(t)$ after the measurement.

Though *in principle* it would be incorrect to regard $E_{\psi(t)}$ as a superposition of all possible E_{ϕ_i} with appropriate weights w_i [since in principle $\psi(t)$ would describe interference effects between the possible states ϕ_i], *in practice* a good measurement is such that the ϕ_i become separated in a way to prevent their future interference with each other.

[6] In case of nonreproducible measurements, ϕ_i is the intermediary state preceding the irreversible change of state at the end of the measurement that makes the measurement nonreproducible, like absorption of light in the nicol in the first example of footnote 3, or the photon emission in the second example of footnote 3.

1.2. First reasons for suggesting hidden-variables theories

The first dissatisfaction about quantum theory comes from people who reason that "*if different members of* E_ψ *are found upon measurement to have different values* A_i *of an observable* A, *then these individual systems must have been in different microstates.*"

The differences between these states then must be more subtle than what can be described by ψ. One would need additional parameters, say ξ, to describe them. According to these people, *the* ψ *and* ξ *together would completely determine the state of an individual system.*[7] By "completely determine" they mean that, if on a thus completely determined physical system a measurement is made which is fully described by giving the complete orthonormal set of eigenfunctions $\{\phi_i\}$ that are the possible results of the measurement,[6] then ψ and ξ together will select from the set $\{\phi_i\}$ unambiguously the state ϕ_n which represents the result of the measurement. This may be described by

$$\psi \to \phi_n \quad \text{with} \quad n = n(\psi, \xi, \{\phi_i\}). \tag{6}$$

The function n simultaneously predicts the result for *every* measurement that could be made, although, of course, for a given individual system this prediction will come true only for the one measurement which we actually choose to make.

The ξ are often called "hidden variables" because they kept themselves hidden for the eyes of those who invented quantum theory. Otherwise the name "hidden" is a misnomer [see Pearle (1968b).] In fact, in Bohm's first theory the hidden variables are observables [Bohm (1952a, b).]

When there is a change (6) of ψ in a measurement, one may well ask whether there is a corresponding change in the additional parameters ξ:

$$\xi \to ?. \tag{7}$$

Different hidden-variables theories may give different answers to this question, if they answer it at all. Most hidden-variables theories assume that the ξ under most circumstances will fluctuate in a rather unpredictable way due to external perturbations like thermal collisions, and that, therefore, for the ξ there will almost always exist uncertainties similar to the "classical" uncertainties in ψ which we discussed in eqs. (3)–(5). In other words, most initial states, even if they are "pure" quantum mechanically, will be "mixed" states with respect to hidden variables. There will be probabilities p_β for hidden-variable values $\xi^{(\beta)}$, and an uncertainty about the outcome (6) of a measurement will be due to the uncertainty of the $\xi^{(\beta)}$ entering for ξ into the unique function $n(\psi, \xi, \{\phi_i\})$.

The uncertainties about the $\xi^{(\beta)}$ are often compared with the uncertainties about the positions and velocities of molecules entering classical kinetic theory of thermal motion. This comparison was stressed by Bohm. [See Bohm (1952b), chap. 7; Bohm (1953a); Bohm and Vigier (1954).] It was one of the reasons for Wiener and Siegel (1953, 1955) to use for the development of their theory the mathematical tool of *differential space* [see Paley and Wiener (1934)], which had been developed for a description of Brownian motion.

[7] Compare, for instance, the first eleven lines of chap. 3 of Gudder (1970).

Like in statistical mechanics for the positions and velocities of molecules, it then is assumed for the hidden variables ξ that their distribution (with probabilities p_β for values $\xi^{(\beta)}$) will rapidly and irreversibly tend toward an "equilibrium distribution" $\bar{p}_\beta \equiv p_\beta\,(t \to \infty)$ which is self-perpetuating:

$$p_\beta(t) \to \bar{p}_\beta \to \bar{p}_\beta. \tag{8}$$

According to this theory, therefore, when an observable A is measured in an ensemble E_ψ corresponding to a pure quantum mechanical state ψ, the spread in results A_n found notwithstanding the uniqueness of the function $n(\psi, \xi, \{\phi_i\})$ in (6) would in most cases be due to the existence of a probability distribution $\bar{p}_\beta$ over the various possible values $\xi^{(\beta)}$ of the hidden variables. If by $\int_{(\beta)}^{(\psi, k)}$ we denote the sum over all β for which, for given state ψ and given index value k we have $k = n(\psi, \xi^{(\beta)}, \{\phi_i\})$, then such a theory would predict a probability

$$\bar{P}_k(\psi, \{\phi_i\}) = \int_{(\beta)}^{(\psi, k)} \bar{p}_\beta \tag{9}$$

for the result ϕ_k, A_k of the measurement. (The bar indicates the equilibrium distribution of the hidden variables.) From the definition of $\int_{(\beta)}^{(\psi, k)}$ we see that the expression (9) depends upon ψ through the ψ that occurs as an argument in the function $n(\psi, \xi^{(\beta)}, \{\phi_i\})$. Through $\{\phi_i\}$ it also depends upon details of the measurement to be made.

1.3. Hidden-variables theories of the first kind

Those who have no other objections against quantum theory than that it was "incomplete" by not providing tools for describing differences between systems in which measurements would have different results, now postulate that the $\bar{p}_\beta$ and the $n(\psi, \xi, \{\phi_i\})$ should be such that

$$\bar{P}_k(\psi, \{\phi_i\}) = w_k(\psi) \equiv |a_k|^2 \equiv |\textstyle\int \phi_k^* \psi|^2. \tag{10}$$

The condition (10) is known as "the *statistical postulate*." It characterizes hidden-variables theories of the first kind. It implies that, in case of an equilibrium distribution of hidden variables, the theory will make exactly the same probability predictions as ordinary quantum theory. Since the latter depend on ψ, it was necessary to have the function n in (6) depend on ψ, so that $\bar{P}_k$ in (9) might depend on ψ.

Note that the $\bar{P}_k(\psi, \{\phi_i\})$, given by the integrals in (9), should according to (10) depend on $\{\phi_i\}$ through the one eigenfunction ϕ_k only. This, however, does *not* imply that the function $n(\psi, \xi^{(\beta)}, \{\phi_i\})$ should not depend on the ϕ_i with i different from the value k which the function n takes. In fact, we shall find in Chapter 3 that it is in general *impossible* for the function $n(\psi, \xi, \{\phi_i\})$ to keep its value k under rotations of the orthonormal set of Hilbert vectors $\{\phi_i\}$ in Hilbert space around ϕ_k.

In this "first" kind of a theory, deviations from quantum theory would occur *only* in *nonequilibrium* distributions of the hidden variables ($p_\beta \neq \bar{p}_\beta$). We will find in Part II that it is easy to perturb the hidden-variables distribution in a predictable way, so that *in principle* it should be possible to investigate experimentally the dependence of the predictions of the theory upon the hidden-variables ξ. *In practice*, however, these experiments are made difficult by the extreme speed with which a perturbed distribution p_β returns to the equilibrium distribution $\bar{p}_\beta$, so that the deviations from quantum theory would disappear before they could be detected. In particular, Papaliolios (1967) found no effects of the difference between p_β and $\bar{p}_\beta$ left after a time lapse of 0.24×10^{-13} s. [See Part II of this book.]

Thus it is not simple at all to distinguish experimentally a hidden-variables theory of the first kind from pure quantum theory. This would easily explain the great success of quantum theory, even if it were true that in principle nature would be governed by a hidden-variables theory of the first kind.

On the other hand, this makes it difficult to prove conclusively by experiments that nature is governed by pure quantum theory if that would be so, because the negative outcome obtained in such case by any attempt at finding deviations from quantum theory after disturbing the hidden-variables distribution could always be explained away by assuming that the relaxation (8) takes place faster than the experiment was performed.

Therefore for a fair decision whether one should prefer quantum theory or a hidden-variables theory of the first kind for explaining the facts of nature, one will have to invoke other principles. Since quantum theory has a simpler formalism than hidden-variables theory, pure quantum theory for reasons of simplicity should have the preference as long as the deviations from quantum theory predicted for $p_\beta \neq \bar{p}_\beta$ have not positively been demonstrated experimentally. This leaves the experimental burden upon hidden-variables theory to prove its validity by disproving simpler quantum theory. This ought to provide for hidden-variables fans of the first kind the stimulus to do faster and faster experiments in the hope of catching up with the speed of relaxation of hidden-variables distributions.

Also, at any stage in this game, the excuse that in the experiments no deviations from quantum theory were found because the relaxation $p_\beta \to \bar{p}_\beta$ took place so fast that the perturbations $(p_\beta - \bar{p}_\beta)$ disappeared before they could cause a measurable effect, will not be an acceptable excuse unless at least qualitative arguments can be given how any reasonable mechanism could *cause* such fast relaxation.

Metaphysical reasons for preferring one choice or the other when the choice between pure quantum theory and hidden-variables theory has to be made in the absence of conclusive experimental evidence are discussed in Section 1.8.

1.4. Incompleteness of some hidden-variables theories

In Sections 1.2 and 1.3 we have recognized a number of the fundamental properties which a satisfactory hidden-variables theory of the first kind should possess. Some theories discussed in the literature do not possess all of these properties.

In all theories the hidden variables are well-defined. (That is, we know what values they can take.) In all theories the shape of the equilibrium distribution $\bar{p}_\beta$ of the hidden variables has been postulated. (See Part II for details.)

In all theories, to some extent, the time dependence of the ξ is explicitly given by an equation for $d\xi/dt$, and it is or can be made plausible why p_β tends to $\bar{p}_\beta$ irreversibly, and why $\bar{p}_\beta$ perpetuates itself. In the theory of Bohm and Bub (1966a), however, there is the possibility of additional but not yet fully known terms in the equation for $d\xi/dt$ active only during measurements. Moreover, no hidden-variables theory so far gives a calculation or a reliable theoretical estimate of the time required for the relaxation process (8).

Most theories provide an algorithm which determines $n(\psi, \xi, \{\phi_i\})$ in (6) for all $\{\phi_i\}$. In Bohm's original theory (1952a, b), however, this algorithm is for electrons unambiguously given only for observables that depend merely on position. For other observables, like momenta, Bohm's $n(\psi, \xi, \{\phi_i\})$ depends intricately upon the hidden variables of the measuring apparatus. (See Part II.) Moreover, this Bohm theory has been formulated as a nonrelativistic theory only. Suggestions have been published how possibly one might try to generalize it to a relativistic theory [Bohm (1953b); Bohm and Vigier (1954)], but such attempts meet various difficulties that so far have not been overcome. (See Part II, Section 2.15.)

All theories of the first kind take care to guarantee the validity of condition (10) for the equilibrium distribution of hidden variables, for all measurements $\{\phi_i\}$ for which the theory provides an algorithm determining $n(\psi, \xi, \{\phi_i\})$. In fact, this is the main criterion by which such theories are of the first kind.

However, not all theories are equally clear about the nature of the process by which ψ changes into ϕ_n by a reproducible measurement. In the original Bohm theory (1952a, b), this transition is an ordinary reduction of the wave packet identical in nature to what is done in pure quantum theory. (See Section 1.1.) That is, it corresponds to a change of the ensemble described by the preparation of the quantum-mechanical state. Since, however, ψ occurs as one of the arguments in the function $n(\psi, \xi, \{\phi_i\})$ that determines the result of future measurements, it then must be (and actually is) shown that this "change of mind" about which ψ shall be used *does not affect* the results predicted by ψ and ξ together for future measurements in those cases where the preceding measurement had the result for which the reduction of ψ was made.

In the theory of Bohm and Bub (1966a), the situation is quite different. In this case, nonlinear terms added to the Schrödinger equation for $d\psi/dt$ provide that during the measurement ψ "condenses onto" ϕ_n. Thus the transition $\psi \rightarrow \phi_n$ in this case is an objective one, and no reduction of the wavefunction is needed.

Siegel and Wiener (1956) try to generalize the theory of measurements of quantum theory for $\psi \rightarrow \phi_n$, but insufficiently clarify how, for not affecting predictions about future measurements, one should "reduce" the ξ in such a voluntary reduction of the wavefunction which the observer could choose or not choose to make after performance of a measurement. [Compare footnote 5 (p. 6). See Section 3.7 of Part II for details.]

Where each of these theories is incomplete in one regard or another, this does not mean that these theories could not be completed by some additional work.

1.5. Hidden-variables theories of the second kind

In the above we assumed that no gross deviations from quantum theory would be found. This was the reason why we postulated in (10) that the predictions of theories of the first kind in the case of an equilibrium hidden-variables distribution would be identical to the predictions of quantum theory.

There are, however, people so much dissatisfied with quantum theory that at the time this is written they are looking for deviations from quantum theory that would exist even if the hidden-variables distribution were an equilibrium distribution.

A special case

Let us look at the aspect of quantum theory which is so much disliked by these people. It is related to the "nonlocality" aspect of the Einstein–Podolsky–Rosen paradox (1935). It is *not* the aspect of incompleteness of quantum theory versus complementarity which at that time was the primary point of discussion in the paper of Einstein *et al.* (1935) and in the reply to it by Bohr (1935). It rather is the aspect lifted up by Furry (1936) and by Bohm and Aharonov (1957). Let us illustrate this point by some remarks about the experiments performed recently by Clauser in Berkeley and by Holt at Harvard University. [See Clauser *et al.* (1969). For more details, see Part III.]

In these experiments, an atom cascades from a state $j = 0$ (Clauser) or $j = 1$ (Holt), via a state $j = 1$, to a state $j = 0$, emitting two photons. When these photons happen to travel in (nearly) opposite directions, they may cause coincidence counts between two photon detectors. We shall take the directions from the source toward these two detectors as our $+z$- and $-z$-axes. In the paths of these two photons, polarization filters F_1 and F_2 may be inserted. It is predicted and found that the number of coincidences is larger for one angle (which we shall call φ_{max}) between the polarization directions of these filters, than for the complementary angle (φ_{min}). Here, $\varphi_{\text{max}} = 0°$ and $\varphi_{\text{min}} = 90°$ for Clauser's 010 case (010 are the successive values of j), and $\varphi_{\text{max}} = 90°$, $\varphi_{\text{min}} = 0°$ for Holt's 110 case. Actually, what quantum theory predicts for the angle φ_{min} between the filters is that *no pair* of photons from one cascading atom can traverse *both* filters, but always *one* from each pair will traverse a filter and the other will not.[8]

If every photon had a definite polarization direction, and these directions for the two photons would be correlated, this would be readily understandable. Quantum theory, however, predicts that the polarization of a photon remains indefinite until the photon meets a polarization filter that distinguishes between two orthogonal polarizations, like linear polarizations along $\hat{x}$ and $\hat{y}$, or linear polarizations along different directions $\hat{x}'$ and $\hat{y}'$ perpendicular to $\hat{z}$, or circular polarizations in right- or left-hand direction. Only then the measurement forces the photon to make up its mind which of the two orthogonal possibilities it chooses. Thus quantum theory predicts for the case of angle φ_{min} between F_1 and F_2 that, of each pair of photons from one cascading atom, the photon reaching F_2 will always make up its mind differently from the photon reaching F_1: If the one traverses the filter, the other will not, and vice versa.[8]

This poses the question how the photon at F_2 can "know" that it should or should not traverse F_2 (depending upon what happened to the photon at filter F_1), if the filters F_1 and F_2 are spatially separated, with F_1 far enough from the source that the time of travel of the first photon from source to F_1 is much larger than the delay in the emission of the second photon as the atom waits before making its second quantum jump.

[8] For details, see Part III.

Since we *do* observe the existence of a correlation between what the two photons do at F_1 and F_2, it then is suggested that the quantum-mechanical picture might be wrong, and that already upon their emission it will be decided for the photons, for whatever filters they later may meet, whether or not they will pass them. (Correlations between such decisions seem feasible for two photons that both are offsprings of the same atom.) However, a polarization preference carried along by a photon may be regarded as a set of hidden variables describing the state of that photon. We then may say that the hidden variables carried by such a particle may be regarded as a *tabulation of instructions* for the particle's future behavior. Thus if and when the particle reaches an apparatus F for measuring a quantity A, with the help of its hidden-variable "manual" it can *locally* answer the question, in which eigenstate of A_{op} shall it appear to the apparatus. (The answer may be determined by some comparing or matching between the particle's "internal" hidden variables and some "external" hidden variables like the parameters describing properties of the apparatus.) The important point here is that the hidden variables of particle 1 at an apparatus F_1 answer this question *without reference to the outcome of any measurement made or not made at a distant apparatus* F_2. The correlations found between the results of measurements made by F_1 on particle 1 and by F_2 on particle 2 are then due to the correlations between the two hidden-variable manuals which the two particles (or photons) carried away from their original interaction or from their common place of birth.

We can now easily see why this (second) kind of a hidden-variables theory may be expected to contradict quantum theory. If the quantum-mechanical two-particles wavefunction $\psi(1, 2)$ of the particles traveling to the two filters is not factorizable into $\psi_1^{(1)}\,\psi_2^{(2)}$, one must expect that quantum theory will predict for particle 1 at apparatus F_1 measuring results that are correlated to and not independent of measuring results for particle 2 at apparatus F_2, as we mentioned above for the experiments of Clauser and of Holt. Therefore if we replace quantum theory by a hidden-variables theory of the second kind which predicts for 1 at F_1 results *without reference* to what 2 does at F_2, such predictions *violate* the predictions of quantum theory. We therefore should not be surprised that the angular dependence predicted for the coincidence rate between the two detectors in the experiments of Clauser and of Holt, as the angle ψ between the polarization directions of the filters F_1 and F_2 is varied, is in any specific hidden-variables theory of the second kind *different* from what it is quantum theory.[8] We thus have here a possibility of verifying experimentally which of the two theories is correct: quantum theory with its correlation predicted in apparent defiance of locality, or a hidden-variables theory that contradicts quantum theory already when its hidden-variables distribution p_β is the stochastic equilibrium distribution $\bar{p}_\beta$. For the latter reason, this hidden-variables theory is not of the first kind. Therefore we call it a theory *of the second kind.*

Some people call this kind of a hidden-variables theory a *local* theory. Against this, the objection has been made that also quantum theory is a "local" theory, notwithstanding the apparent nonlocality in its predictions about the photons at F_1 and at F_2. (Quantum theory is called "local" because its field equations are "local" differential equations.)

In the two-photon case, quantum theory predicts a final state described partially by a two-photon wavefunction. This wavefunction, in Clauser's experiment, is proportional to $\hat{x}_1\hat{x}_2+\hat{y}_1\hat{y}_2 = \hat{x}_1'\hat{x}_2'+\hat{y}_1'\hat{y}_2'$, if $\hat{x}_1$ means a photon traveling in the $+\hat{z}$-direction with polarization along $\hat{x}$, and $\hat{y}_2'$ is a photon traveling in the $-\hat{z}$-direction with polarization along $\hat{y}'$ at angles to $\hat{x}$ and $\hat{y}$. The invariance of this wavefunction under rotation from $\hat{x}, \hat{y}$ to $\hat{x}', \hat{y}'$ is what is meant by saying that the photon has not made up its mind on whether it should choose between $\hat{x}$ and $\hat{y}$ or between $\hat{x}'$ and $\hat{y}'$ before it reaches a filter. Once photon 1 meets a filter $||\hat{x}''$, though, the entire wavefunction becomes $\hat{x}_1''\hat{x}_2''+\hat{y}_1''\hat{y}_2''$, so, if the photon passes that filter, we know the two-photon system has realized a state $\hat{x}_1''\hat{x}_2''$ rather than $\hat{y}_1''\hat{y}_2''$, and therefore the second photon cannot traverse a filter that transmits polarizations $||\hat{y}_2''$. What is nonlocal here is *not the interaction*, but the fact that each of the two terms in ψ between which the measurement by F_1 makes a choice *describes the photon PAIR*, and therefore gives information on the other photon, too.

1.6. Difference between the numbers of distinct hidden-variables states in theories of the first and of the second kind

In general, correlations between the behavior of two distinguishable particles may be expected when the two-particle wavefunction is not factorizable:

$$\psi(1, 2) \neq \phi(1)\, \chi(2). \tag{11}$$

Consider, for instance, an ensemble of two-particle systems in which we do not always make the same measurement on particle 1 but choose for each system one out of R different measurements on particle 1 of which each measurement can have M different results. These possible measurement results are described by R orthonormal sets $\{\phi_i^{(\varrho)}(1)\}$ each of M different eigenfunctions of the appropriate operator, with $i = 1, 2, \ldots, M$, and with $\varrho = 1, 2, \ldots, R$. Simultaneously, let us make one out of S different measurements on particle 2, and let the Hilbert space for particle 2 be N-dimensional, so that these possible measurements are described by orthonormal sets $\{\chi_j^{(\sigma)}(2)\}$ with $j = 1, 2, \ldots, N$, and $\sigma = 1, 2, \ldots, S$. We can combine any of the former R measurements on 1 with any of the latter S measurements on 2, so that we really are considering $P = RS$ different measurements on the pair (1, 2), each measurement in principle with $L = MN$ different possible results $\{\psi_k^{(\pi)}(1, 2)\}$, with $k = 1, 2, \ldots, L$, and $\pi = 1, 2, \ldots, P$. Here k is an abbreviation for the index pair i, j and π for the index pair ϱ, σ, and

$$\psi_k^{(\pi)}(1, 2) = \phi_i^{(\varrho)}(1)\, \chi_j^{(\sigma)}(2). \tag{12}$$

Of course, the state vector $\psi(1, 2)$ may happen to be such that some of the PL wavefunctions $\psi_k^{(\pi)}$ do not occur in some of the P possible different expansions

$$\psi(1, 2) = \sum_{k=1}^{L} C_k^{(\pi)}\, \psi_k^{(\pi)}(1, 2) = \sum_{i=1}^{M} \sum_{j=1}^{N} C_{ij}^{(\varrho, \sigma)}\, \phi_i^{(\varrho)}(1)\, \chi_j^{(\sigma)}(2), \tag{13}$$

because their coefficients $C_k^{(\pi)} = C_{ij}^{(\varrho, \sigma)}$ may happen to vanish. In that case, it means that for that particular pair of measurements (ϱ, σ) the result $\phi_i^{(\varrho)}(1)$ for particle 1 just never would occur simultaneous with the result $\chi_j^{(\sigma)}(2)$ for particle 2.

In general, now,

$$w_k^{(\pi)} \equiv w_{ij}^{(\varrho\sigma)} = |\, C_{ij}^{(\varrho, \sigma)}\, |^2 \tag{14}$$

is the probability that the pair of measurements (ϱ, σ) will have the result $\phi_i^{(\varrho)}$ for particle 1 combined with $\chi_j^{(\sigma)}$ for particle 2.

Now suppose we know the L different coefficients $C_{ij}^{1,1} = C_k^{(\pi)}$ for one of the expansions (13) for $\varrho = \sigma = 1$ $(\pi = 1)$. If we also know the eigenfunctions $\{\phi_i^{(\varrho)}\}$ for all R possible measurements on particle 1, we can calculate the coefficients $\int \phi_i^{(1)*}\phi_{i'}^{(\varrho)}$ in the unitary transformation which expresses the measurement results $\{\phi_{i'}^{(1)}\}$ of any of the other $(R-1)$ possible measurements in terms of the $\{\phi_i^{(1)}\}$. Similarly, let the coefficients $\int \chi_j^{(1)*}\chi_{j'}^{(\sigma)}$ be known in the transformation from $\chi_j^{(1)}$ to $\chi_{j'}^{(\sigma)}$. Then we can calculate all coefficients $C_{k'}^{(\pi)}$ for all P different pairs of experiments from the given $C_k^{(1)}$.

Therefore the *quantum-mechanical* state in general would be determined completely for all measurements by L complex quantities $C_k^{(\pi)}$, i.e., $C_k^{(\pi)}$ with π fixed and $k = 1, 2, \ldots, L$.

A hidden-variables state would require far more information. We could, however, lump together all hidden-variable values that lead to the same predictions for all P different pairs of measurements considered. As each experiment can have $L = MN$ combinations of results (labeled by k or by i, j), the entire set of P experiments could have $L^P = (MN)^{(RS)}$ different results. As all these results should be predicted by ψ together with ξ, what we shall do, for a given ψ, is lumping all ξ values into L^P different heaps, each heap corresponding to a particular set of predictions. We may label these heaps by an index β running from 1 to L^P. In the equilibrium distribution of hidden variables, we can calculate, for each heap labeled by β, the probability $\bar{p}_\beta$ that the hidden variables have values putting them in this particular heap. Then $\bar{p}_\beta$ is the probability that the hidden variables in the equilibrium distribution will simultaneously predict for all P possible combinations of one measurement on 1 with one measurement on 2, the result specified by the label β. (*Different* β then correspond to predictions which for at least one of the P pairs of measurements predict for at least one of the particles a different measurement result.)

The $\bar{p}_\beta$ are all nonnegative. Though they may be related among each other for some specific ψ or for some particular hidden-variables theory, in general for an arbitrary *hidden-variables theory of the first kind* they are L^P *independent quantities*.

A hidden-variables theory of the *second* kind *differs* from this. As it denies the existence of nonlocal quantum mechanical correlations, it is equivalent to changing quantum theory so that the wavefunction ψ for a pair of particles that become separated somehow is changed into a new wavefunction that can be factorized according to

$$\psi(1, 2) = \phi(1)\chi(2) \tag{15}$$

instead of (11). In (13), this means that for any π or (ϱ, σ) we have

$$C_{ij}^{(\varrho, \sigma)} = a_i^{(\varrho)} b_j^{(\sigma)}, \tag{16}$$

so that

$$C_{ij}^{(\varrho, \sigma)} C_{kl}^{(\varrho, \sigma)} = C_{il}^{(\varrho, \sigma)} C_{jk}^{(\varrho, \sigma)}, \tag{17}$$

and

$$w_{ij}^{(\varrho, \sigma)} w_{kl}^{(\varrho, \sigma)} = w_{il}^{(\varrho, \sigma)} w_{jk}^{(\varrho, \sigma)}. \tag{18}$$

In a hidden-variables theory of the second kind, then, the L^P probabilities $\bar{p}_\beta$ are no longer independent, as each of them can be expressed in terms of a smaller number of probabilities.

This is so because a theory of the second kind assumes that the hidden variables can answer *local* questions, about one of the particles at a time. So, for particle 1 the hidden variables can in M^R ways predict for each of R measurements any of M different results, and we can collect the possible values of the hidden variables for particle 1 in M^R heaps, each heap for a particular set of predictions for all R possible measurements. Similarly, the hidden variables making predictions for particle 2 can be sorted out into N^S heaps. Both groups of hidden variables together, therefore, can be sorted out into $(M^R)(N^S)$ heaps in such a way that each of these $(M^R N^S)$ heaps will tell us simultaneously what is predicted for each of the particles separately. Thus a theory of the second kind distinguishes for all possible pairs of experiments on the two particles between $M^R N^S$ heaps, while a theory of the first kind distinguishes $(MN)^{RS}$ heaps as relevantly different. Since, however, $M^R \cdot N^S < (MN)^{RS}$, we need in a hidden-variables theory of the second kind far less information to fully describe the predictions made by the hidden-variables state, than we need in a theory of the first kind. The probabilities of the heaps in this case of the second kind are $(M^R N^S)$ different quantities $\bar{p}_\beta$, with $1 \leqslant \beta \leqslant M^R N^S$.

In Part III we shall derive, from the nonnegative definiteness of the $(M^R N^S)$ probabilities $\bar{p}_\beta$ for any hidden-variables theory of the second kind, certain inequalities holding between linear combinations of probabilities of specific results of measurements, some being measurements made on a pair of particles 1 and 2 simultaneously, and some being measurements made on one of the two particles only. The first of these inequalities was derived by Bell (1964) and another one was derived from the first one by Clauser *et al.* (1969). They are important for understanding the crucial experiments by Freedman and Clauser and by Holt for deciding between quantum theory and any hidden-variables theories of the second kind. In Part III we shall derive the pertinent inequality directly, bypassing Bell's inequality.

The reason why these same inequalities that hold for theories of the second kind do *not* hold for hidden-variables theories of the *first* kind lies in the fact that for the latter the two sides of the inequality would be functions of $(MN)^{RS}$ quantities $\bar{p}_\beta$, between which the relations are missing which must hold in any theory of the second kind, where these $(MN)^{RS}$ quantities become each expressible in terms of a mere $(M^R N^S)$ independent probabilities. By lack of these relations, in a theory of the first kind fewer conclusions can be drawn, and the inequalities are lost.

Thus the crucial experiments which we shall discuss in Part III will decide empirically between the validity of, on the one hand, hidden-variables theories of the second kind, and, on the other hand, either pure quantum theory or hidden-variables theories of the first kind, the latter not being distinguishable from quantum theory in this kind of an experiment. [See below eq. (10) on page 9.] If the results of these experiments would come out in favor of the validity of the inequality, in contradiction of quantum theory and of theories of the first kind, the details of the outcome of these experiments would have an additional importance in that they could tell apart various types of theories of the second kind, of which each must predict the validity of the inequality, but they differ in their predictions for the dependence of both sides of the inequality upon the variable parameters of the experiments.

1.7. Hidden-variables theories of the zeroth kind

Various authors, for instance von Neumann (1932, 1955) and Jauch (1968), have axiomatized quantum theory, and others, like Gleason (1957), have significantly contributed to this theory. The axioms on which such theories are based are, of course, statements of great generality that were part of quantum theory as formulated originally by physicists.

These people have wandered into the domain of hidden-variables theories and have tried to define by some postulates some properties which they thought any hidden-variables theory should possess. They then proved that theories having the properties postulated could not exist.

In each case, however, among the properties postulated there was at least one which the more realistic hidden-variables theories proposed by people more interested in such theories did not possess. Therefore their proofs of impossibility do not apply to hidden-variables theories of the first kind or of the second kind. For having a name for the kind of theories that were disproved, I will call them in the following "hidden-variables theories of the zeroth kind."

One might think at first that hidden-variables theories of the zeroth kind are of no interest. What good is a theory that is impossible?

The interest of hidden-variables theories of the zeroth kind is that they are a *warning*. They are a warning for what one can never expect any realistic hidden-variables theory to accomplish.

In this regard, Gleason's work is of the greatest importance. In a different way, the work of Kochen and Specker (1967) leads to the same conclusions and is somewhat easier to understand. Therefore, in Chapter 3, we shall discuss the work of Kochen and Specker before we turn to the work of Gleason.

1.8. Philosophical, religious, and polemic motives behind the quarrels between proponents of hidden-variables theory and of pure quantum theory

We have seen already in Section 1.3 how difficult it is to distinguish experimentally between hidden-variables theories of the first kind and pure quantum theory. We there concluded that, in absence of positive experimental evidence for hidden-variables theory, our choice should be pure quantum theory for reasons of scientific expediency, that is, for reasons of simplicity.

This "prejudice in favor of quantum theory," no doubt, will have evoked the anger of some of my readers. Some people are deeply involved emotionally in the question whether one should believe in pure quantum theory or whether one should believe in hidden-variables theory. Already the reply of Bohr (1935) to Einstein (1935) in connection to the merits and demerits of quantum theory boiled down mainly to the statement that Bohr *believed* quantum theory to describe nature by ψ *exhaustively* (though not "completely" in Einstein's sense), while Einstein *believed* that the description by ψ is *not* exhaustive. Einstein at later times was more explicit on this mere *difference of opinion*.[9]

[9] See, for instance, Einstein (1949), pages 671–673, where Einstein considers the point of view that ψ describes an ensemble, but that there may not exist any physical hidden variables capable of distinguishing

The main driving force toward a belief that hidden variables should exist, therefore, is in the religious belief[10] that "nature must be deterministic," and that "everything happening in nature must be predetermined by previous happenings in the physical world," even where our own knowledge is too limited for grasping the physical causes of what is happening. It therefore is often said that one would want to believe in hidden variables even if it were fundamentally impossible ever to know their values in advance. In that case, the knowledge that they would predetermine the results of a measurement would, of course, be of little or no *practical* use to us; but it would satisfy our emotional need for believing that *there exist* predetermining causes for all that is happening, even if we cannot measure these causes. (A theory of this kind is sometimes called *cryptodeterministic*.)

Other people do not feel that strongly the need for determinism in science. Where all experimentally known facts point toward the "nonexistence" of hidden variables in the "physical world" (though one cannot deny their existence in the imagination of their proponents), these people would tend to regard an unrestricted belief in determinism as a mere *superstition*.

Then there are those who for religious reasons *do not want* to believe in determinism. For them it is not *a priori* obvious that all happenings in the universe should be ultimately explainable by human reasoning. Anything happening in nature may in one way or another be a manifestation of God acting in this world, but especially those happenings that defy explanation can be "understood" merely as acts of God.[11] Thus the indeterminacy in atomic happenings as predicted by quantum theory is a manifestation of the omnipresence and omnipotence of God. The lack of determinism in quantum theory, to these people, is not only acceptable, but is a reassurance of their religious beliefs. If nature were fully deterministic, there would be no place in it for a God with any freedom to act; the "equations of motion" would determine everything.

Thus the difference in opinion between the convinced believers in determinism and the convinced believers in God is a religious strife in which rationality may be of little help. Those who want to be rational probably will accept the simplest theory that explains all observed facts and then will observe with some curiosity on which side of the dispute between the prejudiced this will land them at any particular time.

Finally, we should not disregard those who have participated in the search for hidden-variables theories for purely polemic reasons. (They are many.) These people invented hidden-variables theories, not because they believed that hidden variables would really govern nature, but because by counterexamples they wanted to expose those who believed that the inventors of theories of the zeroth kind would have proved the impossibility of hidden-variables theories in general.

Also in this group are those who invented hidden-variables theories merely because they

the members of such an ensemble from each other in a way that could *predict* which member will give which measurement result in a subsequent measurement. Einstein then admits that such a point of view is incontestable as a theoretical possibility, but he adds that he does not believe in it, and that he expects that sooner or later one will find the hidden parameters (our ξ) that *would* predetermine the results of measurements.

[10] We use the word "religious" in a rather wide sense. For instance, we would consider dogmatic atheism to be a religious belief.

[11] To some extent this sentence defines the idea of "God." Of course, it does not define specific properties assigned to "God" by specific religions.

wanted to show that they could not be "brainwashed" by what has become known as the *Copenhagen school* in quantum theory. This *Copenhagen* interpretation of quantum theory, like Bohr in 1935, simply states that it believes ψ to give an *exhaustive* description of nature, where "exhaustive" means "the best possible", even though this description is, of course, incomplete in the sense understood by Einstein (1935). This Copenhagen point of view, therefore, is *negative* in as far as it simply decrees that no meaningful parameters ξ exist in nature, so that there would not exist deterministic laws for individual physical systems[11a] by *fiat*. To many, the highhandedness of such a decree was unpalatable, and it pushed them to constructing hidden-variables theories as a protest, even if in the deep of their mind they believed that the Copenhagen school might be right. This "right," however, ought to be justified by showing difficulties with hidden-variables theories, and cannot be justified by simply ignoring the *possibility* of hidden-variables theories.

1.9. The "potential" but "virtual" nature of the predictions made by hidden-variables theories

Not what "is" but what "would be found"

A common misunderstanding about hidden variables is the idea that they might tell us such things as that a certain observable A "has" the value A_n (one of the eigenvalues of A_{op}).

As we mentioned already in Section 1.2, the state vector ψ and hidden variables ξ together *may* tell us by (6) that the result A_k will be found *if* the measurement will distinguish between the members of some specific complete orthonormal set of eigenfunctions $\{\phi_i\}$ of A_{op}, of which ϕ_k is one. If, however, the measurement distinguishes between the members of some different complete orthonormal set of eigenfunctions $\{\phi_i'\}$ of A_{op} of which ϕ_k is a member again (as could happen if A_{op} has some degenerate eigenvalues), then, in general, we cannot expect that $n(\psi, \xi, \{\phi_i\})$ would be equal to $n(\psi, \xi, \{\phi_i'\})$, and therefore it is quite possible that for the same values of ψ and ξ the first measurement might yield A_k and ϕ_k, and the second measurement might yield A_l and ϕ_l' with $\phi_l' \neq \phi_k$.

This at first sight somewhat surprising and at second sight somewhat distressing fact follows for hidden-variables theories proposed in the literature from the "algorithm" they use for determining the function $n(\psi, \xi, \{\phi_i\})$ occurring in (6). As an excuse for this result it is then stated that there is no *a priori* reason why two different measurements (different by the states $\{\phi_i\}$ or $\{\phi_i'\}$ between which they distinguish) should have one and the same result, even when performed on two systems in identical hidden-variable states (ψ, ξ). Yet there are cases in which this oversensitivity of the result of a measurement, for details of the possibil-

[11a] The *statistical* determinism for *ensembles* of physical systems described by a state vector is never questioned. It is expressed by the deterministic Schrödinger equation for ψ. However, we may make observations on a single system, and a single system is not an ensemble, and therefore is in general not described deterministically by any state vector ψ nor by itself uniquely determines the state vector ψ. (The latter depends upon what part of the history of the system is regarded as the common preparation of all members of the ensemble E_ψ in which the probabilities are to be valid that are to be calculated from ψ.)

ities that the measurement does not realize, is at least "not nice." (See Section 3.8.) One therefore would have a tendency to look for alterations of the theory by which this dependence of n on $\{\phi_i\}$ could be avoided.

The importance of the work of Gleason (1957) and of Kochen and Specker (1967) lies in that it shows that under very general circumstances this dependence of n on $\{\phi_i\}$ is not merely a possibility but is an unavoidable *must*. On the other hand, as this work shows where the *reasons* lie for this necessity, it also tells us how in some cases we may avoid *some* of this trouble. (See Section 3.7.) (See also Section 3.6 of Part II.)

One important conclusion we draw from the above is that hidden variables do not tell us what *are* the values of certain observables. They merely tell us what values *would be found* if we would measure these observables in some particular way.[12]

"*Simultaneous*" *values for noncommuting observables?*

A somewhat more primitive objection against hidden variables is that they would be impossible because they would simultaneously predict the values of the coordinates and of the momenta, and "it is well known that these cannot be measured simultaneously."

Again, if the hidden variables would predict $\mathbf{x} = \mathbf{x}^{(o)}$ and $\mathbf{p} = \mathbf{p}^{(o)}$, they do not predict that one measurement would give simultaneously $\mathbf{x}^{(o)}$ *and* $\mathbf{p}^{(o)}$. The prediction $\mathbf{x}^{(o)}$ might be made for a measurement in which the set $\{\phi_i\}$ may be a set of deltafunction eigenfunctions of $\mathbf{x}_{op}$, and the prediction $\mathbf{p}^{(o)}$ might be made for a measurement in which the set $\{\phi_i\}$ might be a set of monochromatic plane waves. Since there are no simultaneous eigenfunctions $\{\phi_i\}$ of $\mathbf{x}_{op}$ *and* $\mathbf{p}_{op}$, it is obvious that the predictions (6) of hidden-variables theory cannot possibly include the prediction of *finding simultaneously* precise values for $\mathbf{x}$ and $\mathbf{p}$.

Instead, the theory can and does *simultaneously predict* precise values, $\mathbf{x}^{(o)}$ for $\mathbf{x}$ when $\{\phi_i\} \approx \left\{\delta(\mathbf{x}-\mathbf{x}_i)^{\frac{1}{2}}\right\}$, and $\mathbf{p}^{(o)}$ for $\mathbf{p}$ when $\{\phi_j\} \approx \left\{V^{-\frac{1}{2}} \exp(i\mathbf{k}_j \cdot \mathbf{x})\right\}$. That is, it simultaneously answers questions about the various *possibilities* for the next measurement which will be made, and which *could* be a measurement of position and which *could* equally well be a measurement of momentum. These are *virtual* predictions for *potentialities*, and are *not* meant to be predictions for some imagined "reality" of a *simultaneous* measurement (which is impossible). [Similarly, in pure quantum theory, the uncertainty relations are

[12] Apparently Bub (1969) is so upset by this aspect of hidden-variables theory that he proposes to reserve the name "hidden-variables theory" for theories of the zeroth kind that try to do what he would like them to do but what is impossible, and he wants to invent a different name (but does not propose one) for realistic theories. The purpose of this change of names is not clear. Moreover, Bub suggests at the top of his page 121 that any attempts at interpreting the distressing fact which above we called "not nice," would necessarily lead to a theory in which the measurement result predicted for a function $f(A)$ of an observable A would differ from $f(A_n)$, if A_n is predicted in the same state (ψ, ξ) to be the result of measurement of A. Assuming that for the measurement of A and of $f(A)$ *the same apparatus is used* (identical sets $\{\phi_i\}$), this claim (which Bub does not substantiate by any explanation) is incorrect, and *must* be incorrect because A and $f(A)$ have the same eigenfunctions, and $n(\psi, \xi, \{\phi_i\})$ merely picks out a particular *eigenfunction* ϕ_n, irrespective of what the operator is of which we regard this ϕ_n to be the eigenfunction. The selected eigenfunction then gives simultaneously the value A_n to A and the value of $f(A_n)$ to $f(A)$. If, on the other hand, for measuring $f(A)$ an apparatus is used for which the set $\{\phi_i\}$ is different, this measurement would perhaps yield $f(A_m)$ with $m \neq n$, but for A itself this apparatus would give A_m.

relations between differences of results of measurements made *on different systems* from the ensemble E_ψ, and do *not directly* tell us anything about the possibility or impossibility of a simultaneous measurement of p and of x on one single system. See Ballentine (1970).]

"Dispersionfree" states

For each of the potential but mutually exclusive measurements separately, the answer to the question what values will be found in such measurements is for given ψ and ξ *"dispersionfree,"* because we *can* ask simultaneously this question about any observable A and its square, and the answer $(A^2)_n$ to the latter question is the square of the answer A_n for the value of A, as both are eigenvalues belonging to the same eigenfunction ϕ_n picked from the $\{\phi_i\}$ by (6). Nevertheless, it is a dangerous play with words if we therefore call a hidden-variables state (as given by ψ and ξ before we pick a measurement $\{\phi_i\}$) a "generally dispersionfree state," as often has been done in the literature. If we use this expression, we should keep in mind that there is a difference between claiming that mutually exclusive measurements each will lead to a state which is dispersionfree for the quantity measured—a claim which hidden-variables theory *has in common with quantum theory*, as seen from the postulate eq. (1) for pure quantum theory—and claiming that there would exist an ensemble E_ψ to *which the rules of quantum theory would be applicable*, and for which yet all observables *simultaneously* would be dispersionless—which is a claim rejected not only by quantum theory, but also by hidden-variables theory, where ψ will never become dispersionless for all observables simultaneously. (See Section 2.3.)

1.10. "Ensembles" corresponding to hidden-variables states and "expectation values" for them

The *only* dispersionfree "ensembles" that hidden-variables theory can construct are trivial ones, consisting of some arbitrary pair of any ψ with any ξ, and *identical copies* of this pair, so that the properties of any such $E_{\psi\xi}$ ensemble are identical with those of a single member of it. These ensembles are in no way comparable with the E_ψ ensembles of quantum theory. They are *not* describable without introducing ξ explicitly, and the way in which they are dispersionless for observables can*not* be understood without considering the function $n(\psi, \xi, \{\phi_i\})$. The value of the *average* $\langle A\rangle_{\psi,\xi}$ of an observable A in such a trivial ensemble (often called the "expectation value" of A for the ensemble) is equal to the value $A_n = A_{n(\psi,\xi,\{\phi_i\})}$ of A for a *single member* of the ensemble:

$$\langle A\rangle_{\psi,\xi} = A_{n(\psi,\xi,\{\phi_i\})} = \text{eigenvalue of } A_{\text{op}}. \tag{19}$$

There are additional differences between the trivial $E_{\psi,\xi}$ ensembles and the ensembles representing quantum mechanical states. For the latter it is useful to introduce a statistical density operator U_{op} of which the matrix elements U_{ji} in any representation have the property that the average value of any observable B over the quantum-mechanical state may be written as

$$\langle B\rangle_{\text{mix}} = \sum_j \sum B_{ij} U_{ji} \,[\equiv \text{Trace}\,(BU)]. \tag{20}$$

In a mixed state with probability p_α for each wavefunction $\psi^{(\alpha)}$, this density operator is easily found to be the operator U_{op} given by

$$U_{op}f(x) = \int_{(\alpha)} p_\alpha \psi^{(\alpha)}(x)\left(\int \psi^{(\alpha)*} f\right), \tag{21}$$

which has the matrix elements

$$U_{ji} \equiv \int \phi_j^* U_{op} \phi_i = \int_{(\alpha)} p_\alpha \left(\int \psi^{(\alpha)*}\phi_i\right)\left(\int \phi_j^* \psi^{(\alpha)}\right) = \int_{(\alpha)} p_\alpha a_j^{(\alpha)} a_i^{(\alpha)*}. \tag{22}$$

The validity of (20) then follows from (4) and (5). We shall find in Section 2.2 that in general there does not exist for the trivial ensemble $E_{\psi,\xi}$ for a hidden-variables states (ψ, ξ) a unique operator V_{op}, the same for all operators A, B, . . ., such that one would have

$$\langle A \rangle_{\psi,\xi} = \sum_i \sum_j A_{ij} V_{ji}, \quad \langle B \rangle_{\psi,\xi} = \sum_i \sum_j B_{ij} V_{ji}, \tag{23}$$

with

$$V_{ji} = \int \phi_j^* V_{op} \phi_i \tag{24}$$

in any representation. This fact, however, does not detract from the value of hidden-variables theory, because density matrices V_{ji} for hidden-variables states are not useful anyhow. They are not needed because we do not need the eqs. (23), as we can always calculate $\langle A \rangle_{\psi,\xi}$ from eq. (19) without using any V_{op} at all.

It is possible to write $\langle A \rangle_{\psi,\xi}$ formally as

$$\langle A \rangle_{\psi,\xi} = \sum_j v_j A_j, \tag{25}$$

with some similarity to $\langle A \rangle_{Av} = \sum_j w_j A_j$ in (4). However, while the quantum-mechanical probabilities

$$w_j = \int_{(\alpha)} p_\alpha \left| \int \phi_j^* \psi^{(\alpha)} \right|^2 \tag{26}$$

for a given mixed ensemble depends merely on one Hilbert vector ϕ_j, and not on a complete set $\{\phi_i\}$, nor on A_{op}, the "truth values"

$$v_j = \delta_{j,\, n(\psi, \xi, \{\phi_i\})} \tag{27}$$

of hidden-variables theory do depend on a *complete* set $\{\phi_i\}$, and not merely on ϕ_j. Since v_j, through $\{\phi_i\}$, is a messy function of A, it follows that the expression (25) is *not* linear in A, so that any results that in quantum theory hold for $\langle A \rangle_\psi$ on account of the linearity of $\int_{(\alpha)} p_\alpha \int \psi^{(\alpha)*} A_{op} \psi^{(\alpha)}$ in A do not have their counterparts for expectation values in hidden-variables states.

Finally, from (19) we see that $\langle A \rangle_{\psi,\xi}$ is always an eigenvalue of A_{op}. Now let A and B be two observables of which the operators do not commute and which have discrete eigenvalues. Then, in general, the eigenvalues $(A+B)_n$ of $(A+B)_{op} \equiv (A_{op}+B_{op})$ cannot be written as the sums of one eigenvalue of A_{op} and some other eigenvalue of B_{op}:

$$(A+B)_n \neq A_{k(n)} + B_{l(n)}. \tag{28}$$

It follows from this that the predictions of hidden-variables theory for the values predicted for A, for B, and for $(A+B)$ in some microstate ψ, ξ *should* be such that in general

$$\langle A+B\rangle_{\psi,\xi} \neq \langle A\rangle_{\psi,\xi}+\langle B\rangle_{\psi,\xi} \tag{29}$$

for the trivial ensemble consisting of identical copies of the one microstate (ψ, ξ). This inequality *is desirable*,[13] but it is distinctly different in form from the rule

$$\langle A+B\rangle_{\text{mix}} = \langle A\rangle_{\text{mix}}+\langle B\rangle_{\text{mix}} \tag{30}$$

which *happens* to be valid in quantum theory, because there $\langle A\rangle_{\text{Av}}$ happens to depend linearly even on noncommuting terms in A_{op}.

We mention all these facts because they often have been overlooked in the literature, sometimes with disastrous results.

[13] We call (29) *desirable* because it is a direct consequence of the primary postulate (6) with (19) of hidden-variables theory, which *must* be postulated in any realistic theory for matching the postulate eq. (1) of quantum theory.

CHAPTER 2

VON NEUMANN'S CLAIM OF IMPOSSIBILITY OF HIDDEN-VARIABLES THEORIES

THE work of von Neumann (1932, 1955) was mainly concerned with the axiomatization of the mathematical methods of quantum theory. His side remarks on hidden variables were merely an unfortunate step away from the main line of his reasoning. This side step was unfortunate because it was not big enough a step. Von Neumann assumed that a hidden-variables theory could be characterized by leaving all his axioms of quantum theory unchanged, *assuming* them to be valid without change also in hidden-variables theory, and merely *adding* one additional postulate for which he chose the assumption that hidden-variables states would be dispersion-free quantum states. He then found that the additional postulate was in contradiction to the axioms he had already introduced for describing quantum theory.

He should have concluded that the kind of hidden-variables theory *defined by him* was an impossible theory—what we called in Section 1.7 a "theory of the zeroth kind." Unfortunately, he did not express himself in this kind of a cautious way. Instead, he claimed[14] "... we need not go any further into the mechanism of the 'hidden parameters,' since we now know that the established results of quantum mechanics can never be re-derived with their help." Though at the present we know very well that this claim is contrary to the results of explicit hidden-variables theories of the first kind [see Section 1.3 below eq. (10)], and that therefore apparently the latter do *not* fall in the category of hidden-variables theories determined by von Neumann's postulates, the authority of von Neumann's overgeneralized claim for nearly two decades stifled any progress in the search for hidden-variables theories. This is especially surprising because of the obviousness of *inapplicability* of one of von Neumann's axioms to any realistic hidden-variables theory of the kind that attempts to explain quantum theory as a special case.

Bell's work

The public recognition that *somehow* von Neumann's formulation of hidden-variables theory was not a fortunate one and could not have the generality and exhaustiveness which von Neumann suggested, goes back at least to Bohm (1952b). The first to publicly *pinpoint*

[14] Von Neumann (1955), page 324, translated from von Neumann (1932), page 171.

the axiom by which von Neumann's formulation violated the elementary principles of any realistic hidden-variables theory was Bell (1966). In the same paper Bell also pinpointed the quantum-theoretical axiom by which Jauch and Piron (1963) made their version of hidden-variables theory into a theory of the zeroth kind.

Moreover, Bell's paper contains an excellent condensation of the argument of Gleason (1957), different from our derivation of this result in Chapter 3 from Kochen–Specker's paradox.

2.1. The source of the trouble with von Neumann's definition of hidden-variables theories

We shall not reproduce here in detail all steps of von Neumann's proof that hidden-variables theories of the kind postulated by him are impossible. An excellent summary of his proof is found on pages 374–375 of a recent paper by Ballentine (1970), who also pinpoints and explains one reason why von Neumann's proof does not apply to hidden-variables theories of any reasonable kind. (See Appendix K for another reason.) This reason for the inapplicability of von Neumann's proof to realistic hidden-variables theories is that one essential property of the hidden-variables theories considered by von Neumann is the assumed validity of

$$\langle aA+bB\rangle_{\psi,\xi} = a\langle A\rangle_{\psi,\xi}+b\langle B\rangle_{\psi,\xi} \tag{31}$$

for the expectation values in the dispersionfree microstates labeled by us by ψ, ξ, for any pair of observables A and B, even if their operators do not commute. For $a = b = 1$, this postulate is in direct disagreement with the inequality (29), which for noncommuting A_{op} and B_{op} with discrete eigenvalues was a necessity because of the more fundamental requirements (6) and (19) which a realistic hidden-variables theory must meet if it is to agree with our postulate eq. (1) for measurements on single physical systems according to pure quantum theory.

Von Neumann apparently copied (31) from the rule

$$\langle aA+bB\rangle_{\psi \text{ or mix}} = a\langle A\rangle_{\psi \text{ or mix}}+b\langle B\rangle_{\psi \text{ or mix}} \tag{32}$$

valid for the ensembles describing pure or mixed quantum states. To hidden-variable theorists, the property (32) for averages over quantum states is a happenstance which is not at all *a priori* obvious because results of measurements of $(aA+bB)$, and of A and of B, require in general entirely different apparatus when A_{op} and B_{op} do not commute, and therefore it is surprising that there should be any relation at all between results of these measurements. In fact, the inequality (29) expresses the absence of such relations for measurements on a single system. It so happens that the other axioms and postulates of quantum theory conspire to make $\langle A\rangle_\psi = \sum_i w_i A_i$ expressible as $\int \psi^* A_{\text{op}}\, \psi$. Since this is linear in A_{op}, the happenstance (32) follows for the ensemble described by ψ. By no means should (32) be regarded as some kind of a definition of how observables should be added. The definition of the sum of observables is naturally comprised in statements like

$$(A+B)_{\text{op}} = A_{\text{op}}+B_{\text{op}} \quad \text{or} \quad (A+B)_{ij} = A_{ij}+B_{ij}. \tag{33}$$

When a physical theory contains valid statements I, II, and III, and it is not only true that III follows logically from I and II, but also I follows from II and III, it is a question of taste whether one wants to treat I and II as axioms, or rather II and III. (The statement left out then becomes a theorem.) In most cases it is physically irrelevant which statement becomes an axiom and which becomes a theorem. We see from the present example, however, that a poor choice may lead to inadequacies when the theory is generalized. If $A+B$ had been defined by $(A+B)_{ij}$ rather than by $\langle A+B\rangle$, there would have been less temptation to take over (32) in the form of (31) into the definition of hidden-variables states, and the violation of the necessary inequality (29) for hidden-variables states might have been avoided.

2.2. The nonexistence of a unique density operator for an ensemble of systems all in the same hidden-variables state (ψ, ξ)

In Section 1.9 we made without proof the claim that in general one cannot invent a unique density operator V_{op} by which one could calculate by eqs. (23)–(24) the values $\langle A\rangle_{\psi,\xi}$, $\langle B\rangle_{\psi,\xi}$, ..., predicted in a hidden-variables state (ψ, ξ) as measurement results for all observables A, B, ...

We might have left our statement that way, leaving the burden of proving the opposite to those who would want to claim the general existence of such a V_{op}. Since, however, much of von Neumann's writing on hidden variables centered around the properties of such an operator V_{op}, it is perhaps for curiosity's sake of some interest to see why in a realistic hidden-variables theory in general no such V_{op} can exist.

This fact is of additional importance because it proves not only that one of the postulates which led von Neumann to the conclusion of existence of a density operator U_{op} in quantum theory cannot be valid in a hidden-variables theory of the first kind, but that also any other set of axioms and postulates that leads to a density operator must contain some assumption not valid in such a hidden-variables theory. In practicular, this must be true for Gleason's proof of the existence of a density operator in quantum theory, which is based upon less far-going assumptions than von Neumann's proof. In Chapter 3 we shall discuss which of Gleason's assumptions does not hold in hidden-variables theories of the first kind.

Since von Neumann started by making the postulate (31) rejected by us, he could prove the existence of a density matrix [expressing it in terms of expectation values of projection operators like

$$P(\phi_n) \equiv |n\rangle\langle n|, \qquad P\left(\frac{\phi_m+\phi_n}{\sqrt{2}}\right) - P\left(\frac{\phi_m-\phi_n}{\sqrt{2}}\right) \equiv |m\rangle\langle n| + |n\rangle\langle m|,$$

and so on, rather than explicitly in terms of the p_α and $\psi^{(\alpha)}$] before he introduced as an afterthought his *hidden-variables postulate of the existence of an all-dispersionfree state*. If the latter idea is accepted as an additional part of the set of axioms and postulates, this set was shown by von Neumann to become *self-contradictory*, so that *any* conclusion drawn from it becomes doubtful. Therefore we must follow a different approach. We shall assume all axioms, postulates, and ensuing theorems that are valid in hidden-variables theories of the first kind, and avoid a self-contradictory set of assumptions by rejecting postulate (31) from the beginning.

Proof

Since we merely want to prove that V_{op} cannot *in general* exist, it is sufficient to consider a simple special case. For this we choose an example in which Hilbert space is two-dimensional. In this Hilbert space we will consider three mutually noncommutative hermitian operators. We shall assume later that ψ is an eigenfunction of one of these operators, which we shall call A_{op}, with eigenfunctions ϕ_i. Any other hermitian operator B_{op} has a set of eigenfunctions $\{\chi_k\}$ which in an N-dimensional Hilbert space could be obtained from $\{\phi_i\}$ by a unitary transformation

$$\chi_k = \sum_{i=1}^{N} T_{ik}\phi_i. \tag{34}$$

We shall confine ourselves to real orthogonal transformations. For $N = 2$, we write (34) as

$$\chi_1 = c\,\phi_1 + s\,\phi_2, \quad \chi_2 = -s\,\phi_1 + c\,\phi_2; \quad c = \cos\theta, \quad s = \sin\theta. \tag{35}$$

When we need later a third operator, we call the new operator B'_{op}, with eigenfunctions χ'_k obtained by changing θ into θ' in (35). Denoting by superscripts A and B the matrix elements of hermitian operators on the two bases $\{\phi_i\}$ and $\{\chi_k\}$, we find that, if V_{op} would exist, then

$$V_{lk}^{(B)} = \int \chi_l^* V_{op}\chi_k = \sum_i \sum_j T_{jl}^* V_{ji}^{(A)} T_{ik}. \tag{36}$$

Now consider a pure state ψ which happens to be an eigenfunction of A_{op} to its smallest eigenvalue $= A_s$. We shall renumber the $\{\phi_i\}$ so that we have $s = 1$, and $\psi = \phi_1$. Then for any ξ the measurement of A shall give A_1 with ϕ_1, so that

$$n(\psi, \xi, \{\phi_i\}) = s = 1 \quad \text{for} \quad \psi = \phi_s = \phi_1 \quad \text{and for any } \xi. \tag{37}$$

It then follows from (19) that, in this case,

$$\langle A\rangle_{\psi,\xi} = A_1. \tag{38}$$

On the other hand, if V_{op} would exist with the property (23),

$$\langle A\rangle_{\psi,\xi} = \sum_k \sum_l A_{kl}V_{lk} = \sum_{j=1}^{N} A_j V_{jj}^{(A)}, \tag{39}$$

where we used the fact that on the basis $\{\phi_i\}$ the matrix of A_{op} is diagonal. Since $A_s \leq$ each A_j, and since the diagonal elements $V_{jj}^{(A)}$ cannot be negative,[15] eq. (39) gives

$$\langle A\rangle_{\psi,\xi} \geqq A_s \sum_{j=1}^{N} V_{jj}^{(A)} = A_s\langle 1\rangle_{\psi,\xi} = A_s = A_1. \tag{40}$$

[15] Consider an operator A'_{op} commutative with A_{op} and having eigenvalues $A'_i = \delta_{ij}$ with fixed j. Then, $\langle A'\rangle_{\psi,\xi} = \sum_i \sum_k A'_{ik}V_{ki} = \sum_i A'_i V_{ii}^{(A)} = V_{jj}^{(A)}$. Since this is the average of measured eigenvalues $\delta_{ij} = 0$ or 1, it follows that $\langle A'\rangle_{\psi,\xi}$ cannot be negative, so that $V_{jj}^{(A)} \geqq 0$.

The inequality (40), however, must be an equality according to (38). Therefore only A_s can contribute to the sum in the last member of (39). It follows that all $V_{jj}^{(A)}$ are zero with the exception of $V_{ss}^{(A)}$. Then $\sum_{j=1}^{N} V_{jj}^{(A)} = \langle 1 \rangle_{\psi, \xi} = 1$ gives

$$V_{jj}^{(A)} = \delta_{js} V_{ss}^{(A)} = \delta_{j1}. \tag{41}$$

If $N = 2$, the matrix $V(\psi, \xi)_{ji}^{(A)}$ then for $\psi = \phi_1$ and for all ξ reduces to

$$V_{ji}^{(A)} = \begin{pmatrix} 1 & v_{12} \\ v_{12}^* & 0 \end{pmatrix}, \tag{42}$$

where $v_{12} = v_{12}(\xi)$ is a function of the hidden variables which is not yet determined by (41).

Then, with the T_{ik} of (34) given by (35), eq. (36) gives

$$V_{11}^{(B)} = c^2 + cs\,(v_{12}+v_{12}^*), \quad V_{22}^{(B)} = s^2 - cs\,(v_{12}+v_{12}^*), \tag{43}$$

and

$$\langle B \rangle_{\psi, \xi} = \sum_k B_k V_{kk}^{(B)} = B_1\, c^2 + B_2\, s^2 + (B_1 - B_2)\, cs\,(v_{12}+v_{12}^*). \tag{44}$$

According to (19), this must be either B_1 or B_2. It follows that

$$\text{either} \quad cs\,(v_{12}+v_{12}^*) = s^2, \tag{45a}$$

$$\text{or} \quad cs\,(v_{12}+v_{12}^*) = -c^2. \tag{45b}$$

According to the statistical postulate (10), we can for a hidden-variables theory of the *first* kind even calculate the probabilities $\bar{P}_1(\psi, \{\chi_k\})$ and $\bar{P}_2(\psi, \{\chi_k\})$ that for the measurement of B (characterized by the orthonormal set $\{\chi_k\}$ of possible results) the hidden variables in an equilibrium distribution will have values making $n(\psi, \xi, \{\chi_k\})$ for our state $\psi = \phi_1$ equal to 1 or to 2. These probabilities should be equal to the quantum-mechanical probabilities $w_1 = |\int \chi_1^* \psi|^2$ and $w_2 = |\int \chi_2^* \psi|^2$. With $\psi = \phi_1$, this yields in our case $\bar{P}_1 = w_1 = c^2$ and $\bar{P}_2 = w_2 = s^2$.

Independent of this result, for *any* realistic hidden-variables theory for our two-dimensional case with $\psi = \phi_1$, we see from (45a, b) that there is some probability P_1 that ξ is such that $\langle B \rangle_{\psi, \xi} = B_1$ and that

$$v_{12}+v_{12}^* = \frac{s^2}{cs} = \tan\theta, \tag{46a}$$

and a probability P_2 that ξ is such that $\langle B \rangle_{\psi, \xi} = B_2$, and that

$$v_{12}+v_{12}^* = \frac{-c^2}{cs} = -\cot\theta. \tag{46b}$$

It does not matter that $v_{12}+v_{12}^*$, as a function of ξ, is necessarily discontinuous where we pass from the ξ values for which (46a) is valid, to the ξ values for which (46b) is valid. In hidden-variables theories, such discontinuities must be expected.

What is disturbing, however, is that $v_{12}+v_{12}^*$ depends on θ. In (42) we defined v_{12} as a function of ξ by

$$v_{12}(\xi) = V(\psi, \xi)_{12}^{(A)} = \int \phi_1^* V(\psi, \xi)_{\text{op}} \phi_2 = \int \phi_1^* V(\phi_1, \xi)_{\text{op}} \phi_2. \tag{47}$$

This depends, besides on ξ, on the eigenfunctions ϕ_i of A_{op}, but there is no dependence at all on B_{op} or on χ_k or on the angle θ in Hilbert space between the orthonormal set $\{\chi_k\}$ and the set $\{\phi_i\}$. In other words, when we vary B_{op} by a further rotation of the orthonormal set $\{\chi_k\}$ in Hilbert space, by changing θ in (35) into θ', it follows from (47) that for given ξ the value of $v_{12}(\xi)$ should not change, while from (46a, b) we see that they do change in a definite way. This contradiction shows that there was an error in our assumption that V_{op} existed in this case. Otherwise expressed, it shows that the existence of a *unique* density operator V_{op} for an ensemble of systems all in the same hidden-variables state (ψ, ξ) (where by "unique" we mean: the same operator V_{op} for all observables A, B, etc.) is in disagreement with the postulates on which realistic hidden-variables theories are based.

Alternative proof

Instead of *starting* with the consideration of the validity of (23) for B_{op}, we shall first apply it to V_{op} itself, or to any observable C for which C_{op} commutes with V_{op} (always starting with the assumption that V_{op} would exist). Then, using the common eigenfunctions of V_{op} and C_{op} as our basis, it follows by the argument of footnote 15 (introducing operators C'_{op}) that $V_{jj}^{(C)} = V_{jj}^{(V)} \geqslant 0$. That is, the eigenvalues of V_{op} should both be positive. These two eigenvalues are easily obtained from $V_{ji}^{(A)}$ given in (42) by solving the secular equation $\begin{vmatrix} 1-V & v_{12} \\ v_{12}^* & 0-V \end{vmatrix} = 0$. This gives two solutions,

$$V_{1,2} = \tfrac{1}{2} \pm \sqrt{\tfrac{1}{4} + |v_{12}|^2},$$

so that $\quad V_{11}^{(C)} = V_1 = \tfrac{1}{2} + \sqrt{\tfrac{1}{4} + |v_{12}|^2}$ and $V_{22}^{(C)} = V_2 = \tfrac{1}{2} - \sqrt{\tfrac{1}{4} + |v_{12}|^2}$.

From $V_{22}^{(C)} \geqslant 0$ it then follows that we must have

$$|v_{12}(\xi)| = 0, \tag{48}$$

so that (42) simplifies for $\psi = \phi_1$ and for all ξ to

$$V_{ji}^{(A)} = \begin{pmatrix} 1 & 0 \\ 0 & 0 \end{pmatrix}. \tag{49}$$

This gives V_{op} the eigenvalues 1 and 0 for the eigenfunctions ϕ_1 and ϕ_2.

We now consider, as another operator for which (23) should be valid, the operator B_{op} considered before. Keeping the old notation (42) for $V_{ji}^{(A)}$, we then from (44) find that $2Re\,\{v_{12}\} = \tan\theta$ or $-\cot\theta$, as in (46a, b). The direct contradiction between this result and (48) then shows that our assumption of existence of V_{op} satisfying (23) for all observables must have been wrong.

2.3. Dispersionfree states in hidden-variables theory

Already in Section 1.8 we alluded to confusion about the meaning of the expression "dispersionfree state" in some of the literature, due again to a failure to distinguish between the meaningful ensembles E_ψ that represent a quantum-mechanical state, and the trivial and useless ensembles of identical copies of a single microstate that are needlessly forced upon the theory when one starts talking about the *average* value $\langle A\rangle_{\psi,\xi}$ in a hidden-variables state which *uniquely* determines the value $A_{n(\psi,\,\xi,\,\{\phi_i\})}$ of A.

We agree, of course, with von Neumann and with Gleason and with all other quantum theorists that there do not exist any pure quantum-mechanical states (ensembles E_ψ), nor any mixed quantum-mechanical states, for which *all* observables A would *simultaneously* be dispersionfree in the sense of

$$\langle A^2\rangle_\psi = (\langle A\rangle_\psi)^2 \quad \text{or} \quad \langle A^2\rangle_{\text{mix}} = (\langle A\rangle_{\text{mix}})^2. \tag{50}$$

On the other hand, it is a trivial consequence of (19), together with the fact that the eigenfunctions ϕ_i of A_{op} are also eigenfunctions of A^2_{op}, that

$$\langle A^2\rangle_{\psi,\xi} = (A^2)_{n(\psi,\xi,\{\phi_i\})} = (A_{n(\psi,\xi,\{\phi_i\})})^2 = (\langle A\rangle_{\psi,\,\xi})^2. \tag{51}$$

There is no contradiction at all between the general validity of (51) and the impossibility of general validity of (50), because (50) is not a consequence of (51). Equation (50), for a particular observable A, imposes upon the density operator U_{op} a severe condition,[16] which makes it not surprising that (50) then cannot be satisfied simultaneously with A replaced by a different matrix B which has no eigenfunctions in common with A. On the other hand, eq. (51) is trivial and does not impose any conditions upon the quantum-mechanical state U_{op} at all. As (51) in view of (19) is not much more than a tautology, no nontrivial conclusions at all can be drawn from (51).

2.4. Von Neumann's three "reasons" for the nonexistence of hidden-variables states

Von Neumann gives three reasons why hidden-variables states should be impossible.

Reason One

In the first place, he claims that the "homogeneous" ensemble E_ψ describing a pure state could not possibly be split up into subensembles $S_{\psi,\xi}$ corresponding to hidden-variables states. He concludes this [17] where he refers to his note 173, by tacit reference to a theorem he

[16] We agree, of course, that for a quantum mechanical state, pure or mixed, there always exists a density operator U_{op} such that (20) is valid for every observable B. [von Neumann (1932, 1955); Gleason (1957).] Then, in a representation diagonalizing the operator of some observable A, eq. (50) requires $\left[\sum_i A_i U_{ii}^{(A)}\right]^2 = \sum_i A_i^2 U_{ii}^{(A)}$. By $\sum_j U_{jj}^{(A)} = \langle 1\rangle = 1$, this requires $\sum_i \sum_j A_i A_j U_{ii}^{(A)} U_{jj}^{(A)} = \sum_i \sum_j \frac{1}{2}(A_i^2 + A_j^2) U_{ii}^{(A)} U_{jj}^{(A)}$ or $\sum_i \sum_j (A_i - A_j)^2 U_{ii}^{(A)} U_{jj}^{(A)} = 0$. Since $U_{ii}^{(A)} \geqslant 0$ for reasons similar to those given in footnote 15 for $V_{jj}^{(A)} \geqslant 0$, it follows that (50) requires that all nonzero $U_{ii}^{(A)}$ be restricted to one single eigenvalue of A_{op}. By $p_\alpha > 0$ in (22) this means that all states ϕ_i for which any of the coefficients $a_i^{(\alpha)}$ of any $\psi^{(\alpha)}$ (with nonzero p_α) does not vanish, must belong to the same eigenvalue A_i.

[17] Von Neumann (1932), page 171, or von Neumann (1955), page 324.

proved a few pages earlier,[18] which considers an ensemble in which the averages of observables are described by a density operator U_{op}, and which is split up into two subensembles with density operators V_{op} and W_{op} which add up to U_{op} and which both are nonnegative definite. He then proves that necessarily V_{op} and W_{op} can differ from U_{op} at most by a normalization factor, so that it would be impossible for V_{op} to describe a generally dispersion-free state, if U_{op} is not generally dispersionfree for all observables.

Our first and least fundamental objection against von Neumann's presentation is that he simply tries to put U_{op} equal to the *sum* of density operators for the ensembles of hidden-variables states that together are supposed to make op E_ψ. This inconsistency is easily remedied. Where von Neumann suggests in his note 173 that one should try to write

$$U(\psi)_{op} = \sum_\beta V(\psi, \xi^{(\beta)})_{op}, \tag{52}$$

where each $V(\psi, \xi^{(\beta)})_{op}$ is defined for a certain interval of ξ which we label by β, we should remember that all density operators are *normalized*, in order that by (20) or possibly (23) they may provide *averages*. Therefore, before they are added, each $V(\psi, \xi)_{op}$ should be multiplied by the probability w_β with which the eigenvalues of observables described by $\xi^{(\beta)}$ occur in the ensemble E_ψ described by $U(\psi)_{op}$. By (10), this is in a theory of the first kind the probability $\bar{p}_\beta$ with which the $\xi^{(\beta)}$ occur in an equilibrium distribution. Thus, instead of (52), we shall try whether we can write U_{op} in the form

$$U(\psi)_{op} = \int_{(\beta)} p_\beta V(\psi, \xi^{(\beta)})_{op}. \tag{53}$$

This improvement, of course, does not make it possible to write U_{op} really as a superposition of density matrices for dispersionfree states (ψ, ξ) of which each could be used to calculate expectation values of all observables in such a dispersionfree state. The reason for the impossibility of making this work, however, is not the reason given by von Neumann. The reason is *not* that dispersionfree hidden-variables states *could not exist* in any *reasonable* hidden-variables theory of the nonzeroth kind. The reason for the impossibility of even the corrected eq. (53) with each $V(\psi, \xi)_{op}$ satisfying (23) for all observables A, B, etc., lies in the *nonexistence of such* $V(\psi, \xi)_{op}$ for such existing states according to Section 2.2.

We saw in Section 2.2 that we *could* define $V(\psi, \xi)_{op}$ if we would relax our requirement of the validity of (23) for all observables, and would require its validity for not more than two noncommutative observables in a two-dimensional Hilbert space, provided we also relinquish the nonnegative definitiness of V_{op}. Let us see what happens to the resolution of U_{op} by (53) in this "mutilated" case. Again we shall consider the example of Section 2.2 in a two-dimensional Hilbert space. We show in Appendix A that, with (23) valid only for the observables A and B considered in eqs. (34)–(46) of Section 2.2, it follows, indeed, that (53) and not (52) is valid, and we also find that in this case the hidden-variables density matrices $V(B_1)_{ji}$ and $V(B_2)_{ji}$ for the two hidden-variables states compatible with our choice $\psi = \phi_1$, $\langle A\rangle_{\psi,\xi} = A_1$ are *not* proportional to the density matrix U_{ji} for this pure state ψ. This shows that in this case von Neumann's theorem about this supposed proportionality is not valid,

[18] Von Neumann (1932), page 170, or von Neumann (1955), page 322.

so that apparently at least one of the assumptions under which he derived this theorem[18] must be violated in our case. It is easily seen (Appendix A) that a condition here violated is von Neumann's requirement that V_{op} should be nonnegative definite. Both $V(B_1)_{op}$ and $V(B_2)_{op}$ here violate this condition.

At the end of Section 2.2, by a different choice of V_{op} we made V_{op} nonnegative definite. For this purpose we had to drop the requirement of validity of (23) for *any* observable B of which ψ is not an eigenfunction. In Appendix A we show that in this case (53) takes the form (A-12), with probability 1 for the state $C = C_1$ with $V_{ji}^{(A)}(C_1) = U_{ji}^{(A)} = \begin{pmatrix} 1 & 0 \\ 0 & 0 \end{pmatrix}$, and probability 0 for the state $C = C_2$ with $V_{ji}^{(A)}(C_2) = \begin{pmatrix} 0 & 0 \\ 0 & 1 \end{pmatrix}$. Here, $w_1^{(C)}\ V_{ji}(C_1)$ and $w_2^{(C)}\ V_{ji}(C_2)$ add up to U_{ji}, and both are this time proportional to U_{ji}, as required by von Neumann, but the latter is proportional to U_{ji} only in a trivial way, by $w_2^{(C)} = 0$, because $V_{ji}(C_2)$ certainly is *not* proportional to U_{ji}.

We thus find no serious fault with von Neumann's mathematical results, but we must object against the conclusion he draws from these results. [See the paragraph below eq. (53).] As no generally applicable V_{op} exist, attempts at decomposition of U_{op} in terms of such V_{op} are doomed to futility. But this is no news. In a hidden-variables theory of the nonzeroth kind, in which we postulate neither (31) nor the existence of $V(\psi, \xi)_{op}$, the nonexistence of $V(\psi, \xi)_{op}$ *is to be expected;* therefore, it also should be impossible to give even (53) a meaning. As discussed at the end of Appendix A as well as below eq. (24) above, this cannot be a criterion by which to disqualify the theory, as the theory *has neither need nor necessity* for assuming relations like (53).

See Appendix K for more on the V_{op} problem.

Reason Two

The second reason given by von Neumann[17] for the nonexistence of hidden-variables states is that they should be quantum-mechanical states represented by ensembles for which his axioms including our eq. (31) would be valid, and for which, additionally, for any observable A the postulate

$$\langle A^2 \rangle_{\psi, \xi} = (\langle A \rangle_{\psi, \xi})^2 \tag{54}$$

would be valid.

As we explained in Section 2.1, any realistic hidden-variables theory drops the postulate eq. (31), thus eliminating the contradiction between assumptions. As, instead, the postulate (6) with (19) is assumed, (54) reduces to the trivial relation (51). Then there is no longer any question of nonexistence of microstates (ψ, ξ) that are dispersionfree in the sense of (54).

Reason Three

As a third reason for the nonexistence of hidden variables, von Neumann[19] alludes to the impossibility of a relation like (31) and to the validity of inequalities like (28) in hidden-variables theory, which he considers to be in conflict with the relations (32) and (30) of

[19] Von Neumann (1932), pages 171 and 167, or von Neumann (1955), pages 325 and 313–314.

quantum theory. We have already told in general terms that the inequality (28), for observables A, B, and $(A+B)$ that are not simultaneously measurable, is not only unavoidable, but also is desirable (see footnote 13), and that the validity of (30) or (32) for quantum theory is not something that should be postulated as if it were an *a priori* obvious triviality, but that it is a happenstance which follows from more elementary parts of the theory which von Neumann unfortunately did not choose to be his fundamental axioms. Though the reader may already have agreed with the general arguments given by us in the preceding in favor of being rather glad than sad about (29) for hidden-variables theory, we want to wash away here any doubt that might still be lingering in any reader's mind about the desirability of (29), by showing, by a very trivial and well-known example taken from the elementary theory of atomic states, that the inequality (29) is a direct consequence of the very fundamentals of quantum theory itself, as soon as it is granted that we may put a label ξ on any individual system for which we claim that somehow we can tell, for whatever measurement the reader plans to make on the system, what the result of the measurement is going to be.[20]

We shall consider here the well-known relation

$$\mathbf{L}^2 = L_x^2 + L_y^2 + L_z^2, \tag{55}$$

and we shall assume that it is possible to measure $\mathbf{L}^2$ or L_x or L_y or L_z, though not necessarily all of them simultaneously. We shall also assume that the atom on which we shall perform the measurements is an atom in a state with quantum numbers $l = 3$ and $m_l = 1$. In this state ($\psi \propto Y_3^1$), we are sure that measurement of $\mathbf{L}^2$ would give the result $12\hbar^2$, while measurement of L_z would give $+\hbar$, so that $L_z^2 = \hbar^2$.

If, therefore, the relation (55) *would* hold between predictions of results of measurements to be taken in the future, with each measurement taken on one single physical system (*no* average over many systems), our predictions for individual measurements of L_y and of L_z would have to satisfy $L_y^2 + L_z^2 = \mathbf{L}^2 - L_x^2 = 11\hbar^2$. On the other hand, whatever the atom's state is, and irrespective of whether we do or do not describe it by hidden variables ξ, quantum theory tells us that we should predict squares of multiples of $\hbar^2$ for both L_y^2 and L_z^2.

There are, however, no two squares of integers that could add up to 11. Consequently, pure quantum theory tells us that the rule (55) cannot possibly hold in this case for any physically possible predictions of results of measurements of L_x^2, L_y^2, L_z^2, and of $\mathbf{L}^2$, notwithstanding the fact that (55) is valid not only in classical mechanics but as an operator equation also between the hermitian operators for these observables.

This impossibility of the general validity of rules like (55) for reasonable predictions of results of individual measurements has never in the past seriously bothered anybody who was studying quantum theory. The fact that somebody labels by values of ξ whatever these predictions of measurement results may be, should not change this lack of concern about the invalidity of (55) for results of measurements of the individual terms in (55) on separate individual physical systems.

[20] Since this is the most fundamental one of all postulates of hidden-variables theory, this is not derived but must be *granted* in any hidden-variables theory.

Then, however, it is equally natural that (55) should not hold when each term is labeled by the symbol $\langle\,\rangle_{\psi,\xi}$, because these averages over trivial ensembles $E_{\psi,\xi}$ of identical measurement predictions are obviously equal to the individual measurement predictions themselves.

This, then, shows that inequalities like (29) should be expected according to the rules of quantum theory itself.

Some other illustrations

What we showed above at the hand of the equality (55) was shown by Bell (1966) at the hand of the equality $A = \alpha + \boldsymbol{\beta}\cdot\boldsymbol{\sigma}$, where $\boldsymbol{\sigma}$ are the Pauli matrices. The example $\sigma_{\hat{n}} \equiv \hat{\mathbf{n}}\cdot\boldsymbol{\sigma}$ would do equally well (where $\hat{\mathbf{n}}$ is some unit vector).

Conclusion about von Neumann's arguments against hidden variables

If the reader is left by the preceding pages with the feeling that we have been diligently explaining the obvious that does not need any explanation at all, this is fine with us. I feel so myself. I always have been puzzled how people could ever have been convinced by von Neumann's arguments that hidden variables could not be introduced. The lack of validity of (31) in any decent hidden-variables theory should have been obvious to anybody by inspection. Von Neumann's choice to pick the quantum-mechanical rule (32) as one of the axioms of quantum theory cannot change this fact and need not trouble us too much as long as this axiom is firmly labeled by subscripts ψ or U_{op} which clarify its inapplicability to states determined by ξ in addition to ψ.

So, what we have been discussing should have been obvious. The truth, however, happens to be that for decades nobody spoke up against von Neumann's arguments, and that his conclusions were quoted by some as the gospel. There must be some magic in his arguments that could fool people into believing that *his* definition of a hidden-variables theory would be the only correct one rather than the obviously inappropriate one.

Trying to dispel the charm of this magic is my excuse for using so many pages for discussing von Neumann's work.

CHAPTER 3

THE WORK OF GLEASON AND OF KOCHEN AND SPECKER

KOCHEN and Specker (1967) claim to have found another proof that hidden-variables theories would all be impossible. Though this conclusion about the meaning of their result is too sweeping, their "proof" is very instructive as a demonstration that hidden variables cannot predict what "is" but only what "would be found." (See Section 1.9.) Moreover, it is a good introduction to the discussion of some conclusions for hidden-variables theory that can be drawn from the earlier and more general work of Gleason (1957).

Kochen and Specker consider a case in which the inequalities (28)–(29) of the end of Chapter 1 happen to be replaced by equalities. It is a case where the operators A and B there added together, though not commutative in general, have some eigenfunctions in common, and where the states ψ considered all are simultaneous eigenfunctions of the operators that are summed, so that for these states the operators behave *as if* they were commutative.

More specifically, consider the relation

$$(\mathbf{J}^2) = (J_x^2)+(J_y^2)+(J_z^2) \tag{56}$$

for a $j = 1$ wavefunction. It really does not matter whether we work with **J** or with **L** or with **S**. In part of their paper they replace **J** by **S** and talk about a triplet state. In general, if ϕ_m is a state with

$$J_z\,\phi_m = m\,\hbar\,\phi_m \quad \text{and} \quad \mathbf{J}^2\,\phi_m = j(j+1)\,\hbar^2\,\phi_m \quad \text{for} \quad |m| \leq j,$$

and if we put

$$\phi_m = 0 \quad \text{for} \quad |m| > j,$$

it is not hard to show that

$$(J_x^2 J_y^2 - J_y^2 J_x^2)\,\phi_m = \hbar^4\,[(1+m)\sqrt{(j+m+2)(j+m+1)(j-m)(j-m-1)}\,\phi_{m+2}$$
$$+(1-m)\sqrt{(j+m)(j+m-1)(j-m+2)(j-m+1)}\,\phi_{m-2}]\cdot \tag{57}$$

For $j = 1$, $\phi_{m\pm 2}$ differs from zero only if $m = \mp 1$, and in that case the factor $(1\pm m)$ in front is zero. Therefore, for $j = 1$, J_x^2 commutes with J_y^2. For reasons of symmetry, then, in that case J_x^2 and J_y^2 and J_z^2 all commute with each other, and, of course, also with $\mathbf{J}^2$. Thus all terms in (56) then commute with each other, and all of them are simultaneously measurable.

In other words, for $j = 1$ the operators J_x^2, J_y^2, and J_z^2 have simultaneous eigenfunctions. When eq. (56) operates on such an eigenfunction, each term of the equation multiplies this eigenfunction by the corresponding eigenvalue, which is $2\hbar^2$ for $\mathbf{J}^2$, and either $\hbar^2$ or 0 for each of the three terms on the right. It follows that in this case *one of the three operators* J_i^2 *must have the value* 0, *and the other two have the value* $\hbar^2$ *each*. (Here, $i = x$ or y or z.) For the simultaneous eigenfunction for which $J_i^2 = 0$, also $J_i = 0$. For the two other components with $j \neq i$, then, we must have $J_j^2 = \hbar^2$, so that this simultaneous eigenfunction must be a linear combination of the eigenfunctions of J_j to the eigenvalues $+\hbar$ and $-\hbar$.

3.1. How to measure simultaneously J_x^2, J_y^2, and J_z^2 in a $j = 1$ state

Kochen and Specker apply the above to orthohelium in its lowest orbital state ($n = 2$, $l = 0$, $s = 1$). A method of simultaneously measuring J_x^2 and J_y^2 and J_z^2 in arbitrarily chosen x,y,z-directions then purportedly is the following. By the application of a small electric field of rhombic symmetry, the energy of the helium atom is perturbed by

$$H_s = aS_x^2 + bS_y^2 + cS_z^2. \tag{58}$$

Here, $\mathbf{S} = \mathbf{J}$. The eigenstates of this perturbation are the same as mentioned above, and the perturbation energy is $(b+c)\hbar^2$ or $(c+a)\hbar^2$ or $(a+b)\hbar^2$ depending on which of the three J_i^2 operators has the value zero. So by measuring the energy of the atom we can determine, for the three orthogonal coordinate axes determined by the rhombic electric field, along *which* of these three axes is $J_i^2 = 0$, and then $J_j^2 = \hbar^2$ along the other two axes.

In as far as we do not tell how the rhombic field is applied, our experiment is a "thought experiment." (Electric fields of rhombic symmetry sometimes are found in certain crystals at the sites of the atoms, but we here assume that we can at will apply this field to a chosen individual atom with the symmetry axes in any directions.)

More specifically, suppose we excite an orthohelium atom to its 2^3P_0 state, from where, in absence of a rhombic field, it could drop down to *any* 2^3S_1 state under emission of a photon. However, we want to measure whether it ends up with its spin in the y,z-plane or in the z,x-plane or in the x,y-plane; so, we want to know for *which* of three perpendicular directions chosen by us, we have $J_n^2 = 0$ and $V_n = 1$ in the final state, where for later convenience we define

$$V_i = 1 - [J_i^2/\hbar^2] \quad (i = x \text{ or } y \text{ or } z). \tag{59}$$

So we apply a small and known electric field with rhombic symmetry to the atom as we excite it, and thus force the atom, as it emits its photon, to choose between the three levels mentioned above for its ground state. Catching the photon in a spectrograph, we find its wavelength and thence the final (perturbed) energy level of the atom, which then tells us which of the three quantities V_x, V_y, and V_z has the value 1, and we then know that the other two will be zero.

If hidden variables exist, knowledge of their value should predict the outcome of this thought experiment.

3.2. The Kochen–Specker paradox

The task of a set of values ξ of the hidden variables, in the above case, is to predict, *for any given orientation of the x,y,z-frame, which one* of the three chosen orthogonal directions is the direction $\hat{n}$ for which $J_n^2 = 0$ and $V_n = 1$.

Kochen and Specker ran into a paradox by making an assumption which at first sight seems harmless but which turned out to lead to contradictions. The assumption was that it would be possible to talk, for any value of ξ and for any direction $\hat{n}$ in space, about the "the" predicted value $J_n^2(\xi)$ or $V_n(\xi)$, without mentioning the other two perpendicular directions that together with $\hat{n}$ make up the set of principal axes of the rhombic field used in the measurement.

For brevity, let us introduce the name "triad" for any set of three *orthogonal* unit vectors. For any cartesian coordinate system, the unit vectors $\hat{x}$, $\hat{y}$, and $\hat{z}$ along the axes would form a triad.

Let T and T' be two triads that have one of their unit vectors (say, $\hat{n}$) in common, so that T' is obtained from T by rotating it around $\hat{n}$. Then, the assumption made by Kochen and Specker was that, in this case, $J_n^2(\xi)$ would have the same value irrespective of whether $\hat{n}$ was considered part of T or part of T'.

If this were true, one could define, *for given* ξ, a set $\mathcal{N}$ of directions $\hat{n}$ so that $J_n^2(\xi) = 0$ and $V_n = 1$ for each direction $\hat{n}$ in $\mathcal{N}$. The complementary set $\mathcal{N}'$ of all remaining directions then should be the set of directions $\hat{n}'$ for which $J_{n'}^2(\xi) = \hbar^2$ and $V_{n'} = 0$.

In the following we shall continue considering the case that a set of hidden variable values ξ is given and kept unchanged during the reasoning, even when we do not repeatedly mention this fact or when we omit the indication (ξ) as argument of J_i^2 in the formulas. So, what Kochen and Specker assumed was that there exists (for each ξ) a set $\mathcal{N}$ of directions so that

$$\left.\begin{array}{llll} V_n = 1 & \text{and} & J_n^2 = 0 & \text{if } \hat{n} \text{ is } \textit{in } \mathcal{N}; \\ V_{n'} = 0 & \text{and} & J_{n'}^2 = \hbar^2 & \text{if } \hat{n}' \text{ is } \textit{outside } \mathcal{N} \text{ (i.e., if } \hat{n}' \text{ is } \textit{in } \mathcal{N}'). \end{array}\right\} \tag{60a}$$

The paradox to which this seemingly harmless assumption (60a) led was that Kochen and Specker then proved that *there does not exist any set* $\mathcal{N}$ *of directions* such that:

(a) the assumptions (60a) are valid;

(b) for any *triad* of directions, one is a direction $\hat{n}$ for which $J_n^2 = 0$, $V_n = 1$, and the other two are directions $\hat{n}'$ for which $J_{n'}^2 = \hbar^2$, $V_{n'} = 0$; (60b)

(c) correspondingly, for any *two* perpendicular directions $\hat{l}$ and $\hat{m}$, *not both* J_l^2 and J_m^2 will be zero simultaneously, but one or both of them must be $= \hbar^2$; that is, *not both* V_l and V_m will be 1, but one or both of them must be zero. (60c)

Since the conditions (60b, c) have to be met according to eq. (56) for $(\mathbf{J}^2) = 2\hbar^2$, Kochen and Specker thought that they had disproved the possibility of the existence of a hidden-variables theory.

What they actually had proved was, of course, that the assumption (60a) was erroneous. That is, hidden variables in general do not assign a unique value to J_n^2 for each given direction $\hat{n}$, but, instead, for a given triad $\hat{l}, \hat{m}, \hat{n}$, they tell which one among J_l^2 and J_m^2 and J_n is $= 0$, and then the other two are $= \hbar^2$. We shall return to this point in Section 3.4. First, we shall indicate some proofs of the impossibility of meeting the conditions (60b, c) under the *assumption* (60a).

3.3. Proofs of the Kochen–Specker paradox

Kochen and Specker themselves proved the incompatibility of (60b, c) with (60a) by considering an ingenious choice of 117 different directions $\hat{n}$ of which many were perpendicular to each other. By repeatedly applying (60b, c) to these 117 directions while assuming the uniqueness (60a) of the value of each J_n^2 (always for given ξ), they showed that it was impossible to avoid contradictions to the rules (60b, c) for some of their directions. For details, see pages 68–70 of the original paper.

For getting some feeling for where the trouble lies, consider the following more illustrative "proof" of the incompatibility of (60a–c). Each direction $\hat{n}$ corresponds to a point on the unit sphere around the origin. We shall color this point red if by (60a) it corresponds to $J_n^2 = 0$, $V_n = 1$, and we shall color it blue if it corresponds to $J_n^2 = \hbar^2$, $V_n = 0$. Since $(-J_n)^2 = J_n^2$, points on the sphere that are each other's antipodes will have the same color.

Any point P and its antipode are the poles of a great circle C representing all directions perpendicular to $\hat{n}$. According to (60c) this circle C for any *red* point P must be a *completely blue* circle. Since, thus, each red point introduces many more blue points, we find that it is impossible to get enough red color onto the sphere so as to make one-third of the sphere red, as condition (60b) suggests we would need.

We could, for instance, start by coloring the inside of the first octant red. Talking in terms of longitudes and latitudes as on the earth, this would be the region between 0° and 90° east on the northern hemisphere. We would also color the adjacent meridians red, including the north pole where they intersect, but then we must color the equator blue, including the part where it borders the red region.

Then, also red will be the antipode octant, between 180° and 90° west on the southern hemisphere, with adjacent meridians and south pole.

Drawing blue great circles C for all points P colored red so far, we find that they cover up all the remainder of the sphere; so we cannot add more red. This is our downfall, because now less than one-third of the sphere is red, and, consequently, there are many triads for which all three directions are blue points on the sphere. For instance, consider the directions with components approximately given by $\left(\sqrt{\tfrac{1}{2}}, \sqrt{\tfrac{1}{2}}, -\varepsilon\right)$, by $\left(\tfrac{1}{2}+\tfrac{1}{2}\varepsilon, -\tfrac{1}{2}+\tfrac{1}{2}\varepsilon, \sqrt{\tfrac{1}{2}}\right)$, and by $\left(-\tfrac{1}{2}+\tfrac{1}{2}\varepsilon, \tfrac{1}{2}+\tfrac{1}{2}\varepsilon, \sqrt{\tfrac{1}{2}}\right)$, where $0 < \varepsilon \ll \tfrac{1}{10}$. If $\varepsilon^2 \approx 0$, these three directions are perpendicular to each other, and they lie in three different blue octants. Thus we have violated the condition (60b).

A more rigorous proof is given in Appendix B. Instead of 117 points on the unit sphere, it uses only eight points. Applying rule (60b) for determining their colors, we find that we can contradict rule (60c).

Still another proof was given already in 1957 by Gleason.[21] We shall outline it in Section 3.6, and in Appendix C we shall reproduce an abbreviated version of this proof given by Bell (1966).

3.4. Conclusion drawn from the Kochen–Specker paradox

As mentioned in Section 3.2, the Kochen–Specker paradox shows that it is not allowed in hidden-variables theory to make the assumption (60a). That is, it is not true that the great circle C for a red point P is necessarily all blue. For instance, if we measure J_z^2 measuring the perturbation energy (58) in a rhombic electric field with principal axes $\hat{x}$, $\hat{y}$, $\hat{z}$, it may be true that a particular set of hidden-variable values ξ predicts $J_z^2 = 0$, so that we may want to color the point (0, 0, 1) on the unit sphere red, and the points (1, 0, 0) and (0, 1, 0) blue. However, when we rotate the rhombic field by an angle ϕ around the z-axis, it is quite possible, according to eq. (6), that the same internal hidden-variable values ξ predict $J_z^2 = \hbar^2$ for this rotated measurement of J_z^2. If so, the point (0, 0, 1) would have turned blue, and one of the perpendicular directions would produce a red point on the equator in the x,y-plane.

3.5. Generalization of the Kochen–Specker paradox to Hilbert space

We shall now generalize the Kochen–Specker paradox as expressed in terms of the quantities V_i defined in eq. (59). In terms of the V_i, this paradox amounted to the following. In a three-dimensional space ($N = 3$) for a given hidden-variables state (ψ, ξ), let a function $v(\hat{n})$ of the variable unit vector $\hat{n}$ be given with the following properties:

(a) Every $v(\hat{n})$ is either 1 or 0; (61a)

(b) For any orthogonal set $\{\hat{n}^{(i)}\}$, with $1 \leq i \leq N$, let

$$\sum_{i=1}^{N} v(\hat{n}^{(i)}) = 1, \tag{61b}$$

so that for $N = 3$ for each triad $\{\hat{n}^{(i)}\}$ one of the three $v(\hat{n}^{(i)})$ will be $= 1$, and the other two $v(\hat{n}^{(i)})$ will be zero.

Then we have seen in Section 3.3 that it is impossible that v will depend on n only. There just does not exist any way of assigning to each direction $\hat{n}$ a value $v = 1$ or $v = 0$ in a way satisfying conditions (61a, b). It *is* possible for each triad $\{\hat{n}^{(i)}\}$ to select the component unit vector $\hat{n}^{(k)}$ privileged by $v(\hat{n}^{(k)}) = 1$; however, when we do so, we know from the Kochen–Specker paradox that necessarily there must be directions $\hat{n}$ such that $v(\hat{n})$ changes to and fro between 0 and 1 when we take a triad with $\hat{n}$ as one of its component unit vectors $\hat{n}^{(k)}$ and then rotate this triad, keeping $\hat{n}^{(k)} = \hat{n}$ fixed and rotating the triad components $\hat{n}^{(j)}$ normal to $\hat{n}^{(k)}$.

For this formulation of our results of Sections 3.3–3.4, the proof in which we painted $v = 1$ red and $v = 0$ blue on the unit sphere remains as valid as it was before. However, in this formulation there is nothing that tells us that $v(\hat{n})$ must mean the V_n of eq. (59). We

[21] Gleason (1957), pages 887–889.

therefore can now forget about the spin of an orthohelium atom, and we can replace $v(\hat{n})$ by any quantity different from (59) that has the same properties (61a, b).

According to Jauch and Piron (1963), we would find such $v(\hat{n})$ in the "truth values" assigned for a given hidden-variables state (ψ, ξ) to "propositions" about Hilbert vectors $\hat{n}$ in a three-dimensional Hilbert space. We shall now first explain this claim.

Truth values of propositions in quantum theory

A "proposition" in quantum theory is a statement that a physical system is in a state corresponding to a Hilbert vector in some given subspace of Hilbert space. This subspace may be a single Hilbert vector or it could be a more-dimensional subspace spanned by several Hilbert vectors. An example of the latter would be the Hilbert subspace corresponding to a degenerate eigenvalue of some operator.

We shall discuss propositions corresponding to more-dimensional subspaces of Hilbert space in Chapter 4 in a discussion of some of the work of Jauch and Piron. For the time being we shall confine ourselves to propositions for single Hilbert vectors, like the proposition "the physical system under consideration is in the quantum state ϕ_n." As applied to reproducible measurements of which in hidden-variables theory we want to predict the outcome, we water down propositions of this kind to statements like: "If we would verify it by any appropriate measurement, we would find the system to be in state ϕ_n."

Jauch and Piron (1963)[22] and Kochen and Specker (1967) have assumed that in *any* hidden-variables theory the hidden variables necessarily should be able to tell uniqely whether such a proposition would be "true" or "false." That is, according to them it should be possible, for any given hidden-variables state (ψ, ξ), to assign to every Hilbert vector ϕ_n a "truth value" v such that $v = 1$ would mean that the proposition mentioned above concerning the system being in state ϕ_n would be true, while $v = 0$ would mean that this proposition would be a false statement.

These v, then, according to them would depend only on ψ, ξ, and on the state ϕ_n about which the statement is made:

$$v = v(\phi_n; \psi, \xi), \tag{62}$$

and it would, for fixed ψ and ξ, satisfy all the conditions (61a, b) if there we read $\hat{n}$ as ϕ_n and $\hat{n}^{(i)}$ as ϕ_i. As some kind of a justification of this claim, which may also be interpreted as a classification of the kind of hidden-variables theories these authors consider, the following argumentation may serve.

When a measurement of some physical observable A distinguishes between an orthonormal set $\{\phi_i\}$ of eigenfunctions of A_{op}, the measurement, if any good, should assign the truth value $v = 1$ to just one of the states ϕ_i, while it should assign $v = 0$ to all other ϕ_i. This would provide the validity of (61a, b) for $v = v(\phi_i; \psi, \xi)$ for this particular type of measurement.

[22] It is open for discussion whether or not Jauch and Piron really meant to make the assumption about truth values ascribed to them in the following paragraphs. Their work, at least, has often been *interpreted* to make this assumption. See Chapter 4 for a more detailed discussion of this historical question.

However, in (62) we assign a value of v to *every* vector ϕ_n in Hilbert space (like Kochen and Specker assigned a value 0 or $\hbar^2$ to J_n^2 for *every* direction in ordinary space). We therefore must reverse the argument, and for this we need a hypothesis which dates back to von Neumann (1932, 1955). This rather serious assumption, criticized by Bohm (1953b), is the following. While it is obvious that any Hilbert vector ϕ_n can be completed to a complete orthonormal set of Hilbert vectors $\{\phi_i\}$ of which ϕ_n is of member, and while it is equally obvious that in infinitely many ways each complete set $\{\phi_i\}$ can be regarded as the complete set of eigenfunctions of some hermitian operator A_{op} (of which the eigenvalues A_i yet may be chosen randomly), it is assumed that

> "one or more of these A_{op} will always correspond to a physical observable A that in principle could be measured, so that the laws of physics about results of measurements would apply to this observable A." (63)

This would guarantee that, for any set $\{\phi_i\}$, we can always find at least one observable A which would allow us to follow the above reasoning, justifying the validity of (61a, b) for the truth values for the propositions concerning each ϕ_i being the result of a measurement of A. However, in order to arrive from this at the conclusion (62), we must make an additional hypothesis, which is that

> "all measurements that, in a given hidden-variables state (ψ, ξ), could verify the value of the truth value $v(\phi_n, \psi, \xi)$ for a chosen state vector ϕ_n (*i*.e. for a chosen proposition about ϕ_n), will predict the same truth value 1 or 0 for $v(\phi_n, \psi, \xi)$, irrespective of the choice of the observable A utilized (as long as ϕ_n is one of its eigenfunctions), and, in particular, irrespective of the choice of an orthonormal set $\{\phi_i\}$ of which ϕ_n is a member." (64)

The condition (64) placed upon the hidden-variables theories considered by Jauch and Piron[22] and by Kochen and Specker would give $v(\phi_n, \psi, \xi)$ the lack of dependence upon the choice of the set $\{\phi_i\}$ of eigenfunctions between which a measurement of A would differentiate, which was the requirement essential for leading to the Kochen–Specker paradox.

In the following we will sometimes drop ψ and ξ as written arguments of the function v, understanding the dependence of v on ψ and ξ tacitly. Then (62) might be written as $v(\phi_n)$, or v_n for short.

Truth values in "realistic" hidden-variables theories

While one may have his doubts about the desirability of the assumption (63) and especially of (64), a confrontation of (62) with our fundamental postulate (6) shows that in "realistic" hidden-variables theories, like the theories of Wiener and Siegel or of Bohm and Bub which will be discussed in Part II, the above claim (62) is not made. As stated in (6), for a given hidden-variables state (ψ, ξ), the ϕ_n which is realized in a measurement (the ϕ_n for which $v = 1$) is picked from a given orthonormal set $\{\phi_i\}$ *provided that this set* $\{\phi_i\}$ *first is given uniquely*. So, if $\{\phi_i\}$ is given, we have

$$v = v(\phi_n; \psi, \xi, \{\phi_i\}) \tag{65}$$

instead of (62). In terms of the function n of (6), we may also write v as a Kronecker delta symbol, as we did in eq. (27) of Section 1.10.

Application of the Kochen–Specker paradox to truth values of propositions

In Appendix D we justify the application of the Kochen–Specker paradox to $v(\hat{n})$ with $\hat{n}$ interpreted as meaning a Hilbert vector ϕ_n in Hilbert space, and with $v(\hat{n})$ satisfying the conditions (61a, b), not only for $N = 3$, and not only for vectors $\hat{n}$ in a real space (for which we provided in Section 3.3 proof of the paradox), but for Hilbert vectors ϕ_n in complex Hilbert spaces with $N(\geqslant 3)$ dimensions. Where in Section 3.3 we would consider a triad $\{\hat{n}^{(i)}\}$ and rotate it by considering transformations like

$$\hat{n}^{(j)\prime} = \sum_{i=1}^{3} X_j^i \hat{n}^{(i)} \quad (j = 1, 2, 3)$$

with real coefficients X_j^i forming an orthogonal matrix, in the generalization we would consider complete orthonormal sets $\{\phi_i(x)\}$ of Hilbert vectors, and would consider transformations

$$\phi_j'(x) = \sum_{i=1}^{N} X_j^i \phi_i(x) \quad (j = 1, 2, \ldots, N), \tag{66}$$

where the X_j^i finally would be generalized to the complex elements of a unitary matrix. Again, it follows that it is impossible to satisfy the conditions (61a, b) if we maintain as in (64) that, for given ψ and ξ, $v(\phi_n)$ shall depend on ϕ_n only. We shall call this impossibility the "generalized Kochen–Specker paradox." Again, this "paradox" will disappear when we allow v to jump to and fro between 1 ("red") and 0 ("blue") for a fixed choice of ϕ_n, when the complete orthonormal set $\{\phi_i\}$ of which ϕ_n is part is unitarily transformed in such a way that ϕ_n is left invariant.

Nonexistence of truth values for propositions

The validity of Kochen and Specker's paradox for the truth values (62) of Jauch and Piron[22] and of Kochen and Specker proves[23] that these truth values (for $N \geqslant 3$) do not exist. This is, of course, exactly what these authors set out to prove. Here we completely agree with their result. What they have shown is that hidden-variables theories that for propositions are required to predict truth values of the kind of (62) are theories of the zeroth kind.

However, nothing is proved that would infer nonexistence of the truth values (65) or (27) for "realistic" hidden-variables theories (of the nonzeroth kind).

[23] At least for the propositions corresponding to single Hilbert vectors considered so far. See Chapter 4 for more general considerations.

Thus all we have shown is the *necessity* of including $\{\phi_i\}$ as an argument of the function n appearing in eqs. (27) and (6). As seen from Appendix D, the proof of this necessity boils down to the proof of the Kochen–Specker paradox. This paradox can also be proved starting from earlier work of Gleason (see Section 3.6). Therefore the proof of the impossibility of assigning truth values of the type (62) for all Hilbert vectors ϕ to hidden-variables states (ψ, ξ) is sometimes called "Gleason's proof of impossibility of hidden-variables theories (of the zeroth kind, of course)." In Appendix C we reproduce a simpler version of Gleason's proof of the Kochen–Specker paradox suggested by Bell (1966).

The price of determinism

Omitting $\{\phi_i\}$ from the function $n(\psi, \xi, \{\phi_i\})$, and therefore also omitting $\{\phi_i\}$ from (63) as attempted by Jauch and Piron and by Kochen and Specker in (62), amounts to denying that the *way* in which a measurement is made could have an effect on the result of that measurement. These authors, of course, do admit that the question whether the measurement will make the system's state condense onto ϕ_n is a meaningless question if ϕ_n is not at all among the set $\{\phi_i\}$ of possible outcomes of the measurement. The Kochen–Specker paradox, however, shows that this is not enough. Whether one would find $J_z^2 = 0$ by the method suggested in Section 3.1 depends, for a given hidden-variables state (ψ, ξ), not only on the z-axis of the rhombic field used in the measurement, but also on the x- and y-axes of this field. In general, the question whether or not ϕ_n is the outcome of a measurement that permits ϕ_n as a possible outcome, depends also on the other outcomes that the measurement permits for different ξ values, if it depends deterministically on any describable initial state (ψ, ξ) at all. This is one price that one has to pay if one wants to save determinism.

The necessity of the restriction imposed by the Kochen–Specker paradox upon hidden-variables theories, that the result $\phi_{n(\psi, \xi, \{\phi_i\})}$ of a measurement must depend on the entire set $\{\phi_i\}$, is also shown by some examples discussed by Turner (1968).

Theories that disregard this restriction are intrinsically inconsistent and therefore are theories of the zeroth kind. An example of such work of the zeroth kind is found in a paper by Misra (1966), who explicitly[24] rejected the above restriction.

3.6. The work of Gleason

We came to the above reasoning by an application of the ideas of Kochen and Specker to Hilbert space. However, the same result is also obtainable from the much earlier work of Gleason (1957).

Contrary to the above, Gleason was not primarily concerned with hidden variables. The problem he set out to solve was a problem in pure quantum theory without hidden variables. He wanted to prove that in ordinary quantum theory probabilities can always be calculated with the help of a density matrix. (An improvement of the work of von Neumann.)

[24] Misra (1966), lower half of page 858.

Gleason's proof of existence of a density matrix for quantum-mechanical states

In the discussion of Gleason's work, we do not talk about truth values v_n, but about the quantum-theoretical probabilities $w_n = w(\phi_n)$ for finding experimentally that a state ϕ_n is realized. Gleason therefore postulates that w as a function of the arbitrary "unit vector" ϕ_n in Hilbert space $\mathfrak{H}$ would be given by a "nonnegative frame function $f(\phi_n)$ of weight $W = 1$."[25] He defines a *frame function of weight W* as a real function $f(\phi)$ of normalized vectors ϕ in $\mathfrak{H}$ such that, if $\{\phi_i\}$ is any orthonormal basis of $\mathfrak{H}$, then $\sum_i f(\phi_i) = W$. Gleason's postulates thus make

$$\sum_i w(\phi_i) = 1 \quad \text{for any complete orthonormal } \{\phi_i\}, \tag{67a}$$

$$0 \leq w(\phi_n) \leq 1. \tag{67b}$$

Gleason then proves in a Hilbert space $\mathfrak{H}$ of three or more dimensions the following theorems. (We just list the more important ones and refer for the proofs to Gleason's original paper.)

(1) Let $\mathfrak{H}_3$ be any three-dimensional subspace of Hilbert space $\mathfrak{H}$, that is, let $\mathfrak{H}_3$ be the space formed by the functions

$$\phi(x) = \sum_{i=1}^{3} a_i \phi_i(x), \tag{68}$$

where ϕ_1, ϕ_2, ϕ_3 are any three orthonormal functions in $\mathfrak{H}$ and where the a_i are complex numbers. Let R_3 be any "real" subspace of $\mathfrak{H}_3$, that is, the space formed by the functions

$$\phi'(x) = \sum_{i=1}^{3} X_i \phi_i'(x), \tag{69}$$

where the $\phi_i'(x)$ are any three orthonormal functions $\phi(x)$ in $\mathfrak{H}_3$, and the X_i are real coefficients. Then Gleason proves a lemma which he neither gives a number nor prints in italics but which says that[26]

> "any nonnegative frame function $w(\phi')$ in R_3 must be a *continuous* function of the coefficients X_i determining ϕ' by eq. (69)." (70)

The proof of this important lemma uses the assumed lack of dependence of each $w(\phi_n)$ upon the other ϕ_j (orthogonal to ϕ_n) which occur as arguments of the w in the sum over i in eq. (67a).

Side remark

Gleason's proof of this lemma remains valid if we replace all w_n in the above by the truth values $v_n = v(\phi_n)$ postulated by Jauch and Piron for the propositions about ϕ_n being a result of a measurement. For these v_n, conditions (67a, b) follow from (61a, b), and the assumed

[25] Gleason (1957), page 893. He makes $W = 1$ for all of $\mathfrak{H}$ only indirectly, by putting $f(\phi_k) = w(\phi_k)$ for some ϕ_k, which achieves the same if $w(\phi_k) \neq 0$.

[26] Gleason (1957), lines 5–4 from bottom of page 889.

lack of dependence of $w(\phi_n)$ upon the choice of a set $\{\phi_i\}$ of which ϕ_n is a member would, according to (62), be matched by the hypothesis (64) about the truth values of propositions. We shall return below to the paradox involved in the validity of the lemma (70) for these truth values that satisfy (62)–(64) in addition to (61a, b). First, we shall quickly complete our outline of Gleason's work.

Back to Gleason

(2) "*Gleason's theorem 2.3*" says that, if the $w(\phi'_n)$ for orthonormal sets $\{\phi'_i\}$ in R_3 depend each only on its own argument ϕ'_n, and if they also satisfy the conditions (67a, b), and are continuous, then $w(\phi')$ must be a quadratic function of the coefficients X_i in (69). [Gleason proves this by showing that $w(\phi')$ could be expanded in terms of spherical harmonics on the unit sphere in R_3.]

(3) By this Theorem 2.3 it follows from the above lemma (70) that any nonnegative frame function in R_3 must be a quadratic function of the X_i. This is Gleason's theorem 2.8.

(4) Let R_2 be any two-dimensional real subspace of any two-dimensional subspace $\mathfrak{H}_2$ of $\mathfrak{H}$. As each R_2 is a subspace of one of the R_3 considered above, it follows that, in any R_2 subspace of $\mathfrak{H}_2$, $w(\phi)$ is a quadratic function of the X_i.

(5) Since[27]

$$w(e^{i\alpha}\phi) = w(\phi) \tag{71}$$

for real α, Gleason then shows that there exists in each $\mathfrak{H}_2$ a self-adjoint 2×2 matrix U so that, for all functions

$$\phi = \sum_{i=1}^{2} a_i\phi_i$$

in $\mathfrak{H}_2$, one has

$$w(\phi) = \sum_{i,j=1}^{2}\sum a_i^* U_{ij}a_j \quad \text{with} \quad U_{ij} = U_{ji}^*. \tag{72}$$

(6) From these 2×2 matrices U_{ij} for subspaces $\mathfrak{H}_2$, Gleason then constructs for $\mathfrak{H}$ a general U matrix so that

$$w(\phi) = \sum_i\sum_j a_i^* U_{ij}a_j \quad \text{for any} \quad \phi = \sum_i a_i\phi_i \tag{73}$$

in all of Hilbert space. In particular, then, $w(\phi_i) = U_{ii}$.

Finally, if P_i is a projection operator in Hilbert space which projects any function ψ onto ϕ_i according to

$$P_i\psi(x) = \phi_i(x)\left(\int \phi_i^*\psi\right), \tag{74}$$

so that the hermitian operator of an observable A with complete orthonormal set of eigenfunctions $\{\phi_i\}$ may be written as

$$A_{\text{op}} = \sum_i A_iP_i, \tag{75}$$

[27] This follows from (67a) in $\mathfrak{H}$ by leaving in $\{\phi_i\}$ all ϕ_i unchanged except one ϕ which is replaced by $e^{i\alpha}\phi$.

Gleason shows that

$$w_i \equiv w(\phi_i) = \text{trace } (U\,P_i), \tag{76}$$

so that

$$\text{trace } (UA_{\text{op}}) = \sum w_i A_i = \langle A \rangle. \tag{77}$$

Thus Gleason proved the result (20) on the basis of lighter assumptions than the ones made by von Neumann, and this was the purpose of Gleason's paper. We, however, want to consider below merely Gleason's first unnamed lemma (70), and draw from it a conclusion which Gleason never printed explicitly in his 1957 paper, but which nevertheless is known as *Gleason's proof of the impossibility of (his kind of) hidden-variables theory*. (Since this kind is characterized by the assumed lack of dependence of the v_n on the $\{\phi_i\}$, it is the zeroth kind.)

"Gleason's proof" (application of Gleason's lemma to truth values)

We mentioned already in our "side remark" above that, if we postulate that the truth values $v_n = v(\phi_n) = v(\psi, \xi, \phi_n)$ which satisfy (61a, b) shall not depend on the choice of the $\{\phi_i\}$ of which ϕ_n is a member, then Gleason's lemma (70) would have to be valid for this function v of ϕ_n because (67a, b) follow from (61a, b) with (62). It thus would follow that v would have to be a continuous function. Then, however, between any ϕ_n for which $v(\phi_n) = 1$ and any $\phi_{n'}$ for which $v(\phi_{n'}) = 0$, there must exist other ϕ for which this function has values *between* 0 and 1. This contradicts our initial assumption that $v(\phi_n)$ would be a truth value always equal to 0 or 1. This suffices to prove the *impossibility* of defining, in a Hilbert space of three or more dimensions, any truth values $v(\phi_n)$ independent of the $\{\phi_i\}$ that would satisfy (61a, b). In other words, we have here another proof of the Kochen–Specker paradox.

As Gleason's proof of his lemma (70) is somewhat messy, we do not give it here. Instead we give in Appendix C the simpler proof given by Bell (1966), which assumes from the beginning that *we* want to apply Gleason's lemma (70) merely to truth values v_i and not to probabilities w_i. Therefore Bell *starts* by assuming that there would exist a truth value $v(\phi_n) = 0$ or 1 which would satisfy (61b) so that it would be a frame function but which would show a discontinuity where v changes from 0 to 1. Then it must be possible to find in R_3 two unit vectors ψ and Φ at some angle $< 26°$ such that $v(\Phi) = 0$ and $v(\psi) = 1$. This is demonstrated by Bell to lead to a contradiction, as he shows that repeated application of (61a, b) in this case yields $v(\psi) = 0$. (See Appendix C.)

Conclusion

We have shown that no possible hidden-variables theory (i.e. no theory of the nonzeroth kind) can assign a set of truth values for all "propositions" of quantum theory to any hidden-variables state. However, "realistic" hidden-variables theories do not even try this. Instead they assign to each hidden-variables state a set of *selection statements* in the form of some algorithm that uniquely determines the function $n(\psi, \xi, \{\phi_i\})$ which in (6) selects the result of a contemplated measurement from a given set $\{\phi_i\}$ of possible results.

The nonexistence of truth values for propositions leads to some troubles with hidden-variables theories which we shall call the *Gleason troubles* of these theories.

3.7. Reducing the impact of the Gleason troubles in hidden-variables theories

If somehow in the future it would become possible to prepare an individual physical system in such a way that we would know not only its quantum mechanical state ψ, but also its hidden variables ξ, we would want the theory to predict unambiguously the result of the next measurement.

The possibility of being able to make a precise prediction exists already in pure quantum theory, namely when ψ is an eigenfunction of the observable to be measured. In this special case, $n(\psi, \xi, \{\phi_i\})$ becomes independent of ξ. That is,

$$n(\phi_k, \xi, \{\phi_i\}) = k, \tag{78}$$

independent not only of ξ, but also of all of $\{\phi_i\}$ except ϕ_k itself.

Also, if $\psi \neq \phi_k$, in any explicit hidden-variables theory the algorithm determining $n(\psi, \xi, \{\phi_i\})$ enables us to *calculate* the selection $n(\psi, \xi, \{\phi_i\})$ for given ψ, ξ and $\{\phi_i\}$.

Our principal worry, however, is whether a given experimental arrangement for making a measurement will enable us to tell with sufficient precision what set $\{\phi_i\}$ describes the possible results of the measurement. There are many cases where there can be doubts.

Manipulating the choice of $\{\phi_i\}$

If we measure an observable A and want to find out whether $A = A_1$ or $A = A_2$, but A_{17} happens to be a threefold-degenerate eigenvalue with infinitely many possible sets $\{\phi_{17,m}\}$ with $m = 1, 2, 3$ spanning the Hilbert subspace corresponding to A_{17}, how do we know which of these sets describes possible results of our measurement?

For instance, in the Bohm–Bub theory, ξ is a vector in Hilbert space. This ξ (luckily) may have a component in the subspace spanned by the $\{\phi_{17,m}\}$. Let ξ_{17} be this component. Now suppose that, among the $\{\phi_{17,m}\}$ which are part of the $\{\phi_i\}$ from which the measurement selects one, there is one $\phi_{17,k}$ which happens to be orthogonal to ξ_{17}. Then this theory predicts for the *slightest* nonzero contribution of $\phi_{17,k}$ to ψ, that the measurement is going to yield $\phi_{17,k}$ even if ϕ_1 and ϕ_2 contribute much more to ψ. [See Part II, Section 3.6, eq. (120).]

The reason why this fact does not lead to a contradiction between hidden-variables theory and quantum theory is that the probability is very small indeed that, for a small $\phi_{17,k}$ component in ψ, the $\phi_{17,k}$ component in ξ will be small enough that we may count it as being zero. [See Part II, eqs. (126)–(127).]

On the other hand, if we just have no information on what the $\{\phi_{17,k}\}$ are, we might think that we could always *choose* them so that one is orthogonal to whatever ξ_{17} happens to be. The above example shows that the predictions of hidden-variables theory are not insensitive to such manipulations of $\{\phi_i\}$ that we do not know. Therefore we should expect that the choice $\{\phi_i\}$ should not be open to arbitrary theoretical manipulation, and should be *uniquely determined by the experimental arrangements made for the measurement.*

The possibility of rotations in Hilbert space as the source of the Gleason troubles

When we review the derivation of the Kochen–Specker paradox [which was at the bottom of all troubles with $v(\psi, \xi, \phi_n)$ in Hilbert space], or if we look at Gleason's proof of his lemma (70), we will find that one basic assumption in these proofs was that a given choice of a basis $\{\phi_i\}$ in Hilbert space could be continuously rotated by orthogonal (or by unitary) transformations into a new position $\{\phi_i'\}$. In Gleason's derivation, the trouble came from the continuity of such a "rotation," as opposed to the discontinuity of v for any of the ϕ_i that make up the basis $\{\phi_i\}$. In the work of Kochen and Specker the trouble arose because for *every* triad $\{\phi_i\}$ we had to assign colors red, blue, blue. We then found that the conditions (61a, b) led to the existence of some blue, blue, blue triads. If, somehow, we could have said that these did not correspond to any possible measurement, the paradox would have disappeared, and it might have been possible to assign truth values to propositions and to avoid a dependence of v_k on ϕ_i with $i \neq k$.

We are, indeed, able to deny the existence of measurements corresponding to most sets $\{\phi_i\}$ that could arbitrarily be chosen. This fact was pointed out already by Bohm (1953b),[28] long before the Gleason troubles had been discovered. We shall illustrate it by the following example.

Absence of Gleason troubles for measurements of position

Consider measurements of position. The possible results of measurement, in a first and mathematically not rigorous approximation, could be represented by wavefunctions $\phi_i(\mathbf{x}) = [\delta_3(\mathbf{x}-\mathbf{x}_i)]^{\frac{1}{2}}$. (We shall not try here to describe the ϕ_i more precisely, by a finite-size wave packet, taking into account uncertainties in the experimental determination of $\mathbf{x}_i$ so that the mathematical position $\mathbf{x}_i$ is no longer the exact observable measured.) Now Gleason or Kochen and Specker would consider other measurements which would have as possible results states $\phi_j'(\mathbf{x})$ obtained by unitary transformation as given by eq. (66). In reality no such measurements exist for arbitrary coefficients X_j^i. They do not exist because a Hilbert vector

$$\tfrac{3}{5}\,[\delta_3(\mathbf{x}-\mathbf{x}_1)]^{\frac{1}{2}} + \tfrac{4}{5}\,\exp(0.13i)\,[\delta_3(\mathbf{x}-\mathbf{x}_2)]^{\frac{1}{2}}$$

does not represent the result of any measurement that can actually be performed.

We pointed at this fact when already in footnote 2 (p. 5) we claimed that a given hermitian operator, contrary to von Neumann's claims, does not necessarily describe an *observable*.[28] The latter fact means that the postulate (63) quite generally used in axiomatic quantum theory may be unacceptable to physicists. Moreover, there is no need for defining truth values $v_n(\psi, \xi, \{\phi_i\})$ for a set $\{\phi_i\}$ that does not correspond to a possible measurement.

[28] Bohm (1953b) notes that, in quantum mechanics, "the transformation theory rests on very weak experimental grounds. For its basic hypothesis; viz., that physical operations exist that would permit the observation of an arbitrary Hermitian operator, has not yet obtained adequate experimental support. In fact, the only observables known so far are...," and he lists them all (*loc. cit.*, page 277). Also: "the usual interpretation" [of quantum theory]... "needs to introduce the postulate that all Hermitian operators are observable; and... the theory contains no prescriptions telling how one might go about trying to verify this postulate" (*loc. cit.*, page 281).

Thus the tools used for deriving mathematically the unavoidability of Gleason troubles are not always tools acceptable to physicists, and there will be many situations in which a physicist may deny the existence of these troubles because of the nonexistence of observables corresponding to the continua of sets $\{\phi_i\}$ used in the above proofs of existence of these troubles.

Degenerate eigenvalues as the real source of Gleason troubles

According to the above, it is easy to deny the relevance of Hilbert bases $\{\phi_j'\}$ formed by (66) from bases $\{\phi_i\}$ with all ϕ_i corresponding to different eigenvalues of the observables measured. Denial of the physical relevance of considering $\{\phi_j'\}$ is, however, impossible when all we are doing in (66) is making linear combinations of eigenfunctions that belong to one and the same eigenvalue of the observable measured.

An example is the Kochen and Specker case of orthohelium. We considered here a continuum of simultaneous eigenfunctions of $\mathbf{J}^2$, J_x^2, J_y^2, and J_z^2. We did, in fact, indicate in Section 3.1 how *in principle* one could make a measurement corresponding to *any* set $\{\phi_j'\}$ from this continuum.

Thus the reality of Gleason troubles *in certain cases* is undeniable.

The excuse we have in this case for the dependence of the prediction for $J_z'^2$ (for given ψ and ξ) upon the direction of x' and y' is that a rotation of x' and y' involves a change in the experimental setup used for the measurement, and this "might affect" the result of the measurement.

In Section 3.8 we shall see that excuses of this type sometimes leave unanswered questions.

Oversensitivity of the result of a measurement for trifling circumstances

Consider a measurement of an observable which to our knowledge is degenerate. We therefore do not really know which $\{\phi_i\}$ to use.

In reality, minute perturbances of wavefunctions by atomic happenstances, impurities, and imperfections may well break up the degeneracy of the quantity really measured, in a way determined by the arrangement, but probably unknown to us. Thus the $\{\phi_i\}$ may be well determined physically,[28a] but uncertain in practice. A very slight change in the perturbations may greatly change $\{\phi_i\}$ under such circumstances, and may affect the result of the measurement. In this case, in a cryptodeterministic way, the microstate (ψ, ξ), together with the apparatus, may predict the outcome of the measurement, but by our lack of knowledge of the detailed perturbations the results look to us unpredictable. The source of the resulting statistical behavior, however, would be not of a quantum-theoretical nature, but of a classical nature.

[28a] An alternative to this method of making $\{\phi_i\}$ unique by happenstances will be discussed in Section 3.6 [eq. (119)] and Appendix J, both of Part II.

The merits of hidden-variables theory

Under the circumstances just described, the merits of hidden-variables theory from an "engineer's" point of view would be nihil, just like the merits of kinetic theory are nihil for describing a thermodynamical measurement. The only merits of hidden-variables theory in such a case would be for the person "believing" in determinism, in reassuring him in his faith.

A decision whether hidden variables are worth the trouble they cause by complicating the formalism of the theory should, of course, be based upon considerations of *experiments* carefully chosen so that hidden-variables predictions on account of the simplicity of the situation become well determined and not subject to uncertainties by Gleason troubles.

Preferred bases $\{\phi_i\}$ for certain observables

Consider again the spin of the orthohelium atom. According to Kochen and Specker, if we simultaneously measure J_x^2 and J_y^2 and J_z^2, or if we simultaneously measure $J_{x'}^2$ and $J_{y'}^2$ and $J_{z'}^2 = J_z^2$, this may result in different outcomes of the measurement of J_z^2 for one and the same microstate (ψ, ξ). Therefore, supposedly, we cannot assign truth values to the propositions $J_z^2 = 0$ and $J_z^2 = \hbar^2$.

However, there are other ways of measuring J_z^2 than by applying an electric field of rhombic symmetry. We could apply a homogeneous magnetic field along the z-axis. This has cylindrical symmetry, so there is no relevant difference anymore between x, y- or x', y'-axes.

Thus we still may assign a truth value to the propositions $J_z = 0$ and $J_z = \hbar$ and $J_z = -\hbar$, and therefore to the propositions $J_z^2 = 0$ and $J_z^2 = \hbar^2$, according to what the state (ψ, ξ) selects from the eigenfunctions of J_z in a magnetic field in the z-direction.

In general, if we are interested in $J_n^2 = (\mathbf{J}\cdot\hat{n})^2$, where $\hat{n}$ has the direction (θ, φ) in spherical coordinates, we may measure this *either* by a rhombic electric field obtained, from one with its symmetry directions along the axes of some x, y, z-frame, by rotations with Euler angles $\alpha = \varphi + \frac{1}{2}\pi$, $\beta = \theta$ and γ arbitrarily chosen; *or* by a magnetic field in the $\hat{n}$-direction θ, φ. In the former case, the triad of eigenfunctions $\{\phi_i\}$ between which the measurement discriminates are the simultaneous eigenfunctions $\phi_{x'}$, $\phi_{y'}$, and $\phi_{z'}$ of $J_{x'}^2$, $J_{y'}^2$, and $J_{z'}^2$ in the x', y', z'-frame with Euler angles α, β, γ. They satisfy

$$J_{i'}^2 \phi_{j'} = (1-\delta_{ij})\,\hbar^2 \phi_{j'}, \quad (i, j = x, y, z), \tag{79}$$

and are listed in Appendix J. In the latter case the triad from which the measurement makes a selection are the eigenfunctions χ'_m of $J_{z'} = J_{n'}$, which satisfy

$$J_{n'}\chi'_m = m\hbar\chi'_m \quad (m = 1, 0, -1). \tag{80}$$

They are related to the ϕ'_j by

$$\chi'_+ = \sqrt{\tfrac{1}{2}}\,(-\phi_{x'} - i\phi_{y'}), \quad \chi'_0 = \phi_{z'}, \quad \chi'_- = \sqrt{\tfrac{1}{2}}\,(\phi_{x'} - i\phi_{y'}), \tag{81a}$$

$$\phi_{x'} = \sqrt{\tfrac{1}{2}}\,(\chi'_- - \chi'_+), \quad \phi_{y'} = i\sqrt{\tfrac{1}{2}}\,(\chi'_- + \chi'_+), \quad \phi_{z'} = \chi'_0, \tag{81b}$$

if we make the choice of the χ'_m dependent upon the Euler angle γ in the way indicated in eq. (J-4) in Appendix J.

If for $J^2_{z'} = \hbar^2$ we do not care whether $J^2_{x'} = \hbar^2$, $J^2_{y'} = 0$, or the other way around, but we would like to know whether $J_{z'} = \hbar$ or $J_{z'} = -\hbar$, the triad $\{\chi'_m\}$ is the preferred set $\{\phi_i\}$ to be used in $n(\psi, \xi, \{\phi_i\})$ for measurement of the observables $J_{z'}$ and $J^2_{z'}$. In this case, there are in practice no Gleason troubles. The only linear conbinations of the $\{\chi'_m\}$ that are obtainable starting from the $\{\chi'_m\}$ themselves by a *continuous* change of the set by unitary transformation, while keeping it a set of eigenfunctions of the operator of a quantity that actually is observable, are the sets $\{\chi'_m\}$ obtained by changing the Euler angles in (J-4). Here, a change of $\alpha = \varphi + \frac{1}{2}\pi$ and of $\beta = \theta$ means a real change of the z'-direction $\hat{n}$, and therefore corresponds clearly to a different measurement (measurement of the component of **J** in a different direction, using a different direction of the magnetic field). A change of the outcome of the measurement under such a real change does not bother anybody.

If we vary γ in eq. (J-4), the resulting change in $\{\chi'_m\}$ is trivial; it merely alters the phase factors in the χ'_m, which are physically undetermined. (The relative phases of the χ'_m among each other are determined merely on the basis of some conventions.) In Part II we will see that arbitrary alteration of such phase factors in the ϕ_i individually in the hidden-variables theories under consideration does not change the function $n(\psi, \xi\{\phi_i\})$ of eq. (6); so the predicted outcome of the measurement is not altered.

If we do not care for the sign of $J_{z'}$, but we would like to know $J^2_{x'}$ and $J^2_{y'}$ simultaneously with $J^2_{z'}$, then $\{\phi_{j'}\}$ of eqs. (J-3) is the preferred set. From (81a, b) we see that $\phi_{x'}$ and $\phi_{y'}$ are not obtainable from χ'_+ and χ'_- by a continuous change of one of the parameters appearing in either (J-3) or (J-4). Of course, mathematically one can dream up a transformation leading continuously from $\{\chi'_m\}$ to $\{\phi_{j'}\}$; for instance, by changing η from 0 to $\frac{1}{4}\pi$ in $\phi_{x'} = \chi'_- \cos\eta - \chi'_+ \sin\eta$, $\phi_{y'} = e^{2i\eta}[\chi'_- \sin\eta + \chi'_+ \cos\eta]$. However, for $0 < \eta < \frac{1}{4}\pi$, these Hilbert vectors would not be eigenfunctions of any hermitian operators that would correspond to quantities that actually would be *observable*, in the physical sense of the word.

The $\{\phi_{j'}\}$, however, can be continuously transformed by a continuous change of the Euler angles, and the set remains descriptive of observable quantities. This corresponds to rotations of the rhombic field. If we look only at $J^2_{z'}$, a change of α and β, as in the case of $\{\chi'_m\}$, corresponds to a change of z'-axis, and nobody is bothered by the fact that the prediction for $J^2_{z'}$ changes when we change the z'-direction. The "Gleason troubles" in this case consisted in the fact that the prediction for $J^2_{z'}$ would occasionally change between 0 and $\hbar^2$ (while the z'-direction was kept fixed) by a change of the Euler angle γ rotating the x'- and y'-directions around the z'-direction. These troubles, however, do not seriously bother us if we are sure that, simultaneous with $J^2_{z'}$, we want to measure $J^2_{x'}$ and $J^2_{y'}$ for one particular choice of the angle γ and not for a different choice.

3.8. Troubles with composite systems in hidden-variables theories of the first kind

For predicting the result $\phi_{n(\psi, \xi, \{\phi_i\})}$ of a measurement, we need knowledge not only of ξ and $\{\phi_i^\dagger\}$, but also of ψ. This may cause additional trouble.

Consider, for instance, a system of two orthohelium atoms which somehow together are in a $^1S_0$ state. The system splits up into two orthohelium atoms each in a $^3S_1$ state. The spins of the two atoms end up in opposite directions, and we may describe the two-particle spin state by

$$\psi(1, 2) = \sqrt{\tfrac{1}{3}}\,[p_1n_2 - z_1z_2 + n_1p_2], \tag{82}$$

if, as in Appendix G, p, z, and n denote states with $j = 1$ and with $m = 1$ or 0 or -1 for the atoms 1 and 2 described by the labels on p, z, and n.

This state (82) cannot be factorized, and the algorithms of existing hidden-variables theories then require that also the hidden variables will describe a two-particle microstate rather than each one-particle state separately.

Therefore we have to know, in addition to $\psi(1, 2)$, also $\xi(1, 2)$ and a unique complete set of states $\{\phi_i(1, 2)\}$ between which the measurement discriminates. If all these things are known, the theory will pick the final state $\phi_n(1, 2)$, which might be, for instance, p_1n_2 or z_1z_2 or n_1p_2, if the measurement consists of measuring J_{z1} and J_{z2} by two magnetic fields, so that the preferred $\{\phi_i\}$ are the nine functions

$$p_1p_2,\ p_1z_2,\ p_1n_2,\ z_1p_2,\ z_1z_2,\ z_1n_2,\ n_1p_2,\ n_1z_2,\ n_1n_2. \tag{83}$$

If, however, we would simultaneously measure J_{z1} and J_{x2}, where J_x has the eigenfunctions $r = \frac{1}{2}p + \sqrt{\frac{1}{2}}z + \frac{1}{2}n$, $m = i\sqrt{\frac{1}{2}}\cdot(n-p)$, and $l = -\frac{1}{2}p + \sqrt{\frac{1}{2}}z - \frac{1}{2}n$ corresponding to $J_x = \hbar$ or 0 or $-\hbar$, we should better write (82) as

$$\psi(1, 2) = \sqrt{\tfrac{1}{12}}\cdot(p_1 + n_1)(r_2 - l_2) + \sqrt{\tfrac{1}{6}}\cdot[i(n_1 - p_1)m_2 - z_1(l_2 + r_2)], \tag{84}$$

while the $\{\phi_i\}$ basis is now

$$p_1r_2,\ p_1m_2,\ p_1l_2,\ z_1r_2,\ z_1m_2,\ z_1l_2,\ n_1r_2,\ n_1m_2,\ n_1l_2. \tag{85}$$

Now it is possible for this ψ that a given $\xi(1, 2)$ selects from the $\{\phi_i\}$ given by (83) the result z_1z_2, but that from the $\{\phi_i\}$ given by (85) the same ξ will pick the result p_1r_2. Now suppose we really measure merely J_{z1}, and do not try to measure either J_{z2} or J_{x2}. Will we find $J_{z1} = 0$ (corresponding to the factor z_1 in the state z_1r_2), or will we find $J_{z1} = 1$ (corresponding to p_1 in the state p_1r_2) or perhaps $J_{z1} = -1$ (corresponding to a factor n_1 in some state picked from the $\{\phi_i\}$ used in a simultaneous measurement of J_{z1} with J_{y2})? (For a possible, but somewhat paradoxical answer to this question, see Appendix K of Part II.)

Also, if we *do* measure J_{z1} and J_{x2} simultaneously, but the two apparatus are spatially separated, how will the apparatus measuring J_{z1} know whether we are simultaneously measuring J_{x2} or J_{z2}? So how does it know whether it should yield a state z_1 or p_1 for particle 1? The latter question shows that the predictions of hidden-variables theories of the first kind, like pure quantum theory, have aspects of nonlocality that are not easily answered and that lead people to the invention of hidden-variables theories of the second kind.

Finally, there is an unanswered question related to the "reduction of ξ." When in the above two-particle case particle number 2 gets lost, and on particle 1 we measure $J_{z1} = 1$, from (82) as initial state we will reduce the wave packet to $\psi(1) = p_1$, and assume that this will suffice as a starting point for future predictions on this one particle that did not get lost.

However, for predictions by hidden-variables theory for particle 1 alone, we need not only $\psi(1)$, but also $\xi(1)$, and no procedure so far has been suggested for deriving $\xi(1)$ by some kind of "reduction" from $\xi(1, 2)$ in a way that never will discontinuously alter the predictions of the theory made by the algorithm that determines the function $n(\psi, \xi, \{\phi_i\})$, at the instant that the reductions of ψ, ξ, and $\{\phi_i\}$ from two-particle to one-particle functions are made.

CHAPTER 4

THE WORK OF JAUCH AND PIRON

THE work of Jauch and Piron (1963)[29] on hidden-variables theories starts out from some axiomatic considerations about quantum theory that are more general than anything we consider in our survey. As formulated by them, quantum theory assigns probabilities to propositions that form a "lattice" (see Section 4.2). In Section 3.5 we defined "propositions" in terms of the Hilbert-space formulation of states. Much of the generality of the work of Jauch and Piron stems from the fact that they do not postulate the existence of a Hilbert space. We will not try here to keep this generality, for which the reader is referred to the original work. Thus we will continuously "illustrate" Jauch and Piron's work by "projecting it onto Hilbert space," that is, by assuming that a Hilbert space exists and by applying their work in this case. Even then, their propositions are more general than the ones we discussed in Chapter 3 because, if in the Hilbert-space language of Section 3.5 we say that a proposition a is a statement of the kind of "the system is found to be in a state described by a state vector inside a subspace α of Hilbert space $\mathfrak{H}$," we discussed in Chapter 3 merely propositions for which this subspace was one-dimensional (containing just products of one single Hilbert vector with arbitrary complex numbers), while in the work of Jauch and Piron we also consider more-dimensional subspaces α of Hilbert space, spanned by several Hilbert vectors. This is useful if, for instance, a measurement of an observable A would find $A = A_k$, where A_k is a degenerate eigenvalue of A_{op}.

The proposition indicated above for a Hilbert subspace α, we called a. Similarly, b is the proposition corresponding to a Hilbert subspace β, and so on. By the assignment of Hilbert subspaces α, β, etc., to propositions a, b, etc., we lose some of the generality of Jauch and Piron's original work. On the other hand, to readers trained in understanding quantum theory in terms of Hilbert space, it makes it easier to understand the applicability of Jauch and Piron's considerations to this more familiar kind of theory.

In absence of the assumption that each proposition corresponds to a subspace in Hilbert space, Jauch and Piron need a number of axioms in order to establish some properties of propositions (see Section 4.3). Since we apply their theory while assuming a correspondence between propositions and subspaces in Hilbert space, some of their axioms in our eyes will look like theorems which we can prove or make plausible by considerations in Hilbert space. [See, for instance, the "proof" of axiom (4)° given in Section 4.3, or of axiom (P), and of a similar assumption (P′) in Appendix F.]

[29] Essentially the same reasonings are found in Jauch (1968).

4.1. Jauch–Piron's hidden-variables theory

Let $w^{[\beta]}(a)$ represent the probability that a proposition a is true in a quantum-mechanical state $[\beta]$ which may be a mixed state with probability $p_\sigma^{[\beta]}$ for the state vector $\psi^{(\sigma)}$. If $w^{(\sigma)}(a)$ is the corresponding probability in the pure state $\psi^{(\sigma)}$, we would expect

$$w^{[\beta]}(a) = \int_{(\sigma)} p_\sigma^{[\beta]} w^{(\sigma)}(a); \tag{86}$$

but Jauch and Piron do not use this relation explicitly.

Next, Jauch and Piron consider the ensemble E of physical systems described by some arbitrary quantum-theoretical state (pure state ψ, or mixed state with probabilities p_τ for state vectors $\psi^{(\tau)}$). Let $w(a)$ be the probability (among measurements upon the systems contained in E) that a measurement of A shall give the result described by the proposition a.

Jauch and Piron now define their hidden-variables theories, in imitation of some ideas of von Neumann, as quantum theories in which for any quantum-mechanical state represented by such an ensemble E it is possible to write $w(a)$ in the form[30]

$$w(a) = \int_{[\beta]} p_{[\beta]}\, w^{[\beta]}(a) \quad \text{with} \quad p_{[\beta]} > 0 \quad \text{and} \quad \int_{[\beta]} p_\beta = 1, \tag{87}$$

$$\text{with} \quad w^{[\beta]}(a) = \quad 0 \text{ or } 1 \quad \text{for all propositions } a. \tag{88}$$

This restriction of $w^{[\beta]}(a)$ to 0 or 1 for all (a) is equivalent to saying that the quantum mechannical state $[\beta]$ is *dispersionfree*, that is, that the state $[\beta]$ has zero dispersion[31] on all propositions a.

Comparison with "realistic" hidden-variables theories

Jauch–Piron's formulation (87) never mentions the hidden variables ξ themselves explicitly. It is therefore often assumed tacitly by readers of this work that, where Jauch and Piron write[32] $w^{[\beta]}(a)$ (where $w^{[\beta]}$ would be a function of a state $[\beta]$ and a proposition a only), they must have meant $v(\psi, \xi^{(\beta)}, a)$, with β referring to a hidden-variables state, and that they omitted ξ from their notation for brevity only. Indeed, with the help of eq. (27) of Section 1.10, we may write the definition (9) of the equilibrium hidden-variables probability of a state ϕ_k as

$$\bar{P}_k(\psi, \{\phi_i\}) = \int_{(\beta)} \delta_{k,\, n(\psi,\, \xi^{(\beta)},\, \{\phi_i\})} \bar{p}_\beta = \int_{(\beta)} \bar{p}_\beta\, v_k(\psi, \xi^{(\beta)}, \{\phi_i\}), \tag{89}$$

where v_k is the truth value of the selection statement "the hidden-variables state $(\psi, \xi^{(\beta)})$ selects state ϕ_k from the set $\{\phi_i\}$" and therefore is either 0 or 1. The similarity of this formula (89) valid for realistic hidden-variables theories to Jauch–Piron's formula (87) is even more striking when we remember from Section 1.3 that in theories of the first kind the left-hand member of (89) is equal to the quantum-mechanical probability w_k which for $a \equiv \phi_k$ is also the left-hand member of (87). Therefore, Jauch–Piron's eq. (87) is naturally

[30] Jauch and Piron (1963), page 835.
[31] Jauch and Piron (1963), bottom of page 833.
[32] $w_\xi(a)$ in their notation.

mistaken for a shorthand version of (89), where $w^{[\beta]}(a)$ is written for a truth value like $v_k(\psi, \xi^{(\beta)}, \{\phi_i\})$, with the proposition for ϕ_k generalized to the proposition a for the Hilbert subspace α. Correspondingly, we might generalize the set $\{\phi_i\}$ in the argument of v to a set of Hilbert subspaces $\{\alpha_i\}$ of which α_k is one, and one might read $w^{[\beta]}(a_k)$ as a shorthand for the truth value $v_k(\psi, \xi^{(\beta)}, \{\alpha_i\})$ of the selection statement "the hidden-variables state ψ, $\xi^{(\beta)}$ selects subspace α_k from the set $\{\alpha_i\}$ of subspaces between which the measurement A differentiates." This way of understanding (87) we shall call the *reinterpretation* ($w \rightarrow v$) of the work of Jauch and Piron.

From this point of view it is disturbing that eq. (87) does not mention $\{\alpha_i\}$ as an argument of w. One would have expected, as a shorthand for $\{\alpha_i\}$, at least an indication of the dependence of w (or rather v) on A. However, Jauch and Piron do not explicitly mention A as an argument of their function w. If, however, Gleason's reasoning is valid (that is, in cases where rotations in Hilbert space lead continuously from one set of possible results of a measurement, to another set which may again be regarded as a set of possible results of a different measurement, as discussed in Section 3.7), then v should depend on $\{\phi_i\}$ or more generally on $\{\alpha_i\}$ in a relevant way, as v would have to change occasionally discontinuously upon continuous rotation in Hilbert space of this set of possible measurement results describing the measurement A in detail. The absence of A as an argument in the work of Jauch and Piron makes their w, in a given state, merely a function of the proposition a, as in pure quantum theory, and in contrast to the v of more realistic hidden-variables theories which by eq. (27) of Section 1.10 or by a generalization thereof should be functions of a and A together. Therefore the reinterpretation ($w \rightarrow v$) of Jauch and Piron's work would give the truth values v of their theory the character of the v_n of eq. (62), which is equivalent to the postulate (64) that truth values would not depend on A in addition to their dependence upon a.

We have already concluded at the end of Section 3.6 that the adoption of this postulate is by itself sufficient reason for considering any resultant hidden-variables theory a theory of the zeroth kind.

Always assuming that we may accept Gleason's work or the Kochen–Specker paradox, there is no way out of this conclusion by simply regarding Jauch and Piron's a as an abbreviated notation for a and A together, because it is an essential feature of their work that they treat their a as forming a *lattice;* and, while it is possible to treat the propositions a as a lattice, it is not clear how a lattice should be formed out of what we called "selection statements" at the conclusion of Section 3.6, which replace the propositions in hidden-variables theories based upon algorithms of the nature of eq. (6). [The latter make the truth values functions of A as well as of a, as in eq. (27).]

A second and even more serious difficulty with the reinterpretation ($w \rightarrow v$) of Jauch and Piron's work is that where they write w and one would be reading it as v, they also assign properties to the w which the v could not possibly possess.

Therefore, for an understanding of what they really proved, it is preferable first to forget about hidden variables altogether and to follow Jauch and Piron's reasoning as it would apply to pure quantum theory, reading w and not v for their w. Then, using *their* definition of what *they* call "hidden-variables theory," we will find that they proved the *impossibility* of such a theory, and therefore that they proved themselves that their theory is "of the zeroth kind." Finally, and only as an afterthought, we shall consider what happens to their derivations if one *tries* to read their probabilities w as truth values v. In that last part of the discussion we will overlook the fact that the specifications A or $\{\phi_i\}$ of the measurements were not given, reasoning as if Gleason's arguments would not necessarily apply.

4.2. Notation used by Jauch and Piron

Jauch and Piron use an abbreviated notation in which propositions are indicated by a single letter. For instance, consider the measurement of an observable A with operator A_{op} which has eigenvalues A_i, and let α_i in Hilbert space $\mathfrak{H}$ be the subspace containing all eigenfunctions ϕ_{ij} that belong to A_i. (If A_i is degenerate, α_i is more-than-one-dimensional.) Then the proposition "measurement of A yields the result A_i" is denoted by a_i. In case the measurement is reproducible, and if a_i is true, the measurement would project ψ into the Hilbert subspace α_i. On the other hand, if ψ lies entirely within one of the subspaces α to start with, one would expect the proposition a_i to have a 100% probability.

In the following we often will not write explicitly the index (i) by which for the measurement of A the various propositions are distinguished from each other.

For measurements of different observables, B, C, ..., we would similarly indicate the propositions by b, c, ..., and the corresponding subspaces in Hilbert space by β, γ....

The collection of all possible propositions a, b, c, ..., is called $\mathcal{L}$. By defining for it two operations, $\wedge$ and $\vee$, we make it a "lattice."

The operation $\wedge$ is called the *intersection*. Jauch and Piron write $\cap$ for it. Following their notation, we write $(a_i \cap b_j)$ for the proposition that a simultaneous measurement of A and B would lead simultaneously to the results A_i *and* B_j, with the resulting state in the intersection $(\alpha_i \cap \beta_j)$ of α_i and β_j in Hilbert space. If A_{op} and B_{op} have no common eigenfunction, this is, of course, impossible, as $(\alpha_i \cap \beta_j)$ then is an empty subspace. If $\gamma = \alpha \cap \beta$ is *empty*, $c = a \cap b$ is called an *absurd* proposition, which we denote by $\varnothing$. Conversely, if α is *all* of Hilbert space $\mathfrak{H}$, the corresponding proposition a is called the *trivial* proposition I.

By $(a_i \vee b_j)$ we refer to some kind of measurement which (if reproducible) would project ψ into the *sum* of the Hilbert subspaces α_i and β_j. The *sum* $(\alpha \vee \beta)$ of Hilbert subspaces α and β is defined as the space containing *all linear combinations* of the Hilbert vectors in α and in β. (Jauch and Piron write $\cup$ for $\vee$; we prefer to use the more conventional notation $\vee$ for the *sum*.)

4.3. Axioms and postulates of Jauch and Piron

With the above definitions, the distributive laws which in the theory of sets hold for the operations of intersection $\cap$ and of union $\cup$ are not valid for the operations of intersection $\cap$ and of sum $\vee$ on the lattice $\mathcal{L}$ of propositions. This is so because $\alpha \vee \beta$ (contrary to $\alpha \cup \beta$) contains Hilbert vectors which are contained neither in α nor in β. Thus, for instance, if a, b, and c are three propositions, we may have

$$a \cap (b \vee c) \neq (a \cap b) \vee (a \cap c), \tag{90}$$

as easily seen from examples (see Appendix E). Jauch and Piron therefore replace the distributive law by a new axiom. After defining in a peculiar way "compatibility" of two propositions (see Appendix F), they claim that the needed axiom is:

$$\text{(P)} \qquad \text{“If } a \subseteq b, \text{ then } a \text{ and } b \text{ are compatible”} \qquad (91)$$

[“compatible” in the sense of eq. (F-1) of the Appendix].

Next, Jauch and Piron consider the ensemble E of physical systems described by the quantum theoretical (pure or mixed) state considered. They define $w(a)$ to be the probability (among measurements upon the systems contained in E) that a measurement of A shall give the result described by the proposition a. They postulate for $w(a)$ the following properties:

$$(1) \quad 0 \leqslant w(a) \leqslant 1; \qquad (2) \quad w(\varnothing) = 0, \quad w(I) = 1. \qquad (92)$$

(3) If a and b are compatible, then

$$w(a)+w(b) = w(a \cap b)+w(a \vee b). \qquad (93)$$

$$(4)^\circ \quad \text{If} \quad w(a) = 1 \quad \text{and} \quad w(b) = 1, \quad \text{then} \quad w(a \cap b) = 1; \qquad (94)$$

(5) If $a \neq \varnothing$, there exists a state such that $w(a) \neq 0$.

For pure quantum theory, these postulates make good sense. In particular it is important to see why postulate (4)° is made. If $w(a) = 1$, it follows that all $\psi^{(\sigma)}$ for which $p_\sigma \neq 0$ must lie inside the Hilbert subspace α. When also $w(b) = 1$, these $\psi^{(\sigma)}$ must also lie in β. Therefore they lie in the intersection $\alpha \cap \beta$. This, in the first place, proves that $\alpha \cap \beta$ is not empty in this case. Moreover, from the fact that all $\psi^{(\sigma)}$ with $p_\sigma \neq 0$ lie in $(\alpha \cap \beta)$, it follows that the corresponding proposition $(a \cap b)$ is true for all physical systems in the ensemble E, so that $w(a \cap b) = 1$.

This reasoning remains valid if a and b are incompatible (see Appendix G for an example showing that two propositions may be “incompatible” in the sense of invalidity of (F-1), even though they might have probabilities 1 simultaneously.)

For part of their further derivations, Jauch and Piron use an axiom slightly stronger than (4)°. It is

$$(4) \quad \text{If} \quad w(a_i) = 1 \quad \text{for} \quad i \subset I, \quad \text{then} \quad w\left(\bigcap_{i \subset I} a_i\right) = 1.$$

4.4. Conclusions of Jauch and Piron

Jauch and Piron now give two different proofs that *no dispersionfree states can exist.* The first proof[33] uses the stronger axiom (4) above. We refer for it to the original publication. The proof uses the fact that in quantum theory the lattice of propositions is of a particular kind. In the simplest case,[34] the lattice has the property that to each proposition (except $\varnothing$ and I) it contains some other proposition which is incompatible with it (see Appendix H). The lattice then is called *coherent*.[34] It then is shown that on such a proposition system no dispersionfree states [with $w(a) = 0$ or 1 for all a] can exist.[35] Applying this to the states

[33] Jauch and Piron (1963), page 834.

[34] Jauch and Piron go beyond this in their consideration of “superselection rules,” which for simplicity we shall ignore here.

[35] This is Jauch and Piron's “corollary 1.”

$[\beta]$ occurring in (87), we conclude that the assumption (88) was impossible, which completes[34] the first proof of Jauch and Piron that their hidden-variables theory is of the zeroth kind.

The second proof[36] uses axiom (4)° instead of axiom (4). It starts by showing that the assumption (88) that the states $[\beta]$ would be dispersionfree leads by (92)–(94) to the validity of

$$w(a)+w(b) = w(a \vee b)+w(a \cap b) \tag{95}$$

for any of the states $[\beta]$ {label $[\beta]$ for brevity omitted in (95)} and for any pair of propositions a and b. Next, the validity of (95) follows also for the w in the left-hand member of (87). Finally, Jauch and Piron derive from (95) by (4)° that

$$w((a \cap b')\vee b) = w((b \cap a')\vee a) \quad \text{for all } w, \tag{96}$$

that is, for all states w entering eq. (87) on the left. Here, a' means the *negation* of a, defined in Appendix F. From (96), they conclude that

$$(a \cap b')\vee b = (b \cap a')\vee a, \tag{97}$$

that is, any two propositions a and b would be compatible in the sense of (F-1). Since we know that the latter result is false, it follows that the original assumption (88) must have been false.

Their application of the axioms (92)–(94) in these proofs to the $w^{[\beta]}(a)$ in the right-hand member of (87) means that Jauch and Piron themselves cannot possibly have regarded their $w^{[\beta]}(a)$ in (87) to be the v of (89), as we will see in Section 4.5 that the reinterpretation ($w \rightarrow v$) of their eq. (87) makes the derivation of their corollary 1 and of their eq. (95) utterly nonsensical.

From a mathematical point of view, Jauch and Piron's proofs are an improvement over the work of von Neumann. On the other hand, the result of Jauch and Piron is not very useful. They merely showed that "their" hidden-variables theory (which does not differ much from von Neumann's hidden-variables theory) is a theory of the zeroth kind and therefore not worth further consideration. As their postulate (87) (featuring w and not v) definitely differs from the formula (89) used in "more realistic" theories, they have proved nothing about the latter theories.

4.5. Reinterpretation ($w \rightarrow v$) of the work of Jauch and Piron

Bell (1966), criticizing Jauch and Piron's work, has remarked quite properly that Jauch and Piron's postulate (4)° is applicable to quantum-mechanical probabilities w, but that the same postulate cannot be generally valid for truth values v. Bohm and Bub (1966b) quite properly added to this that the reason for the inapplicability of (4)° to truth values v lies in the fact that $a \cap b$ may be empty even in cases where for some particular hidden-variables state (ψ, ξ) both a and b happen to be true. Examples of this are plentiful. One was already given by Bell (1966). Another example is given at the end of our Appendix I. On the other hand, Jauch and Piron (1968), quite understandably, regarded Bohm and Bub's remarks as a misinterpretation of their statements about w and not about v.

[36] Jauch and Piron (1963), pages 835–836.

The inapplicability of $(4)^\circ$ to truth values v is easily seen from the fact that, whenever $a \cap b$ happens to be empty (as is, for instance, the case when α and β are single Hilbert vectors which do not coincide), it follows that $v(a \cap b)$ must be zero, in disagreement with the fact that $(4)^\circ$, *if* applicable to v instead of w, would predict $v(a \cap b)$ to be 1 when $v(a) = v(b) = 1$. [If ψ is not normal to either α or β, there will always exist some ξ values for which $v(a) = v(b) = 1$.]

We thus see that the assumption $(4)^\circ$ and (4) made by Jauch and Piron for their w, and applied in their proofs of eqs. (95) and (96) and in proving corollary 1, would no longer remain valid when the w in (87) would be changed into v. As, consequently, Jauch and Piron's conclusions drawn from (87) are no longer valid for theories that replace (87) by (89), we conclude that Jauch and Piron's reasoning cannot be generalized by a reinterpretation ($w \to v$) to a proof of impossibility of more realistic hidden-variables theories. In fact, with v replacing w, many results of Jauch and Piron are easily shown to become false. For instance, (95) loses by this change its validity, as it can easily be shown by explicit examples that it is quite possible that for certain hidden-variables states

$$v(a)+v(b) \neq v(a \cap b)+v(a \vee b). \tag{98}$$

A case of this is the example given in Appendix I, where $v(a) = v(b) = 1$ and where $a \cap b = \varnothing$, so that $v(a \cap b) = 0$. Then, (98), if it had been an equality, would have given $v(a \vee b) = 2$, which is, of course, impossible.

4.6. Gudder's reformulation of postulate $(4)^\circ$

In an attempt to make Jauch and Piron's axioms for w applicable to v, too, Gudder (1968) has suggested to water down the axiom $(4)^\circ$ by adding to it the condition "if $(a \cap b) \neq \varnothing$." In a quantum-mechanical state, this adds nothing, since $a \cap b \neq \varnothing$ follows from $w(a) = w(b) = 1$. In a hidden-variables state of nonzeroth kind, Gudder's suggestion would change (94) into

$$\text{If} \quad v(a) = v(b) = 1 \quad \text{and} \quad a \cap b \neq \varnothing, \quad \text{then} \quad v(a \cap b) = 1. \tag{99}$$

With this alteration, the inequality (98) would become an equality for those cases for which $a \cap b \neq \varnothing$. This, however, would not make (96) valid with v replacing w, because Jauch and Piron's derivation of (96) uses (95) with a and b replaced once by $(a \cap b')$ and b, and once by a and $(a' \cap b)$, and $(a \cap b') \cap b$ and $a \cap (a' \cap b)$ both are always empty, so that the modified (99) would not apply. Moreover, in existing hidden-variables theories, like those of Wiener and Siegel or of Bohm and Bub, the axiom (99) is not generally valid. [That is, the algorithm (6) in these theories may lead to results contrary to Gudder's suggestion.]

APPENDICES TO PART I

APPENDIX A

AN ATTEMPT AT WRITING THE QUANTUM-MECHANICAL DENSITY OPERATOR $U(\psi)_{op}$ IN A SIMPLE CASE AS A SUPERPOSITION OF WEIGHTED MICROSTATE DENSITY OPERATORS $V(\psi, \xi)_{op}$

WE CONSIDER the special example discussed in Section 2.2. We there considered a pure state $\psi = \phi_1$ in a two-dimensional Hilbert space. Then the quantum mechanical density matrix U_{op} is given in the notation of Section 2.2 by

$$U_{ji}^{(A)} = \begin{pmatrix} 1 & 0 \\ 0 & 0 \end{pmatrix}. \tag{A-1}$$

Thence, on the basis of eigenfunctions of the operator B_{op} of Section 2.2, we obtain by a transformation similar to (36) with T_{ik} from (34)–(35),

$$U_{ji}^{(B)} = \begin{pmatrix} c^2 & -cs \\ -cs & s^2 \end{pmatrix}. \tag{A-2}$$

We found in (2.2) that an operator V_{op} with

$$V_{ji}^{(A)} = \begin{pmatrix} 1 & v_{12} \\ v_{12}^* & 0 \end{pmatrix} \tag{A-3}$$

was able to yield by (23) not only $\langle A \rangle_{\psi, \xi}$, but also $\langle B \rangle_{\psi, \xi}$ for the observable with eigenfunctions given by (35), provided we would choose

$$\left.\begin{aligned} v_{12}+v_{12}^* = 2\,Re\{V(\psi, \xi)_{12}^{(A)}\} &= \tan\theta \text{ for states } \xi \text{ for which } \langle B \rangle_{\psi, \xi} = B_1, \\ &= -\cot\theta \text{ for states } \xi \text{ for which } \langle B \rangle_{\psi, \xi} = B_2. \end{aligned}\right\} \tag{A-4}$$

Let $V(B_1)$ mean the density operator V_{op} for a hidden-variable state predicting $B = B_1$. Then, with the abbreviation $t = \tan\theta = s/c$, and with f and g real but so far undetermined, we have, from (A-4),

$$V_{ji}^{(A)}(B_1) = \begin{pmatrix} 1 & if+t/2 \\ -if+t/2 & 0 \end{pmatrix}, \quad V_{ji}^{(B)}(B_2) = \begin{pmatrix} 1 & ig-1/2t \\ -ig-1/2t & 0 \end{pmatrix}. \tag{A-5}$$

From this we obtain by (36)

$$V_{lk}^{(B)}(B_1) = \begin{pmatrix} 1 & if-t/2 \\ -if-t/2 & 0 \end{pmatrix}, \quad V_{ji}^{(B)}(B_1) = \begin{pmatrix} 0 & ig-1/2t \\ -ig-1/2t & 1 \end{pmatrix}. \tag{A-6}$$

We notice that if we choose

$$c^2 f + s^2 g = 0 \tag{A-7}$$

we find

$$c^2\, V_{ji}^{(A)}(B_1) + s^2\, V_{ji}^{(A)}(B_2) = U_{ji}^{(A)}, \tag{A-8}$$

$$c^2\, V_{lk}^{(B)}(B_1) + s^2\, V_{lk}^{(B)}(B_2) = U_{lk}^{(B)}, \tag{A-9}$$

which, by $w_1 = c^2$ and $w_2 = s^2$ for the probabilities of B_1 and B_2 in the ensemble E_ψ for the state $\psi = \phi_1$, is in agreement with our suggestion (53) for the equilibrium distribution of hidden variables in a theory of the first kind. [Compare the paragraph in Section 2.2 between eqs. (45b) and (46a).] That is, if we identify $V(B_1)$ with von Neumann's V, and $V(B_2)$ with his W, we find

$$U = w_1\, V(B_1) + w_2\, V(B_2) = c^2 V + s^2 W \tag{A-10}$$

rather than von Neumann's $U = V+W$.

Comparing (A-5) with (A-1), or (A-6) with (A-2), we also see that the matrices $V = V(B_1)$ and $W = V(B_2)$ are *not* proportional to the matrix U, so that some of the assumptions made by von Neumann when he proved this proportionality must have been violated in our example. We easily find that one of the conditions violated is the nonnegative definite character of V and W postulated by von Neumann as a desirable property of any reasonable density matrix. The $V(B_1)$ and $V(B_2)$ found above by the method of Section 2.2 do not have this property. This is easily verified by calculating the eigenvalues of $V(B_i)$ by solving the secular equations for these matrices (A-5) and (A-6). This gives the eigenvalues

$$\begin{aligned} &\tfrac{1}{2}|c|\left\{|c| \pm \sqrt{1+c^2 f^2}\right\} \quad \text{for} \quad c^2 V(B_1), \\ &\tfrac{1}{2}|s|\left\{|s| \pm \sqrt{1+s^2 g^2}\right\} \quad \text{for} \quad s^2 V(B_2). \end{aligned} \tag{A-11}$$

Since $|c| = |\cos\theta|$ and $|s| = |\sin\theta|$ lie both between 0 and 1, we see from (A-11) that (for any f and g, including $f = g = 0$) each of these two matrices for $\theta \neq 0$ has always one positive eigenvalue and one negative eigenvalue. This proves the lack of nonnegative definiteness of these matrices.

Even without calculating the eigenvalues of V explicitly, the indefiniteness of the matrices $V(B_1)$ and $V(B_2)$ is seen immediately from the fact that their determinants have negative values $= -|v_{12}|^2$. (The determinant is equal to the product of the eigenvalues.)

At the end of Section 2.2 ("Alternative proof") we tried to make V_{op} nonnegative definite by making (23) valid for operators A_{op} and C_{op} instead of A_{op} and B_{op}, with C_{op} by postulate commutative with V_{op}. This led to the requirement $v_{12} = 0$ or $V_{12}^{(A)} = 0$, which makes $V_{op} = U_{op}$, so that C_{op} commutes with A_{op}. Then, in the state $\psi = \phi_1$ of Section 2.2, the quantum-mechanical probability distribution between the eigenvalues of C_{op} becomes $w_1 = 1$ for the state with eigenfunction ϕ_1, and $w_2 = 0$ for the state ϕ_2, and the superposition (53) becomes

$$U = 1.V(C_1)+0.V(C_2). \tag{A-12}$$

Again, we find that there is no difficulty in writing U_{op} as a superposition (53) of density matrices V_{op} as long as we dilute the meaning of these density matrices V_{op} sufficiently to make their existence possible, by requiring (23) to be valid for small enough a number of noncommutative observables. As shown in Section 2.2, we make the existence of V_{op} and of the expansion (53) impossible as soon as we postulate the validity of (23) for too large a number of observables. Especially, if V_{op} is to be nonnegative definite, it can be used to calculate $\langle B\rangle_{\psi,\xi}$ by (23) only for observables B of which ψ is an eigenfunction.

Thus for given ψ and ξ it is always possible to find some observables B' for which $\langle B'\rangle_{\psi,\xi} \neq \sum_m \sum_n B'_{mn} V_{nm}$, so that our task to find density matrices $V(\psi, \xi)_{op}$ for *all* hidden-variables states is already a hopeless one for a *single* one such state.

We thus agree with von Neumann on the point of impossibility of defining density operators for the trivial ensembles $E_{\psi,\xi}$ representing hidden-variables states, but we disagree with him that this would be a sign that all hidden-variables theories would be impossible. This impossibility followed for von Neumann's hidden-variables theories of the *zeroth* kind from his making our eq. (31) an axiom which he "proved" to lead automatically to the possibility of constructing a V_{op} giving $\langle B\rangle_{\psi,\xi}$ for all B by (23). (We put "proved" between quotation marks because his set of postulates for a hidden-variables theory was self-contradictory, as he showed himself.) Since we dropped this axiom (31) for all hidden-variables theories of the nonzeroth kind, in order to avoid a self-contradictory set of postulates, von Neumann's objection is not a valid one for the hidden-variables theories of the first kind with which we will be concerned in Part II of this book in which the statistical postulate [our eq. (10)] will guarantee that quantum theory will become part of hidden-variables theory as the special case of an equilibrium distribution of the hidden variables.

APPENDIX B

A SIMPLE NEW PROOF OF THE KOCHEN–SPECKER PARADOX

As in Chapter 3, when below we talk about red (or blue) points on the unit sphere, we mean directions $\hat{n}$ for which $J_n^2 = 0$, $V_n = 1$ (or $J_n^2 = \hbar^2$, $V_n = 0$).

We first note that there certainly should exist red points on the sphere which are separated by distances of between 71° and 90° of arc. If this were not so, all red points would be contained in two areas, antipodes to each other, of which each would be a "shrunk octant" of which the edges would be arcs of less than 71° (instead of 90°). Thus less than one-quarter of the sphere would be red; much less than one out of every three points, as would be needed for having on every triad one red point to every two blue points.

Take now two such red points, R_1 and R_2. Take the z-axis toward R_1; so, it has coordinates $(0, 0, 1)$. Take the y, z-plane through R_2; so R_2 has coordinates $(0, s, c)$, where $c = \cos\theta$, $s = \sin\theta$, with $71° < \theta < 90°$, so, $0 < \cot^2\theta < \frac{1}{8}$.

R_1 and R_2 are poles of two blue great circles, $C_1(z = 0)$ and $C_2(sy+cz = 0)$. On C_1 take a blue point P_1 with coordinates $(\Gamma, \Sigma, 0)$ with $\Gamma = \cos\varphi$, $\Sigma = \sin\varphi$. (We shall choose φ later.) On C_2, take the "corresponding" blue point P_2 with coordinates $(Nc\Sigma, -Nc\Gamma, Ns\Gamma)$, where $N = (1-s^2\Sigma^2)^{-\frac{1}{2}} = (c^2+s^2\Gamma^2)^{-\frac{1}{2}} = (\Gamma^2+c^2\Sigma^2)^{-\frac{1}{2}}$ for placing P_2 onto the unit sphere. We see at once that P_1 and P_2 are in perpendicular directions $(x_1x_2+y_1y_2+z_1z_2 = 0)$. The point P_3 with coordinates $(-Ns\Gamma\Sigma, Ns\Gamma^2, Nc)$ completes the triad, since $x_ix_3+y_iy_3+z_iz_3 = \delta_{i3}$ for $i = 1, 2, 3$.

Since P_1 and P_2 are blue, P_3 must be red. We obtain another red point, P_3', by repeating the same construction, starting with P_1' at a different angle (φ' instead of φ) from the z-axis. We shall show that we can contradict the condition (60c) by making the directions of P_3 and of P_3' perpendicular to each other.

For this we need

$$0 = x_3x_3'+y_3y_3'+z_3z_3' = NN'\,[s^2\Gamma\Sigma\Gamma'\Sigma'+s^2\Gamma^2\Gamma'^2+c^2],$$

that is

$$\Gamma\Gamma'(\Sigma\Sigma'+\Gamma\Gamma') = -c^2/s^2,$$

or

$$\cos\varphi\,\cos\varphi'\,\cos(\varphi-\varphi') = -\cot^2\theta. \tag{B-1}$$

The left-hand member of this equation, as a function of φ and φ', can take all values between 1 and $-\frac{1}{8}$. (Its maximum 1 is attained when φ and φ' are zero or multiples of 180°, while its minimum value $-\frac{1}{8}$ is attained, for instance, at $\varphi = 60°$ with $\varphi' = 120°$). Since we had $0 < -\cot^2\theta < -\frac{1}{8}$, we are certain that there are angles φ and φ' solving eq. (B-1).

This completes our proof of the paradox, as we have shown that there exist red points P_3 and P_3' in perpendicular directions, in contradiction to eq. (60c).

The "Kochen–Specker diagram" for the above proof is shown in Fig. B.1.

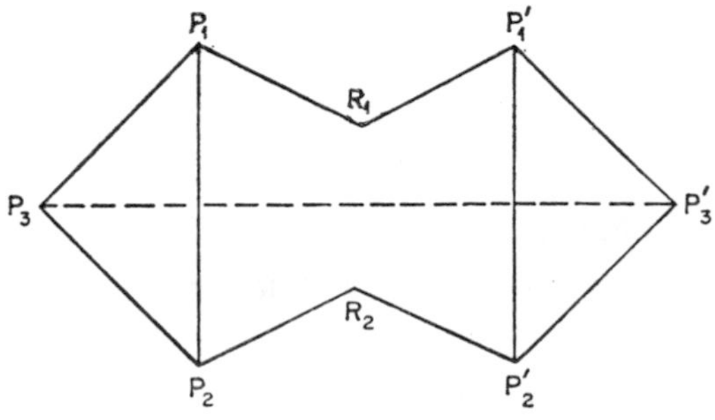

FIG. B.1. Kochen–Specker diagram of Belinfante's proof. [Compare Kochen and Specker (1967), pages 68–69.] Each vertex represents a point on the unit sphere. Each line connects two perpendicular directions. We choose R_1 and R_2 red. Then, P_1, P_2, P_1', and P_2' are blue, so, P_3 and P_3' are red. However, the latter can be chosen perpendicular to each other, in contradiction to rule (60c).

APPENDIX C

BELL'S VERSION OF GLEASON'S PROOF OF THE KOCHEN–SPECKER PARADOX

In Section 3.6 we mentioned a lemma (70) of Gleason according to which in a real three-dimensional space "all nonnegative frame functions are continuous." That is, if $w(\hat{n})$ is a function of the direction of the unit vector $\hat{n}$, and if for all triads of orthogonal directions $(\hat{n}^{(1)}, \hat{n}^{(2)}, \hat{n}^{(3)})$ the sum $\sum_{i=1}^{3} w(\hat{n}^{(i)})$ has the same value (for which we will choose the value 1), then w must depend continuously on $\hat{n}$.

At the end of Section 3.6 we applied this to "truth values" $v(\hat{n})$ which we assumed naturally to satisfy the conditions (61a, b). For making Gleason's lemma applicable we also had to assume (62) that $v(\hat{n})$ would not depend on the triad $\{\hat{n}^{(i)}\}$ of which $\hat{n}$ was a member. Then, however, the continuity of $v(\hat{n})$ following from (70) contradicted the discontinuity of $v(\hat{n})$ following from (61a). We concluded that it is impossible to satisfy (61a, b) and (62) simultaneously, which is nothing else than the Kochen–Specker paradox.

Since the proof of Gleason's lemma (70) is somewhat complicated, Bell (1966) modified and simplified the above reasoning as follows by turning it around.

Like Kochen and Specker (1967), he *starts* by assuming that there *would* exist a frame function $v(\hat{n})$ taking values 0 and 1 only, satisfying $\Sigma v(\hat{n}^{(i)}) = 1$ (61b), and depending on $\hat{n}$ itself only. As in Section 3.3, we shall call $\hat{n}$ "red" or "blue" depending upon whether $v(\hat{n}) = 1$ or $v(\hat{n}) = 0$. Near the borderline between red and blue regions it then must be possible to find, arbitrarily close together, two vectors, which Bell calls ψ and Φ, such that ψ is red $[v(\psi) = 1]$ and Φ is blue $[v(\Phi) = 0]$.

For simplicity, Bell takes the arguments of v off the unit sphere, generalizing $v(\hat{n})$ to a function $v(\mathbf{n})$ of an arbitrary vector $\mathbf{n}$ by putting $v(\mathbf{n})$ by definition equal to $v(\hat{n})$ for $\hat{n} = \mathbf{n}/|\mathbf{n}|$. Bell now shows that the assumptions (61a, b) about this function v lead to a contradiction, if the angle between ψ and Φ is chosen to be less than $\arctan \frac{1}{2} \approx 26°$ (which it certainly can be). He starts by somewhat reformulating the requirements (61a, b) imposed upon a frame function with $v(n) = 0$ or 1 and with $\sum_i v(\hat{n}_i) = 1$ for triads. Instead of (61a, b), he assumes:

(A) if $v = 1$ for one of three orthogonal directions, then $v = 0$ for both directions normal to it; and

(B) if for *two* orthogonal directions $\hat{n}_1$ and $\hat{n}_2$ we have $v = 0$, then $v = 0$ for all other directions in the plane of $\hat{n}_1$ and $\hat{n}_2$.

Clearly, (A) is a consequence of (61a, b), but the validity of (A) does not require the invariance of v under rotation, as in (A) we talk merely about one single tetrad. On the other hand, (B) is a consequence of (61a, b) combined with the invariance (62) of $v(\hat{n})$ under rotation around $\hat{n}$, as seen by the following three steps. First, it follows from (61b) that for the direction $\hat{p}$ normal to $\hat{n}_1$ and $\hat{n}_2$ we must have $v(\hat{p}) = 1$; then it follows from

(62) that after rotation of the tetrad around $\hat{p}$ we still will have $v(\hat{p}) = 1$; finally, it follows from (61a, b) or from (A) that for the rotated directions $\hat{n}_1'$ and $\hat{n}_2'$, normal to $\hat{p}$ and in the plane of $\hat{n}_1$ and $\hat{n}_2$, we still must have $v = 0$.

Bell starts by constructing a triad of orthogonal unit vectors ψ, ψ', and ψ'', where ψ is the "red" direction for which $v(\psi) = 1$, so that ψ' and ψ'' are "blue" directions $[v(\psi') = v(\psi'') = 0]$, while ψ' is chosen so that the "blue" direction Φ lies in the ψ, ψ' plane. So, he puts $\Phi = N\cdot(\psi+\varepsilon\psi')$, where $\varepsilon = \tan\angle(\psi,\Phi) < \frac{1}{2}$, and, since $v(\Phi/N) = v(\Phi)$, he can replace Φ by a new Φ given by

$$\Phi = \psi+\varepsilon\psi' \tag{C-1}$$

and still have Φ blue. Obviously, Φ is normal to ψ''.

Then he introduces four vectors, which we shall call ϕ_1, ϕ_2, χ_1, and χ_2, by

$$\left.\begin{aligned}\phi_1 &= \Phi+\tfrac{1}{2}(1+\sigma)\psi'' = \psi+\varepsilon\psi'+\tfrac{1}{2}(1+\sigma)\psi'',\\ \phi_2 &= \Phi-\tfrac{1}{2}(1+\sigma)\psi'' = \psi+\varepsilon\psi'-\tfrac{1}{2}(1+\sigma)\psi'',\end{aligned}\right\} \tag{C-2}$$

$$\left.\begin{aligned}\chi_1 &= -\varepsilon\psi'+\tfrac{1}{2}(1-\sigma)\psi'',\\ \chi_2 &= -\varepsilon\psi'-\tfrac{1}{2}(1-\sigma)\psi'',\end{aligned}\right\} \tag{C-3}$$

where he chooses

$$\sigma = \sqrt{1-4\varepsilon^2}, \tag{C-4}$$

which is real for $\varepsilon < \frac{1}{2}$. [Bell writes $\varepsilon\gamma$ for our $\pm\frac{1}{2}(1-\sigma)$.] We notice at once that ϕ_1 and ϕ_2 lie in the plane of the two blue vectors Φ and ψ'' which were normal to each other, so that by assumption (B) also ϕ_1 and ϕ_2 are blue. Similarly, χ_1 and χ_2 both lie in the plane of the two blue and normal vectors ψ' and ψ'', so that also χ_1 and χ_2 are blue.

Since ψ, ψ', and ψ'' were normal unit vectors, the dot products $(\phi_1\cdot\chi_1)$ and $(\phi_2\cdot\chi_2)$ are easily found to vanish on account of (C-4). Thus, again by (B), any vector in the plane of ϕ_1 and χ_1 is blue, and so is any vector in the plane of ϕ_2 and χ_2. In particular, the two vectors

$$\eta_1 \equiv \phi_1+\chi_1 = \psi+\psi'' \quad\text{and}\quad \eta_2 \equiv \phi_2+\chi_2 = \psi-\psi'' \tag{C-5}$$

are both blue. However, they are normal to each other, so that, by (B) again, any linear combination of them should be blue, including half their sum, which is ψ. This contradicts the original assumption that ψ was red.

(Of course, there is *no* contradiction if the color of ψ may depend upon whether it is taken to be part of the triad ψ, ψ', ψ'', or part of the triad $\psi, \eta_1\sqrt{\frac{1}{2}}, \eta_2\sqrt{\frac{1}{2}}$.)

Thus Bell has shown that, in case $v(\hat{n})$ is zero or one and satisfies (A) and (B), it is impossible that a "red" direction ψ (with $v = 1$) and a "blue" direction Φ (with $v = 0$) would make an angle less than 26° with each other. This again would exclude the discontinuities in $v(\hat{n})$ which, on the other hand, *must* exist where v jumps from 0 to 1. This final contradiction completes Bell's proof of the Kochen–Specker paradox.

In Fig. C.1 we give the Kochen–Specker diagram for Bell's proof. In it, the vectors p, π_1, and π_2 represent the red "poles" of the all-blue "equators" of $(\psi'', \Phi, \phi_1, \phi_2)$, of (ϕ_1, χ_1, η_1), and of (ϕ_2, χ_2, η_2), as Bell's requirement (B) is represented in the Kochen–Specker diagram by the fact that there exists a common pole of $(\hat{n}_1, \hat{n}_2)$ and of some rotated $\hat{n}_1'$ direction in the $\hat{n}_1, \hat{n}_2$ plane.

Comparison of Fig. C.1 with Fig. B.1 shows the somewhat greater simplicity of Belinfante's derivation of the Kochen–Specker paradox as given in Appendix B, while the derivation we gave in Section 3.3 is far simpler yet. (The simplicity goes at the expense of rigor, though.)

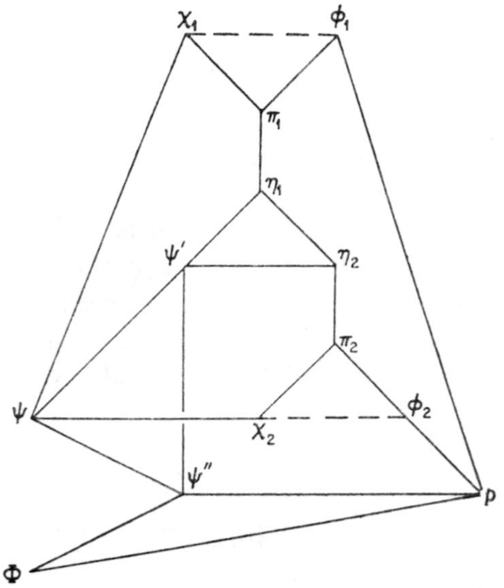

FIG. C.1. Kochen–Specker diagram of Bell's proof. We assume ψ is red and Φ is blue with an angle $< 26'$ between them, as should be possible for a discontinuous function $v(\hat{n})$ that is here red and there blue, but we show that it leads to a contradiction. Red ψ makes ψ' and ψ'' blue and therefore χ_1 and χ_2 blue; Φ and ψ' blue make (p red and) ϕ_1 and ϕ_2 blue; ϕ_i and χ_i blue make (π_i red and) η_i blue; η_1 and η_2 blue make (ψ' red and) ψ blue. This contradicts (ψ' blue and) ψ red. [The broken lines rely on eq. (C-4), which is possible for $\varepsilon < \frac{1}{2}$ only.]

APPENDIX D

GENERALIZATION OF THE KOCHEN–SPECKER PARADOX TO COMPLEX HILBERT VECTORS IN HILBERT SPACES OF THREE OR MORE DIMENSIONS

We shall here consider in more detail the generalization of the Kochen–Specker paradox discussed in Section 3.5, from real space of three dimensions, to complex Hilbert space of N dimensions (with $N \geqslant 3$).

In the Hilbert space $\mathfrak{H}$ of the possible states of a system S, consider an orthonormal set of N Hilbert vectors ϕ_i which span an N-dimensional subspace $\mathfrak{H}_N$, such that the state vector ψ of S in $\mathfrak{H}$ assigns zero probability to S being outside this $\mathfrak{H}_N$. We define a "real" subspace R_N of $\mathfrak{H}_N$ as the space of all Hilbert vectors ϕ'_j which can be obtained from the ϕ_i chosen above by orthogonal transformations (66) in which the X^i_j are real constants that form an orthogonal matrix. [Note that we call this R_N "real" even though the ϕ_i and therefore the ϕ'_j are in general complex functions. However, all products $(\phi'_j, \phi''_k) \equiv \int \phi'^*_j \phi''_k$ in R_N obviously are real, on account of $(\phi_i, \phi_l) = \delta_{il}$.]

We may geometrically visualize the transformations (66) as rotations by imagining that $\{\phi_i(x)\}$ would be a real orthogonal set of unit vectors $\{e_i\}$ in N-dimensional space; then the $\phi'_j(x)$ would form a new orthogonal set of unit vectors e'_j. Thus $\{\phi_i\}$ and $\{\phi'_j\}$ form in Hilbert space two orthogonal coordinate systems of which these unit vectors form the axes.

Temporarily we shall confine ourselves to the consideration of observables A of which the operators have N orthonormal eigenfunctions that happen to lie in R_N. We also will assume in R_N the validity of (63), so that, for any set $\{\phi'_j\}$ obtainable in R_N by (66), there would exist a measurable observable A' with $\{\phi'_j\}$ as its eigenfunctions. Then, for given ψ and ξ (which we will not keep mentioning as arguments of v), there would exist truth values $v(\phi'_j)$ or for short v'_j which in imitation of (61a, b) would have the properties that

$$\text{each } v'_j \text{ is either equal to 1 or to 0;} \tag{D-1}$$

$$\text{for each orthonormal set } \{\phi'_j\} \text{ in } R_N, \quad \sum_{j'=1}^{N} v'_j = 1. \tag{D-2}$$

For $N = 3$ we know already from the Kochen and Specker paradox that the existence of such v'_j satisfying (D-1) and (D-2) for all orthonormal sets obtainable from each other by (66) is *impossible* if we postulate (64) that each v'_k shall not depend on the choice of the set $\{\phi'_j\}$ of which ϕ'_k is a member. That is, the existence of truth values is impossible unless we allow v'_k to depend on the choice of $\{\phi'_j\}$, allowing v'_k to change from 1 to 0 and vice versa under transformations (66) that leave ϕ'_k invariant.

We shall now prove that the same holds true also for $N > 3$.

We prove this by induction. That is, we assume that it is true for an $(N-1)$-dimensional real Hilbert space R_{N-1}, and then derive it for R_N.

For this proof, take in R_N some direction ϕ_n for which $v_n = 0$. Renumber the axes so that it becomes direction number N, so that $v_N = 0$. The directions perpendicular to ϕ_N then form the basis for a real subspace R_{N-1}. For the hidden-variables state ψ, ξ under discussion, we then from (D-2) find that

$$\sum_{i=1}^{N-1} v_i = 1 - v_N = 1 - 0 = 1, \tag{D-3}$$

with all v_i satisfying (D-1). This is our problem in $(N-1)$ dimensions. We assumed that we knew that in R_{N-1} it is impossible to assign v'_j values to all Hilbert directions ϕ'_j and yet to satisfy (D-1) and (D-3) for all orthonormal sets, unless some v'_n may change value from 0 to 1 and back, under rotation of the other ϕ'_j around that ϕ'_n. Thus the necessity of this nonuniqueness of the truth values of propositions in the subspaces R_{N-1} of R_N proves the same nonuniqueness in R_N itself.

Having shown this nonuniqueness in real subspaces of Hilbert spaces of N dimensions ($N \geqslant 3$), we want to generalize our result to complex Hilbert spaces $\mathfrak{H}_N$, by allowing the X_j^i in (66) to form unitary instead of real orthogonal matrices. Since R_N is a subspace of $\mathfrak{H}_N$, it is clear that if, under condition (64), it is already impossible to assign for given ψ and ξ unique v_n to state vectors in R_N, it certainly cannot be possible in the complex Hilbert space $\mathfrak{H}_N$ of which R_N is part.

APPENDIX E

INVALIDITY OF THE DISTRIBUTIVE LAW FOR QUANTUM-THEORETICAL PROPOSITIONS

Consider a spin $\frac{1}{2}$ system and let b and c be the propositions $S_z = +\frac{1}{2}\hbar \left[\text{spin function} \begin{pmatrix}1\\0\end{pmatrix}\right]$ and $S_z = -\frac{1}{2}\hbar \left[\text{spin function} \begin{pmatrix}0\\1\end{pmatrix}\right]$ for a measurement of S_z. Then, $(b \vee c) = I$ is the trivial proposition that the result of the measurement is a state with the spin either up or down.

If a is the proposition $S_x = +\frac{1}{2}\hbar \left[\text{spin function} \begin{pmatrix}\sqrt{\frac{1}{2}}\\ \sqrt{\frac{1}{2}}\end{pmatrix}\right]$, then

$$a \cap (b \vee c) = a \cap I = a, \tag{E-1}$$

corresponding to

$$\alpha \cap (\beta \vee \gamma) = \alpha \cap \mathfrak{H} = \alpha$$

in Hilbert space. On the other hand, $(\alpha \cap \beta)$ and $(\alpha \cap \gamma)$ are empty, because S_x and S_z for spin $\frac{1}{2}$ have no eigenfunctions in common. Therefore, also $(\alpha \cap \beta) \vee (\alpha \cap \gamma)$ is empty and

$$(\alpha \cap b) \wedge (\alpha \cap c) = \varnothing. \tag{E-2}$$

Since $a \neq \varnothing$, the inequality (90) follows for this example from (E-1) and (E-2), and therefore the distribution law cannot generally be valid for propositions.

APPENDIX F

COMPATIBILITY OF PROPOSITIONS ACCORDING TO JAUCH AND PIRON

Jauch and Piron introduce the *negation* a' of proposition a. It is the proposition which states that the result of a certain experiment is finding the observed system in an unspecified state which we merely know to be orthogonal to *any* of the states in which the system might have been found by a measurement for which the proposition a would be true. Thus the Hilbert subspaces α and α' corresponding to a and a' have the *sum* $\alpha \vee \alpha' = \mathfrak{H}$, and have an empty intersection $\alpha \cap \alpha'$.

Jauch and Piron then call propositions a and b *compatible* if the relation

$$(a \cap b') \vee b = (b \cap a') \vee a \tag{F-1}$$

would happen to be valid.

The propositions a and b of Appendix E are an example of *in*compatible propositions. In this case we have $b' = c$ and $a' =$ the proposition $S_x = \frac{1}{2}\hbar$. Then

$$(a \cap b') = \varnothing \quad \text{and} \quad (b \cap a') = \varnothing \tag{F-2}$$

are both absurd propositions corresponding to empty subspaces in $\mathfrak{H}$. Therefore, the left-hand member of (F-1) is b and its right-hand member is a. Since these differ from each other, we conclude that $S_x = \frac{1}{2}\hbar$ and $S_z = \frac{1}{2}\hbar$ are what Jauch and Piron call *incompatible* propositions.

Note that, according to the definition by Jauch and Piron, *a proposition a and its negation a' are always compatible* [Piron (1963)], because (F-1) for $b = a'$ becomes $a \vee b = b \vee a$.

In Jauch and Piron's axiom P [eq. (91)], by $a \subseteq b$ we mean that the corresponding Hilbert subspaces are related by $\alpha \subseteq \beta$. This axiom (P) is a plausible one, since correspondingly in Hilbert space it is easily seen that, if $\alpha \subseteq \beta$, then $\alpha \cap \beta' =$ empty, and then β is the *sum* of α and of the part of α' that lies inside β, so that in this case $(\alpha \cap \beta') \vee \beta = \beta = \alpha \vee (\alpha' \cap \beta)$.

We easily find the following counterpart to axiom (P):

(P') If $a \subseteq b'$, then a and b are compatible. (F-3)

As an explanation, consider the following.

In Hilbert space the statement $a \subseteq b'$ is equivalent to $\alpha \subseteq \beta'$, which tells us that α is normal to β, and therefore $\beta \subseteq \alpha'$. Therefore $a \subseteq b'$ implies $b \subseteq a'$.

From $a \subseteq b'$ we conclude that $a \cap b' = a$, and from $b \subseteq a'$ we conclude that $b \cap a' = b$. Therefore in this case both sides of (F-1) are equal to $a \vee b$. This completes the proof that a and b are compatible when $a \subseteq b'$ or when $b \subseteq a'$.

With a and b compatible if $a \subseteq b$ or $a \subseteq b'$ or if $b \subseteq a$ or if $b \subseteq a'$, we may ask *under what circumstances are a and b not compatible?* Usually a and b will be not compatible if a contains parts which are part neither of b nor of b', and/or if b contains parts which are part neither of a nor of a'. That is, if α contains Hilbert vectors which are linear combinations of a Hilbert vector in β and of a Hilbert vector perpendicular to β (both with nonzero coefficient), a and b will *not* be compatible.

An instance of this is the example given in Appendix E. Here, $c = b'$, and the spin function a was a linear combination of b and b'.

In this particular example, not only the propositions $(a \cap b')$ and $(b \cap a')$ were absurd, but also $(a' \cap b') = \varnothing$ *and*

$$(a \cap b) = \varnothing. \tag{F-4}$$

However, these two additional relations are mere happenstance. For instance, in Appendix G we shall give an example of two incompatible propositions a and b such that $(\alpha \cap \beta)$ is not empty, so that $(a \cap b) \neq \varnothing$. On the other hand, a and b may be compatible even if $(a \cap b) = \varnothing$. An example of this was the case $b = a'$ mentioned above. Therefore, the question whether or not two propositions are compatible is unrelated to the question whether or not the corresponding Hilbert subspaces have any Hilbert vectors in common.

APPENDIX G

VALIDITY OF JAUCH AND PIRON'S POSTULATE (4)° FOR INCOMPATIBLE PROPOSITIONS

We shall give here an example of two propositions a and b such that $w(a) = w(b) = 1$ even though a and b are not compatible, and we shall show that in this case $w(a \cap b) = 1$ anyhow. In the following we shall write $\mathbf{L}$ for $\mathbf{L}/\hbar$.

Consider the four-dimensional Hilbert space of the $(1+3)$ states of a spinless particle which can be in states with $l = 0$ or 1. Denote by s the s-state with $\mathbf{L}^2 = 0 = L_x = L_y = L_z$. For $l = 1$, $\mathbf{L}^2 = 2$, we denote by p, z, n (for "positive," "zero," "negative") the three states with $L_z = +1$ or 0 or -1. Similarly, by r, m, l (for "right," "middle," "left") we denote the three $\mathbf{L}^2 = 2$ states with $L_x = +1$ or 0 or -1. We can express r and m and l in terms of p and z and n, by

$$\left.\begin{aligned} r &= \tfrac{1}{2}(n+p)+z\sqrt{\tfrac{1}{2}}, \\ m &= i(n-p)\sqrt{\tfrac{1}{2}}, \\ l &= -\tfrac{1}{2}(n+p)+z\sqrt{\tfrac{1}{2}}. \end{aligned}\right\} \tag{G-1}$$

By e, let us denote the Hilbert vector

$$e = \sqrt{\tfrac{1}{3}}\cdot z-\sqrt{\tfrac{2}{3}}\cdot p = \sqrt{\tfrac{2}{3}}\cdot l-i\sqrt{\tfrac{1}{3}}\cdot m, \tag{G-2}$$

so that $e \subset (p \vee z)$ and $e \subset (m \vee l)$.

Now define the Hilbert subspaces

$$\alpha = s \vee n \quad \text{and} \quad \beta = s \vee l \vee m, \tag{G-3}$$

so that their "negations" are $\alpha' = z \vee p$ and $\beta' = r$. We then easily find $(\alpha' \cap \beta) = e$ and $(\alpha \cap \beta') =$ empty. Therefore

$$\alpha \vee (\alpha' \cap \beta) = s \vee n \vee e \neq s \vee l \vee m = \beta = \beta \vee (\beta' \cap \alpha), \tag{F-4}$$

so, a and b in this example are not compatible. Yet

$$(\alpha \cap \beta) = s, \quad \text{so,} \quad (a \cap b) \neq \varnothing. \tag{G-4}$$

Now, consider the s-state of this particle ($\psi = s$). In that case, $\psi \subseteq \alpha$ and $\psi \subseteq \beta$, so, $\psi \subseteq (\alpha \cap \beta)$. Consequently, in this state $w(a) = 1$ and $w(b) = 1$, and $w(a \cap b) = 1$ as a consequence.

Thus we see that compatibility of a and b in Jauch's and Piron's sense is not necessary for the existence of common Hilbert vectors which α and β share, and that then for $\psi \subseteq (\alpha \cap \beta)$ two incompatible propositions may simultaneously have 100% probability to be true, while postulate (4)° is applicable in this case.

APPENDIX H

COHERENCE OF LATTICES OF PROPOSITIONS IN QUANTUM THEORY

In Appendix F we have seen that two propositions a and b usually are not compatible when α contains a Hilbert vector which has nonzero projections on both β and β', or if β contains a Hilbert vector which has nonzero projections on both α and α'.

Let a be a proposition different from $\varnothing$ and from I, so that the same holds for its negation a'. Let $\{\alpha_i\}$ and $\{\alpha_j'\}$ be orthonormal bases of α and of α'. Let $\beta_\pm = \sqrt{\tfrac{1}{2}} \cdot (\alpha_1 \pm \alpha_1')$. Let $\tilde{\alpha}$ be the Hilbert subspace spanned by all α_i except α_1, and let $\tilde{\alpha}'$ be spanned by all α_j' except α_1'. Then,

$$\beta_+' = \tilde{\alpha} \vee \tilde{\alpha}' \vee \beta_-,$$

$$\alpha \cap \beta_+' = \tilde{\alpha}, \quad (\alpha \cap \beta_+') \vee \beta_+ = \tilde{\alpha} \vee \beta_+,$$

$$\beta_+ \cap \alpha' = \text{empty}, \quad (\beta_+ \cap \alpha') \vee \alpha = \alpha = \tilde{\alpha} \vee \alpha_1.$$

Thence

$$(\alpha \cap \beta_+') \vee \beta_+ \neq (\beta_+ \cap \alpha') \vee \alpha, \tag{H-1}$$

so that the propositions a and b_+ corresponding to α and β_+ are not compatible in the sense of (F-1).

Because we thus can construct in the Hilbert space of quantum theory, for every a different from $\varnothing$ or I, a proposition b_+ not compatible with a, we call the lattice of propositions of quantum theory *coherent*.

An *exception* to the above must be made when it is impossible to make the linear combination $\beta_{\pm}$. This is so when there is a *superselection rule* which tells us that linear combinations of Hilbert vectors in α and Hilbert vectors in α' are meaningless. [See footnote 34 (p. 58).]

APPENDIX I

AN EXAMPLE HOW $v(a) = v(b) = 1$ DOES NOT EXCLUDE $v(a \cap b) = 0$

Consider two propositions a and b for which $(\alpha \cap \beta)$ is empty so that $(a \cap b)$ is absurd. Below we shall give a simple example of such a case.

It is clear that, in this case, there does not exist a state vector ψ that lies both in α and in β. Therefore $w(a)$ and $w(b)$ cannot both be 100% in this case, and postulate (4)° does not apply.

We may, however, have a state vector ψ in a. Then $w(a; \psi) = 1$. While certainly $w(b; \psi) < 1$ in this case, most likely also $w(b; \psi) > 0$. If that is so, among all individual systems in $E(\psi)$ there must be *some* for which the hidden variables ξ make $v(b)$ equal to 1 instead of 0. Take these hidden-variables values. For them not only $v(b) = 1$, but also $v(a) = 1$ because $w(a) = 1$ means that $v(a) = 1$ for all ξ-values. Thus we have, for these hidden-variables values, $v(a) = v(b) = 1$.

Since, however, $(a \cap b)$ was absurd, we have in this case $w(a \cap b) = 0$ for this ψ, and, therefore, $v(a \cap b) = 0$ for any ξ. Thus we have constructed a case where $v(a) = v(b) = 1$ does not exclude $v(a \cap b) = 0$.

For a *specific example*, consider an atom in a state ψ with $l = 3$ ($\mathbf{L}^2 = 12\hbar^2$) and with $L_x = \hbar$. Using Clebsch–Gordon coefficients, we can express this ψ in terms of the eigenfunctions Y_l^m of $\mathbf{L}^2$ and L_z, and find $\sqrt{(15/64)}$ as the coefficient of the $l = 3$, $m_l = 3$ state. Therefore, if a is the proposition $L_x = \hbar$, and b is the proposition $L_z = 3\hbar$, then our ψ gives $w(a) = 1$, $w(b) = 15/64$. However, there exists no Hilbert vector for which both $L_x = \hbar$ *and* $L_z = 3\hbar$, so that $\alpha \cap \beta =$ empty, $a \cap b = \varnothing$, and $w(a \cap b) = 0$. In this state, all systems in $E(\psi)$ have $v(a) = 1$ and $v(a \cap b) = 0$, and a fraction 15/64 of all systems in $E(\psi)$ have hidden variables ξ for which, additionally, $v(b) = 1$. This shows that the correct axiom (4)° becomes wrong when w in it is replaced by v.

APPENDIX J

ANGULAR EIGENFUNCTIONS FOR $j = 1$

We shall give here the eigenfunctions ϕ'_j and χ'_m used in the last part of Section 3.7. We introduce the following abbreviations:

$$\left.\begin{aligned} C &= \cos\beta = \cos\theta, \quad S = \sin\beta = \sin\theta,\\ \Gamma &= \cos\gamma, \quad \Sigma = \sin\gamma, \quad v = \sqrt{\tfrac{1}{2}}. \end{aligned}\right\} \tag{J-1}$$

With Euler angles α, β, γ for the x',y',z'-frame, we have

$$\left.\begin{aligned} J'_x &= \begin{pmatrix} S\Sigma & ve^{-i\alpha}(\Gamma - iC\Sigma) & 0 \\ ve^{i\alpha}(\Gamma + iC\Sigma) & 0 & ve^{-i\alpha}(\Gamma - iC\Sigma) \\ 0 & ve^{i\alpha}(\Gamma + iC\Sigma) & -S\Sigma \end{pmatrix},\\ J'_y &= \begin{pmatrix} S\Gamma & -ve^{-i\alpha}(\Sigma + iC\Gamma) & 0 \\ -ve^{i\alpha}(\Sigma - iC\Gamma) & 0 & -ve^{-i\alpha}(\Sigma + iC\Gamma) \\ 0 & -ve^{i\alpha}(\Sigma - iC\Gamma) & -S\Gamma \end{pmatrix},\\ J'_z &= \begin{pmatrix} C & ivSe^{-i\alpha} & 0 \\ -ivSe^{i\alpha} & 0 & ivSe^{-i\alpha} \\ 0 & -ivSe^{i\alpha} & -C \end{pmatrix}. \end{aligned}\right\} \tag{J-2}$$

Here, rows and columns are labeled by the quantum number $m(= 1, 0, -1)$ in the unprimed x,y,z-frame. We shall give the ϕ'_j and χ'_m in this same unprimed representation:

$$\phi'_x = \begin{pmatrix} v(iC\Sigma - \Gamma)\, e^{-i\alpha} \\ S\Sigma \\ v(iC\Sigma + \Gamma)\, e^{i\alpha} \end{pmatrix}, \quad \phi'_y = \begin{pmatrix} v(iC\Gamma + \Sigma)e^{-i\alpha} \\ S\Gamma \\ v(iC\Gamma - \Sigma)e^{i\alpha} \end{pmatrix}, \quad \phi'_z = \begin{pmatrix} -ivSe^{-i\alpha} \\ C \\ -ivSe^{i\alpha} \end{pmatrix}. \tag{J-3}$$

$$\chi'_+ = e^{-i\gamma}\cdot\begin{pmatrix} \tfrac{1}{2}(C+1)e^{-i\alpha} \\ -ivS \\ \tfrac{1}{2}(C-1)e^{i\alpha} \end{pmatrix}, \quad \chi'_0 = \begin{pmatrix} -ivSe^{-i\alpha} \\ C \\ -ivSe^{i\alpha} \end{pmatrix}, \quad \chi'_- = e^{i\gamma}\cdot\begin{pmatrix} \tfrac{1}{2}(C-1)e^{-i\alpha} \\ -ivS \\ \tfrac{1}{2}(C+1)e^{i\alpha} \end{pmatrix}. \tag{J-4}$$

Note that $e^{i\alpha} = ie^{i\varphi}$ and $e^{-i\alpha} = -ie^{-i\varphi}$, since $\varphi = \alpha - \frac{1}{2}\pi$. The phases in (J-3) and (J-4) were chosen here in such a way that

$$\left.\begin{aligned} &\chi'_m = \chi_m \quad \text{for} \quad \alpha = \beta = \gamma = \theta = 0, \quad \varphi = -\tfrac{1}{2}\pi;\\ &\phi'_z = \chi'_0; \end{aligned}\right\} \tag{J-5}$$

$$J'_x\phi'_y = -J'_y\phi'_x = i\phi'_z, \quad J'_y\phi'_z = -J'_z\phi'_y = i\phi'_x, \quad J'_z\phi'_x = -J'_x\phi'_z = i\phi'_y; \tag{J-6}$$

$$\left.\begin{aligned} J'_x\chi'_+ &= v\chi'_0, \quad J'_x\chi'_0 = v(\chi'_- + \chi'_+), \quad J'_x\chi'_- = v\chi'_0,\\ J'_y\chi'_+ &= iv\chi'_0, \quad J'_y\chi'_0 = iv(\chi'_- - \chi'_+), \quad J'_y\chi'_- = -iv\chi'_0. \end{aligned}\right\} \tag{J-7}$$

APPENDIX K

VON NEUMANN'S PROOF OF IMPOSSIBILITY OF DISPERSIONFREE STATES

Von Neumann's proof of nonexistence of generally dispersionfree states is fine when applied within the framework of pure quantum theory, as in quantum theory the assumptions upon which this proof is based are all valid. Apparently, however, it does not apply to dispersionfree hidden-variables states in theories of the nonzeroth kind, so that the question may be asked which one(s) of his various assumptions are not valid in such theories.

This question has been discussed by various authors. Since Bell (1966), the major culprit among von Neumann's axioms has been recognized in our eq. (31), assumed by von Neumann to be valid for hidden-variables states just like our eq. (32) is valid for quantum states.

We can easily trace how von Neumann's reasoning can be broken when the assumption (31) is dropped. At the hand of the excellent summary of von Neumann's proof published by Ballentine (1970), we shall show below where the break occurs. However, it is interesting to go a bit deeper and to see how the inapplicability of (31) in this case is caused by one of the most fundamental assumptions of hidden-variables theories of the nonzeroth kind, namely, the assumption (19), which for a hidden-variables state s we shall write here as

$$\langle A \rangle_s = A_{n(s,\{\phi_i\})}. \tag{K-1}$$

That is, the value of A in the state s is an eigenvalue A_j of A_{op} with j given as a function n of the state s and of a selected complete set $\{\phi_i\}$ of eigenfunctions of A_{op}. (That is, the subscript s on the left is not sufficient to describe $\langle A \rangle$ for a hidden-variables state if A_{op} is degenerate, and more precisely one should in that case write $\langle A \rangle_{s,\{\phi_i\}}$.)

The set $\{\phi_i\}$ represents the possible results of the measurement which is to be used for finding the value of A. The necessity of a dependence of $\langle A \rangle_s$ on this set is discussed in Chapter 3. Because of this dependence of $\langle\,\rangle_s$ on $\{\phi_i\}$, a combination of $a\langle A \rangle_s$ and $b\langle B \rangle_s$ to $\langle aA+bB \rangle_s$ as occurs in (31) becomes ostentatively unlikely when there is no common set $\{\phi_i\}$ of eigenfunctions of the observables A, B and $(aA+bB)$, as then in (31) one would be trying to combine unlike operations $\langle\,\rangle_s$. A more naive and therefore easier-to-understand reason for the breakdown of (31) on account of (K-1) was given already in Section 1.10 in the discussions around eq. (28). That is, the breakdown of the *general validity* of (31) is unavoidable because any $\langle A \rangle_s$ for a hidden-variables state s must by (K-1) be an eigenvalue of A_{op}.

Since this merely shows the breakdown of the *general validity* of (31), it still remains to be shown why (31), on account of (K-1), is invalid *in the specific application of (31)* made by Von Neumann in giving his "proof." This is the purpose of this appendix. It somehow *shifts the reason* for the breakdown of von Neumann's proof from the lack of general validity of (31) to the assumed validity of (K-1) for hidden-variables states of the nonzeroth kind.

As explained by Ballentine (1970), what von Neumann did was showing the appearance of a contradiction if the formula

$$\langle A\rangle_s = \sum_m \sum_n A_{mn} U[s]_{nm} \equiv \text{trace}\ \{A \cdot U[s]\} \tag{K-2}$$

would be combined with the assumption that A would be dispersionfree in state s if A were the "observable" that would correspond to the hermitian projection operator

$$P_\phi \equiv |\phi\rangle\langle\phi| \quad (\text{with } \langle\phi|\cdot|\phi\rangle = (\phi|\phi) = 1) \tag{K-3}$$

for any normalized Hilbert vector ϕ in Hilbert space. Ballentine writes ϱ for our U and R for our P_ϕ, and does not explicitly label the hidden-variables state by s. We shall refer to equations of his paper by prefixing his equation numbers by Ba.

Let us briefly summarize how von Neumann and Ballentine construct a contradiction out of (K-2) together with the assumption

$$\langle P_\phi^2\rangle_s = \langle P_\phi\rangle_s^2. \tag{K-4}$$

Equation (K-3) gives $P_\phi^2 = P_\phi$; thence (K-4) gives

$$\langle P_\phi\rangle_s = 0 \quad \text{or} \quad 1. \tag{K-5}$$

By (K-2) for $A = 1$, we find

$$\text{trace}\ U[s] \equiv \sum_n U[s]_{nn} = 1. \tag{K-6}$$

For $A = P_\phi$, we find by (K-2)

$$\begin{aligned}\langle P_\phi\rangle_s &= \sum_m \sum_n (P_\phi)_{mn}\,(U[s])_{nm}\\ &= \sum_m \sum_n \langle\varphi_m|\cdot|\phi\rangle\langle\phi|\cdot|\varphi_n\rangle\,(U[s])_{nm}\\ &= \sum_n \sum_m (\phi|\varphi_n)\,U[s]_{nm}\,(\varphi_m|\phi) = (\phi|U[s]|\phi).\end{aligned} \tag{K-7}$$

From (K-5) and (K-7) we then can conclude that $(\phi|U[s]|\phi)$ would for any ϕ be $= 0$ or 1 [see Ba(6.8)], but it would be continuous as a functional of ϕ, and therefore should have the same value (either 0 or 1) for *all* ϕ. This would make either $U[s] = 0$, or $U[s] = 1$ as an operator. [This is Ba(6.9).] Thence $U[s]_{mn} = 0$ or δ_{mn}, which contradicts (K-6).

Therefore, for breaking von Neumann's proof we should show that (K-2) does not apply to $A = P_\phi$ given by (K-3). Our reasoning in Sections 2.2 and 2.4 stresses the *nonexistence* of a density operator $U[s]$ for a hidden-variables state s (we there write V for $U[s]$), but von Neumann claims he can *prove* the existence of $U[s]$; so, our task is to find where the latter proof breaks down.

This proof is given by Ballentine in equations Ba(6.2) through (6.6). Ballentine follows von Neumann in defining hermitian operators $u^{(n)}$, $v^{(nm)}$, and $w^{(nm)}$ by

$$\left.\begin{aligned} u^{(n)} &= |n\rangle\langle n|,\\ v^{(nm)} &= |m\rangle\langle n| + |n\rangle\langle m| = v^{(mn)},\\ w^{(nm)} &= i(|m\rangle\langle n| - |n\rangle\langle m|) = -w^{(mn)};\end{aligned}\right\} \tag{K-8}$$

so, with $R(A_{mn}) = \frac{1}{2}(A_{nm}+A_{mn})$ and $I(A_{mn}) = \frac{1}{2}i(A_{nm}-A_{mn})$, we may write any hermitian operator A as

$$A \equiv \sum_m \sum_n |m\rangle A_{mn} \langle n| = \sum_n u^{(n)} A_{nn} + \sum\sum_{m<n} \{v^{(nm)} R(A_{mn}) + w^{(nm)} I(A_{mn})\}. \tag{K-9}$$

Now *if we assume* that $u^{(n)}$, $v^{(nm)}$, and $w^{(nm)}$ are observables for which it makes sense to talk about the average result $\langle \dots \rangle_s$ of their measurement in the state s, we can *define* the operator $U[s]$ by

$$U[s] = \sum_n \langle u^{(n)} \rangle_s u^{(n)} + \tfrac{1}{2} \sum\sum_{m<n} \{\langle v^{(nm)} \rangle_s v^{(nm)} + \langle w^{(nm)} \rangle_s w^{(nm)}\}, \tag{K-10}$$

and we obtain, from (K-10) and the matrix elements of (K-8), for $k < l$,

$$\left.\begin{aligned} U[s]_{ll} &= \langle u^{(l)} \rangle_s, \quad U[s]_{kl} = \tfrac{1}{2}\{\langle v^{(lk)} \rangle_s + i\langle w^{(lk)} \rangle_s\}, \\ U[s]_{lk} &= \tfrac{1}{2}\{\langle v^{(lk)} \rangle_s - i\langle w^{(lk)} \rangle_s\}, \end{aligned}\right\} \tag{K-11}$$

so, for $m < n$,

$$\left.\begin{aligned} \langle u^{(n)} \rangle_s &= U[s]_{nn}, \quad \langle v^{(nm)} \rangle_s = U[s]_{nm} + U[s]_{mn}, \\ \langle w^{(nm)} \rangle_s &= i\{U[s]_{nm} - U[s]_{mn}\}. \end{aligned}\right\} \tag{K-12}$$

Then we find from (K-9), *if (31) would be applicable to it*,

$$\begin{aligned} \langle A \rangle_s &= \sum_n U[s]_{nn} A_{nn} + \sum\sum_{m<n} \{U[s]_{nm} + U[s]_{mn}\} R(A_{mn}) + \sum\sum_{m<n} i\{U[s]_{nm} - U[s]_{mn}\} I(A_{mn}) \\ &= \sum_n U[s]_{nn} A_{nn} + \sum\sum_{m<n} \{U[s]_{nm} A_{mn} + U[s]_{mn} A_{nm}\} = \sum_m \sum_n A_{mn} \cdot U[s]_{nm} = \text{trace}\,\{A \cdot U[s]\}, \end{aligned} \tag{K-13}$$

which is eq. (K-2). This is von Neumann's proof of the validity of (K-2). We see that it breaks down if (31) is *not* applicable to (K-9), but it also requires an assumption above eq. (K-10). It now is our task to understand (1) *why the assumption made above (K-10) cannot be generally valid,* and (2) *why in particular for (K-9) the validity of (31) should break down,* all on account of the postulate (K-1).

In order to give $\langle u^{(n)} \rangle_s$, $\langle v^{(nm)} \rangle_s$, and $\langle w^{(nm)} \rangle_s$ in (K-10) any meaning under the assumption (K-1) whatsoever, we must be defining $\langle \dots \rangle_s$ here in each case for a selected set of eigenfunctions of the operator ... of which the value $\langle \dots \rangle_s$ is calculated. A possible simultaneous set of eigenfunctions of all operators $u^{(n)}$ is given by $\{|j\rangle\}$. They are eigenfunctions to the eigenvalues δ_{nj}. For $n \neq j \neq m \neq n$, $|j\rangle$ is also an eigenfunction of $v^{(nm)}$ and of $w^{(nm)}$ to the eigenvalue zero. However, (for $m \neq n$) the eigenfunctions of $v^{(nm)}$ to the eigenvalues ± 1 are $\sqrt{\frac{1}{2}}(|n\rangle \pm |m\rangle)$, and the eigenfunctions of $w^{(nm)}$ to the eigenvalues ± 1 are $\sqrt{\frac{1}{2}}(|n\rangle \pm i|m\rangle)$. There are, therefore, *no* simultaneous eigenfunctions of *all* operators (K-8); in particular, any $|j\rangle$ is not an eigenfunction of $v^{(jk)}$ or $v^{(kj)}$ or $w^{(jk)}$ or $w^{(kj)}$ for any $k \neq j$.

This by itself is enough to make (K-10) meaningless as a "definition" of $U[s]$ in a hidden-variables state s. That is, (K-1) tells us that $\langle \dots \rangle_s$, which is really an abbreviation for $\langle \dots \rangle_{s,\{\varphi_i\}}$, means picking the eigenvalue of ... which would be the result of the measurement of ... specified by the fact that it discriminates between the set $\{\varphi_i\}$ of eigenvalues of ...; on the other hand, there exists no set $\{\varphi_i\}$ that can be used as specification of $\langle \dots \rangle_s$ in all terms of (K-10), so that there is no way of defining the meaning of $\langle \dots \rangle_s$ for each term separately so that $\langle \dots \rangle_s$ would mean the same in all terms of (K-10).

Moreover, for applicability of (K-12), that same $\langle\ldots\rangle_{s,\{\varphi_i\}}$ should appear also on the left in (K-13), which means that the set $\{\varphi_i\}$ which specifies the meaning of $\langle\ldots\rangle_s$ in the definition (K-10) of the density operator $U[\sigma]$, *should also be a set of eigenfunctions of* A_{op}. This could only be so if the $|j\rangle$, which form the simultaneous eigenfunctions at least of the $u^{(n)}$ for all n, would simultaneously be eigenfunctions of A_{op}. That is, the A_{kl} then should form a *diagonal matrix*, with $A_{nm} = A_{mn} = 0$ for $m < n$. [Otherwise, the derivation of (K-2) by the application (K-13) of (31) to (K-9) makes no sense.]

This, however, leads at once to the recognition of a *special case* in which the *validity of (K-13) and of (K-2) can be rescued for a particular observable A*. If A_{kl} is diagonal, the off-diagonal elements of $U[s]$ in (K-13) have vanishing coefficients and may be ignored. We may use the eigenfunctions of A_{op} in this case as the set $\{\varphi_i\}$ specifying the meaning of $\langle\ldots\rangle_s$ wherever it appears. This, of course, makes $\langle v^{(nm)}\rangle_s$ and $\langle w^{(nm)}\rangle_s$ nonsensical, but we can in this case *redefine* a density operator

$$V^{(A)}[s] = \sum_n \langle u^{(n)}\rangle_{s,\{\varphi_i\}}\, u^{(n)} \tag{K-14}$$

simply omitting the meaningless terms. That is, as the off-diagonal elements of the density matrix do no longer appear in (K-13), we need not calculate them through (K-12) from (K-11), and may as well replace them by zero, as we do in V^A given by (K-14). In that case,

$$\langle A\rangle_{s,\{\varphi_i\}} = \text{trace}\,\{A\cdot V^A[s]\} \tag{K-15}$$

for the value of A predicted by the hidden-variables state s as the result of a measurement discriminating between the states φ_i.

The definition (K-14) of $V^{(A)}[s]$ gives, on the basis $\{\varphi_i\}$, the density matrix

$$V^{(A)}[s]_{kl} = \delta_{kl}\langle |k\rangle\langle k|\rangle_{s,\{|j\rangle\}} = \delta_{kl}\langle P_{|k\rangle}\rangle_{s,\{|j\rangle\}}\,. \tag{K-16}$$

Since the eigenvalues of the projection operator are 0 or 1, and $\langle P_{|k\rangle}\rangle_{s,\{|j\rangle\}}$, therefore is zero or one depending upon whether k is the label of the eigenstate of A picked for the measurement of A by the state s, we may use for it the notation $\delta_{k,n(s,\{|j\rangle\})}$ used in eq. (27) of section 1.10, and write

$$V^{(A)}[s]_{kl} = \delta_{kl}\,\delta_{k,n(s,\{|j\rangle\})}\,. \tag{K-17}$$

Equation (K-15) then yields

$$\langle A\rangle_{s,\{\varphi_i\}} = A_{n(s,\{\varphi_i\})}, \tag{K-18}$$

as the sets $\{\varphi_i\}$ and $\{|j\rangle\}$ here both were taken to be the same set of eigenfunctions of A_{op}. Thus, with a tremendous effort, we simply have found back the result that was already postulated in eq. (19). Similar results are obtained by different reasoning in Section 2.2, and are reviewed in the first part of Section 2.4, where, too, we find that the only density matrices that can be constructed for hidden-variables states are density matrices $V^{(A)}$ that can serve in (31) for the calculation of the hidden-variables value $\langle A\rangle_s$ of merely one particular observable A_{op}, and where, too, we find that there does not exist one general V_{op} of which the V^A_{kl}, V^B_{mn}, etc., would be matrix elements on different bases.

PART II

THEORIES OF THE FIRST KIND

FOREWORD TO PART II

In Part I we defined hidden-variables theories of the first kind as hidden-variables theories constructed with the following two primary purposes in mind: (1) the theory shall not be self-contradictory, and (2) the theory shall yield all results of pure quantum theory for all ensembles of physical systems in which the distribution of the hidden-variable values is a certain equilibrium distribution. By (1), these theories differ from theories of the zeroth kind, which we discussed in Part I; by (2), they differ from theories of the second kind, which we will discuss in Part III of this book.

Part II starts with a discussion of Bohm's theory of 1951, which is largely a reinterpretation of ordinary, nonrelativistic wave mechanics as a hidden-variables theory. A typical achievement of Bohm's theory is that it allows us in a two-slit interference experiment to tell through which slit a particular particle reached the screen, without destroying the interference pattern.

The main trouble with this theory of Bohm is that its relativistic generalization meets difficulties.

Next we will discuss the theory of Wiener and Siegel. From this theory we obtain the theory of Bohm and Bub by first omitting from the Wiener–Siegel theory all parts that are irrelevant for its application and by then adding Bohm–Bub's ideas about additional terms in the Schrödinger equation effective only during measurements. We will, in this connection, also consider some of Tutsch's more general, ideas related to the latter point.

Finally, we shall discuss possible experiments aimed at finding deviations from quantum theory which theories of the first kind predict for measurements made in rapid succession. We shall calculate the effects predicted by the theories of Wiener and Siegel and of Bohm and Bub, and by a variation thereof. Some of these experiments have already been performed in 1967 by Papaliolios, and gave negative results, which impose an upper limit upon the relaxation time in which in these experiments a purposely perturbed hidden-variables distribution returned to an equilibrium distribution.

We discuss a method of removing the ambiguity of the set $\{\phi_i\}$ of possible measurement results in the case of degenerate observables, but we find that even thus there are cases where application of the theory of Wiener–Siegel or Bohm–Bub leads to paradoxes.

CHAPTER 1

INTRODUCTION

IN PART I we discussed in Section 1.2 the general structure of hidden-variables theories of the nonzeroth kind, and in Section 1.3 of Part I we told what special property is the characteristic of theories of the first kind [see (I:10).[1]]. In Section 1.4 of Part I we pointed out that some theories that have been published are not complete theories, and leave certain questions unanswered. (In Section 3.8 of Part I, however, we suggested that even so-called "complete" theories leave a number of practical questions unanswered about the application of such theories.)

As we often in Part II want to refer back to these discussions, we shall very briefly summarize some of their main points, but we shall start with some clarification of the difference between "hidden-variables states" and "quantum states."

1.1. Hidden-variables states (microstates) and quantum states

Technically, a hidden-variables state is distinguished from a quantum state by specifying certain additional variables ξ in addition to the variables ψ that describe the quantum state (see Section 1.2). In order to understand what the problems are in experimenting with hidden-variables states, we shall make here a comparison of the "microstates" of hidden-variables theory, with the "microstates" of statistical mechanics. Though in many regards the comparison is not a happy one, yet something can be learned from it.

Consider a gas in a vessel. It is easy to prepare it in a state with given volume and shape of the vessel, given pressure and temperature. (Possibly even with given macroscopic currents in the gas.) However, the exact thermal motion of each individual molecule is unknown. The macrostate of the gas thus determined, we want to compare here to the quantum state of an atomic system. The state vector ψ describes the system, according to hidden-variables theory, only incompletely, like macroscopic parameters describe a gas only incompletely.

Microstates of the gas might be described by points in at least $6N$-dimensional phase space of all N molecules. It is neither customary nor useful to assign to such a microstate a macrostate parameter like the temperature. The latter describes the average energy $\overline{\mathcal{E}}$ in the ensemble representing the macrostate. One *could*, of course, assign to a microstate of

[1] By I: we refer to equation numbers of Part I of this book.

energy $\mathcal{E}$ a "temperature" [*sic*] equal to that of an ensemble whose $\bar{\mathcal{E}}$ equals the microstate's $\mathcal{E}$, but, in that case, all microstates that are members of the ensemble that describes a macrostate of temperature T would have temperatures which are different among each other and different from T. For avoiding this, macrostate parameters are not assigned at all to microstates.

In hidden-variables theory, we postulate things to be different. Each microstate is said to be described not merely by "microscopic" variables ξ, but it is assumed that the "macroscopical" parameter ψ that describes the quantum state (the ensemble to which the microstate or "hidden-variables state" belongs) is also a good additional parameter for describing the microstate. Contrary to what we find about the "temperatures" [*sic*] of microstates of a gas, we postulate in hidden-variables theories that each physical system in the ensemble E_ψ describing the quantum state ψ *shall have the same* ψ *as the ensemble has.*

Thus a microstate is in hidden-variables theory described not by ξ, but by ψ and ξ together, and we shall see that ψ here plays a necessary and *very important* part in the description of this state. In Bohm's theory (Chapter 2), ψ helps determine how the microstate evolves ($d\xi/dt$ depends on ψ); in all hidden-variables theories which we shall discuss, it helps determine for given ξ the outcome of any measurement.

The fact that the ensemble parameter ψ is also "used" as a parameter for microstates causes, of course, a problem. One and the same microstate could be part of more than one ensemble. In quantum theory the "reduction of a wave packet" by (or after) a measurement is often regarded as such a change of ensemble made voluntarily by the physicist for the sake of future convenience. (Of course, it is *not* an *arbitrary* change of ψ and of ensemble, but one that is justified by the result of an objective measurement; see Section 2.11.) The question then has to be considered whether such a change of ψ is allowable in the description of a microstate, without discontinuously changing the "predictions for the future" that are determined by ψ and ξ together. We shall meet this question in the discussion of the 1951 theory of Bohm (see Section 2.5). The Bohm–Bub theory avoids the problem by replacing a discontinuous reduction of a wave packet as an act of wisdom, by a continuous change of ψ described by a modification of the Schrödinger equation. Wiener and Siegel (Section 3.7) answer the question incompletely.

Another great difference between microstates in hidden-variables theory and microstates in statistical mechanics is that in the latter case the number of variables used to describe a macrostate is far and far less than the number of variables needed to describe a microstate, while this is not necessarily so in hidden-variables theory. For instance, in the theory of Bohm and Bub the microstate may be described by giving a second state vector in the Hilbert space in which the quantum state was already given by a first Hilbert vector ψ.

Can hidden-variables states be prepared experimentally?

We might say that, for a gas, p, T, etc., describe the state in which the gas is "prepared" to be. Similarly, one says that the quantum state ψ is given by the "preparation" of a system on an atomic scale.

It is practically impossible to "prepare" a microstate of a gas. However, one could (by perturbations) create situations which are not well represented by an ensemble describing statistical equilibrium. Take, for instance, a very wide and high vessel filled with an extremely rarified Knudsen gas. Then inject a small amount of gas at the top. These molecules will drop down without collisions, so that, lower down, before the molecules have hit bottom or walls, these additional molecules will have a Maxwell distribution from which velocities below a certain minimum have been cut off. This distribution could not be maintained. Pretty soon such a distribution would "relax" and change into an ordinary Maxwell distribution. But one might imagine doing some experiments which demonstrate the curious nonMaxwellian distribution while it lasts (before many incoming molecules have hit the walls of the vessel).

Similarly, a quantum state ψ will normally be a "mixed state as regards the possible hidden-variables states", and is well describable by an ensemble of states (ψ, ξ) with an "equilibrium" distribution of the ξ in which the "probability" p_β of the hidden-variable value $\xi^{(\beta)}$ is some definite function of β or of $\xi^{(\beta)}$; say, $\bar{p}_\beta$. This distribution normally is self-perpetuating, like a Maxwell distribution of molecular velocities in a gas. But we could by some trick make p_β temporarily different from the equilibrium $\bar{p}_\beta$. In Chapter 5 we shall describe how this can easily be done according to the hidden-variables theories of Wiener and Siegel and of Bohm and Bub. These deviations lie at the bottom of all attempts of showing experimentally the existence of hidden-variables states if nature would be ruled by a theory of the first kind. (We shall discuss in Part III how one can use much cruder experiments for verifying whether or not nature is ruled by a theory of the *second* kind.)

The short duration of the deviation of p_β from $\bar{p}_\beta$ is the source of the difficulty of such experiments.

1.2. The main parts of a theory of the first kind

We shall now enumerate the main parts of hidden-variables theories of the first kind, as we have learned them from Sections 1.2–1.3 of Part I:

(A) We must know what values ξ can take.

(B) We must know its equilibrium distribution, that is, the equilibrium probabilities $\bar{p}_\beta$ for hidden-variable values $\xi^{(\beta)}$. In practice, we need one or many *continuous* parameters β to label the possible ξ, and the $\bar{p}_\beta$ are replaced by products of probability densities and differential volume elements in β-space.

(C) One should make it plausible that p_β, when subjected to random alterations of some reasonable kind, would tend toward $\bar{p}_\beta$ when differing from it, or at least that it would stay equal to $\bar{p}_\beta$ when it was initially equal to $\bar{p}_\beta$.

(D) An *algorithm* should be given which determines for given ψ and ξ what the result will be of a measurement of which it is given which complete orthonormal set of Hilbert vectors $\{\phi_i\}$ represents the possible results of the measurement. The algorithm has the form of a selection statement. From the set $\{\phi_i\}$ it picks a result ϕ_n, where n is a function of ψ, ξ, and $\{\phi_i\}$ [see eq. (I:6)]. Thus the state vector in case of a reproducible measurement would be reduced to ϕ_n if the measurement would be carried out:

$$\psi \rightarrow \phi_n \quad \text{with} \quad n = n(\psi, \xi, \{\phi_i\}). \tag{1}$$

Consequently, if a quantity A is measured (and therefore the ϕ_i all must be eigenfunctions of A_{op}), the result of the measurement would be

$$A_n \quad \text{with} \quad n = n(\psi, \xi, \{\phi_i\}), \tag{2}$$

[see (I:19)]. For possible "Gleason troubles" when some A_i are degenerate, see Section 3.7 of Part I. A way out of such troubles, if any arise, will be suggested in eq. (119) of Section 3.6 below.

In Bohm's theory we shall find n in eqs. (1)–(2) to depend not merely upon the hidden variables ξ of the system on which the measurement is made, but also, in a way that in practice is impossible to describe, upon the hidden variables that describe the microstate of the measuring apparatus.

(E) Combining the postulated distribution $\bar{p}_\beta$ of (B) with the algorithm eq. (1) of (D), it should be *proved* that it *follows* that the theory predicts the same probability densities as ordinary quantum theory predicts. The distributions which the hidden-variables theory predicts for the various states ϕ_k in the set $\{\phi_i\}$ are [see (I:9) and (I:89)]

$$P_k(\psi, \{\phi_i\}) = \int_{(\beta)} p_\beta \, \delta_{k,\, n(\psi,\, \xi^{(\beta)},\, \{\phi_i\})}. \tag{3}$$

It should be shown that, upon substitution of $\bar{p}_\beta$ from (B) for p_β, the integral in (3) gives

$$\bar{P}_k(\psi, \{\phi_i\}) = w_k(\psi) \equiv |\alpha_k|^2, \tag{4}$$

where

$$\alpha_k = \int \phi_k^* \psi \tag{5}$$

is the quantum-mechanical probability amplitude.

The Kronecker symbol in (3) is the truth value of the selection statement "k is chosen by ψ and $\xi^{(\beta)}$ from $\{\phi_i\}$." In Part I we used for it the symbol v:

$$v_k(\psi, \xi, \{\phi_i\}) = \delta_{k,\, n(\psi,\, \xi,\, \{\phi_i\})}. \tag{6}$$

In Bohm's theory, the explicit dependence of n in (3) on the ξ is not known. Yet the validity of (6) is guaranteed there as Bohm borrows his theory of measurements from ordinary quantum theory.

(F) The theory should make it clear in what way ψ becomes changed into ϕ_n as a consequence of the measurement discussed in eqs. (1)–(2). In some theories, this may be a Copenhagen-style ("voluntary") reduction of the wave packet. If so, it should be explained why in this case the prediction by the algorithm eq. (1) for the *next* measurement is *independent of whether or not* for our single system described by ξ *the reduction from* $\psi(t)$ *(at time t after the first measurement) to* ϕ_n *is performed.*

In other theories such a proof may be made redundant by making the reduction from $\psi(t)$ to ϕ_n *unvoluntary*, that is, by giving an equation for $d\psi/dt$ which leads to an *automatic* change of $\psi(t)$ into ϕ_n as time progresses.

(G) The theory should determine $d\xi/dt$ in absence of measurements. This is often provided by some plausible *postulates*, like, for instance, that, for measurements of constants of motion, for given initial ψ and ξ, the n predicted by (1) shall be independent of time. In Bohm's 1951 theory, $d\xi/dt$ is determined otherwise.

(H) In theories which alter the Schrödinger equation for ψ by nonlinear terms which are effective during measurements and which reduce ψ to ϕ_n, the question should be discussed whether or not the equation for $d\xi/dt$ should be altered in a similar way. Different proposals for (H) will lead to different predictions of the theory for experiments in which measurements follow each other so rapidly that the perturbed (biased) distribution p_β resulting from the first measurement has no time to relax to $\bar{p}_\beta$ before the second measurement is performed.

Before a theory as complicated as a hidden-variables theory of the first kind becomes a honest candidate for being generally accepted as a reasonable description of the happenings in nature, one may want to demonstrate experimentally the existence of the deviations from quantum theory which such a theory predicts for measurements performed in rapid succession (see the last part of Section 1.3 of Part I). Then measurements of this kind may also decide between various simple forms of the equation for $d\xi/dt$ during a measurement (see Section 5.7.)

Finally, it would be desirable, if any hidden-variables theory can be taken serious, to get a better understanding of the relaxation mechanism which changes p_β into $\bar{p}_\beta$. This understanding should enable us to *estimate numerically the time required* for a perturbed p_β to return to the canonical distribution $\bar{p}_\beta$ to a sufficient degree as to guarantee the validity of quantum-theoretical predictions. As one would be able to calculate approximately this relaxation time, one would be better able to devise experiments for investigating whether or not quantum theory has the deficiency ascribed to it by hidden-variables theory.

1.3. Theories discussed in Part II

In Chapter 2 we shall discuss the 1951 theory of Bohm, its relation to work of De Broglie, and its application to various problems (partially in appendices). As the theory is incomplete, we discuss possibilities of its completion. In this connection, we discuss Bell's model of a theory of spin $\frac{1}{2}$ and some generalizations of it. The problem of generalizing Bohm's theory to a relativistic theory leads to more serious difficulties.

In Chapter 3 we discuss the theory of Wiener and Siegel, based upon Wiener's "differential-space theory." This theory is much more complete, but some of the mathematical tools it uses give physicists the feeling that its "hidden variables" are rather artificial; in fact, so much so that one of its authors has proposed to use the name "hidden-variables state" no longer for states with given values of the quantities orginally introduced as hidden variables, but rather for states characterized by a set of predictions made on the basis of a set of values for the original quantities.

In Chapter 4 we present the theory of Bohm and Bub and some related work of Tutsch. We show how the Bohm–Bub theory is naturally obtained in two steps. The first step consists of eliminating from the Wiener–Siegel theory all mathematical details that are irrelevant for its application. This gives the theory an appearance that is much more acceptable for a physicist. The second step consists of adding to the theory a mechanism that replaces the voluntary reduction of wave functions. As there is ambiguity in this step, Tutsch has investigated some possible forms for it, and has concluded that the Bohm–Bub formalism is the simplest one that can do the job. It is likely to be an oversimplification.

In Chapter 5 we ask what predictions the Wiener–Siegel/Bohm–Bub theory can make that are experimentally verifiably different from predictions made by quantum theory. Since experiments using the polarization of light are easier to perform than experiments using spin polarizations of electrons, we discuss the application of the theory to photons. We also reconsider the time dependence of the hidden variables as it influences the results predicted for the experiment. We then calculate these predictions for a few simple experiments.

One of these experiments has already been tried by Papaliolios (1967). The results were negative. This *may* mean that the best hidden-variables theory of the first kind so far in existence is wrong. It is safer to say that it means that, so far, we have *no need* for this theory. [See Section 1.2, below (H).] Since it has not yet been established how fast one may theoretically expect biased hidden-variables distributions to relax, it is not certain at this stage whether or not crucial experiments between the Wiener–Siegel or Bohm–Bub theory and pure quantum theory will be feasible in the future. If a lower limit can be set to the time needed for relaxation, such a crucial experiment might or might not be found possible. If an upper limit to this relaxation time can be found that is obviously too short to be beaten experimentally, we can establish that there will never be a need for a hidden-variables theory of the first kind, as pure quantum theory will do all jobs of application of the theory in much simpler a way.

CHAPTER 2

BOHM'S 1951 THEORY

THIS nonrelativistic theory, originated in 1951 by Bohm (1952a, b), had its starting point in some considerations which were proposed already by Madelung (1926).

2.1. Ensemble velocity field and particle velocities

The basic idea is the following. Consider an electron in a potential field $V(\mathbf{x})$. If no measuring apparatus disturbs the electron in some hard-to-describe fashion, the electron is described quantum mechanically by the Schrödinger equation

$$i\hbar \frac{\partial \psi}{\partial t} = \left[-\frac{\hbar^2}{2m} \nabla^2 + V(\mathbf{x}) \right] \psi(\mathbf{x}, t) \tag{7}$$

and the conjugate equation for $\partial\psi^*/\partial t$. These two equations combine to a continuity equation

$$\frac{\partial(\psi^*\psi)}{\partial t} = \operatorname{div} \left\{ \frac{i\hbar}{2m} [\psi^* \nabla \psi - (\nabla \psi^*)\psi] \right\}. \tag{8}$$

Regarding

$$\varrho = \psi^*\psi \tag{9}$$

as the probability density of the ensemble of electrons described by ψ in $\mathbf{x}$-space, we see that we may write (8) as

$$\frac{\partial \varrho}{\partial t} = -\operatorname{div} (\varrho \mathbf{v}) \tag{10}$$

if we define the local ensemble velocity $\mathbf{v}$ as a function of $\mathbf{x}$ by

$$\mathbf{v} = \frac{\hbar}{2im} \frac{\psi^* \nabla \psi - (\nabla \psi^*)\psi}{\psi^*\psi}. \tag{11}$$

This allows us in nonrelativistic wave mechanics to make an assumption which goes beyond the Copenhagen interpretation[2] of the wavefunction.

[2] The basic interpretation of quantum theory is what Ballentine (1970) calls the *statistical interpretation.* By "Copenhagen interpretation" *we* mean essentially the statistical interpretation, together with the belief that it is not advantageous to introduce hidden variables as long as there is no experimental evidence necessitating their introduction.

In x,y,z,t-space we visualize the "sample particles" of which at each time t the ensemble E_ψ is made up[3], as points distributed on the corresponding surface $t =$ constant at positions $\mathbf{x} = \boldsymbol{\xi}$. We then can connect these points from surface to surface by worldlines with a slope corresponding to the velocity $\mathbf{v}$ given by eq. (11). The result will look like a picture of the worldlines of the particles of a fluid, though, of course, in reality it may have been the existence of one single electron that made us write down the Schrödinger equation (7). Because of the continuity equation (10), this space–time picture of the ensemble is consistent without worldlines stopping dead or starting from nothing.

We then might *imagine* (though at the present no experiments exist which could verify or falsify this) that any single particle from the ensemble described by ψ would perform on its own a well-defined continuous motion by following one of the worldlines just described, if somehow its initial (or final!) position were known. *That this would be so is the basic claim made by Bohm's 1951 theory.*

2.2. The imagined or actual particle momentum and the quantum-mechanical potential

In imitation of nonrelativistic classical mechanics we shall define for our particle a quantity which we shall provisionally call its *imagined* momentum $\boldsymbol{\pi}$:

$$\boldsymbol{\pi} \equiv m\, d\boldsymbol{\xi}/dt = m\mathbf{v} \quad [\text{that is, } \boldsymbol{\pi}(\mathbf{x}) = m\mathbf{v}(\mathbf{x})]. \tag{12}$$

According to Bohm's theory, this $\boldsymbol{\pi}$ would be not just "imagined" but would be the actual particle momentum. By (11),

$$\boldsymbol{\pi}(\mathbf{x}, t) = \frac{\hbar}{2i} \nabla \ln (\psi/\psi^*) = Re\left\{\frac{\hbar}{i} \nabla \ln \psi\right\}. \tag{13}$$

Correspondingly, the imagined (or actual) force $\mathbf{f}$ is defined at $\mathbf{x}$ by

$$f(\mathbf{x}, t) = \frac{d\boldsymbol{\pi}(\mathbf{x}(t), t)}{dt} = \left[\frac{\partial}{\partial t} + \mathbf{v}\cdot\nabla\right] \boldsymbol{\pi}(\mathbf{x}, t). \tag{14}$$

Calculations are simplified if we put

$$\psi = R \exp (iS/\hbar), \tag{15a}$$

$$R = |\psi| = \sqrt{\varrho}, \quad S = (\hbar/2i) \ln (\psi/\psi^*). \tag{15b}$$

Then (13) becomes

$$\boldsymbol{\pi} = \nabla S. \tag{16}$$

As shown in Appendix A, eqs. (12)–(16), with the help of the Schrödinger equation (7), give

$$\mathbf{f} = \frac{d\boldsymbol{\pi}}{dt} = m\frac{d^2\boldsymbol{\xi}}{dt^2} = -\nabla(V+U), \tag{17}$$

where the right-hand member is to be taken at $\mathbf{x} = \boldsymbol{\xi}$ and where

$$U = -\frac{\hbar^2}{2m}\frac{\nabla^2 R}{R} = -\frac{\hbar^2}{4m}\left[\frac{1}{\varrho}\nabla^2\varrho - \frac{1}{2}\left(\frac{\nabla\varrho}{\varrho}\right)^2\right]. \tag{18}$$

[3] By E_ψ we refer to the ensemble described by ψ. See Part I. See also Ballentine (1970).

This U has been named by Bohm (1952a) the *quantum-mechanical potential*. Correspondingly, we shall call $-\nabla U$ the quantum-mechanical force. By (17), it is responsible for the difference between the (imagined or actual) motion of a particle according to this Bohm interpretation of nonrelativistic wave mechanics, and the motion of a particle according to classical mechanics. Note that U does not depend on the phase $S/\hbar$ of the wavefunction ψ, and that it depends only locally on the quantum-mechanical ensemble density ϱ of the ensemble E_ψ, and its first two derivatives. Moreover,

U is not changed when the ensemble density ϱ is multiplied by a numerical constant factor (19)

as in a renormalization of the wavefunction.

What, from the point of view of pure quantum theory, are just *imagined* motions of particles with positions $\boldsymbol{\xi}(t)$, velocities $\mathbf{v}$, and momenta $\boldsymbol{\pi}$, motions that are not verifiable but that are not in disagreement with the predictions of wave mechanics for the probability distribution $\varrho(\mathbf{x}, t)$, these same motions are from this 1951 Bohm point of view the possible motions of the *actual particles* that might be described by ψ.

In some simple cases the equations of motion (17) can be integrated. Examples are given in Appendices E–G. In most cases, however, the term $-\nabla U$ in (17) causes a motion which is wildly complicated.

Moreover, for integrating the equation (17) for $\boldsymbol{\xi}(t)$ in a known $\psi(\mathbf{x})$ field, we should first know the initial value of $\boldsymbol{\xi}$. [The initial value of $d\boldsymbol{\xi}/dt = \boldsymbol{\pi}/m$ then follows from (16) at $\mathbf{x} = \boldsymbol{\xi}$.] So far, however, no methods are known of measuring $\boldsymbol{\xi}$ without disturbing $\psi(x)$. [See Section 2.11.] Therefore, the $\boldsymbol{\xi}(t)$ are for us as "*hidden*" as are the motions of individual molecules in a gas. Like the latter in kinetic theory, therefore, for given ψ the $\boldsymbol{\xi}(t)$ can only be used *statistically*, by averaging over all possible motions.

2.3. Possible alteration of Bohm's theory

We repeat here four essential assumptions made in the above:

(1) The ordinary Schrödinger equation holds for $\psi(\mathbf{x}, t)$.
(2) The particles cannot have arbitrary velocities. Their initial velocities are determined by $\boldsymbol{\pi} = (\nabla S)_{\mathbf{x}=\boldsymbol{\xi}}$.
(3) The velocities then change according to eq. (17), thus maintaining $\boldsymbol{\pi} = \nabla S$ at all times.
(4) The probability density of $\boldsymbol{\xi}$, which we may call $P(\boldsymbol{\xi})$, is equal to $\varrho(\mathbf{x}) \equiv |\psi(\mathbf{x})|^2$ at $\mathbf{x} = \boldsymbol{\xi}$; this is so initially, and remains so by eq. (16), as seen from (8).

The theory determined by these four rules then will yield all results of ordinary wave mechanics.

We see that ψ satisfies a field equation (the Schrödinger equation) like the electromagnetic field does (which satisfies Maxwell's equations). The ψ field exerts a force on a particle (by $-\nabla U$) like an electromagnetic field does (by $\varepsilon\mathbf{E}+\varepsilon\mathbf{v}\times\mathbf{B}/c$). However, the initial velocity of a particle does not depend on the electromagnetic field, so the assumption (2) that it would depend by (16) or (11) on the ψ field is not satisfactory.

Therefore Bohm has suggested that $\boldsymbol{\pi} = \nabla S$ be merely an approximation, and that under circumstances which make us doubt the validity of wave mechanics (as when distances of 10^{-13} cm or times of 10^{-23} s are involved), this equation may be violated. He suggests adding to $-\nabla U$ in (17) a term which makes $d\boldsymbol{\xi}/dt$ tend to $\nabla S/m$ when it is different from it. If L is a length of around 10^{-13} cm, this could be achieved by adding to (17) on the right a term $(c/L)(\nabla S - m\, d\boldsymbol{\xi}/dt)$. (Here, $1/L$ may be some function of $\boldsymbol{\xi}$ or ψ and their derivatives.) Then the initial $d\boldsymbol{\xi}/dt$ can be arbitrary, and will rapidly "decay" to the value $\nabla S/m$.

Under these circumstances, the current density $\varrho\, d\boldsymbol{\xi}/dt$ will, of course, not satisfy a continuity equation with ϱ. So Bohm suggests that in this generalization of the theory the ensemble of $\boldsymbol{\xi}$ values at a given time t shall have a density $P(\boldsymbol{\xi})$ different from $\varrho(\mathbf{x})$ at $\mathbf{x} = \boldsymbol{\xi}$, so that the continuity equations shall hold between $P(\boldsymbol{\xi})$ and $P(\boldsymbol{\xi})\, d\boldsymbol{\xi}/dt$. After $d\boldsymbol{\xi}/dt$ has decayed to $\nabla S/m$, we shall show in Appendix B that $P(\boldsymbol{\xi})$ in most cases will decay to $\varrho(\boldsymbol{\xi})$, with certain exceptions that are meaningful. (See the last part of the appendix.)

This proof, however, uses the Schrödinger equation (7). Bohm does consider the possibility that even this equation will have to be modified either by terms that vanish for $\nabla S = m\, d\boldsymbol{\xi}/dt$, or by terms that are nonlinear in ψ and that become important only in problems involving distances $\lesssim 10^{-13}$ cm, or both. However, he does not make any specific suggestions about these changes.

Bohm then claims[4] that, if the correct theory would incorporate some of these changes, it may be possible to invent methods, using the altered theory, for measuring $\boldsymbol{\xi}$ without disturbing $\psi(\mathbf{x})$; but he does not indicate any realistic way of doing so.

More convincingly, he remarks that, if $P(\boldsymbol{\xi}) \neq |\psi(\boldsymbol{\xi})|^2$, it should be possible to discover deviations from quantum theory experimentally.[5]

In the following we shall consider the unaltered theory which should yield all results of ordinary wave mechanics while giving a model in which each particle would simultaneously have a precise position $\boldsymbol{\xi}$ and a precise momentum $m\, d\boldsymbol{\xi}/dt = \boldsymbol{\pi}$, contrary to the beliefs of the Copenhagen school.[6]

Also, we shall in the following write simply $\mathbf{x}$ for the particle's $\boldsymbol{\xi}$, but we shall maintain the distinction between the particle's momentum $\boldsymbol{\pi} = m\mathbf{v}$, and the quantum theoretical observable $\mathbf{p}$ described by the operator $(\hbar/i)\nabla$ [see Section 2.11].

2.4. Comparison with De Broglie's work

Most of the formulas of Sections 2.1 and 2.2 are found already in old work of De Broglie (1926, 1927a, b).[7] For De Broglie, however, the ψ wave was not the "real thing." The ψ wave satisfying the Schrödinger equation (7) and describing by $\varrho = \psi^*\psi$ both a probability density and by (18) a quantum-mechanical potential, was merely a "pilot wave" telling the particle

[4] See Bohm (1952b), sec. 6, where his reasoning on p. 185 is too speculative to be very convincing.

[5] This idea is a forerunner of thoughts about the Bohm–Bub theory (Chapter 4) which led to the experiments of Papaliolios (Chapter 5 of this Part II).

[6] See Part I, Section 1.8.

[7] For a more detailed account of De Broglie's ideas of 1924–27, and for the reasons why he dropped them in 1927, and why he took them up again in 1951 [see De Broglie (1951)] together with Vigier (1952) after he had seen a preprint of Bohm's work, see De Broglie (1953).

how to move. According to De Broglie, the "real thing" was another wave, which not only would tell the particle how to move, but which would also tell where the particle is located. This wave,

$$u = R' \exp(iS/\hbar), \tag{20}$$

would have the same phase $S/\hbar$ as the ψ-wave, but would have an amplitude R' with a singularity at the actual position of the particle. In 1924–7, De Broglie thought that u could be just a different solution of the same Schrödinger equation (7). Therefore he called his theory (which used both u and ψ) the *theory of the double solution*. In 1951, however, De Broglie realized that u, if it existed, would have to satisfy a slightly different, *nonlinear* equation [De Broglie (1952)], and was reinforced in this idea by Vigier (1953). Any specific equation for u, or its solution or application, to my knowledge was never published.

2.5. Solving for the possible trajectories of a particle

For finding the possible trajectories of a particle according to Bohm's theory in a given potential field $V(\mathbf{x})$, we first solve the Schrödinger equation (7) for $\psi(\mathbf{x}, t)$, using appropriate boundary conditions. Then we could calculate $U(\mathbf{x}, t)$ from $\varrho = |\psi|^2$ by (18), and, for a given initial position $\mathbf{x}$ of the particle, we could find its initial imagined momentum from (13). We then could integrate the equation of motion (17), as one would do in classical mechanics, except for our present inclusion of U as an additional potential.

However, instead of using this second-order differential equation for $\mathbf{x}(t)$, it is simpler to integrate the first-order differential equation

$$d\mathbf{x}/dt = m^{-1} \nabla S(\mathbf{x}, t) \tag{21}$$

obtained from (12) and (16), with S given by (15b). This is equivalent to calculating $\mathbf{v}$ from (11). Thus we obtain, at time $(t+dt)$, the new position $\mathbf{x}+d\mathbf{x} = \mathbf{x}+\mathbf{v}\,dt$ of the particle, completely bypassing the "dynamics" of eqs. (17)–(18).

2.6. In what way is Bohm's 1951 theory a hidden-variables theory?

Let us see now to what extent does Bohm's theory meet the requirements (A)–(G) of Section 1.2.

We met *requirement (A)* by identifying the hidden[8] variable ξ of a particle with its position $\boldsymbol{\xi}$ or $\mathbf{x}$. In a quantum-mechanical one-particle state given merely by ψ, the position $\mathbf{x}$, indeed, is a parameter which picks in Bohm's theory the worldline of an individual particle from the ensemble E_ψ of particles described by ψ.

We meet *requirement (B)* by identifying the equilibrium distribution $\bar{P}(\mathbf{x})$ of the hidden variable $\mathbf{x}$ with the wave-mechanical probability density:

$$\bar{P}(\mathbf{x}) = \varrho(\mathbf{x}) \equiv |\psi(\mathbf{x})|^2. \tag{22}$$

Discussing *requirement (C)*, we ask why $P(\mathbf{x})$ should tend toward $\bar{P}(\mathbf{x})$. In his later papers, Bohm makes it successively more plausible that random interactions (collisions) will make

[8] Remember that the name "hidden" is a misnomer. [See Part I, Section 1.2, and Pearle (1968b).]

the ratio $P(\mathbf{x})/\varrho(\mathbf{x})$ approach a constant value independent of $\mathbf{x}$. This constant ratio then must be $= 1$, if we normalize both $P(\mathbf{x})$ and $\psi(\mathbf{x})$. (See Appendix B.)

As to *requirement (D)*, Bohm's theory in its original form is an incomplete theory. For "observables" $F(\mathbf{x})$ that merely depend upon the particle's position, the algorithm is simply that, for given initial value $\mathbf{x}(0)$ of the hidden variable, the predicted result of a measurement of F at time t will be $F(\mathbf{x}(t))$, with $\mathbf{x}(t)$ calculated as in Section 2.4.

For the particle's *spin*, Bohm's original theory has no prescription at all. We shall consider in Section 2.14 a method suggested by Bell (1966) of treating the spin of an electron.

As to the momentum, we must distinguish here between the particle's momentum $\boldsymbol{\pi}$, which from Bohm's point of view is a physically significant property of the system, and the quantum-theoretical observable $\mathbf{p}$. The algorithm for determining $\boldsymbol{\pi}$ is simply $\boldsymbol{\pi} = \nabla S$ [eq. (16)], which in terms of ψ and ξ is given by (13):

$$\pi(\psi, \xi) = (\hbar/2i)\,(\partial/\partial\boldsymbol{\xi})\ln\,[\psi(\boldsymbol{\xi})/\psi^*(\boldsymbol{\xi})]. \tag{23}$$

However, just as we have no way of *measuring* $\boldsymbol{\xi}$, we have no way either of measuring $\boldsymbol{\pi}$, so, saying that (23) is the predicted outcome of such a measurement is not very meaningful.

On the other hand, when ψ is given, we *can* measure the *observable* $\mathbf{p}$. This observable, though, is not really a property of the physical system by itself. As Bohm has expressed this in his textbook on quantum theory,[9] observables are merely "potentialities" of the system, and can be brought out only by the measuring apparatus in its uncontrollable microstate. As we shall see in Section 2.11, where we discuss the theory of measurements for Bohm's theory, the eigenvalue Q_n of an observable Q that is found by a measurement depends in Bohm's theory not merely on ψ and Q and the hidden variables ξ of the system on which Q is being measured, but it depends also upon the hidden variables ξ_Q of the apparatus by which the measurement of Q is made. As the latter will never be known, in Bohm's theory the function $n = n(\psi, \xi, \xi_Q, Q)$ which determines the label n of the eigenvalue Q_n found will remain unknown even if and when at some future time it will have become possible to measure the system's hidden variables ξ.

We see now at once why von Neumann's arguments against hidden-variables theories do not apply to Bohm's theory. Bohm readily agrees[10] that in his theory *one* set of hidden variables of a system cannot assign precise values to *two* noncommuting observables Q and P simultaneously. In his theory, the precise value of Q is a function of the combined hidden variables ξ of the system *and* ξ_Q of the apparatus measuring Q, while the value of the other observable P is a function of ξ together with ξ_P. (Both, of course, are functions of ψ.)

On the other hand, von Neumann's arguments do not apply to $\boldsymbol{\pi}$, because $\boldsymbol{\pi}$ is not an observable in the quantum theoretical sense. (No $\boldsymbol{\pi}_{\mathrm{op}}$ is defined in Hilbert space.) Therefore it lies beyond reach of von Neumann's considerations.

[9] See Bohm (1951), chap. 6, sec. 9. Note that this book predates Bohm's hidden-variables theory. In fact its chap. 22, sec. 19, purports to show that hidden-variables theories are impossible.

[10] Bohm (1952b), sec. 9.

Inapplicability of Gleason's arguments to Bohm's theory

While we are at it, we might mention here, too, why "*Gleason's proof*" (page 46) does not apply to Bohm's theory. It does not apply to allocation of dispersionfree values to the hidden variables $\boldsymbol{\xi} = \mathbf{x}$ because linear superpositions of eigenfunctions of $\mathbf{x}_{op}$ are eigenfunctions of hermitian operators that do *not* correspond to observables. [See Part I, Section 3.7, page 48.] It does not apply to the predictions of Bohm's theory for measured $\mathbf{p}$-values for a similar reason. It does not even apply to predictions of Bohm's theory for measurements of the angular momentum $\mathbf{L}$ because Bohm does not assign an L_n value to a state (ψ, ξ), but leaves L_n undetermined as long as the hidden variables of the apparatus are undetermined. If for L_n in one direction $\mathbf{n}$ the apparatus hidden variables *would* be known, they certainly would be different for the angular momentum component $L_{n'}$ in a different direction $\mathbf{n}'$, as $L_{n'}$ would be measured by a *different* apparatus. Therefore, the arguments of Chapter 3 of Part I, where we considered the predictions for components L_n in a three-dimensional space all for one and the same hidden-variables state (ψ, ξ), cannot apply here, where the ξ include at one time the ξ_{L_n}, and at another time the $\xi_{L_{n'}}$.

The algorithm (D) for $\boldsymbol{\Lambda} = \boldsymbol{\xi} \times \boldsymbol{\pi}$ is, of course, that it has the value of $\mathbf{x} \times \nabla S(\mathbf{x})$ at $\mathbf{x} = \boldsymbol{\xi}$. For the observable $\mathbf{L} = \mathbf{x} \times \mathbf{p}$, with operator $\mathbf{L}_{op} = \mathbf{x}_{op} \times (\hbar/i)\nabla$, the situation is the same as for the observable $\mathbf{p}$: the result of the measurement is determined as discussed in Section 2.11, and the result is an unpredictable function of the hidden variables of the measuring apparatus together with the particle's own hidden variables (ψ, ξ).

Requirement (E) is for the observable $\mathbf{x}$ satisfied trivially by eq. (22). For other observables it is satisfied practically without considering the hidden variables at all because of the identity between Bohm's theory of measurement for observables and the conventional theory of measurement for quantum theory (see Section 2.11).

For $\boldsymbol{\pi}$ there is no requirement (E) because pure quantum theory does not make probability predictions for this quantity that is not an observable.

Reduction of wave packets

Before we can discuss *requirement (F)*, we must first remark that in Bohm's theory the reduction of the wavefunction $\psi(x)$ upon a measurement is performed exactly like in pure wave mechanics [see Section 2.11]. As a typical example, consider a particle scattering on some target [compare Section 2.10]. The scattered wave can be written as a superposition $\sum_i c_i \psi_i$ of different kinds of waves ψ_i. In and near the target there are intricate interference phenomena between these waves ψ_i. This causes wild motions of the particles due to the effect of $-\nabla U$ in eq. (17). It is now important that the incident wave describing an individual particle approaching the scattering center is *not* the infinitely extended plane monochromatic wave of given momentum $\mathbf{p}$ which theoreticians for the sake of simple calculations often postulate. The correct description of the incident wave is a finite wave packet aimed so as not to bypass the target. This finite incident wave gives rise to finite secondary waves which in the case of inelastic scattering may be spherical waves of different radial velocity depending upon the amount of energy lost in the collision. These spherical waves after a certain time become separated by spaces in which $\psi = 0$, and, though a particle could cross such regions at infinite speed [see Section 2.8], it is more plausible that the particle, when present in one of these spherical wave packets, will stay inside it. If by any observation we find out in which of these wave packets the particle is trapped, this observation will introduce in that wave packet a factor which will depend on additional variables and which will make this part of the wavefunction orthogonal to the remainder of ψ. [See Section 2.10 for an example.] When people want to ignore such dependence of ψ on additional variables, they often simulate its effects by putting an uncontrollable phase factor $e^{i\alpha}$ in the part of the wavefunction that has interacted with some gadget enabling us to observe the presence or absence of the particle in that wave packet.[11] The effect in any case is that terms in ψ with and without such factors, or with different factors of this nature, can no longer give rise to observable effects of interference among each other. If we indicate such factors by f, the result is that the wavefunction ψ which originally was a superposition $\psi = \sum_i c_i\psi_i$ changes by the possibility of observation into

$$\psi = \sum_i f_i c_i \psi_i, \tag{24}$$

so that the probability distribution changes from $\sum_i\sum_j c_i^* c_j \psi_i^* \psi_j$ to $\sum_i\sum_j (f_i^* f_j) c_i^* c_j \psi_i^* \psi$, where, upon repeating the experiment many times for observing the presence or absence of interference, the factor $(f_i^* f_j)$ for $i \neq j$ either is zero by orthogonality as a function of other variables, or averages to zero because $f_j = \exp(i\alpha_j)$ with α_j varying arbitrarily from one measurement to the next. Thus the probability distribution effectively becomes

$$(|\psi|^2)_{\mathrm{Av}} = \sum_i |c_i|^2 |\psi_i|^2 = \sum_i w_i |\psi_i|^2 \tag{25}$$

if we assume normalization of f_i so that $(f_i^* f_i) = 1$. The probability distribution (25), however, is identical with the distribution given by a "mixed state" where there is a probability $w = |c_i|^2$ for a pure wavefunction ψ_i. Replacing, therefore, the pure (superposition) state $\psi = \sum f_i c_i \psi_i$ by this mixed state, the probabilities w_i become classical probabilities, so that our knowledge from the observation that the particle is in wave packet ψ_n allows us to replace the $w_i = |c_i|^2$ by $w_i = \delta_{in}$ as better describing the latest information we have,[12] and the mixed state with distribution (24) becomes again a pure state described by a new wavefunction $= \psi_i$ called the "reduced" wavefunction.[11]

Omitting the above explanations, we may obtain the final result ψ_n from the original $\psi = \sum_i c_i \psi_i$ in two steps. Firstly, we *omit* from ψ all terms except the term $c_n \psi_n$ corresponding to the packet in which the scattered particle is found. Secondly, we *renormalize* the result $c_n \psi_n$ thus obtained by multiplying it by a new normalization factor c_n^{-1} in order to have again a normalized wavefunction ψ_n.

An important feature in the above is that, *due to the finite size of the incident wave packet*, the interference phenomena causing wild motion of the scattered particle do not last forever but quiet down as the different wave packets separate from each other.[13]

[11] See Bohm (1951), chap. 6, secs. 5–6, and chap. 22, secs. 8–11.

[12] Bohm (1951), chap. 6, sec. 4, page 126.

[13] See Bohm (1952a), pages 176–177, and Bohm (1952b), pages 191–192. Also De Broglie, according to his reply to Pauli at the 1927 Solvay Conference, seems vaguely to have recognized this point [see Solvay (1928), pages 282–283], but apparently he did not convince himself. [See De Broglie (1953), pages 12–13.]

The reduction of the wavefunction drastically changes $|\psi|^2$ in the regions where the other wave packets ψ_i for $i \neq n$ are different from zero, and therefore will affect $U(\mathbf{x})$ of eq. (18) in those regions, but this change does not affect the further motion of the scattered particle by eq. (17) because the particle supposedly stays in the wave packet ψ_n that is left, and does not appear at the locations where U is altered by the reduction.

In the region of wave packet ψ_n, after "step one" (omission of the other parts of ψ), the remaining wavefunction $c_n\psi_n$ is not different from the original ψ, and therefore also U there is unchanged. In "step two" we have to replace this $c_n\psi_n$ by the renormalized wave-function ψ_n. According to (19), however, this does not affect $U(\mathbf{x})$ either, and the further motion of the scattered particle as predicted by the equation of motion (17) (or by $d\mathbf{x}/dt = m^{-1}\nabla S$, for that matter) is the same *after* the reduction of the wave packet as it would have been *without* this reduction. Thus the requirement (F) is met.

For all of this, it is important that the reduction of the wavefunction will change ψ, *along the trajectory* $\mathbf{x}(t)$, *merely* by multiplication by the new normalization factor c_n^{-1}. Therefore it is necessary that the regions where the omitted wave packets ψ_i for $i \neq n$ do not vanish, will not overlap with the region where the remaining wave packet ψ_n does not vanish. All of this would not have been possible without introducing new variables upon which ψ might depend[14] if the wave packets would not have separated. Failure to see that this separation can be achieved by keeping the incident wave finite was apparently the main reason why De Broglie abandoned his theory of the pilot wave (Section 2.4) from 1928 until 1951.

Requirement (G) of Section 1.2 is satisfied obviously by the fundamental assumption eqs. (12)–(13) of Bohm's theory.

2.7. How to tell through which slit a particle came that is observed in the interference pattern behind several slits

An amusing part of Bohm's theory is that in principle it can tell us through which slit a particle came as it hits a screen in one of the constructive-interference fringes behind a system of slits.

The Copenhagen version of quantum theory claims that the question through which slit such a particle came has no answer, and therefore should be a (physically) meaningless question. More correctly, quantum theory—according to its statistical interpretation—says that any answer given to the above question is not experimentally verifiable without destroying the interference pattern.[15]

Bohm's theory has no objections against the latter more correct statement and shows the inaccuracy of the first sloppy claim to which a reply on its own level might be the question:

[14] If in $\mathbf{x}$-space the wave packets ψ_i do not vanish at $\psi_n \neq 0$, after observing the particle is in ψ_n we still can follow the above reasoning if in (24) above we use the factors f_i that depend on the additional variables introduced for describing the observation. If they are $\mathbf{y}$, we should consider the system's trajectory a curve in $\mathbf{x},\mathbf{y}$-space (see Section 2.9). The wave packets that overlap in $\mathbf{x}$-space will not overlap in $\mathbf{x},\mathbf{y}$-space because they are separated in $\mathbf{y}$-space.

[15] Any particle detector placed at any one of the slits supposedly perturbs the phase of the wave through that slit so that the interference pattern is washed out. Compare Bohm (1951), chap. 6, secs. 2–5.

How can Copenhagen forbid us to answer the original question by *any* answer we please to give? Thus Bohm's theory gives to the original question a *theoretical* answer, and, since *no experimental verification* is involved and no gadgets are placed at the slits, the interference pattern is not washed out by the act of just *claiming* through which slit the particle came.

For a more specific model of the interference experiment we are talking about, assume we have a screen with slits of atomically exactly known properties and geometry, onto which a low-intensity beam of incident particles is falling so that never more than one particle is near the slit system. We assume that we can describe this beam by a steady incident wavefunction $\psi(\mathbf{x}, t)$ in front of the split system for which we have found boundary conditions that are consistent with the fact that it really is a superposition of an incident wave and a reflected wave. For simplicity let us assume that this incident (and reflected) wave is monochromatic, and depends on time merely by a factor $\exp(-2\pi i \nu t)$.

Given the slit system, in principle we can integrate the Schrödinger equation through the slits into the region behind the slits and up to a screen set far behind the slits, where we assume that the wave is completely absorbed. (No reflection.) This is, of course, an idealization, and this author does not claim that it would be easy to really execute this program (say, on a computer). For the sake of the argument, let us assume that it can be done.

We thus have calculated De Broglie's *pilot wave* (see Section 2.4). Thence at any point between the slits and the absorbing screen we can calculate the "imagined velocity" or "actual particle velocity" $\mathbf{v}$ by eq. (11).

When a particle at time t hits the screen at a point $\mathbf{x}(t)$, we now integrate eq. (23) backward in time until we have traced back from which slit the particle came.[16] Of course, we need not wait for a particle to arrive. Once and for all we can determine for all imagined particle trajectories emerging from each slit, where, according to the integrated eq. (11), they are going to hit the screen in the back. Thus we can mark off on that screen which regions will receive the particles from each of the separate slits.

It is, of course, true that the interference pattern thus subdivided into regions for particles from each slit will completely change when we tamper with any of the slits, either by closing one, or by perturbing the wave through that slit by some particle detector. We then have to start our computing program from scratch, as we have to solve the Schrödinger equation behind the slits under the new circumstances. We will find a different $\psi(\mathbf{x})$; thence, a different $v(\mathbf{x})$ field, a different $U(\mathbf{x})$, and different $d^2\mathbf{x}/dt^2$ if we care for the latter. Thence, particles from each slit will now reach different parts of the screen if they move along the imagined trajectories of the theory of Madelung, De Broglie, and Bohm.

2.8. What keeps particles absent from regions where $\psi = 0$?

For finding the particle trajectories in an interference experiment, we suggested in Section 2.7 first to solve the Schrödinger equation behind the slits. Suppose this makes us find regions of destructive interference in which $\psi = 0$, even though $V \neq \infty$. (Probably, $V = 0$).

[16] Since the differential equation (17) does not involve "uncertainties" of either $\mathbf{x}$ or $\boldsymbol{\pi}$, the impossibility in wave mechanics to trace back the motion from the interference pattern to the slits [see Bohm (1951), chap. 6, sec. 2] does not apply to our present problem.

Then, dynamically speaking, what keeps particles from spending time in these "forbidden" regions?

Bohm (1952a, page 174) has suggested that the absence of particles at $\psi = 0$ would be due to the appearance of $R = |\psi|$ in the denominator of the expression (18) for the quantum-mechanical potential U. If, therefore, $U \to +\infty$, the particle would bounce back from the forbidden region; if $U \to -\infty$, the particle would be infinitely attracted and would spend no time in the region $R = 0$ as it would pass through it at an infinite speed. (Remember that Bohm's theory is a nonrelativistic theory.)

In Appendix C we shall show that the latter is not a likely thing to happen; that is, $U \to -\infty$ is not likely to occur in regions where not also $V \to +\infty$, and $(U+V)$, which determines the particles' motion by (17), is more likely to stay finite.

2.9. Generalization of Bohm's theory to interacting particles

Bohm's theory of "imagined" or "actual" trajectories of particles described quantum-mechanically by a wavefunction is easily generalized to more than one particle. See Appendix D for the mathematical details. For N particles, ψ, and correspondingly V and U, are functions of $3N$ coordinates. The trajectory now becomes a curve in $3N$-dimensional space, telling simultaneously the motion of all N particles as a function of one single time coordinate t. The "imagined" or "actual" momenta $\boldsymbol{\pi}_j$ of all particles become the gradients of $\hbar$ times the common phase $\{S(\mathbf{x}_1, \mathbf{x}_2, \ldots, \mathbf{x}_n)/\hbar\}$ of the N-particle wavefunction eq. (D-1). We find that $\sum_k \boldsymbol{\pi}_k$ is conserved when V and U (or V and ψ) depend merely on the differences $(\mathbf{x}_i - \mathbf{x}_j)$ between particle positions, while, in a stationary state, the fixed and conserved energy $\mathscr{E}$ is the sum of the imagined potential energy $(V+U)$ and the imagined kinetic energy $\sum_j \boldsymbol{\pi}_j^2/2m_j = \sum_j \frac{1}{2} m_j \mathbf{v}_j^2$.

The exclusion principle

Since $\psi(\mathbf{x}_1, \mathbf{x}_2, \ldots)$ for fermions is antisymmetric between pairs of particles, it is zero when two positions coincide.[17] By the argument of Appendix C, generalized to $3N$-dimensional space, the trajectory of the N-particle system therefore will spend zero time at regions where two particles would have the same position.

2.10. Inelastic scattering

Using the generalization of Bohm's theory to many-particle systems discussed in Section 2.9, we can better understand Bohm's treatment of scattering processes to which we alluded already in Section 2.6 while discussing the problem of the reduction of a wavefunction.

We shall briefly discuss here the problem of inelastic scattering, as occurs in the Franck–Hertz experiment. [Bohm (1952a), sec. 7.]

[17] We are here ignoring spin.

Suppose we have an initial state with an atom in the ground state $\psi_o(\mathbf{x})$ at the origin, an incident wave packet with wavefunction

$$f_o(\mathbf{y}, t) = \int d^3\mathbf{k} f(\mathbf{k}-\mathbf{K}) \exp(i\mathbf{k}\cdot\mathbf{y} - i\hbar k^2 t/2m) \tag{26}$$

heading toward the atom according to $\mathbf{y} \approx \hbar\mathbf{K}t/m$ [if $f(\mathbf{k}-\mathbf{K})$ is maximal at $\mathbf{k} = \mathbf{K}$] and going to hit it at $t \approx 0$, and suppose we have no photons present, which we shall indicate by a wavefunction $\psi^{(F)}_{\text{vac}}(q_f)$ for the radiation field. (See Appendix H.) Thus the initial wavefunction is

$$\Psi_o(\mathbf{x}, \mathbf{y}, q_f, t) = \psi_o(\mathbf{x}) \exp(-i\mathcal{E}_o t/h) f_o(\mathbf{y}, t)\, \psi^{(F)}_{\text{vac}}. \tag{27}$$

We should add to (27) additional factors describing the initial states for any other variables that at a later time might interact with the system we are considering here.

It is easily seen that, as long as Ψ can be factorized into separate factors describing separate particles, each of these particles will satisfy its own eqs. (D-1) through (D-19) without involving the other particles, as long as no interaction terms like $V(\mathbf{x}_1, \mathbf{x}_2, \ldots)$ start mixing up the Ψ field so that it no longer can be factorized in this way.

When, just before $t = 0$, the wave packet (26) comes close to the atomic wavefunction $\psi_o(\mathbf{x})$, such perturbation of (27) starts taking place. We can use ordinary time-dependent perturbation theory or any other means of integrating the time-dependent Schrödinger equation for $\Psi(\mathbf{x}, \mathbf{y}, q_f, t)$, taking into account the interaction between $\mathbf{x}$ and $\mathbf{y}$, and for $t > 0$ we will find

$$\Psi(\mathbf{x}, \mathbf{y}, q_f, t) = c_o\Psi_o + \sum_n \psi_n \exp(-i\mathcal{E}_n t/\hbar) f_n(\mathbf{y}, t)\, \psi^{(F)}_{\text{vac}}(q_f), \tag{28}$$

where the asymptotic form of $f_n(\mathbf{y}, t)$ is

$$f_n(\mathbf{y}, t) \to \int d^3\mathbf{k} f(\mathbf{k}-\mathbf{K})\, g_n(\theta, \phi, \mathbf{k})\, r^{-1} \exp(ik_n r - i\hbar k_n^2 t/2m), \tag{29}$$

where r, θ, ϕ are the spherical coordinates of the point $\mathbf{y}$, and where k_n is defined by

$$\hbar^2 k_n^2/2m = \hbar^2 k^2/2m + \mathcal{E}_o - \mathcal{E}_n. \tag{30}$$

The $g_n(\theta, \phi, \mathbf{k})$ can be calculated by considering the simpler case that just a plane monochromatic wave $\exp(i\mathbf{k}\cdot\mathbf{y} - i\hbar k^2 t/2m)$ would have been incident.

The various n-values correspond to different possible excitations of the atom at the origin. As each wave packet $f_n(\mathbf{y}, t)$ has its k_n values centered around K_n given by replacing k by K in (30), it will expand radially at a speed $\sim \hbar K_n/m$, and, for t larger than a certain positive value, the wave packets for different n will start to separate.

The excited atom, in the meantime, interacts with the radiation field, and, according to quantum electrodynamics (compare Appendix H), will spontaneously return to the ground state ψ_o under emission of a photon of energy $\mathcal{E}_n - \mathcal{E}_o$ in some arbitrary direction. We might indicate this final state of the radiation field by $\psi^{(F)}_{f(n)}$, where the index $f(n)$ denotes the polarization, energy, and momentum (photon direction) of the radiation field containing this photon. Thus, after a while, (28) is changed into

$$\Psi(\mathbf{x}, \mathbf{y}, q_f, t) = c_o\Psi_o + \psi_o(\mathbf{x}) \sum_n \exp(-i\mathcal{E}_n t/\hbar) f_n(\mathbf{y}, t) \sum_{f(n)} c_{f(n)} \psi^{(F)}_{f(n)}(q_f), \tag{31}$$

where the $\exp(-i\mathcal{E}_n t/\hbar)$ combines the time dependence of $\psi_o(\mathbf{x})$ and of $\psi^{(F)}_{f(n)}(q_f)$.

In (31) each term of different n-value features a different radiation field wavefunction, and therefore all these terms now are orthogonal among each other, even if the $f_n(\mathbf{y}, t)$ would still overlap. Therefore these terms in ψ can no longer interfere in $\int |\Psi|^2 \, dq_f \ldots$, when one asks for $|\Psi|^2$ at some given $\mathbf{y}$-value, but integrated over the radiation field variables. Therefore, as a function of the position $\mathbf{y}$ of the scattered particle alone, (31) is equivalent to a *mixture* of states (29). That is, we may from here on assume that $\psi(\mathbf{y})$ is purely any *one* of the states (29), with a classical probability w_n obtained by integrating $|f_n(\mathbf{y}, t)|^2$ over $\mathbf{y}$-space (over the region covered by the nth wave packet). If, additionally, we are lucky enough to observe the emitted photon, from its wavelength we can determine the value of n realized in the final state, and replace the corresponding w_n by 1, and we have a reduced single pure wavefunction for the scattered particle.

If now this scattered wave hits a second atom, which, in turn, is excited and then emits another photon, by a reasoning similar to the one above we can conclude that there will be no interference between the second-scattered wave and the parts of the first-scattered wave that missed the second atom. Also, a second reduction of the wavefunction of the scattered particle will become possible after observation of the second photon and the location of its source,[18] leaving as twice-reduced wave the spherical wave packet of appropriate radial velocity around the second atom.

Relation to Bohm's theory

The above is all very fine, but it is pure quantum theory, and we do not need Bohm's theory for it. What, then, is added conceptually to the above, if we make Bohm's hypothesis of precise locations and velocities of particles described by wavefunctions?

This contribution is not very exciting, and the main reason for giving above in some detail qualitatively the wave-mechanical treatment is for giving the reader an opportunity to verify that it contains *nothing that would contradict* Bohm's theory. (As mentioned earlier,[13] Pauli convinced De Broglie in 1927 that something would be wrong with this kind of a theory when applied to problems like the one above.) All we can say about the motion of the scattered particle in the above case is that its motion will be pretty violent[19] in regions where wave packets for different n-values overlap in (28). Really, we have one single trajectory of a point P in $\mathbf{x}$, $\mathbf{y}$, q_f-space. The different $\psi_n(\mathbf{x})$, however, overlap, so that from the projection of the $\mathbf{x}$, $\mathbf{y}$, q_f-point onto $\mathbf{x}$-space we cannot determine the value of n. Also, by the time the excited atom emits its photon, the excited one-photon state in q_f-space is, according to Appendix H, obtained from $\psi^{(F)}_{\text{vac}}$ by multiplication by one of the q_f [as the Hermite polynomial $H_1(q_f) = 2q_f$], and thus also $\psi^{(F)}_{f(n)}$ and $\psi^{(F)}_{\text{vac}}$ overlap in q_f-space, and observation of the one single q_f-value of the point P at a given time cannot tell us to which wave packet in q_f-space the system belongs.[20]

[18] This will not always be possible, but we might consider the hypothetical case that only a few atoms at known positions are around, and that these atoms are spaced far enough apart that approximate observation of the momentum and localization of the photon can tell us which atom emitted that photon, by tracing back the track of the photon wave packet. [Compare eqs. (H-36) through (H-43).]

[19] The quantum-mechanical potential U may take very large positive or negative values where $|\psi|^2$ in the interference pattern becomes small. The corresponding large gradients of U may cause tremendous accelerations $m^{-1} \, d\boldsymbol{\pi}/dt$ of the particles $\mathbf{x}$ and $\mathbf{y}$ even if their mechanical interaction by V is weak.

[20] The system is completely described not by the point P in $\mathbf{x}$, $\mathbf{y}$, q_f-space alone, but by a combination of

Therefore, from Bohm's point of view, if somehow we would be able to follow the trajectory of the descriptive point P in $\mathbf{x}$, $\mathbf{y}$, q_f-space, we would recognize the state n to which the atom was excited only at the time that in $\mathbf{y}$-space the nth wave packet has separated from the other wave packets. Then, projection of P onto $\mathbf{y}$-space will tell us in which wave packet the system got trapped. Plausibly, P then would stick to that wave packet, and not at infinite speed cross regions $\psi = 0$ in $\mathbf{y}$-space to a different wave packet. Only then, knowledge of the $\mathbf{y}$ coordinates of P alone will suffice to allow us to reduce the wave function.

This is the reason for the importance attached by Bohm to description of the initial state by a finite wave packet.

2.11. Bohm's theory of measurement of observables

The theory of measurements in Bohm's theory, as far as applied to reproducible[21] measurements of observables,[22] is identical with the theory of measurements in pure quantum theory,[23] which we shall here briefly summarize. As in the case of scattering theory (Section 2.10), consideration of Bohm's hypothesis merely allows us to add some comments without changing anything that was already said.

Reproducible measurements

For a reproducible measurement, the influence of the apparatus should not alter the value of the observable measured. The interaction $\mathcal{H}_i$ between the apparatus and the system observed therefore should have no matrix elements between different eigenvalues q of the observable Q observed, and therefore should commute with Q_{op}. We therefore put

$$\mathcal{H}_i = F[Q(\mathbf{x}, \mathbf{p}), y, p_y], \tag{31}$$

where $\mathbf{x}$ is the coordinate of the observed system and y the coordinate of the apparatus

Let the initial state of the observed system, expanded in terms of eigenfunctions $\phi_q(\mathbf{x})$ of Q_{op}, be given by

$$\psi_o(\mathbf{x}) = \sum_q c_q \phi_q(\mathbf{x}), \tag{32}$$

and let the initial state of the apparatus be given by a normalized wave function $g_o(y)$, so that

$$\Psi_o(\mathbf{x}, y) = \sum_q c_q \phi_q(\mathbf{x})\, g_o(y). \tag{33}$$

After a time t, this will have changed into

$$\Psi(\mathbf{x}, y, t) = \sum_q f_q(y, t)\, \phi_q(\mathbf{x}); \quad [f_q(y, t) = \int \phi_q^*(\mathbf{x})\, \Psi(\mathbf{x}, y, t)\, d^3\mathbf{x}]. \tag{34}$$

P with $\Psi(\mathbf{x}, \mathbf{y}, q_f)$. The knowledge of the $\psi^{(F)}$ factor in Ψ *will* tell us what photon has been emitted, in Bohm's theory as well as in ordinary quantum electrodynamics. Thus if the wave packets do not separate in $\mathbf{y}$-space, the value of n is still determined by Ψ alone, even when P alone will not provide this information.

[21] Compare Part I, Section 1.1.

[22] Compare in Section 2.6 the discussion of the requirement (D) of Section 1.2 for the observable $\mathbf{p}$.

[23] Compare Bohm (1951), chap. 22, secs. 5–11.

We either consider an impulsive measurement[24] in which $\mathcal{H}_i$ is very large but lasts very briefly, or the above formulas are to be understood in an interaction picture.[25] In either case, we have

$$i\hbar\, \partial\Psi(\mathbf{x}, y, t)/\partial t = \mathcal{H}_i \Psi(\mathbf{x}, y, t), \tag{35}$$

that is

$$i\hbar\, \partial f_q(y, t)/\partial t = F[q, y, -i\hbar\partial/\partial y]\, f_q(y, t). \tag{36}$$

This equation holds from the start of the interaction at $t = 0$, when $f_q(y, 0) = c_q g_o(y)$, until the end of the interaction at some time Δt. Integrating (36), which is linear homogeneous in f_q, we find

$$f_q(y, \Delta t) = c_q g(q, y, \Delta t) \tag{37}$$

at the end of the interaction. Here, for hermitian $\mathcal{H}_i$, $g(q, y, \Delta t)$ will be a wavefunction normalized like $g(q, y, 0) = g_o(y)$ was.

In a ballistic measurement, $|g(q, y, \Delta t)|^2$ may be of the form of a peaked function of y where the peak lies at $y = V(q)\cdot\Delta t$. For a static measurement, the peak of $|g|^2$ may lie at $y = Y(q)$, and the width of each peak may be shrinking as Δt increases.

However, $\Psi(\mathbf{x}, y, t)$ is the superposition (34) of different wave packets for different q-values. If the eigenvalues of Q are discrete, these wave packets may cease to overlap either for large enough Δt, or, in a ballistic measurement, at large enough a time after the end of the interaction at Δt. In that case, the situation is like in the scattering problem of Section 2.10. As the narrowness of the peaked functions $|g|^2$ allows us to make a reading ($\langle y \rangle = Y$ or $\langle dy/dt \rangle = V$) of the apparatus, we may reduce the wavefunction to one single term from (34) which will leave the system in the reduced state $\psi(\mathbf{x}) = \phi_q(\mathbf{x})$, which shows that *a reproducible measurement can be used for preparing a system with a known wavefunction* ϕ_q.

The factor c_q in (37) shows us that the probability of the measuring result $Q = q$ is equal to the probability amplitude $|c_q|^2$ that wave mechanics assigns to the eigenvalue $Q = q$ at $t = 0$ on account of (32). This probability does not change during the interactions, since $\mathcal{H}_i$ commuted with Q. This conclusion, however, is valid only if the unperturbed Hamiltonian $\mathcal{H}_o(\mathbf{x}, \mathbf{p})$ of our system commutes with Q. If it does not, Q would be continuously changing with time anyhow, and we could at best measure its *instantaneous* value, by making Δt above so short that the change $(i\hbar)^{-1}\,[Q\mathcal{H}_o - \mathcal{H}_o Q]\,\Delta t$ of Q during the time Δt is negligible.

Relation to Bohm's theory

The above is pure quantum theory. What comments upon the above may be made when Bohm's hypotheses are made?

There is, indeed, very little comment on the above that is not trivial. Again, the point P in $\mathbf{x}$, y-space representing simultaneously the microstate of system and of apparatus will move wildly between the various wave packets in the sum (34) as long as the wave packets overlap. During this time, no conclusive reading of the apparatus is yet possible. A unique reading is possible after the packets have sufficiently separated so we can tell by looking at

[24] Bohm (1951), chap. 22, sec. 5, page 591.
[25] See Merzbacher (1970), chap. 18, sec. 7.

$\langle y \rangle_\varphi$ in which q's wave packet y has ended up. That is, like in the scattering problem, the point P finally will be trapped in one of the packets, as seen by projecting on y-space; then, the other packets can be ignored. But, because the wave packets are labeled by q, this also tells us which $\phi_q(\mathbf{x})$ is the final reduced wave function of the system observed.

Nonreproducible measurements

The above reasoning was for generality kept rather abstract, and we did not even give examples of the function F in (31). It would be nice to give some specific examples for measurements of some fundamental observables, like, for instance, $\mathbf{x}$ and $\mathbf{p}$. One would like to give examples that are meaningful to the experimenter who must make the measurement. However, most examples that come first to mind are examples of *nonreproducible* measurements, and it must be granted that those are just as good. As a matter of fact, Ballentine (1970) makes it a point that by "measurement" one ought to understand as a matter of principle a *nonreproducible* measurement which tells us something about the state of the system *before* the measurement (at least, if the measurement is repeated often enough to enable us to find the probabilities $|c_q|^2$ of the various possible results), while what we above called "reproducible measurements" should rather be called "state preparations."[26]

Let us therefore briefly consider some methods of measuring $\mathbf{x}$ and $\mathbf{p}$ that may be *non*-reproducible, but let us start this time by defining the instrumentation, rather than giving an abstract formula like (31) for $\mathcal{H}_i$.

Measurements of position

We may measure positions by a scintillation screen or by a photographic plate. If the latter absorbs the particle, the black dot we see is the particle's tombstone, showing where it is now buried in a bound state. If the screen or plate does not keep the particle, the observed scintillation or blackening is like a Kilroy-was-here sign, and we have no idea where the particle went afterwards.

The $\mathbf{x} = \boldsymbol{\xi}$ measured is in this case simultaneously an observable and a particle property.

Measurements of momentum

Consider the measurement of the instantaneous momentum of an electron inside an atom by a Compton effect. A known photon hits the atom; a photon leaves the atom and hits a plate (which determines the direction of the secondary photon) and there causes some secondary process which allows us to calculate the energy of the secondary photon; the electron from the atom enters a cloud chamber with a magnetic field determining the electron's momentum. Combining these data we find the initial momentum of the electron inside the atom by the law of conservation of momentum.

[26] Ballentine (1970), page 366.

From the theory of the Compton effect we know that what we are measuring here is the momentum $\mathbf{p}$ (not $\boldsymbol{\pi}$) of the electron inside the atom. The conservation of momentum in a Compton effect is easy to understand when also the initial electron is free; it is due to the vanishing of the matrix element if the $\exp(i\mathbf{k}\cdot\mathbf{x}+i\mathbf{p}\cdot\mathbf{x}/\hbar)$ factor of the electromagnetic wave and of the electron before and after the collision do not have the same value for $\mathbf{k}+\mathbf{p}/\hbar$.

If the initial state of the electron is a bound state, we expand it as $h^{-3/2}\int A(\mathbf{p})\times \exp(i\mathbf{p}\cdot\mathbf{x}/\hbar)\,d^3\mathbf{p}$. We then interpret $|A(\mathbf{p})|^2\,d^3\mathbf{p}$ as the probability for a momentum inside $d^3\mathbf{p}$, and, when such a momentum is realized, use the $\exp(i\mathbf{p}\cdot\mathbf{x}/\hbar)$ wave for calculating what will happen. That is, the $\mathbf{p}$ that enters the law of conservation of momentum as the momentum of the initial electron (and which the measurement described above gives as the result of the measurement) is the $\mathbf{p}$ of the Fourier component of the initial bound state, and occurs with the probability predicted for the value $\mathbf{p}$ of the observable with eigenfunctions $\exp(i\mathbf{p}\cdot\mathbf{x}/\hbar)$ according to the rules of wave mechanics. That is, the observable that has $(\hbar/i)\nabla$ as its operator. That is $\mathbf{p}$, not $\boldsymbol{\pi}$.

As shown in Appendix F, the $\boldsymbol{\pi}$ distribution for an electron bound in an atom is quite different from the $\mathbf{p}$ distribution for such an electron. For instance, an electron in an s-state is always at rest, with the Coulomb force $-\nabla V$ exactly balanced by the quantum-mechanical force $-\nabla U$. While in this case $\boldsymbol{\pi}=0$, yet the electron's $\mathbf{p}$ distribution is not concentrated at $\mathbf{p}=0$ at all. [In the ground state of the hydrogen atom, there is a probability given by eq. (F-7) that $|\mathbf{p}|$ is larger than a fraction f of the momentum $\hbar/r_o$, where r_o is the first Bohr radius.]

Differences between $\mathbf{p}$ *and* $\boldsymbol{\pi}$

The above is just one of many cases in which $\mathbf{p}$ and $\boldsymbol{\pi}$ have different probability distributions. In Appendix G we calculate these two distributions in the Gaussian wave packet for a free electron, and also discuss Bohm's particle trajectories in such a wave packet. If the packet has its smallest size at time $t=0$, it is usually assumed that its later spread is caused by the dispersion in the initial momentum distribution, but that any particle in the ensemble described by the wave packet ψ will have a constant momentum, as the particle is free. This all is true if we interpret $\mathbf{p}$ as the momentum. However, we find that things are different when we interpret with Bohm $\boldsymbol{\pi}$ as the particle momentum. At time $t=0$, the packet has no $\boldsymbol{\pi}$ dispersion at all. The trajectories that come in from the left do *not* continue straight through and leave toward the right, but they bounce back toward the left, repelled by the quantum-mechanical force $-\nabla U$. Thus, the particle's $\boldsymbol{\pi}$ is not conserved at all. See the Appendix for further details.

We also distinguish the "observable" angular momentum $\mathbf{L}$ from the "imagined" or "actual" particle angular momentum $\boldsymbol{\Lambda}$:

$$\mathbf{L}=\mathbf{x}\times\mathbf{p} \quad \text{with} \quad \mathbf{L}_{\text{op}}=\mathbf{x}\times(\hbar/i)\nabla, \quad \text{and} \quad \boldsymbol{\Lambda}=\boldsymbol{\xi}\times\boldsymbol{\pi}=\mathbf{x}\times m\mathbf{v}. \tag{38}$$

In Appendix E we discuss this distinction in atomic stationary states and in the calculation of an atom's magnetic moment. We find that, in an atomic state with given quantum numbers l and m_l, the Bohm orbits of electrons are parallel circles of arbitrary radius around the

z-axis, at arbitrary elevation above (or below) the x,y-plane. For a superposition of stationary states, the trajectory is too complicated for discussion.

Algebraically, the difference between **p** and $\boldsymbol{\pi}$ is seen by calculating $\mathbf{p}_{op}\psi = \mathbf{p}_{op}[R \exp(iS/\hbar)]$. We find

$$\mathbf{p}_{op}\psi = (\hbar/i)\nabla\psi = [R \nabla S + (\hbar/i) \nabla R] \exp(iS/\hbar) = \boldsymbol{\pi}\psi + (\hbar/i)(\nabla \ln |\psi|) \psi. \quad (39)$$

In a comparison of **p** and $\boldsymbol{\pi}$ for a particle described by ψ, though, we must keep in mind that **p** has no definite value for such a particle. We can at most compare **p**-distributions in the ensemble E_ψ with $\boldsymbol{\pi}$-distributions, or we could compare some particular $\boldsymbol{\pi}$ with the average $\langle \mathbf{p} \rangle_\psi$.

Only when ψ is a monochromatic plane wave (eigenfunction of $\mathbf{p}_{op}$), so that $\nabla \ln |\psi| = 0$, no distinction between **p** and $\boldsymbol{\pi}$ is needed.

Stationary state of electron in one-dimensional infinite square well

For a particle in an infinite one-dimensional square well of width L, the stationary states are given by

$$\psi_n = L^{-\frac{1}{2}} \sin(n\pi x/L) \exp(-it\, n^2\pi^2\hbar/2mL^2).$$

In this case, $\nabla S = 0$, and according to Bohm the particle stands still wherever it is located. The **p** distribution, however, is 50% various p toward the left and 50% the same p toward the right.

If we suddenly drop the infinite potential walls, half of the ψ-wave will run out toward the left and half toward the right. The **p** distribution is not changed, but the $\boldsymbol{\pi}$ distribution outside will become equal to the **p** distribution. Failure to appreciate this kept N. Rosen in 1945 from pursuing a proposed theory essentially identical with De Broglie's theory of the pilot wave and with Bohm's theory.[27]

2.12. The Einstein–Podolsky–Rosen paradox explained by Bohm's theory

The "paradox" of Einstein *et al.* (1935) has served people to discuss two aspects of quantum theory—its incompleteness in Einstein's sense, and the apparent nonlocality it causes in correlations between spatially separated measurements on systems that interacted in the past. While often the first aspect has drawn primary attention, since Bohr's reply (1935) confined itself almost exclusively to it, we shall here, as in Part I, Section 1.5, consider merely its nonlocality aspect.

Consider a system of two unlike particles of which the initial preparation is described by a state vector $\psi(1, 2)$ or $\psi(x, y)$. Let E be the ensemble of similarly prepared two-particle systems. Suppose on particle 1 we measure a quantity A with eigenfunctions $\phi_i(x)$, while particle 2 may be at a spacelike distance from particle 1. We may expand $\psi(1, 2)$ according to

$$\psi(x, y) = \sum_i \phi_i(x)\, \chi_i(y) \quad \text{with} \quad \chi_i(y) \equiv \int d^3x \phi_i(x)^* \psi(x, y). \quad (40)$$

[27] See Bohm (1952b), appendix B and sec. 5.

If $A = A_k$ is the result of the measurement, we may reduce the wavefunction, replacing it by the single term

$$\psi_k(x, y) = N\phi_k(x)\chi_k(y), \tag{41}$$

where N is a normalization constant for χ_k. The new wavefunction describes the statistics for future measurements on systems of the new ensemble E_k obtained by adding to the preparation of the system a check-up that measurement of A yielded A_k as a result (the system being rejected when this is not so).

The change from ψ to ψ_k influences the probability distribution of predictions for measurements upon particle 2, even though it is at a spacelike distance, so that no signals from measurement 1 can reach particle 2 for telling it how it should react to measurements on it. Thus are the facts of quantum theory. Those who do not like it suggest replacing quantum theory by a hidden-variables theory of the second kind. (See Part III.) Here, however, we will assume that we see nothing wrong with these predictions of quantum theory, and we are merely curious how Bohm's theory deals with these facts.

For simplicity we shall assume here that the measurements on the two particles both are measurements of position (see Section 2.6). Different possible results of such a measurement then correspond to different trajectories of the particles. In this case, these trajectories are described by a single curve in six-dimensional configuration space. The measurement on particle 1 determines the projection upon x_1,y_1,z_1-space of a point of this curve. This excludes from the original collection of curves all those that have the wrong projection. This reduction of the number of trajectories that need consideration as possibilities for a measurement on particle 2, corresponds to the reduction of ψ to ψ_k.

Nonlocal quantum-mechanical forces on particles in Bohm's theory

By inserting $\boldsymbol{\pi}_k = m_k \mathbf{v}_k = m_k\, d\mathbf{x}_k/dt$ in eq. (D-9), we obtain

$$m_k\, d^2\mathbf{x}_k/dt^2 = -\nabla_k V - \nabla_k U. \tag{42}$$

When particles 1 and 2 are so far apart that they do no longer interact mechanically, $-\nabla_k V$ for each particle becomes a function of that particle's position *independent of the location of the other particle*. The same, however, is not true for U, when ψ is not factorizable as a product of wavefunctions each depending on one particle only. This is seen from eq. (D-10), where, for a nonfactorizable ψ, each term in the sum over j will be a function of all particle positions $\mathbf{x}_1, \mathbf{x}_2, \ldots, \mathbf{x}_n$.

Thus eq. (42) shows how particle k may be accelerated in a way depending upon the position of particle j, through the term $-\nabla_k U$ in the imagined force on particle k, even if particle j is at a spacelike distance from particle k. This quantum-mechanical instantaneous action-at-a-distance between separate particles is avoided only when the wavefunction ψ is factorizable into wavefunctions for each particle separately. In that case, in each term in (D-10), all factors in R not belonging to particle j will cancel out in the quotient $(\nabla_j^2 R)/R$, and $U(\mathbf{x})$ becomes a sum $\sum_j U_j(\mathbf{x}_j)$, so that $\nabla_k U(\mathbf{x}) = \nabla_k U_k(\mathbf{x}_k)$ depends on $\mathbf{x}_k$ only.

If the particles did interact in the past, their initial ψ will *not* be factorizable in this way.

The resulting quantum-mechanical action-at-a-distance force by the one particle on the other explains why behavior of the one particle should depend upon the location of the other particle.

Bell's comment on Bohm's explanation of this apparent nonlocality of quantum theory

Bell (1966) remarks about the above feature of Bohm's theory: "... the Einstein–Podolsky–Rosen paradox is resolved in the way which Einstein would have liked least." This, of course, is not so much an objection against Bohm's theory as it is an expression of regret about a property of pure quantum theory that Bohm's theory does not try to contradict. The removal of the nonlocality would have required a theory of the second kind (Part I, Section 1.5), and a violation of pure quantum theory.

2.13. Treatment of photons in Bohm's theory

In both of his first two papers on the subject, Bohm (1952a, b) discusses some applications of his theory to problems of a quantum-electrodynamical nature. His treatment of photons [appendix A of Bohm (1952b)] differs drastically from attempts of De Broglie who, without justification, tried to apply the electron formula $\mathbf{v} = m^{-1}\nabla S$ to photons, using $m = h\nu/c^2$.

Bohm's treatment is based on the treatment of radiation as a system of oscillators of which the coordinates q_f, like the positions $\mathbf{x}_i$ of particles, are treated as hidden variables. Quantization of the radiation field is described by a wavefunction ψ which is a function of the q as well as of the $\mathbf{x}_i$, and which is written as $\psi = R \exp(iS/\hbar)$. Now the $\partial S/\partial q_f$ can be regarded as velocities in the q_f-direction, of a fluid in $\mathbf{x}_i,q_f$-space, of density $|\psi|^2$, representing the ensemble of systems described by ψ. The velocity $\mathbf{v}_i$ of this fluid in the $\mathbf{x}_i$-direction this time, however, turns out to be given by $m_i \mathbf{v}_i = \pi_i - (\varepsilon_i/c)\mathbf{A}(\mathbf{x}_i)$ with $\pi_i = \partial S/\partial \mathbf{x}_i$ and with the vector potential $\mathbf{A}$ in the radiation gauge. [This gauge is essential for making the continuity equation for the "ensemble fluid" hold on account of the Schrödinger equation. See eqs. (H-50)–(H-56) of Appendix H.]

The coordinates q_f of the oscillators in this theory cannot be the coordinates q used by Pauli (1933) in his famous *Handbuch* article on wave mechanics and later in 1937 used by Kramers (1957). Instead, the q_f used by Bohm are the (real and imaginary parts of the) coefficients of the Fourier expansion of the magnetic field (or of the vector potential in the radiation gauge). These q_f and the conjugated p_f differ by a canonical transformation from those used by Pauli and Kramers. Use of the latter would have been equivalent to introducing in the Schrödinger equation interaction terms bilinear in the momenta of particles and of radiation. If this would not make the theory impossible, it would certainly greatly complicate it.

For proofs of these statements, and for a mathematical formulation of the theory, see Appendix H. For applications of the theory, see Bohm's original 1952 papers.

As the imagined "hidden-variables point" in $\mathbf{x}_i,q_f$-space wanders around, it describes by its projection onto $\mathbf{x}_i$-space the trajectories of all electrons considered, and by its projection onto x_f-space the varying modes of vibration of the radiation field. An ensemble of such points moving in $\mathbf{x}_i,q_f$-space, with an initial density distribution $|\psi(0)|^2$ at time $t = 0$, will automatically have the density distribution $|\psi(t)|^2$ at later times, with $i\hbar\ \partial\psi/\partial t$ given by ordinary quantum electrodynamics. Thus, for an ensemble of systems, Bohm's theory predicts all the results of (nonrelativistic) quantum electrodynamics. (Self-energy problems are ignored.)

If the matter–radiation interaction is ignored, there are stationary states of the radiation field which have ψ equal to the product of real eigenfunctions of the harmonic-oscillator problem, times a phase vector exp $(\mathcal{E}t/i\hbar)$. In this case, all $\partial S/\partial q_f$ vanish, and the radiation field is in a stationary mode of vibration $(dq_f/dt = 0)$, in which the energy lies by multiples of $h\nu_f$ above the vacuum energy level. These states correspond to standing electromagnetic waves and may represent possible stationary states in black-body radiation in a cavity. Free photons are described by linear combinations of these states in which the real and imaginary parts of the complex Fourier components of $\mathbf{A}(\mathbf{x})$ are simultaneously excited with a particular phase relation [see eq. (H-35)]. Moreover, realistic photons, as are emitted by atoms and as undergo Compton scattering as observed in a cloud chamber, are described not by plane waves, but by wave packets [see eq. (H-36)]. The Compton effect described by Bohm (1952b) in his appendix A is one in which the initial electron is bound in an atom.

2.14. Bell's hidden spin variable for an electron

Bohm's theory in its original form describes merely nonrelativistic spinless particles. Bell (1966) has shown one way in which a theory of this kind could be generalized so it could distinguish "spin-up" states from "spin-down" states.[28] We shall here present this theory.

Suppose we have a one-electron state described by the product ψ of a wavefunction $\phi(\mathbf{x})$ scalar under spatial rotations, and a normalized two-component spin function χ that is scalar under translations, but which transforms as a spinor under spatial rotations. As our theory is nonrelativistic, we do not consider here Lorentz transformations.

In Bell's formalism, an individual electron has its spin described by a scalar hidden spin variable which he calls λ and which can take any value between $-\frac{1}{2}$ and $+\frac{1}{2}$. This parameter λ together with the known spin function χ determines the outcome of any measurement involving the electron spin, by the following steps.

We first rotate the coordinate axes so that the spin function $\chi = \begin{pmatrix}\chi^+\\ \chi^-\end{pmatrix}$ in the new coordinate system takes the form $\begin{pmatrix}1\\ 0\end{pmatrix}$. Even without knowing λ, we then know that the spin component in the new z-direction will be positive ("up") and not negative ("down").

Other spin-dependent observables can always be written in the hermitian form

$$A = \begin{pmatrix}\alpha+\beta_z & \beta_x-i\beta_y\\ \beta_x+i\beta_y & \alpha-\beta_z\end{pmatrix} = \alpha+\boldsymbol{\beta}\cdot\boldsymbol{\sigma}. \tag{43}$$

[28] Bell did not consider the $\mathbf{x}$-dependence of the spinor describing the spin state imagined, and he did not consider N-particle spin states. Below we shall generalize his ideas. In Section 3.8 we explain how the treatment of the spin in the theories of Wiener–Siegel and Bohm–Bub may be put into the form of Bell's theory.

The eigenvalues of this matrix are $\alpha \pm |\boldsymbol{\beta}|$, so, measurement of A should give either $\alpha+|\boldsymbol{\beta}|$ or $\alpha-|\boldsymbol{\beta}|$. Bell's algorithm is that the result of the measurement will be[29]

$$A_\lambda = \alpha+|\boldsymbol{\beta}| \text{ sign } (\lambda|\boldsymbol{\beta}|+\tfrac{1}{2}|\beta_z|) \text{ sign } \beta_z, \tag{44}$$

if $\boldsymbol{\beta}_z$ is measured along the z-axis which gives χ the form $\begin{pmatrix}1\\0\end{pmatrix}$, or, in any frame,

$$A_\lambda = \alpha+|\boldsymbol{\beta}| \text{ sign } (\lambda|\boldsymbol{\beta}|\chi^\dagger\chi+\tfrac{1}{2}|\boldsymbol{\beta}\cdot\chi^\dagger\sigma\chi|) \text{ sign } (\boldsymbol{\beta}\cdot\chi^\dagger\sigma\chi). \tag{45}$$

Verification of Bell's proposal

Equation (44) fulfills requirement (D) of Section 1.2. We satisfy requirement (B) by postulating that all λ-values between $-\frac{1}{2}$ and $+\frac{1}{2}$ shall by equally probable. It then is our task to verify that requirement (E) is satisfied. We therefore must show that the λ intervals that give A_λ by (44) each of its two possible values are equal to the quantum-mechanical probabilities $w_\pm$ for the eigenvalues $A_\pm = \alpha \pm |\boldsymbol{\beta}|$ in the state given by $\psi = \begin{pmatrix}1\\0\end{pmatrix} \phi(\mathbf{x})$ in our new coordinate system.

In the following we shall omit the invariant factor $\phi(\mathbf{x})$ from our writing. We shall also write β for $|\boldsymbol{\beta}|$.

We find easily for the normalized eigenfunctions $\chi_\pm$ of A

$$\chi_\pm = \begin{pmatrix}[(\beta\pm\beta_z)/2\beta]^{\frac{1}{2}} \\ \pm(\beta_x+i\beta_y)/[2\beta(\beta\pm\beta_z)]^{\frac{1}{2}}\end{pmatrix}. \tag{46}$$

Thus, the state $\chi = \begin{pmatrix}1\\0\end{pmatrix}$ is decomposed into $C_+\chi_+ + C_-\chi_-$ with

$$C_\pm = [(\beta\pm\beta_z)/2\beta]^{\frac{1}{2}}, \tag{47}$$

and the quantum-theoretical probabilities for $A = A_\pm$ are

$$w_\pm = |C_\pm|^2 = \tfrac{1}{2}(1\pm\beta_z/\beta). \tag{48}$$

On the other hand, from (44) we see that

$$\left.\begin{aligned} A_\lambda = \alpha+\beta \quad &\text{if} \quad -\tfrac{1}{2}(\beta_z/\beta) < \lambda < \tfrac{1}{2} \quad \text{for} \quad \beta_z > 0, \\ \text{or} \quad &\text{if} \quad -\tfrac{1}{2} < \lambda < \tfrac{1}{2}(\beta_z/\beta) \quad \text{for} \quad \beta_z < 0. \end{aligned}\right\} \tag{49}$$

Either of these two intervals has a width $\frac{1}{2}(1+\beta_z/\beta)$. Then the remaining λ-interval of width $\frac{1}{2}(1-\beta_z/\beta)$ will predict $A_\lambda = \alpha-\beta$.

Thus the probabilities that λ will predict $A_\lambda = A_\pm$ are equal to the quantum-theoretical probabilities $w_\pm$ of eq. (47), and requirement (E) is met.

[29] For the special case $\beta_z = 0$, Bell gives a different formula, in which the last factor in (44) is changed into sign β_x, or, if also $\beta_x = 0$, then into sign β_y. This is to make A_λ unique also in those special cases. We shall in the following ignore these cases in which $\boldsymbol{\beta}\cdot\chi^\dagger\boldsymbol{\sigma}\chi = 0$.

Measurement of spin in a different direction

As an application, consider the measurement of the spin in the direction of the unit vector **n** so that $\alpha = 0$ and $\boldsymbol{\beta} = \frac{1}{2}\hbar\mathbf{n}$. We obtain $w_+ = \frac{1}{2}(1+\cos\theta) = \cos^2(\frac{1}{2}\theta)$ and $w_- = \frac{1}{2}(1-\cos\theta) = \sin^2(\frac{1}{2}\theta)$ as the probabilities for $S_n = \pm\frac{1}{2}\hbar$, where θ is the angle between **n** and the z-axis for which $\chi = \begin{pmatrix}1\\0\end{pmatrix}$. This agrees with the quantum-theoretical prediction.

Generalizations of Bell's method

Correlations between spin and position

Consider the one-electron state described by the two-component wavefunction $\psi(\mathbf{x}) = \psi_+(\mathbf{x})\begin{pmatrix}1\\0\end{pmatrix}+\psi_-(\mathbf{x})\begin{pmatrix}0\\1\end{pmatrix}$. As hidden variables, use the particle's position **x** and the scalar spin variable λ (with $-\frac{1}{2} \leqslant \lambda \leqslant \frac{1}{2}$). Suppose we measure the quantity $A(\mathbf{x}) = \alpha(\mathbf{x})+\boldsymbol{\beta}(\mathbf{x})\cdot\boldsymbol{\sigma}$ at the position of the particle. We then may generalize Bell's algorithm (44) by postulating that the result of the measurement will be

$$\begin{aligned}A(\mathbf{x}, \lambda) = \alpha(\mathbf{x})+|\boldsymbol{\beta}(\mathbf{x})|\ \mathrm{sign}\ \{\lambda\,|\boldsymbol{\beta}(\mathbf{x})|\,\psi^\dagger(\mathbf{x})\,\psi(\mathbf{x}) \\ +\tfrac{1}{2}\,|\psi^\dagger(\mathbf{x})\,\boldsymbol{\beta}(\mathbf{x})\cdot\boldsymbol{\sigma}\psi(\mathbf{x})|\}\ \mathrm{sign}\ \{\psi^\dagger(\mathbf{x})\,\boldsymbol{\beta}(\mathbf{x})\cdot\boldsymbol{\sigma}\psi(\mathbf{x})\}.\end{aligned} \tag{50}$$

N-Particle spin states

For an N-electron state, the wavefunction has 2^N components $\psi_{s_1\ldots s_N}(\mathbf{x}_1, \ldots, \mathbf{x}_N)$, and we expect to have hidden variables $\mathbf{x}_1, \ldots, \mathbf{x}_N$ and $\lambda_1, \ldots, \lambda_N$. We would want to postulate an algorithm for the finding of the eigenvalues of observables like

$$A = \alpha+\sum_{i=1}^{N}\boldsymbol{\beta}_i\cdot\boldsymbol{\sigma}_i+\sum_{i=1}^{N}\sum_{j=1}^{N}\boldsymbol{\sigma}_i\cdot C_{ij}\cdot\boldsymbol{\sigma}_j + \text{etc.} \tag{51}$$

For $N = 2$ and $C_{ij} = 0$ the four eigenvalues of A are $A = \alpha\pm|\boldsymbol{\beta}_1|\pm|\boldsymbol{\beta}_2|$, and the two signs may be determined by a generalization of (50). For $C_{ij} \neq 0$ or for $N > 2$, the problem of predicting the outcome of a measurement depending simultaneously on the spins of several electrons has not yet been solved. For any suggested algorithm, the requirement (E) of Section 1.2 is to be verified. Moreover, one should have algorithms for observables different from (51) by lack of symmetry between the points $\mathbf{x}_i$. For instance, for two particles, at positions $\mathbf{x}_1$ and $\mathbf{x}_2$, one might want to measure the spin of the particle at $\mathbf{x}_1$, while leaving the spin of the other particle undetermined. It would be nice if the predictions for the results of such measurements would be consistent with simultaneous measurements of the spins of *both* particles.

As we said earlier, Bohm's theory is not yet a complete theory, and more work is needed for its completion. The Wiener–Siegel theory (1953, 1955) may suggest a solution. (See Section 3.8 for the one-particle case.)

2.15. Relativistic generalizations of Bohm's theory

When we attempt to generalize Bohm's theory to make it applicable to Dirac electrons, we must start from scratch. In this case,

$$\varrho = \psi^\dagger \psi \equiv i\bar{\psi}\gamma^0\psi \tag{52}$$

satisfies the continuity equation $\partial\varrho/\partial t = -\operatorname{div} \mathbf{J}$ with

$$\mathbf{J} = ic\bar{\psi}\boldsymbol{\gamma}\psi \equiv c\psi^\dagger \boldsymbol{\alpha}\psi. \tag{53}$$

This suggests[30] defining the hidden velocities of electrons and of positons by

$$d\mathbf{x}/dt \equiv \mathbf{v} = \mathbf{J}/\varrho. \tag{54}$$

This immediately poses a great number of questions, which we shall only briefly discuss, and which are the reasons why Bohm's theory so far has not been worked out to everybody's complete satisfaction. A suggestion different from (54) made by De Broglie (1952b) is discussed in Appendix I.

Charge interpretation of ϱ and J

The conventional treatment of Dirac electrons avoids densities including infinite numbers of electrons in negative-energy states by replacing (52) by a second-quantized and normal-ordered operator[31]

$$\varrho_\varepsilon/\varepsilon = :\boldsymbol{\psi}^\dagger\boldsymbol{\psi}: \equiv \sum_{A=1}^{4} \{\boldsymbol{\psi}_A^*\boldsymbol{\psi}_A^{(+)} - \boldsymbol{\psi}_A^{(-)}\boldsymbol{\psi}_A^*\}, \tag{55}$$

where ϱ_ε is the charge density operator, ε the electron charge, $\boldsymbol{\psi}^{(+)}$ is the part of the electron field operator $\boldsymbol{\psi}(\mathbf{x})$ that annihilates an electron, and $\boldsymbol{\psi}^{(-)}$ is the part that creates positons. Then, $\langle\varrho\rangle$ would not be positive definite, and could even be zero where $\langle\mathbf{J}\rangle \neq 0$. This interpretation of ϱ, therefore, is not well adapted to use in eq. (54).

Without quantizing the ψ field, the difficulty with using (52) in (54) is seen as follows. For c-number ψ, the following identity due to Darwin (1928) and Fock (1929) is a tautology when written out explicitly in terms of the sixteen components of either $\psi_A^*\psi_B$ of $\bar{\psi}_A\psi_B$:

$$(\bar{\psi}\gamma_\nu\psi)(\bar{\psi}\gamma^\nu\psi) = (\bar{\psi}\psi)^2 - (\bar{\psi}\gamma_5\psi)^2, \tag{56}$$

where we use the conventional notation for Dirac electron theory.[32] Therefore, (52)–(53) yields

$$\varrho^2 - \mathbf{J}^2/c^2 = -j_\nu j^\nu = (\bar{\psi}\psi)^2 - (\bar{\psi}\gamma_5\psi)^2. \tag{57}$$

The right-hand side shows that the left-hand member can be positive or negative or zero, so that there is no guarantee that the four-vector j^ν is timelike. Thus, spacelike four-vectors j^ν with vanishing component $j^0 = \varrho$ are not excluded.

[30] Compare Vigier (1952) and Bohm (1953b).

[31] See, for instance, Schweber (1961), chap. 8. See also Furry and Oppenheimer (1934) and see Wick (1950).

[32] Here, $\bar{\psi} = \psi^\dagger\beta$, $\dot{\mathbf{x}}_{\text{op}} = c\boldsymbol{\alpha}$, $\boldsymbol{\alpha}^\dagger = \boldsymbol{\alpha}$, $\beta^\dagger = \beta$, $x^0 = -x_0 = ct$, $\alpha_x\alpha_y = i\sigma_z$, $\boldsymbol{\alpha} = -\gamma_5\boldsymbol{\sigma}$, $\alpha^0 = 1 = -\alpha_0$, $\gamma_\mu = -i\beta\alpha_\mu$, $\gamma_\mu\gamma_\nu + \gamma_\nu\gamma_\mu = 2g_{\mu\nu}$. With $\gamma_4 = i\gamma^0$, then, eq. (56) reads $\sum_{j=1}^{5} (\bar{\psi}\gamma_j\psi)^2 = (\bar{\psi}\psi)^2$.

Particle interpretation of ϱ and **J**

For Bohm's interpretation of ϱ and **J**, we would prefer an interpretation of ψ, ϱ, and **J** in particle configuration space. For nonrelativistic electron theory, where there are no positon problems, the relation between the second-quantized version of the theory and the formulation using antisymmetric wavefunctions in configuration space was worked out by Fock (1932). For Dirac's theory of electrons and positons, this theory was worked out by Belinfante (1949) in the 1940s, but its publication in connection with its application to vacuum polarization and the Uehling effect in the 1950s[33] never got beyond the preprint state. Luckily, the essentials of it were then published by Wightman and Schweber.[34]

In configuration space, the state vector is a product

$$\psi = \Psi_{(\varepsilon-)}\Psi_{(\varepsilon+)} \tag{58}$$

of a negative electron wavefunction $\Psi_{(\varepsilon-)}$ antisymmetric in the negatons present, and a positon wavefunction $\Psi_{(\varepsilon+)}$ antisymmetric in the positons present, so that in a state with n negatons and p positons present we can define the negaton particle probability density $\varrho_{(-)}$ and the positon particle probability density $\varrho_{(+)}$ and have them both positive definite. For instance, in a given point **x** in three-dimensional space, $\varrho_{(-)}(\mathbf{x})$ would be obtained by integrating

$$n\Psi_{(\varepsilon-)}(\mathbf{x}, \mathbf{x}_1, \ldots, \mathbf{x}_{n-1})^\dagger\, \Psi_{(\varepsilon-)}(\mathbf{x}, \mathbf{x}_1, \ldots, \mathbf{x}_{n-1})$$

over all values of $\mathbf{x}_1, \mathbf{x}_2, \ldots, \mathbf{x}_{n-1}$. Similarly, we could define separate "particle number probability current densities" $\mathbf{J}_{(-)}$ and $\mathbf{J}_{(+)}$ for negatons and positons, by inserting $c\boldsymbol{\alpha}$ between $)^\dagger$ and Ψ in the middle. Since states with different numbers of particles present are orthogonal to each other, and at this point we do not consider operators that change the numbers of particles, for the sake of defining densities and current densities we may treat superpositions of wavefunctions (58) for different numbers n and p here as "mixed states."

As in eq. (D-3), we may prefer to define a particle probability density ϱ in $3(n+p)$-dimensional configuration space by

$$\varrho = \psi^\dagger\psi = \{\Psi_{(\varepsilon-)}(\mathbf{x}_1 \ldots \mathbf{x}_n)^\dagger\, \Psi_{(\varepsilon-)}(\mathbf{x}_1 \ldots \mathbf{x}_n)\}\{\Psi_{(\varepsilon+)}(\mathbf{y}_1 \ldots \mathbf{y}_p)^\dagger\, \Psi_{(\varepsilon+)}(\mathbf{y}_1 \ldots \mathbf{y}_p)\},$$

where the sum over all $n!$ permutations of $\mathbf{x}_1 \ldots \mathbf{x}_n$ and all $p!$ permutations of $\mathbf{y}_1 \ldots \mathbf{y}_p$ in $\varrho(\mathbf{x}_1 \ldots \mathbf{x}_n, \mathbf{y}_1 \ldots \mathbf{y}_p)\, d^3\mathbf{x}_1 \ldots d^3\mathbf{x}_n\, d^3\mathbf{y}_1 \ldots d^3\mathbf{y}_p$ would be the probability of simultaneous presence of one negaton in each of the volume elements $d^3\mathbf{x}_1$ through $d^3\mathbf{x}_n$, and one positon in each of the $d^3\mathbf{y}_1$ through $d^3\mathbf{y}_p$.

For free particles, each of the two antisymmetric wavefunctions $\Psi_{(\varepsilon+)}$ and $\Psi_{(\varepsilon-)}$ may be an expansion in terms of Slater determinants built up from one-free-particle wavefunctions.

One-particle states and a relativistic uncertainty relation

The one-free-particle wavefunctions of which $\Psi_{(\varepsilon-)}$ and $\Psi_{(\varepsilon+)}$ are made up all are *positive-energy* wavefunctions, and therefore $\Psi_{(\varepsilon+)}$ and $\Psi_{(\varepsilon-)}$ each are superpositions of states which *do not form complete sets*. [The negaton states together with the charge conju-

[33] See Belinfante (1953, 1954). See also footnotes 12 and 47 of Wightman and Schweber (1955).

[34] Wightman and Schweber (1955), especially the summarizing tables on pages 815 and 829–830, and their discussion in chap. 3.

gates[35] of the positon states would form a complete set.] When the second-quantized quantum field operator $\psi^{(+)*}(\mathbf{x})$ acts upon the antisymmetric electron state $\Psi_{(\varepsilon-)}(\mathbf{x}_1, \mathbf{x}_2, \ldots, \mathbf{x}_n)$, the result is a numerical factor C_n times an antisymmetric state with $(n+1)$ electrons,[36]

$$\psi^{(+)*}(\mathbf{x})\, \Psi_{(\varepsilon-)}(\mathbf{x}_1, \mathbf{x}_2, \ldots, \mathbf{x}_n) = C_n \Sigma_P\, \varepsilon_P\, f(\mathbf{x}_1-\mathbf{x})\, \Psi(\mathbf{x}_2, \mathbf{x}_3, \ldots, \mathbf{x}_{n+1}), \tag{59}$$

where Σ_P is a sum over permutations of $1, 2, 3, \ldots, n, n+1$, and where $\varepsilon_P = +$ (or $-$) for even (or odd) permutations. However, while in the nonrelativistic theory of second quantization, according to Fock (1932), $f(\mathbf{x}_1-\mathbf{x})$ would be the Dirac delta-function $\delta_3(\mathbf{x}_1-\mathbf{x})$ appearing in (59) as the wavefunction of an electron created at the field point $\mathbf{x}$, the function f is naturally a different function in second-quantized negaton-positon theory. The delta-function appearing for $f(\mathbf{x}_1-\mathbf{x})$ in Fock's nonrelativistic theory may be expressed in terms of any complete orthonormal set $\{\phi_n(\mathbf{x})\}$ of functions by

$$f(\mathbf{x}_1-\mathbf{x}) = \sum_n \phi_n(\mathbf{x}_1)\, \phi_n^*(\mathbf{x}). \tag{60}$$

Equation (60) remains valid in the relativistic theory; only the sum is no longer over a complete set of functions but merely over all positive-energy states. (The "sum", in infinite space, becomes an integral.) One thus finds

$$f(\mathbf{x}) = S_+(\mathbf{x}, t=0)\, \gamma_0 \equiv [\{\partial_0 - \boldsymbol{\alpha}\cdot\nabla - i(mc/\hbar)\beta\}\, \Delta_+(x)]_{t=0}, \tag{61}$$

which is a function of *finite extension*. The region around $\mathbf{x}$, where $f(\mathbf{x}_1-\mathbf{x})$, as a function of $\mathbf{x}_1$, differs appreciably from zero, has a radius of the order of magnitude of *half the Compton wavelength*. This means that, according to the relativistic theory, *a single electron or positon cannot be localized more precisely than within a distance $h/2mc$* ($\approx 10^{-10}$ cm). This is not surprising, as measurements trying to make the uncertainty Δx smaller than $h/2mc$ would require $\Delta p > 2mc$, for instance by probing the position of the electron by photons of energy $> 2mc^2$, which, however, could cause *pair production*, so that one may be creating a three-particle state while investigating a one-particle state.

The *mathematical* necessity for this relativistically needed additional uncertainty relation $\Delta x \gtrsim \hbar/2mc$ is seen when we remember that a precisely determined position of a particle would require a wavefunction like $\delta_3(\mathbf{x}_1-\mathbf{x})$, which by eq. (60) requires for its expansion a *complete* set of functions $\phi_n(\mathbf{x}_1)$, while only positive-energy functions are considered physically allowable as particle wavefunctions in the relativistic theory.

In view of this *fuzziness in the localizability* of electrons in the relativistic theory, description of the "hidden" motion of individual electrons by *sharp worldlines* $\mathbf{x}(t)$, with $d\mathbf{x}/dt$ possibly given by (54), loses some of its attractiveness.

Limited validity of the continuity equation for particle numbers

While $\partial\varrho_\varepsilon/\partial t = -\operatorname{div}\mathbf{J}_\varepsilon$ holds strictly for the charge density and the charge current density, for the particle number probability density $\psi^\dagger\psi = (\Psi^\dagger_{(\varepsilon-)}\Psi_{(\varepsilon-)})(\Psi^\dagger_{(\varepsilon+)}\Psi_{(\varepsilon+)})$ in $3(n+p)$-dimensional configuration space (when n negatively charged and p positively charged

[35] Schweber (1961), chap. 4.
[36] Schweber (1961), chap. 6g, eq. (98). In our (59), $C_n = n^{1/2}/n!$

electrons are present), together with a particle current density $\mathbf{J} = c\psi^\dagger\left(\sum_{j=1}^{n+p} \alpha_j\right)\psi$, a continuity equation holds only when the particles are strictly free and do not interact with each other and the electromagnetic field. When these interactions are taken into account, pair creations and annihilations spoil the validity of the continuity equation. The only way to re-establish the conservation law for the number of particles is by counting antiparticles by *negative* numbers of particles, but this brings us back to the interpretation of ϱ as proportional to a charge density with its possibility of being zero where the current does not vanish. This would make $\mathbf{v} = \infty$ at the creation of a pair, and this is not nice in a relativistic theory.

Lack of invariance of number of particles under Lorentz transformations in the presence of interactions

The difficulties with the use of particle number densities for electrons and positons, when their interactions are taken into account, start already at energies insufficient for creation of pairs. Even if pairs cannot be created *actually*, they would be created *virtually*. That is, the state vector will contain time-dependent admixtures of states in which pairs may have been created (or annihilated). The result of this is that the splitting of the electron field operator $\boldsymbol{\psi}(\mathbf{x})$ into $\boldsymbol{\psi}^{(+)}$ and $\psi^{(-)}$ [as in eq. (55)], and the corresponding splitting of electron states into states for positive-energy negative-charge electrons ($\Psi_{(\varepsilon-)}$) and states for positive-energy positons charge-conjugate to negative-energy negaton states [as needed in defining ψ by (58)], *both are not invariant procedures* in the presence of interactions. This lack of invariance is due to the fact that, in the wedge-shaped space-time region between the hypersurface $t =$ constant in one Lorentz frame, and $t' =$ constant in a different Lorentz frame, pairs may be created or annihilated virtually if not really. Therefore, the separation of the *undor*[37] (= Dirac wave function) $\boldsymbol{\psi}$ into *undors* $\boldsymbol{\psi}^{(+)}$ and $\boldsymbol{\psi}^{(-)}$, and the splitting of a set of solutions $\{\phi_n\}$ of the Dirac equations into positive- and negative-energy solutions in one Lorentz frame do not correspond to similar separations in another Lorentz frame, unless the $\boldsymbol{\psi}$ field and the solutions ϕ_n satisfy the free-electron Dirac equation *without* electromagnetic interaction terms. That is, the positon-and-negaton formalism really makes sense only in the *interaction picture* of quantum field theory.[38]

Crudely speaking in the language of classical physics, this means the following. Let P be a point in space–time on the worldline of an electron. If, on one hypersurface $t =$ constant through P, this electron is the only electron present, then, on a different hypersurface $t' =$ constant through the same space–time point P, there may be additional pairs of electrons present, when interaction is taken into account. Therefore, whether at P the wavefunction $\psi = \Psi_{(\varepsilon-)}\Psi_{(\varepsilon+)}$ is a one-particle or a three-particle wavefunction depends on the Lorentz frame in which we calculate ψ.

[37] See Belinfante (1939a).
[38] Schwinger (1949); Merzbacher (1970), chap. 18, sec. 7.

What happens at pair creation and annihilation?

Bohm's worldlines picture of wave mechanics meets a special problem where pairs are created and annihilated. Here, at a given time, the number of worldlines increases or decreases by two, so that no continuity equation can hold for the *number* densities.

From the point of view where we count *charge*, or where we count antiparticles negatively, we would give worldlines a time direction,[39] toward the future for negatons, and toward the past for positons, so that the worldline merely curves back to an opposite time direction. When a pair is created at *P*, a charge density ϱ_ε, originally zero, suddenly splits up into positive and negative regions by moving charge from one side to the other. This is a start from $\varrho_\varepsilon = 0$ by $\mathbf{J}_\varepsilon \neq 0$, which suggests instantaneously $\mathbf{v} = \infty$, as would correspond to a worldline that has a U-shape rather than a V-shape.

On the other hand, a V-shape tangential to the light cone would suffice to give the pair initially the infinite kinetic energy needed to overcome at *P* the initial infinite negative Coulomb potential of attraction between the positon and the negaton. Relativity theory takes more kindly to worldlines corresponding to $v \leqslant c$. However, the V-shape has the disadvantage of a discontinuity in direction at *P*.

Thus, either way, the worldline picture is problematic.

Relations between velocity, momentum, and ∇S

If something like (54) is adopted, and yet we want to relate $\mathbf{v}$ to the gradient of some phase S obtained from a factor $\exp(iS/\hbar)$ to be split off from ψ, we meet two problems. One is to relate $\mathbf{J}$ to gradients of ψ, and the other is to determine what phase factor $\exp(iS/\hbar)$ should be used when writing

$$\psi_A(x) = \phi_A(x) \exp(iS/\hbar). \tag{62}$$

Attempts at removing the ambiguity of S

The latter problem cannot be solved by making $\phi_A(x)$ real, like we did in the nonrelativistic case, because the ratios of the four components of ψ in general are complex, and we want only one single scalar function $S(x)$, and not four functions $S_A(x)$. An attempt at defining $S(x)$ uniquely was made by De Broglie (1952b) and is discussed in Appendix I. The differential equation for S to which it leads, however, seems to have no solutions (see the last paragraph of the appendix). Let us tentatively assume here that the problem of determining that S in (62) uniquely has *some* solution may be different from De Broglie's suggestion. (We will show below that $S = 0$ might be alright.)

[39] Feynman (1949).

Relating the particle's momentum to the gradient of S

The connection between **J** and $\nabla\psi$, then, was made by Vigier (1952) by eqs. (I-8) through (I-10) of the appendix. Inserting (62) in these equations, Vigier finds

$$j_\mu = \frac{\bar{\phi}\phi}{mc}\left\{\partial_\mu S - \frac{\varepsilon}{c}A_\mu + \frac{\hbar}{2i\bar{\phi}\phi}[\bar{\phi}\,\partial_\mu\phi - (\partial_\mu\bar{\phi})\phi] - \frac{\hbar}{4i\bar{\phi}\phi}\partial_\alpha(\bar{\phi}I^{\alpha}_{.\mu}\phi)\right\} \tag{63}$$

with $I^{\alpha}_{.\mu} = \gamma^\alpha\gamma_\mu - \gamma_\mu\gamma^\alpha$. He then puts

$$P_\mu = \frac{\varepsilon}{c}A_\mu + \frac{\hbar}{2i\bar{\phi}\phi}[(\partial_\mu\bar{\phi})\phi - \bar{\phi}\partial_\mu\,\phi] + \frac{\hbar}{4i\bar{\phi}\phi}\partial_\alpha(\bar{\phi}I^{\alpha}_{.\mu}\phi) \tag{64}$$

and

$$M_o = m\,[\bar{\phi}\phi]^{-1}[(\bar{\phi}\gamma_\alpha\phi)(\bar{\phi}\gamma^\alpha\phi)]^{\frac{1}{2}}, \tag{65}$$

$$u_\mu = cj_\mu[-j_\alpha j^\alpha]^{-\frac{1}{2}} = ic\,(\bar{\phi}\gamma_\mu\phi)\,[(\bar{\phi}\gamma_\alpha\phi)(\bar{\phi}\gamma^\alpha\phi)]^{-\frac{1}{2}}. \tag{66}$$

Then, (63) takes the form

$$M_o u_\mu = \partial_\mu S - P_\mu. \tag{67}$$

Vigier then identifies u^μ with $dx^\mu/d\tau$ along the worldline of a particle, and correspondingly uses $df(x)/d\tau = u^\alpha\,\partial_\alpha f(x)$. Also using $u^\alpha u_\alpha = -c^2$, and therefore $u^\alpha\,\partial_\mu u_\alpha = 0$, he finds

$$\begin{aligned} d(M_o u_\mu)/d\tau &= u^\alpha\,\partial_\alpha(\partial_\mu S - P_\mu) = u^\alpha\,\partial_\mu(\partial_\alpha S - P_\alpha) + u^\alpha(\partial_\mu P_\alpha - \partial_\alpha P_\mu) \\ &= -\partial_\mu M_o c^2 + u^\alpha(\partial_\mu P_\alpha - \partial_\alpha P_\mu), \end{aligned} \tag{68}$$

and he claims that this shows that the particle moves as if it has a "mass" M_o and is acted upon by forces derived from a scalar potential $M_o c^2$ and a four-vector potential P_μ. Finally, M_o can be simplified by the identity (56), which yields

$$M_o = m[1 - (\bar{\phi}\gamma_5\phi)^2/(\bar{\phi}\phi)^2]^{\frac{1}{2}}. \tag{69}$$

Satisfying requirements (B) and (E) of Section 1.2

In order to satisfy requirement (B) of Section 1.2 we want to identify the density $\bar{P}(\mathbf{x})$ of the ensemble of hidden-variables states in the equilibrium situation with the quantum mechanical probability density ϱ of finding a particle. If we *ignore* the difficulties explained earlier which are related to positon theory, we therefore would want to put

$$\bar{P}(x) = \varrho_\varepsilon/\varepsilon = \varrho = \psi^\dagger\psi = i\bar{\psi}\gamma^0\psi. \tag{70}$$

(According to Dirac's hole theory of positons, really $\varrho_\varepsilon = \varrho - \varrho_{\text{vac}}$, and, as a particle density, this has the wrong sign when only positons are present.) Equation (70) differs from the proposal (I-23) made by De Broglie (1952b), of which it is not clear how it would jibe with requirement (B).

We see from (66) that Vigier (1952) chooses the four-vector u^μ parallel to the four-vector j^μ, and normalizes it according to $u^\mu u_\mu = -\,c^2$ in order to be able to identify u^μ with $dx^\mu/d\tau$, and therefore $\mathbf{u}/u^0$ with $\mathbf{v}/c$. Therefore,

$$\mathbf{v} = c\mathbf{u}/u^0 = c\mathbf{j}/j^0 = \mathbf{J}/\varrho = \mathbf{J}/\bar{P}. \tag{71}$$

That is, we may then identify the quantum-mechanical current density $\mathbf{J} \equiv \mathbf{J}_\varepsilon/\varepsilon$ with the current density $\bar{P}\mathbf{v}$ in the ensemble of hidden-variables states, and the continuity equation $\partial_\mu j^\mu = 0$ becomes the continuity equation

$$\partial\bar{P}/\partial t = -\operatorname{div}(\bar{P}\mathbf{v}) \tag{72}$$

for the ensemble of hidden-variables states $\mathbf{x}$. As in the nonrelativistic case, this guarantees that velocities $dx/dt = \mathbf{v}$ of the hidden variables $\mathbf{x}$ will leave $\bar{P} = \varrho$ valid at later times, when $\bar{P} = \varrho$ initially. (Compare also Appendix B.)

For observables that are merely functions of $\mathbf{x}$, requirement (E) is satisfied trivially by $\bar{P} = \varrho$, as in the past. For other observables, this requirement again is satisfied by the identity of Bohm's theory of measurement with the quantum-mechanical theory. (See Section 2.11.)

Remember that this all ignores the problems with the charge interpretation of ϱ and $\mathbf{J}$ mentioned above.

Choosing S

Since ψ may be substituted for ϕ in (65) and (66), the ambiguity of S does not affect the definitions of M_o and of u_μ, and it only affects the way in which $M_o u_\mu$ is split up in (67) into $\partial_\mu S$ and P_μ. A different splitup acts on P_μ like a gauge transformation, and does not affect the curl of P occurring in the last term of (68). This means, of course, that one might as well take $S = 0$ and $\phi = \psi$.

Unsolved problems ignored in Vigier's theory

The main problem with (65) and (66) is that it ignores the fact that in Dirac's electron theory there is no guarantee that j^μ is timelike. Since (66) makes u^μ parallel to j^μ, then also there is no guarantee that u^μ will be timelike, as one would desire for the worldline of a particle.

Where j_μ becomes spacelike, one could at least save the *reality* of u_μ by replacing $[-j_\alpha j^\alpha]^{-1/2}$ by $|j_\alpha j^\alpha|^{-1/2}$. A similar change then should be made in the formula for M_0^{-1}. This change may affect eq. (68) where $u_\alpha u^\alpha$ passes from -1 to $+1$. The particle worldlines now are no longer always timelike.

Also, the identification $\bar{P} = \varrho = \varrho_\varepsilon/\varepsilon$ seems dubious for positons, or for cases where both positons and negatons are present.

2.16. Evaluation of Bohm's theory

We have seen in Section 2.15 that the generalization of Bohm's theory to a relativistic theory meets complications that so far have not yet satisfactorily been resolved. Therefore, we shall consider here only the value of Bohm's theory *as a nonrelativistic approximation.*

The value of the theory as a mere illustration of pure wave mechanics

If in talking about $\mathbf{v}$, $\boldsymbol{\pi}$, and $\mathbf{f}$ we use the epitheton "imagined" velocity, momentum, or force, and we do not pretend that these imaginations are related to some kind of reality, there is nothing in Bohm's theory that is objectionable to the pure quantum theoretician. Bohm's trajectories then merely illustrate a way in which the ensemble E_ψ *could* depend on time with conservation of the description of the quantum-mechanical probability predictions by this picture of the ensemble.

De Broglie's objection against Bohm's interpretation of the pilot wave

When Bohm in his 1951 theory revived De Broglie's 1927 theory of the pilot wave without attaching to it a theory of the double solution (see Section 2.4), De Broglie (1951) objected against this by arguments that might have been presented by a pupil of the Copenhagen school.

We shall illustrate his argument at the hand of Bohm's interpretation of the Gaussian wave packet (see Appendix G). Consider a particle which, from a hidden-variables point of view, at $t = 0$ passes some specific point $\mathbf{x}_1$ with a velocity $\mathbf{v}_1 = \hbar\mathbf{k}_1/m$. Suppose we got some approximate knowledge of this fact, thinking it passed through some point near $\mathbf{x}_o$ with a speed approximately equal to $\mathbf{v}_o = \hbar\mathbf{k}_o/m$. After the measurement which provided us with this knowledge, and which either gave rise to the hidden-variables state $\mathbf{x}_1$, $\mathbf{v}_1$, or did not disturb it, we try to assign to the particle a reduced wavefunction describing the preparation of the state as *we* know it. This may be done best by one of the Gaussian wavefunctions of eq. (G-1), with $\mathbf{x}$ replaced by $\mathbf{x}-\mathbf{x}_o$, and $\mathbf{k}$ replaced by $\mathbf{k}_o$. We still have σ_o at our disposal, for describing which value we trust better, $\mathbf{x}_o$ (then choose σ_o small), or $\mathbf{k}_o$ (then choose σ_o large).

According to Bohm, this choice is no good, because this choice (G-1) of ψ would imply that at $t = 0$ the actual particle momentum $\boldsymbol{\pi}$ would be *exactly* $\hbar\mathbf{k}_o$, while actually it was $\hbar\mathbf{k}_1$, which is, of course, unknown to the observer described above. Though this by itself would seem to make the practicality of Bohm's theory doubtful, we can find a way out of this difficulty by using the ideas of Section 2.3, where we allowed that the momentum $\boldsymbol{\pi}$ of a particle *may* deviate from the value ∇S predicted by ψ, and then very speedily will adjust itself to the theoretical prediction by an additional term in the particle's equation of motion. Thus the observer can change the particle's future *actual* velocity by merely interpreting the result $\hbar\mathbf{k}_o$ of his momentum measurement as the exactly correct value of ∇S.

Here the question arises to *which* value of $\hbar\mathbf{k}_o$ the actual momentum $\boldsymbol{\pi}$ would quickly adjust itself when two observers quarrel about the value of $\hbar\mathbf{k}_o$ measured.

So far, we considered merely the initial adjustment of $\hbar\mathbf{k}_1$ to $\hbar\mathbf{k}_o$ within the first 10^{-22} s after the observer invented the wavefunction. Now look at the predictions of eqs. (G-12) through (G-24) for the motion during the next few seconds.

At time t, the particle supposedly has moved to the point

$$\mathbf{x}(t) = \mathbf{x}_o + (\hbar\mathbf{k}_o/m)t + (\mathbf{x}_2 - \mathbf{x}_o)\,[1 + (\hbar t/2m\sigma_o^2)^2]^{\frac{1}{2}}, \tag{73}$$

where $\mathbf{x}_2$, near $\mathbf{x}_1$, is the point reached by the actual particle in its first 10^{-22} or so seconds.

Notice, again, the dependence of (73) upon both $\mathbf{x}_o$ and σ_o. Again, where the particle actually will be after time t, depends on the accuracy with which the observer chose his ψ. The momentum of the particle now has changed, according to eq. (G-23), to

$$\pi(t) = \hbar\mathbf{k}_o + (\mathbf{x}_2 - \mathbf{x}_o)\, m(\hbar/2m\sigma_o^2)^2\, t[1+(\hbar t/2m\sigma_o^2)^2]^{-\frac{1}{2}}. \tag{74}$$

It has changed due to the repulsive force exerted by the wave packet $|\psi|^2$. Whether it has increased toward the left or toward the right depends upon whether the observer's guess at the position $\mathbf{x}_o$ was lying toward the right or toward the left of the unknown exact position of the particle. Again, what is the particle to do when two observers do not agree about the measurement of $\mathbf{x}_o$ or the estimate of σ_o?

De Broglie, therefore, concluded that the wavefunction ψ was merely an *approximation* for a "physically real" wave u which would carry a singularity that represented the actual particle (see Section 2.4).

Bohm's refutation of De Broglie's objections

Bohm turns De Broglie's arguments around. If it is true that the actual particle is accelerated by a quantum-mechanical force that depends on $|\psi|$, then $|\psi|$ just like an electromagnetic field is a physical reality, and if the above observer is not able to correctly guess its actual value, that is just too bad for him, and, when two observers quarrel about $|\psi|$, obviously at least one of them must be wrong. It is not the observer's guess at $|\psi|$, but the physically realized $|\psi|$, that determines how the particle is going to move. Our present tools at measuring $|\psi|$ for an individual particle are about as poor as our ways of measuring $\mathbf{x}$ or π; maybe we should consider the ψ-field a hidden variable itself. In certain cases, however, ψ is well predictable as the reduced wavefunction after a reproducible measurement of an observable preparing the state of the particle. (See Section 2.11.)

Impossibility of verifying Bohm's hypothesis with present techniques

Whether it is true or false that particles' accelerations depend on a quantum mechanical force $-\nabla U$ can be determined experimentally only when it is possible to make at least three successive exact measurements of the particle's position, not only without disturbing the particle in an uncontrollable way, but also without disturbing the wavefunction ψ which was determined by a preceding reproducible measurement of some other observable. Unfortunately, present measurements of positions, as measurements of the *observable* $\mathbf{x}$, cause a reduction of the wave packet or disturb it in an often unpredictable way. (See Section 2.11.) Thus Bohm's hypothesis at the present is unverifiable.

Copenhagen point of view

The Copenhagen point of view, if not purely positivistic, is at least strongly influenced by positivism, which is the philosophy which calls anything hogwash that cannot be verified. Enough said.

If some *specific* complaints against predictions by the Bohm theory are wanted, the following remark on Bohm's theory of the Gaussian wave packet might suffice. If the wave packet is to describe our initial uncertainty of both **x** and **p** of a particle *assumed to be moving freely*, it may be just a poor choice of ψ at $t = 0$ that makes the packet describe *exact* initial π-values; but, in whatever way we modify ψ to remove this inadequacy of the Gaussian packet from Bohm's point of view, we will keep the fact that the ψ-wave will repel free particles on its outskirts. (This follows from the dissipation of wave packets, as well as by generalization of the reasoning of Section 2.8.) Therefore Bohm predicts accelerations of free particles. With no experimental evidence to support such a contradiction in terms, there is no sense in upholding such a claim.

Who is right?

We presented in this section various mutually contradictory points of view. We leave it to the reader to make his choice. Additional possibilities of choice are offered by different hidden-variables theories presented in Chapters 3 and 4. As discussed in Chapter 5, these other theories may be a bit easier to test experimentally.

CHAPTER 3

THE THEORY OF WIENER AND SIEGEL

3.1. The mathematical apparatus

The Wiener–Siegel hidden-variables theory[40] is based upon some mathematical ideas developed by Wiener,[41] and on the invariance of a Gaussian isotropic probability distribution in complex space under unitary transformations. The latter is to be understood as follows. In a space with complex coordinates $\Delta_1, \Delta_2, \ldots, \Delta_N$ consider a probability distribution given by[42]

$$d\bar{P} = \left(\frac{N}{\pi}\right)^N \exp\left(-N\sum_{\mathrm{i}=1}^{N}|\Delta_{\mathrm{i}}|^2\right) d\delta_1\, d\delta_2 \ldots d\delta_{2N}, \tag{75}$$

where we split the Δ_{i} into their real and imaginary parts by

$$\Delta_{\mathrm{i}} = \delta_{2\mathrm{i}-1}+i\delta_{2\mathrm{i}}. \tag{76}$$

We may write (75) also as

$$d\bar{P} = \prod_{\iota=1}^{2N}\left\{\sqrt{\frac{N}{\pi}}\exp\left(-N\delta_\iota^2\right) d\delta_\iota\right\}, \tag{77}$$

so that it is a product of Gaussian distributions of dispersion $(2N)^{-1/2}$ for all coordinates of a $2N$-dimensional real δ_ι-space.

Now consider a unitary transformation in complex Δ_{i}-space. This leaves $\sum_{\mathrm{i}=1}^{N}|\Delta_{\mathrm{i}}|^2 = \sum_{\iota=1}^{2N}\delta_\iota^2$ invariant; so, it is a rotation in δ_ι-space. Moreover, the jacobian of such transformation is unity both between the old and the new Δ_{i}, and between the old and new δ_ι. Thence the distribution (75) or (77) is form-invariant under this transformation.

In fact, the isotropic Gaussian distribution is the *only* distribution that is factorizable into separate factors for each coordinate, simultaneously in all N-dimensional complex spaces that are obtainable from each other by unitary transformations. This fact is the reason why Wiener and Siegel use the distribution (75) for satisfying requirement (B) of Section 1.2, using it as the basis of their equilibrium distribution of hidden variables. (See Sections 3.2 and 3.5.)

[40] See Wiener and Siegel (1953, 1955); Siegel (1962, 1966).

[41] See Paley and Wiener (1934), chap. 9, and see Wiener *et al.* (1966) for later developments.

[42] In our formulas we change some conventions and correct some errors in factors 2 of the original papers of Wiener and Siegel. See footnote 22 on page 994 of Wiener and Siegel (1955).

Shortcut formulation of the Wiener–Siegel theory

The remainder of this Section 3.1 deals with some mathematical tricks which need be mentioned when the reader wants to understand the mathematical framework of the Wiener–Siegel theory *as published*. There exists, however, a short-cut method which bypasses much of the formalism of Wiener and Siegel, throwing it out as deadwood, while proceeding immediately to the essentials. (Compare the second half of Section 3.10.)

The reader who does not care for cute formulas and wants to take the short-cut road probably can omit the remainder of this Section 3.1; read Section 3.2 to just above eq. (82); skip Section 3.3 altogether; read Section 3.4 up to eq. (100) only, and then read eq. (105) without its middle member as a consequence of (100); then in Section 3.5 skip eq. (117) and take eq. (118) as a *definition* of Wiener's and Siegel's equilibrium distribution of the hidden variables $\Xi(x)$ as expressed by (105) in terms of the ξ_i; and then start reading regularly from Section 3.6 on. In Section 3.9 all paragraphs about $\Psi(\alpha)$ then can be skipped.

As far as I can see, by this short cut, no physics is lost. What is lost is an appreciation of Wiener's mathematical ingenuity and an understanding of why Wiener and Siegel (1953, 1955) claim that for their theory they use "differential space." The understanding how considerations about Brownian motion led Wiener and Siegel to the discovery of their formalism is greatly lost in my description of their work, anyhow, as for brevity I only occasionally refer to it in the sections skipped by the short-cut reader.

Fusion of variables

Another mathematical idea used in the Wiener–Siegel theory is the mapping of $2N$ real variables $\alpha_1, \alpha_2, \ldots, \alpha_{2N}$, all defined in the interval from 0 to 1, upon one single real variable α in this same interval, in an *almost*[43] everywhere continuous one-to-one and measure-conserving manner. Thus, when α is given, all variables $\alpha_1, \alpha_2, \ldots, \alpha_{2N}$ can be retrieved from it. A change of α smaller than ε will almost[43] always yield a change of each α_ι by an amount less than $\varepsilon^{1/2N}$. There is in the α-interval from 0 to 1 a subset of measure zero, of points where some α_ι as functions of α are not continuous. The existence of these exceptional α-values is disregarded in the development of the theory, even though these points α where some α_ι make jumps lie everywhere dense in the interval from 0 to 1. (They are the α-values that are rational numbers of which the denominator is some power of 2.)

This mapping of all α_ι into a single α is easily achieved by writing all α_ι as well as α as binary fractions. If $\alpha_{\iota,n}$ is the nth binary digit (0 or 1) behind 0. in the binary presentation of α_ι, we simply write α as

$$\alpha = 0.a_{1,1}\, a_{2,1}\, a_{3,1} \ldots a_{2N,1}\, a_{1,2}\, a_{2,2} \ldots a_{2N,2}\, a_{1,3} \ldots \tag{78}$$

We retrieve from this the value of α_ι by merely looking at the ιth, the $(2N+\iota)$th, the $(4N+\iota)$th ... digits of α behind 0. Since 0.01011 in binary is identical with 0.010101111111 ..., the α-values with finite numbers of binary digits are the points where

[43] Most statements made in the theory of Wiener and Siegel require the qualification "almost". This "almost" means "except for a subset of relatively zero measure."

there is a discontinuity of the α_ι and where two different sets $\{\alpha_\iota\}$ correspond to the same value of α.

The relation (78) between α and the set $\{\alpha_\iota\}$, and the inverse relation, we shall denote by

$$\alpha = \text{fuse}\,(\{\alpha_\iota\}_1^{2N}), \quad \alpha_\iota = \text{split}_{(2N),\,\iota}(\alpha). \tag{79}$$

Paley and Wiener (1934) show that this fusion of the set $\{\alpha_\iota\}_1^{2N}$ into one variable almost everywhere *conserves measure*. That is, ignoring the exceptional points, any interval $d\alpha$ somewhere between 0 and 1 corresponds to either a single volume element, or a set of volume elements, in $\alpha_1, \alpha_2, \ldots, \alpha_{2N}$ space such that the sum of the latter volume elements $d\alpha_1\, d\alpha_2 \ldots d\alpha_{2N}$ equals the original $d\alpha$:

$$d\alpha = d[\text{fuse}\,(\{\alpha_\iota\}_1^{2N})] = (\Sigma) \prod_{\iota=1}^{2N} d\alpha_\iota\,. \tag{80}$$

For a further discussion of this mapping, see Paley and Wiener (1934).

Differential space

A *real* Δ_i-space was originally introduced by Wiener for describing the displacements $\Delta_i \equiv \Delta_i x$ of a particle in Brownian motion during the ith time interval Δt considered. {That is, $\Delta_i = x(i\,\Delta t) - x([i-1]\,\Delta t)$.} Therefore, Wiener called this Δ_i-space "differential space," and considered Gaussian probability distributions for the values of each Δ_i. In their hidden-variables theory, the *complex* Δ_i had properties rather similar to the *real* Δ_i considered in the theory of Brownian motion. Therefore they took over the name "differential space" for their complex Δ_i-space.

Simplifying assumptions

For a simplified presentation of Wiener–Siegel's hidden-variables theory, we shall follow Wiener and Siegel (1953, 1955) in replacing the entire space of independent variables upon which the state vector ψ depends, by a single real variable x that runs from 0 to 1 only, and by representing the range of this continuous variable x by a set of N discrete points lying at distances $1/N$ from each other, so that, for n particles with spin,

$$\int_{-\infty}^{+\infty} \cdots \int_{-\infty}^{+\infty} d^3\mathbf{x}_1 \ldots d^3\mathbf{x}_N \sum_{m_{s1}} \cdots \sum_{m_{sN}} \quad \text{is represented by} \quad \int_0^1 dx = \frac{1}{N}\sum_{i=1}^{N}\,. \tag{81}$$

This simplified treatment can be generalized to a more realistic one. [See Wiener *et al.* (1966).]

3.2. The role of the Ξ_i and of Wiener and Siegel's Δ_i or δ_ι or α_ι or α as hidden variables

As seen from the title of this section, the Wiener–Siegel theory features a whole array of variables, dependent upon each other, of which each may be regarded as the hidden variables which in our Part I we denoted by ξ. We shall keep using this name "hidden variables" for any of these parameters, even though Siegel (1966) later suggested reserving this name for something else. (See Section 3.10.)

There is a close relation between the Δ_i of Wiener and Siegel, and the "Hilbert[44] vectors" Ξ used by Bohm and Bub (1966a) in their hidden-variables theory discussed in our Chapter 4. In order to bring out this relation from the beginning, we shall introduce here Wiener's and Siegel's hidden variables Δ_i, etc., not in the way they did themselves, but *via* the Ξ of Bohm and Bub. This possibly somewhat pins down the "meaning" of the Δ_i, and thus our treatment loses some generality. (If we had started from the Δ_i of Wiener and Siegel as basic quantities, and reconstructed Ξ from it, there would not have been a guarantee that Ξ would lie in the same[44] Hilbert space as ψ.) We pay this price of loss of generality for the sake of a simpler presentation of the theories of Wiener–Siegel and of Bohm–Bub, and their mutual relationship, and, for practical applications of the theory to experimental questions (as in our Chapter 5), nothing really is lost.

The second Hilbert vector Ξ and the Δ_i of differential space

The Hilbert vector $\psi(x)$ used as state vector in quantum theory does not describe a single physical system, but describes an ensemble E_ψ of systems identically prepared. When subjected to one and the same measurement, these systems may give different results. This might be partially due to differences unknown to us in the "microstate" of the measuring apparatus, as Bohm assumed in his theory of measurement. (See Section 2.11.) In the theories we shall consider now, the differences in outcome are ascribed to differences between the microstates of *the observed systems themselves*, and it is assumed that these microstates can be described by hidden variables.

In the Bohm–Bub theory (Chapter 4), the hidden variables can be described by a second state vector, $\Xi(x)$, which lies in a space of which the Hilbert space of the state vectors $\psi(x)$ is a part.[44] (One might call ψ the quantum-mechanical state vector, and Ξ either the microstate vector or the hidden-variables state vector.) This second state vector $\Xi(x)$ is normalized like $\psi(x)$ is, in the theory of Bohm and Bub. In the theory of Wiener and Siegel, we assume that $\Xi(x)$ usually is *not* normalized. We shall now discuss how to obtain from this $\Xi(x)$ the Δ_i of Wiener and Siegel in the simplified description of their theory proposed at the end of Section 3.1.

[44] Bohm and Bub needlessly complicate matters by making Ξ lie in a space of which the conjugated Hilbert space $\mathfrak{H}^*$ in which ψ^* lies is a subspace. So, we take our Ξ conjugate to theirs. If the Hilbert space in which ψ lies is a function space in which the allowable functions must satisfy certain conditions of decent behavior, we may want to drop some of such conditions for the Ξ. Therefore the function space of the Ξ may be larger than the Hilbert space for ψ, but the latter space is certainly *contained* in the space of Ξ functions.

As we replace the x interval by N discrete points x_i in the interval from 0 to 1, and thus $\psi(x)$ by $\psi_i \equiv \psi(x_i)$, we will replace $\Xi(x)$ by $\Xi_i \equiv \Xi(x_i)$. Now consider systems from the ensemble E_ψ in different microstates Ξ. We may distinguish the different functions $\Xi(x)$ by a label α, by writing $\Xi_\alpha(x)$. Correspondingly we could write $\Xi_{\alpha, i}$ for $\Xi_\alpha(x_i)$, but more often we will write $\Xi_i(\alpha)$ for it.

We mentioned already that the Wiener–Siegel theory does not restrict the rather arbitrary functions $\Xi(x)$ by a normalization condition. As we shall see later, this generalization of Ξ somewhat simplifies certain calculations. (See the last part of Section 4.1.) It means, however, that in the Wiener–Siegel theory Ξ and $C\Xi$ will describe the same microstate, for any value of C. This equality in the physical meanings of Ξ and $C\Xi$ in the Wiener–Siegel theory will become evident in Section 3.6, when we give the algorithm telling how Ξ predicts the outcome of the measurement of an observable A.

When x-space is represented by N discrete points spaced out evenly between 0 and 1, we define the N complex quantities Δ_i describing a state α by

$$\Delta_i(\alpha) = \Xi_i(\alpha)/N \quad [\equiv \Xi_\alpha(x_i)/N]. \tag{82}$$

Thus the exact state of a system taken from E_ψ is (by postulate) determined by $\psi(x)$ (i.e. the N complex quantities ψ_i) together with the N complex quantities Δ_i or the $2N$ real quantities δ_ι. By this postulate we meet requirement (A) of Section 1.2. We meet requirement (B) by postulating that under ordinary circumstances, in the ensemble E_ψ, *the hidden variables Δ_i or δ_ι will be distributed according to eqs. (75) or (77).* [We see at once that this postulate of the Wiener–Siegel theory makes it *impossible* to normalize Δ_i or Ξ_i or therefore $\Xi(x)$ in any way.]

The α variable of Wiener and Siegel

We mentioned above how some label α might be used to distinguish one microstate from another. We shall now explain how Wiener and Siegel (1953, 1955) chose for α a specific parameter in the interval $0 < \alpha < 1$. Their choice makes the δ_ι and Δ_i functions of α that are *explicitly known once and for all*, so that the value of the one label α, by (82), immediately will determine the Ξ_i characterizing the Bohm–Bub Hilbert vector $\Xi(x)$. (Bohm and Bub's normalized Ξ can always be found from any nonnormalized Ξ of Wiener and Siegel by simply normalizing the latter Ξ.)

From this unique dependence of Ξ upon α, it is clear that also α may serve as the hidden variable determining the microstate of the system.

The real hidden variable α is found by fusion from $2N$ real variables α_ι, each in the interval $0 < \alpha_\iota < 1$, by the procedure discussed in eqs. (78)–(79). Here each of the α_ι is defined in terms of the corresponding δ_ι by the equation

$$\alpha_\iota = \tfrac{1}{2} + \tfrac{1}{2} \operatorname{erf}(\delta_\iota \sqrt{N}) \equiv \int_{-\infty}^{\delta_\iota} [N/\pi]^{\frac{1}{2}} \exp(-Ny^2)\, dy. \tag{83}$$

This definition guarantees $0 < \alpha_\iota < 1$ for $-\infty < \delta_\iota < +\infty$. The inverse of (83) is given by

$$\delta_\iota = N^{-\frac{1}{2}} \operatorname{erf}^{-1}(2\alpha_\iota - 1), \tag{84}$$

which for given N and α allows us to calculate $\Xi_i(\alpha)$ from (α) by (79), (84), and (76):

$$\Xi_i(\alpha) = N\,\Delta_i(\alpha) = N^{\frac{1}{2}}\{\mathrm{erf}^{-1}\,[2\,\mathrm{split}_{(2N),\,2i-1}\,(\alpha)-1]+i\,\mathrm{erf}^{-1}\,[2\,\mathrm{split}_{(2N),\,2i}\,(\alpha)-1]\}. \quad (85)$$

This dependence of Ξ_i on α is pretty horrible. We shall explain below why this does not harm us.

The equilibrium distribution of the hidden variable α

When α would be used as the hidden variable, according to requirement (B) of Section 1.2 we should know its equilibrium distribution. This distribution is easily found as follows.

The differential of (83) is

$$d\alpha_\iota = [N/\pi]^{\frac{1}{2}} \exp\,(-N\delta_\iota^2)\, d\delta_\iota\,. \quad (86)$$

Thence, by (77) and by (80), we obtain

$$d\bar{P} = \prod_{\iota=1}^{2N} d\alpha_\iota = d\alpha. \quad (87)$$

This shows that equal α intervals have equal probabilities, and, since all microstates fall inside the interval $0 < \alpha < 1$, we need not even a normalization factor for obtaining the equilibrium probability $d\bar{P}$ for some chosen interval $d\alpha$.

This relation (87) between the parameter α and probability allows us to write in a particularly simple way the average over hidden-variables states of any function $F(\psi, \alpha)$ of the microstate of a system, when the macrostate (quantum state) ψ is given, and when the hidden-variable distribution is the canonical one given by (77) or (87):

$$\langle F(\psi, \alpha)\rangle_{\mathrm{Av}} = \int F(\psi, \alpha)\, d\bar{P} = \int_0^1 F(\psi, \alpha)\, d\alpha. \quad (88)$$

[We will later have to show that this average is equal to the quantum-theoretical average $\langle F\rangle_\psi$, if $F(\psi, \alpha)$ is the result of the measurement of the observable F in the state (ψ, α). See Section 3.7.]

Triviality of the result

The result (88) may seem impressive, but really it is trivial. Suppose we had not specified in detail the equilibrium distribution (77), but, instead, we had merely claimed that $d\bar{P} = \prod_\iota \{f_\iota(\delta_\iota)\, d\delta_\iota\}$. We would have written $\alpha_\iota = \int_{-\infty}^{\delta_\iota} f_\iota(y)\, dy$ instead of (83), and (87)–(88) would still have been valid, even without attempts at determining $f_\iota(\delta_\iota)$.

Use made of the α

The horrible form of the relation (85) poses the question what good is the variable α for practical calculations. The answer is: none. Wiener and Siegel (1953) do use α often as a simple label for a hidden-variables state without trying to calculate its value. Formal derivations may yield formulas in which α, like in (88), appears as the integration variable.

For evaluating such an integral in practice, we would invert the reasoning which led from (77), by (86), to (87). Thus integrals over α can be expressed in terms of integrals over the δ_ι. In Section 3.5 we shall express $d\bar{P}$ in terms of other variables yet, and replace integrals over δ_ι by integrals over variables η_λ.

By avoiding the use of the α in practical calculations we also avoid the use of the manipulations "fuse" and "split," which may be well defined mathematically but which give physicists the creeps.

3.3. Some properties of the $\Delta_i(\alpha)$

When, below eq. (82), we postulated that the δ_ι would ordinarily be distributed according to eq. (77), we made $d\bar{P}$ a product of separate factors each for a different δ_ι. This suggests the assumption that the various δ_ι distributions be really independent of each other, without any correlation. With such lack of coherence, we may asssume[45]

$$\langle\delta_\iota\delta_\varkappa\rangle_{\mathrm{Av}} \equiv \int_0^1 \delta_\iota(\alpha)\delta_\varkappa(\alpha)\,d\alpha = 0 \quad \text{for} \quad \iota \neq \varkappa, \tag{89}$$

and therefore also

$$\langle\Delta_{\mathrm{i}}^*\,\Delta_{\mathrm{k}}\rangle_{\mathrm{Av}} \equiv \int_0^1 \Delta_{\mathrm{i}}(\alpha)^*\,\Delta_{\mathrm{k}}(\alpha)\,d\alpha = 0 \quad \text{for} \quad \mathrm{i} \neq \mathrm{k}. \tag{90}$$

For $\iota = \varkappa$ we obtain by (86) and (83)

$$\langle\delta_\iota^2\rangle_{\mathrm{Av}} \equiv \int_0^1 \delta_\iota(\alpha)^2\,d\alpha_\iota = \int_{-\infty}^{+\infty} [N/\pi]^{\frac{1}{2}} \exp(-N\delta_\iota^2)\,\delta_\iota^2\,d\delta_\iota = 1/2N. \tag{91}$$

Thence, for i = k, by (91),

$$\langle|\Delta_{\mathrm{i}}|^2\rangle_{\mathrm{Av}} = \langle\delta_{2\mathrm{i}-1}^2\rangle_{\mathrm{Av}} + \langle\delta_{2\mathrm{i}}^2\rangle_{\mathrm{Av}} = 1/N. \tag{92}$$

We combine (90) and (92) to

$$\langle\Delta_{\mathrm{i}}^*\,\Delta_{\mathrm{k}}\rangle_{\mathrm{Av}} \equiv \int_0^1 \Delta_{\mathrm{i}}^*(\alpha)\;\Delta_{\mathrm{k}}(\alpha)\,d\alpha = \delta_{\mathrm{ik}}/N. \tag{93}$$

Since i and k represent points x_{i} and x_{k} in the space of the coordinates x on which ψ depends, we may interpret (93) as stating that

$$\langle\Delta^*(x_{\mathrm{i}})\Delta(x_{\mathrm{k}})\rangle_{\mathrm{Av}} \equiv \int_0^1 \Delta_\alpha^*(x_{\mathrm{i}})\Delta_\alpha(x_{\mathrm{k}})\,d\alpha = N^{-2}\delta(x_{\mathrm{i}} - x_{\mathrm{k}}),$$

where for large N we identified

$$\delta_{\mathrm{ik}} \to N^{-1}\delta(x_{\mathrm{i}} - x_{\mathrm{k}}) \tag{95}$$

because $\sum_{\mathrm{k}} \delta_{\mathrm{ik}} f(x_{\mathrm{k}}) = f(x_{\mathrm{i}}) = \int \delta(x_{\mathrm{i}} - x_{\mathrm{k}}) f(x_{\mathrm{k}})\,dx_{\mathrm{k}}$ is by (81) equivalent to $N^{-1}\sum_{\mathrm{k}} \delta(x_{\mathrm{i}} - x_{\mathrm{k}}) f(x_{\mathrm{k}})$.
By (82), we may write (94) also as

$$\langle\Xi^*(x)\,\Xi(x')\rangle_{\mathrm{Av}} = \int_0^1 \Xi_\alpha^*(x)\,\Xi_\alpha(x')\,d\alpha = \delta(x - x'). \tag{96}$$

[45] This was one of the fundamental properties of the distribution of the coordinate differences Δ_ι in the theory of Brownian motion which served Wiener as a model for his theory.

3.4. Transformation of $\Delta_i(\alpha)$ from x_i-space to A_i-space

In Part I of this book we discussed that on account of "Gleason troubles" (see the end of Section 3.6 of Part I) hidden variables *cannot* generally tell us uniquely the values of observables, but they *can select*, from a *given* complete orthonormal set $\{\phi_i(x)\}$ of possible results of a measurement of some observable A (of which the ϕ_i are eigenfunctions), the one particular state $\phi_n(x)$ which the measurement would realize when actually performed. This is exactly what the hidden variables in the theories of Wiener–Siegel and of Bohm–Bub are postulated to do. The algorithm telling how they do it will be discussed in Section 3.6. For its understanding, however, we first must transform our set of hidden variables $\{\Delta_i\}$ into a new set $\{\xi_i\}$. For this transformation, we use the set $(\phi_i(x)\}$ mentioned above.

We shall represent the $\Delta_i(\alpha)$ here by the $\Xi_i(\alpha)$ which by (82) are proportional to them. Using Dirac bra–ket notation, we use the

$$\phi_i(x) \equiv \langle x \mid i \rangle = \langle i \mid x \rangle^* \tag{97}$$

as the transformation matrix for a unitary transformation from the representation

$$\langle x \mid \alpha \rangle \equiv \Xi_\alpha(x) \tag{98}$$

of $\mid \alpha \rangle$ in x-space to its representation[46]

$$\langle i \mid \alpha \rangle \equiv \xi_i(\alpha) \tag{99}$$

in i-space,[47] according to

$$\langle i \mid \alpha \rangle = \int \langle i \mid x \rangle \, dx \, \langle x \mid \alpha \rangle = \xi_i(\alpha) = \int \phi_i(x)^* \Xi_\alpha(x) \, dx. \tag{100}$$

Representing x-space by N discrete points x_i, we may rewrite this by (81) and (82) as

$$\xi_i(\alpha) = \sum_{\mathrm{i}} \phi_i^*(x_\mathrm{i}) \cdot \Delta_\mathrm{i}(\alpha). \tag{101}$$

The unitarity of (97) is a consequence of the fact that we assumed the $\{\phi_i(x)\}$ not only to be orthonormal according to

$$\int \langle i \mid x \rangle \, dx \, \langle x \mid k \rangle = \int \phi_i^*(x) \, \phi_k(x) \, dx = \delta_{ik} = \langle i \mid k \rangle, \tag{102}$$

but also to be complete according to

$$\int \langle x \mid i \rangle \, di \, \langle i \mid x' \rangle \equiv \sum_i \phi_i(x) \, \phi_i^*(x') = \delta(x - x') = \langle x \mid x' \rangle, \tag{103}$$

or, equivalently by (95),[48]

$$\sum_i \phi_i(x_\mathrm{j}) \, \phi_i^*(x_\mathrm{k}) = N \delta_{\mathrm{jk}}. \tag{104}$$

[46] Wiener–Siegel and Bohm–Bub would write ξ^* for our ξ. Compare footnote 44. The only equation in which their distinction of ξ from ξ^* looks perhaps nicer than ours, is eq. (114) below.

[47] This i-space is the one-dimensional space in which the coordinate i by its value picks an eigenfunction $\phi_i(x)$ of the observable A. If the eigenvalues A_i of A_{op} are not degenerate, we may use the eigenvalues A_i for labeling the points in i-space. Thence the A_i in the title of this section.

[48] From (104) we see that, in discrete x_i-space, the unitary transformation between the i-representation and the i-representation is given not by $\langle i \mid x_\mathrm{i} \rangle = \phi_i^*(x_\mathrm{i})$, but by $U_{i\mathrm{i}} = N^{-1/2} \phi_i^*(x_\mathrm{i})$, which then by (101) transforms $N^{1/2} \Delta_\mathrm{i}(\alpha)$ into $\xi_i(\alpha)$. Compare eq. (106).

By (101) and (104) we find

$$\sum_i \phi_i(x_j)\,\xi_i(\alpha) = N\Delta_j(\alpha) = \Xi_j(\alpha), \tag{105}$$

which is the inverse $(\int\langle x_j\,|\,i\rangle\, di\,\langle i\,|\,\alpha\rangle = \langle x_j\,|\,\alpha\rangle)$ of the transformation (100). Multiplying (105) by $\Delta_j^*(\alpha)$ and summing over j, we find by the conjugate of (101)

$$N\sum_i |\Delta_i(\alpha)|^2 = \sum_i |\xi_i|^2, \tag{106}$$

which we will need in the following sections.

Transformation from x-space to α-space

Wiener and Siegel like to consider not only the $\langle x\,|\,i\rangle = \phi_i(x)$ and $\phi_i^*(x) = \langle i\,|\,x\rangle$ as matrices for unitary transformations between x-space and i-space, but they also like to consider the $\langle x\,|\,\alpha\rangle = \Xi_\alpha(x)$ and $\Xi_\alpha^*(x) = \langle\alpha\,|\,x\rangle$ as the matrices for some kind of transformation between x-space and α-space.[49] The $\Xi_\alpha(x)$ do form an orthonormal set of functions of the hidden variables α, labeled by x (or by x_i in the discrete version of x-space), as seen from (96) using the notation (98):

$$\int \langle x'\,|\,\alpha\rangle\, d\alpha\,\langle\alpha\,|\,x\rangle = \langle x'\,|\,x\rangle. \tag{107}$$

However, they do not form in α-space a complete set; that is,

$$\begin{aligned}\int \langle\alpha\,|\,x\rangle\, dx\,\langle x\,|\,\alpha'\rangle &= N^{-1}\sum_i \Xi_i^*(\alpha)\,\Xi_i(\alpha')\\ &= N\sum_i \Delta_i^*(\alpha)\,\Delta_i(\alpha') = \textit{unknown}.\end{aligned} \tag{108}$$

We may also express this by saying that the Hilbert vectors $\Xi_\alpha(x)$ representing different microstates labeled $\alpha, \alpha', \ldots$, need not be orthogonal among each other, and in the Wiener–Siegel theory certainly are not normalized.

Therefore when we use the $\langle\alpha\,|\,x\rangle$ and $\langle x\,|\,\alpha\rangle$ as transformation matrices, these transformations are not unitary. They are what Wiener and Siegel call *semiunitary*. That is, when we transform $|\,i\rangle$ from the x-representation to the α-representation by

$$\langle\alpha\,|\,i\rangle = \int\langle\alpha\,|\,x\rangle\, dx\,\langle x\,|\,i\rangle = \xi_i^*(\alpha) = \int \Xi_\alpha^*(x)\,\phi_i(x)\,dx, \tag{109}$$

the $\xi_i^*(\alpha)$ in α-space acquire the orthonormality which the $\phi_i(x)$ had in x-space, according to

$$\begin{aligned}\int\langle i\,|\,\alpha\rangle\, d\alpha\,\langle\alpha\,|\,k\rangle &= \int \xi_i(\alpha)\,\xi_k^*(\alpha)\,d\alpha = \langle\xi_i\xi_k^*\rangle_{\text{Av}}\\ &= [\text{by (101), (93) and (81)}] = \sum_i\sum_k \phi_i^*(x_i)\,\langle\Delta_i\Delta_k^*\rangle_{\text{Av}}\phi_k(x_k)\\ &= \int \phi_i^*(x)\,\phi_k(x)\,dx = \delta_{ik} = \langle i\,|\,k\rangle.\end{aligned} \tag{110}$$

On the other hand, while the $\phi_i(x)$ formed a complete set in x-space, the $\xi_i^*(\alpha)$ do *not* form

[49] Actually, they use the $\Delta_i(\alpha) = \Xi_\alpha(x_i)/N$ instead of the $\Xi_\alpha(x_i)$ themselves.

a complete set in α-space, as seen from

$$\int \langle \alpha | i \rangle \, di \, \langle i | \alpha' \rangle = \sum_i \xi_i^*(\alpha) \, \xi_i(\alpha') = [\text{by (101) and (104)}] = N \sum_j \Delta_j^*(\alpha) \, \Delta_j(\alpha') = \textit{unknown} \tag{111}$$

as in (108).

The state vector $\psi(x)$ in hidden-variables representation

Wiener and Siegel (1953) also apply the above semiunitary transformation to the quantum-mechanical state vector $\psi(x)$. This yields the hidden-variables representation of ψ given by

$$\Psi(\alpha) = \int \Xi_\alpha^*(x) \, \psi(x) \, dx = \sum_i \Delta_i^*(\alpha) \, \psi(x_i). \tag{112}$$

If $\psi(x)$ is expanded in terms of the $\phi_i(x)$ by

$$\psi(x) = \sum_i a_i \cdot \phi_i(x), \quad a_i = \int \phi_i^*(x) \, \psi(x) \, dx, \tag{113}$$

then (109) and (112) give

$$\Psi(\alpha) = \sum_i a_i \xi_i^*(\alpha). \tag{114}$$

Therefore, eq. (110) gives

$$\langle |\Psi|^2 \rangle_{\mathrm{Av}} = \int |\Psi(\alpha)|^2 \, d\alpha = \sum_i |a_i|^2 = 1. \tag{115}$$

3.5. Equilibrium hidden-variables distribution in ξ-space

We assumed in Section 3.2 that the equilibrium distribution of the hidden variables in complex Δ_i-space was given by eq. (75). By (101) a distribution in complex Δ_i-space generates a corresponding distribution in complex ξ_i-space. We shall split up each ξ_j into its real and imaginary part according to

$$\xi_j = \eta_{2j-1} + i\eta_{2j}. \tag{116}$$

By (76) and (106) we then have

$$N \cdot \sum_{\iota=1}^{2N} \delta_\iota^2 = N \cdot \sum_{i=1}^{N} |\Delta_i|^2 = \sum_{i=1}^{N} |\xi_i|^2 = \sum_{\lambda=1}^{2N} \eta_\lambda^2. \tag{117}$$

Thus the unitary transformation (101) between the $(N^{1/2}\Delta_i)$ and the ξ_i (see footnote 48) yields an orthogonal transformation between the δ_ι and the η_λ. As the Jacobian of an orthogonal transformation is unity, it follows that the distribution generated by (75) or (77) in complex ξ_i-space or in real η_λ-space is given by[50]

$$d\bar{P} = \pi^{-N} \exp\left(-\sum_{i=1}^{N} |\xi_i|^2\right) d\eta_1 \, d\eta_2 \, \dots \, d\eta_{2N}$$

$$= \prod_{\lambda=1}^{2N} \left\{\pi^{-\frac{1}{2}} \exp\left(-\eta_\lambda^2\right) d\eta_\lambda\right\}. \tag{118}$$

[50] For writing the δ_ι- or Δ_i-distribution in the form (75)–(77) we had to assume that i took N (or ι took $2N$) discrete values, so that x-space contained N discrete points. As the $\{\phi_i(x)\}$ form a complete set in this space of N points, there must be N functions in this set. Therefore $1 \leqslant i \leqslant N$ and, consequently, $1 \leqslant \lambda \leqslant 2N$.

3.6. How the hidden variables determine the result of a measurement in the theory of Wiener and Siegel

Avoiding troubles for measurements of "degenerate" observables

When an observable A is measured, one might hope that knowledge of the hidden variables and of ψ will tell us which eigenvalue A_n will be realized. If the Hilbert subspace S_n spanned by the eigenfunctions $\phi_{n,j}$ of A_{op} belonging to A_n is more-than-one-dimensional, the eigenfunction resulting as reduced wavefunction after the measurement (here assumed to be of the reproducible kind) should, according to von Neumann, be the *orthogonal projection of the original state vector ψ upon that subspace S_n*.

Thus if some atom is in an $l = m_l = 1$ state and we measure its energy without disturbing its angular momentum state, we expect the result to be an $l = m_l = 1$ state, and not, for instance, a state with $l = 1$ which is a superposition of states with different values of m_l.

If, however, there is in the measuring apparatus something that would make the angular momentum during the energy measurement choose between the possible eigenvalues of L_y, we would expect that this final state would be a simultaneous eigenfunction of L_y and the energy operator and the possible results $\{\phi_i\}$ of the measurement would no longer have $m_l = 1$.

In general, the experimental setup will largely determine the possible results $\{\phi_i\}$ for the reduced wavefunction immediately after the (supposedly reproducible) measurement, but the ϕ_i need not all be *uniquely* determined by the apparatus alone. There still may be two-or-three-or-more-dimensional subspaces S in Hilbert space in which *any* Hilbert vector seems to be a candidate for being a possible result of the measurement.

When this is so, we know from Gleason's work that we are in trouble with predicting the results of the measurement on the basis of the values of the hidden variables, if any of the S is of dimension 3 or more.[51]

Wiener and Siegel (1955) realized that their method of determining the result of the measurement (by the "*polychotomic algorithm*" discussed below) would cause trouble already in cases where some S is two-dimensional. They demonstrated this by an example given in an appendix. In Appendix J we give an example of it which is rather similar to theirs.

Wiener and Siegel (1955) therefore stressed that the dimensionality of the subspaces S of degeneracy of the measurement should be reduced by describing each measurement as a simultaneous measurement of a sufficient number of mutually commutative observables so that each set of simultaneous eigenvalues of all these observables (like a "complete" set of quantum numbers) would uniquely determine a one-dimensional subspace in Hilbert space (a single Hilbert vector ϕ_i, or, of course, this ϕ_i times a numerical complex factor).[52] In that case, not only Gleason troubles[51] can be avoided, but also the additional troubles caused by degeneracy in application of Wiener and Siegel's "algorithm". (Compare Appendix J.)

[51] See Part I, Section 3.6.

[52] See Wiener and Siegel (1955), pages 989–990, and again on page 997.

Tutsch's rule for making the set $\{\phi_i\}$ unique

We can avoid Wiener and Siegel's requirement of reducing to 1 the dimensionalities of all Hilbert subspaces S in which every Hilbert vector seems to be a common eigenfunction of all observables measured simultaneously. This is done by imposing an additional requirement upon the choice of the set $\{\phi_i\}$. [Tutsch (1969).]

If any of these subspaces S has a dimensionality $n > 1$, we can postulate[53] that we must choose in S, as one of the ϕ_i admitted as a possible result of the measurement, the *normalized orthogonal projection ϕ_s of ψ onto S*. Then, choose an orthonormal set of $(n-1)$ additional ϕ_i in S orthogonal to ϕ_s, and thus also orthogonal to ψ, so that for *any* values of the hidden variables there is a zero probability that any of these additional ϕ_i will be realized.

If $n-1 > 1$, there still will be a lack of uniqueness in these ϕ_i that are normal to ψ, but this does not harm anybody.

In the first place, *this* ambiguity in the complete set $\{\phi_i\}$ does not cause any Gleason troubles.[51] Gleason's unnamed lemma [see (I:70)] told us merely that any nonnegative frame function $w(\phi)$ of the Hilbert vectors in a three-dimensional "real"[51] subspace R_3 of Hilbert space [such that, for any orthonormal "triad" $\{\phi_i\}$ in R_3, $\sum_i w(\phi_i) =$ invariant under rotations of the triad in R_3] must be a *continuous* function of the ϕ in R_3. From this, we derived in Section 3.6 of Part I the impossibility of existence of frame functions that were truth values [$w(\phi) = v(\phi) = 0$ or 1] by considering the case $\sum_i w(\phi_i) = 1$ in R_3, in which $w(\phi) = 0$ or 1 would lead to jumps of $w(\phi)$ between 0 and 1 in contradiction to continuity. But, in the present case, in the $(n-1)$-dimensional part of S normal to ϕ_s, we consider merely the case $\sum_i w(\phi_i) = 0$ and all $w(\phi_i) = 0$ (for $\phi_i \perp \phi_s$), so that the continuity of $w(\phi) = 0$ in this subspace is trivial and leads to no contradictions.

In the second place, this ambiguity, remaining in the part of the set $\{\phi_i\}$ that in the expansion of ψ occurs with coefficients $a_i = 0$, does not cause any ambiguity in the predictions of the Wiener–Siegel theory either, as we will find below that their "polychotomic algorithm" eq. (120) will *never* predict that the outcome of any reproducible measurement would be a state ϕ_i normal to the initial ψ, and that these states ϕ_i with $a_i = 0$ will have no influence whatsoever upon the determination, by the algorithm, of the state ϕ_n that *would* be the result of that measurement.

Consequently, the part of the complete orthonormal set $\{\phi_i\}$ that is *relevant* for the application of the algorithm then consists merely of one Hilbert vector ϕ_s in every Hilbert subspace S corresponding to each eigenvalue A_s of the observable measured—provided we adopt Tutsch's rule[53] which adds to the theory a prescription for the choice of the set $\{\phi_i\}$ which states that not only shall each ϕ_i be an eigenfunction of each of the observables simultaneously measured, but also,

> where this first prescription leaves subspaces S of degeneracy of the measurement which have a dimensionality > 1 in Hilbert space, the set $\{\phi_i\}$ should contain in each such S one vector ϕ_s which is the normalized orthogonal projection of ψ upon S, (119)

[53] See Tutsch (1969), page 1117, top half of second column.

while the set as a whole should be a complete orthonormal set. Thus in the expansion $\psi = \sum a_i\phi_i$ only one ϕ_i in each subspace S of degeneracy will occur with a nonvanishing coefficient a_i.[54]

Remaining of Gleason or Kochen–Specker troubles

As, for any observable A, Tutsch's rule makes the relevant part of the set $\{\phi_i\}$ unique for any given state vector ψ, Gleason's proof[51] that there should exist *continuous* frame functions $w(\phi_i)$ breaks down for those cases where *degeneracy of* A_{op} was taken in Part I as a justification for considering all Hilbert vectors in some three-or-more dimensional Hilbert subspace as possible results of some measurement.

This, however, does not remove the Kochen–Specker paradox or break the validity of the reasoning based upon Gleason's unnamed lemma, in those cases where the Hilbert vector in a three-or-more-dimensional Hilbert subspace obviously all *are* possible results of measurements as in the case of measurements of the spin of an orthohelium atom. (See Part I, Chapter 3.) In this Kochen–Specker example, for each measurement with a given rhombic electric field, the set $\{\phi_i\}$ was always uniquely given to start with, and Tutsch's rule was not needed for making the $\{\phi_i\}$ unique. Nevertheless, we found it impossible to assign truth values to "propositions" such as that a certain component of the spin for given ψ and ξ would be zero.

For proving the nonexistence of unique truth values for such propositions, we did not even need Gleason's three-dimensional continuum of Hilbert vectors that could be results of measurements. Kochen and Specker (1967), as we mentioned early in Section 3.3 of Part I, succeeded in proving the impossibility of unique truth values by considering merely 117 different Hilbert vectors representing states with spin component zero or with spin component nonzero, in 117 different directions in space. In Part I, Appendix B, we succeeded in simplifying this proof using spin components in merely eight directions in space. This example shows that the impossibility of assigning to propositions unique truth values $v_n(\psi, \xi)$ that depend on the microstate (ψ, ξ) only can occur already when the propositions for just a few appropriately chosen Hilbert vectors are considered, and there is in that case no need for considering rotations in a Hilbert subspace, so that Gleason's unnamed lemma for this case is irrelevant. Therefore it is in this case more appropriate to talk about "Kochen–Specker troubles" than about "Gleason troubles" when referring to this impossibility of assigning unique truth values to propositions.

We talk here about "troubles" because it sounds paradoxical to those accustomed to pure quantum theory that in a given microstate (ψ, ξ) it still would depend upon the complete orthonormal set $\{\phi_i\}$ of possible measurement results instead of merely upon the state ϕ_n itself whether the proposition "measurement will find the system in state ϕ_n" would be true or false. Since, however, this dependence of the truth value v_n not only on ψ and ξ but also on the complete orthonormal set $\{\phi_i\}$ is accepted by hidden-variables enthusiasts as a necessary evil, we merely call it a "trouble" and not a "proof of impossibility" of hidden-variables theory.

[54] The alternative to postulating (119) was discussed in Section 3.7 of Part I under the heading "Oversensitivity of the result of a measurement for trifling circumstances."

Since in the Kochen–Specker example the axes of the rhombic field uniquely determine the set $\{\phi_i\}$, Tutsch's rule is not needed and it does not take away the "paradox" or "trouble." If we would use a cylindrical instead of rhombic field, singling out only one instead of three preferred axes, but yet the measurement would determine only the *square* of the spin component in the preferred directions, we could use Tutsch's rule to restore for each given ψ the uniqueness of $\{\phi_i\}$ lost by this additional symmetry of the electric field applied. As uniqueness of $\{\phi_i\}$ was no guarantee against Kochen–Specker troubles, we should not be surprised when, even where Tutsch's rule is applicable, paradoxes occur.

We conclude that Tutsch's rule is useful for taking the arbitrariness out of $\{\phi_i\}$ where such arbitrariness exists. We show in Appendix S that for the "polychotomic algorithm" of the Wiener–Siegel theory defined in eq. (120) below, Tutsch's rule leads to a generalization eq. (S-15) of this algorithm for degenerate observables, which was already proposed by Wiener and Siegel in 1953. However, the theory, thus made unique by Tutsch's rule (119), still leads to paradoxes, of which some are worse than the Kochen–Specker troubles in that they are much less (if at all) justifiable as a "result of the influence of the measuring arrangement." We discuss two types of such "new" paradoxes in Appendix K.

The polychotomic algorithm

Having determined the set $\{\phi_i\}$ of possible final wavefunctions at the end of the reproducible measurement, we now need a prescription to select from the set $\{\phi_i\}$ the one measurement result ϕ_n, where n should be a unique function of $\psi(x)$ and of $\Xi(x)$ or Δ_i or α. In any case, for given $\{\phi_i\}$ we can express the microstate in terms of the $\{\xi_i\}$, since the set $\{\xi_i\}$ follows from $\Xi(x)$ by (100), from Δ_i by (101), and from α by (85) followed by (101). We can also express $\psi(x)$ in terms of the set $\{a_i\}$ given by (113).

The algorithm finally chosen by Wiener and Siegel for determining the choice ϕ_n for given sets $\{a_i\}$ and $\{\xi_i\}$ was called by them the *polychotomic algorithm*.[55] It makes n depend upon the set of quotients[56] $\{|a_i/\xi_i|\}$. If among all $|a_i/\xi_i|$ the largest quotient is $|a_k/\xi_k|$, the polychotomic algorithm states that $n = k$, and that ϕ_k will be realized by the measurement. If NO(k) means *any* index *except* k, this algorithm thus makes the selection

$$\{\phi_i\} \to \phi_k \quad \text{when} \quad |a_k/\xi_k| > |a_{\mathrm{NO}(k)}/\xi_{\mathrm{NO}(k)}|. \tag{120}$$

We shall not bother here to invent a special rule for the special case that two members of the set $\{|a_i/\xi_i|\}$ *share* the property of being the largest of the set, as the probability that this would happen is zero.

[55] In their first paper, Wiener and Siegel (1953) proposed, as a possible alternative to the polychotomic algorithm, the successive application of a "dichotomic" algorithm, which was to make finer and finer subdivisions of the set $\{\phi_i(x)\}$ until only one $\phi_n(x)$ would be left. Later it was pointed out by Siegel (1966) that this alternative fails to yield statistical results agreeing with quantum theory because of a bias introduced in the hidden-variables distribution each time the algorithm is applied.

[56] In case a_i vanishes, we put $|a_i|/|\xi_i| = 0$ by definition, even if also ξ_i would vanish.

3.7. Verification that the Wiener–Siegel theory meets the requirements of a hidden-variables theory of the first kind

We have so far satisfied several of the requirements of Section 1.2. We satisfied requirements (A) and (B) already in Section 3.2 and by eq. (118).

As to requirement (C), neither Wiener and Siegel nor Bohm and Bub make much of an attempt at explaining how the equilibrium distribution $\bar{P}$ of hidden variables is re-established when somehow it is disturbed. They do explain (see Section 3.9 below) how this distribution perpetuates itself in absence of perturbations once it has been established. Then, by a reasoning similar to the one we have met in Bohm's theory, it should be *plausible* that distributions P different from $\bar{P}$, which are not stable, have a tendency of approaching by perturbations toward the only stable distribution. As thus the tendency $P \rightarrow \bar{P}$ is plausible without a real effort at explaining the *mechanism* of this transition of P into $\bar{P}$, nobody has carefully looked into the details of this mechanism. In Chapter 5 we will find that an understanding at least of the *speed* with which this "relaxation" of any distribution P into the canonical distribution $\bar{P}$ takes place would help us drawing conclusions from the results of experimental attempts at proving or disproving these hidden-variables theories.

Requirement (D) was that some algorithm be given for predicting results of measurements for given ψ, ξ, and $\{\phi_i\}$. The polychotomic algorithm (120) satisfies this requirement.

Verification that the polychotomic algorithm is alright

Requirement (E) is that we should be able to prove that the canonical distribution $\bar{P}$ together with the algorithm will lead to the same statistical predictions about the results of measurements as made by quantum mechanics. Before we prove this, first a few general remarks.

Consider N-dimensional complex ξ_i-space. The unit vectors along the axes are the eigenfunctions ϕ_i, and the point Ξ in it has the complex coordinates ξ_i. In this same space we have a fixed point ψ with complex coordinates a_i. While ψ is a single point, we consider an ensemble of points Ξ, distributed according to (118).

To each point Ξ we assign one of the axes (say, axis ϕ_k out of the set of axes $\{\phi_i\}$) by comparing the ratios $|a_i/\xi_i|$ for all i as in (120). Below, we will count the number of points Ξ in the ensemble to which *some particular axis* ϕ_k is assigned. Think of all those points Ξ being painted red. (The remainder we then could paint blue, as we did in Part I, Section 3.3.)

We must scan over all points Ξ for deciding about each point separately whether it should be painted red or blue. We can do the scanning along all radii from the origin (as if we would be using polar coordinates in our space, as in fact we will in Section 4.1), or we can scan parallel to the axes (as the use of cartesian coordinates ξ_i would suggest.) We will do here both.

We do first the radial scanning for deriving a result that we will need in Section 4.1. Next we do the cartesian scanning because it simplifies the actual count of the red Ξ points in the ensemble.

We define the "radial distance" r of a point Ξ by $\mathrm{r}^2 = \sum |\xi_i|^2 = \int |\Xi(x)|^2 dx$.

Radial scanning

Compare each point Ξ with the point Ξ_1 on the same radius at unit distance (r = 1) from the origin. The ξ_i values at Ξ then are r times the ξ_{1i} of the point Ξ_1:

$$\xi_i = \mathrm{r}\,\xi_{1i}.$$

Therefore, each ratio in the set $\{a_i/\xi_i\}$ of ratios for the point Ξ is r^{-1} times the corresponding a_i/ξ_{i1} for Ξ_1. Then, if a_k/ξ_{1k} is (or is not) larger than all other a_i/ξ_{1i} in the set for Ξ_1, it follows that also a_k/ξ_k is (or is not) larger than all other a_i/ξ_i in the set for Ξ. That is, if Ξ_1 is red (or blue), then also Ξ is red (or blue).

We thus have shown that *radii in our space all have one single color throughout*.

Since the distribution (118) is spherically symmetric in our space, *each radius contains an equal number of points of the Ξ ensemble*.

Therefore, the *probability* $\bar{P}_k$ in our ensemble (118) for a point Ξ being red, is equal to the *fraction of all radii* that is red. It thus is equal to the ratio of the red solid angle to the overall solid angle.

As it may be difficult to measure solid angles in a *complex* N-dimensional space, in Section 4.1 [in eq. (176′)] we will translate this result into *real* solid angles in $2N$-dimensional η_λ-space.

However, we want a formula for this solid angle in terms of the coordinates a_i of the point ψ, which by (120) determines which Ξ points will be red and which will be blue. This formula is easier to derive if we use "cartesian scanning." The simplification comes from the fact that (118) can be factorized into Gaussian distributions for each coordinate ξ_i by itself.

Cartesian scanning

We give here the simplified calculation of $\bar{P}_k$ (and proof of its equality to w_k) given by Wiener and Siegel (1955) in their second paper.

We first introduce polar coordinates r_i and θ_i in the complex plane of each ξ_i, so that

$$\eta_{2i-1} = r_i \cos\theta_i, \quad \eta_{2i} = r_i \sin\theta_i, \quad \xi_i = r_i \exp(i\theta_i). \tag{121}$$

We also introduce new variables

$$X_i \equiv r_i^2 = |\xi_i|^2 = \eta_{2i-1}^2 + \eta_{2i}^2. \tag{122}$$

Then (118) becomes

$$d\bar{P} = \prod_{i=1}^{N} \{\exp(-X_i) r_i\, dr_i\, d\theta_i/\pi\}. \tag{123}$$

Integration over all θ_i from 0 to 2π gives

$$d\bar{P} = \prod_{i=1}^{N} \{\exp(-X_i)\, dX_i\}. \tag{124}$$

Now put

$$w_k \equiv |a_k|^2. \tag{125}$$

Then the conditions determining k in (120) may be rewritten as

$$X_i > X_k w_i / w_k \quad \text{for all } i \neq k. \tag{126}$$

The sum $\bar{P}_k$ of all probabilities $d\bar{P}$ for hidden-variable values for which (126) holds is given by

$$\bar{P}_k = \underset{\text{conditions (126)}}{\int \dots \int} dP = \int_0^\infty \exp(-X_k)\, dX_k \cdot \prod_{i(\neq k)} \left\{ \int_{X_k w_i/w_k}^\infty \exp(-X_i)\, dX_i \right\}$$

$$= \int_0^\infty dX_k \exp(-X_k/w_k) = w_k, \tag{127}$$

where we used $\sum w_i = 1$. The result $\bar{P}_k = w_k$ proves that the polychotomic algorithm (120) predicts the same probability for finding the result ϕ_k out of a set of possible results $\{\phi_i\}$, as quantum mechanics predicts. Thus, the requirement (E) of Section 1.2 is satisfied.

Also, if $F(\psi, \xi)$ is the result of a measurement of the observable F for a system in the microstate (ψ, ξ), the average of $F(\psi, \xi)$ over all hidden-variables states in the canonical distribution $\bar{P}$ of these states is by $\bar{P}_k = w_k$ equal to the quantum-mechanical "expectation value" $\langle F \rangle_\psi = \sum_k w_k F_k$ of F in the quantum-mechanical state ψ:

$$\langle F(\psi, \xi) \rangle_{\mathrm{Av}} = \langle F \rangle_\psi. \tag{128}$$

This proves the claim made in Section 3.2 below eq. (88).

Functional relations between observables

Note also that, if there exists a functional relation $B = f(A)$ between two observables, the set $\{\phi_i\}$ of Hilbert vectors that are the possible results of a reproducible measurement of A in a state $\psi(x)$ is also the set of possible results of a reproducible measurement of B. Therefore, for given $\Xi(x)$ and $\psi(x)$, both measurements feature the same ξ_i and a_i, and the algorithm (120) yields for both the same value of k and the same ϕ_k. Therefore, the functional relation $B_k = f(A_k)$ in this case exists between the predictions for these two observables.

Continuous spectra

In the above we have merely considered the case of discrete eigenvalues A_k of the observable measured, since we treated the index i in $\phi_i(x)$ as an index taking discrete values. The mathematical complications introduced by the consideration of observables with continuous spectra can be dealt with in a variety of ways, treating continuous spectra as limiting cases of discrete spectra. The physical ideas remain the same. [See Wiener *et al.* (1966).]

External coordinates

Suppose we measure an observable A of which the eigenvalue problem $A_{\mathrm{op}} \phi_i(x) = A_i \phi_i(x)$ determines a nondegenerate spectrum $\{A_i\}$, so that A_{op} by $\{\phi_i\}$ determines a unique set of directions in the Hilbert space of functions of x. Suppose, however, that the system on which A is measured has been interacting with a different system described by a coordinate y, so that the state vectors ψ and Ξ are functions of y as well as of x, and can be expanded in terms of the $\phi_i(x)$ by

$$\psi(x, y) = \sum_i a_i(y)\phi_i(x), \quad a_i(y) = \int \phi_i^*(x)\psi(x, y)\, dx, \tag{129}$$

$$\Xi(x, y) = \sum_i \xi_i(y)\phi_i(x), \quad \xi_i(y) = \int \phi_i^*(x)\Xi(x, y)\, dx. \tag{130}$$

We show in Appendix L that, then, by (119), the polychotomic algorithm predicts that the result of the measurement is A_k if the ratio

$$\left\{\int |a_i(y)|^2\, dy\right\} \Big/ \left|\int \xi_i^*(y)a_i(y)\, dy\right| \tag{131}$$

is larger for $i = k$ than it is for any other value of i. (The ratio is counted to be zero if its numerator is zero, even if its denominator would vanish, too.) The numerator in this ratio,

$$w_i = \int |a_i(y)|^2\, dy, \tag{132}$$

is the quantum-mechanical probability for a result $A = A_i$ in an ensemble for which $\psi(x, y)$ is given by (129). In this application of the polychotomic algorithm it was assumed that in x, y-space eigenfunctions χ_{im} of A_{op} were chosen in agreement with Tutsch's rule (119), and that the coefficients ξ_{im} in the expansion of $\Xi(x, y)$ are canonically distributed according to (118) in the ensemble described by $\psi(x, y)$.

In case the set $\{\phi_i(x)\}$ is *not* uniquely determined by A_{op}, problems arise that have not been solved. (See Appendix L.)

A slight variation of the method discussed above leads to a prescription given by Wiener and Siegel (1953) for calculating by the polychotomic algorithm a unique value for the probability of finding in a measurement a particular eigenvalue of an observable, even if it is degenerate. The prescription is to pick, for an arbitrary complete orthonormal set $\{\phi_{im}\}$ of eigenfunctions given by $A_{op}\phi_{im}(x) = A_i\phi_{im}(x)$, the largest *not* among the $|a_{im}/\xi_{im}|$, but among the quantities

$$\left(\sum_m |a_{im}|^2\right) \Big/ \left|\sum_m \xi_{im}^* a_{im}\right|. \tag{131'}$$

See Appendix S for details.

This method is another application of Tutsch's rule (119), and therefore suffers from the paradoxes of Appendix K. Wiener and Siegel do not mention that the use of (131′) may be justified by postulating that the only possible results of a reproducible measurement can be the orthogonal projections of ψ upon the Hilbert subspaces S_i belonging to the degenerate eigenvalues A_i of the observable measured.

Reduction of the state vector ψ, and possibly of the hidden-variables state or distribution, after a measurement

If we measure A reproducibly, and, whenever the result is A_k, we follow up by measuring B, then the statistics about the results of the measurement of B can be understood quantum mechanically if we assume that finding A_k changes ψ into ϕ_k as the initial wavefunction for the measurement of B. Then the statistics for B can be understood by the polychotomic algorithm if we assume that *not only* is ψ changed into ϕ_k by the measurement of A, but also the canonical distribution (118) of hidden variables is restored before the measurement B is made.

The latter might be just a case of "relaxation" $P \to \bar{P}$ of the hidden-variables distribution, although there is a problem with the canonical distribution that we discuss in Appendices M and N.

Requirement (F) of Section 1.2 was the need for an understanding of why or how ψ would be "reduced" to ϕ_k during or after the measurement $A = A_k$.

Siegel and Wiener (1956), in an attempt to discuss the theory of measurement within the framework of their theory, give for the reduction of ψ *itself* no other reasons then the ones given usually in quantum theory. See their section 1. In their section 2 they add a discussion of how a final hidden-variables state of an *observed object* might be derived for a given outcome of the observation from a given hidden-variables state of the *composite system* of object and measuring apparatus. Their treatment (see our Appendix N) seems at first sight plausible because of the fact that, if the reduction of ψ changes the a_{im} in the expansion (S-2) into $\delta_{iI}a_{Im}$ when A_I is observed, the quantity (S-15) used in the polychotomic algorithm *for that observable A* does no longer depend upon the ξ_{im} for $i \neq I$. Therefore Siegel and Wiener also reduce Ξ by omitting its part normal to the Hilbert subspace S_I corresponding to the observed eigenvalue A_1. However, they do not show that this reduction would not affect predictions made by the polychotomic algorithm for *subsequent measurements of other observables B*. An attempt to investigate this question in the first part of Appendix N is unsuccessful probably because of oversimplification of the model there used. [See below eq. (N-10).] The calculations of Chapter 5, however, certainly suggest that this kind of a reduction of Ξ is not allowable.

In practice, usually the evolution or reduction of the Ξ vector is irrelevant because relaxation changes the Ξ state into a mixed state given by a canonical distribution. We then are left with the problem of reducing the canonical distribution of the composite system of object and instrument, to a canonical distribution of the object alone [see eq. (N-16)], though the two distributions are in spaces of different dimensionality, and such a reduction therefore looks incomprehensible if one would assign "physical reality" of some sort to the hidden variables. (Compare Appendix M.) The only justification for *this* kind of a reduction can lie in its effectiveness in giving the correct statistical predictions.

The quantum-mechanical reasons for $\psi \to \phi_k$ which Siegel and Wiener (1956) give in section 1 of their paper are essentially the ones we gave in Sections 2.10 and 2.11. The essential two points in a reduction are (1) a changeover from a pure state of the composite system of instrument and object of measurement to a mixed state of the object alone as a "satisfactory" substitute when interference effects between different results of the measurement have ceased either by orthogonality of instrument states, or by loss of coherence; and (2) a changeover from this mixed state to a pure state corresponding to the observed result of the measurement by inserting values 1 or 0 for classical probabilities. It is well known that such a reduction of the state vector is not a real necessity (compare Appendix N), and it must be admitted that in a *rigorous* application of quantum theory it therefore can be avoided. However, if reduction is not performed, the description of the object of measurement becomes more and more encumbered by the inclusion of coordinates of instruments that interacted with the system some time in the past. [See Everett (1957), Wheeler (1957), Shimony (1963), Pearle (1967), DeWitt (1970).] For completeness sake, also coordinates of instruments that might in the future interact with the system should then be carried along from the beginning [like our $\psi_3^{(o)}(x_3)$ in eq. (N-2) of the Appendix]. *In principle* this is beautiful. *In practice* it causes unnecessary complications of the formalism. [Compare eq. (N-7) to eq. (120).] Therefore the "reduction" procedure is used as a device for efficiency. It cannot really be justified otherwise except that one can show that it is a very good approximation when it is performed after not only we know the result of the past measurement, and not only has interaction with the instrument ceased, but also all coherences of state have been lost between instrument and object. [See Bohm (1951), chap. 22.]

However, such justifications are based upon statistical arguments.

In hidden-variables theory, therefore, "reduction" may be an acceptable tool in statistical considerations of ensembles described by state vectors ψ and canonical distributions of hidden variables. In Appendix N, at the hand of a simple idealized example, we show how to handle those canonical distributions in such a reduction. Again, the procedure proposed is proposed merely because it is effective; there is no logical justification.

An attempt in Appendix N at finding out whether any reduction procedure could be used in the *nonstatistical* case of an assumedly *known* hidden-variables state fails because in the oversimplified example considered even the "rigorous" treatment without reductions breaks down and predicts nothing. [See eq. (N-10).] (With no result to compare with, it is impossible to judge whether any proposal at "reduction" in such a case does or does not give acceptable results.)

Time dependence of the hidden variables

The questions raised above about the measurement problem require for their satisfactory discussion a precise knowledge of how the hidden variables depend on time. The need of knowledge about this point, *between* measurements and *during* measurements, were the subject of requirements (G) and (H) of Section 1.2. As we shall see in Section 3.9, requirement (G) is pretty well satisfied in the Wiener–Siegel theory, as we will find it necessary to require that between measurements $\Xi(x)$ shall satisfy the same[59] Schrödinger equation as $\psi(x)$.

As to the equations of motion of [$\psi(x, t)$ and of] $\Xi(x, t)$ during measurements, Wiener and Siegel have not proposed any particular alteration of the Schrödinger equation for (ψ and) Ξ during such times.

Thus the Wiener–Siegel theory seems to be a rather complete hidden-variables theory of the first kind except for a so far incomplete treatment of some problems centering around questions about the reduction of states after a measurement (see Appendix N).

Paradoxes in the Wiener–Siegel theory

We want to finish this section by pointing out paradoxes that are found when one tries to apply the polychotomic algorithm to a composite system for which the state vectors ψ and Ξ are not factorizable. An example of this is given in Appendix K. It is found, that, for a given hidden-variables state, the result of a simultaneous measurement of A on part 1 and of B on part 2 may differ from the results of separate measurements of A and B, even if the two parts are so far apart that there can be no interactions between the two measurements [Belinfante (1972)]. The paradox is not comparable with the Einstein–Podolsky–Rosen paradox, since it is based upon an assumed perfect knowledge of the initial hidden-variables state. Rather, it is closely related to the "Kochen–Specker troubles" mentioned in Section 3.6.

We also mention in Appendix K paradoxes of this type that may exist already for single systems. We discuss an example first signaled by Tutsch (1969) in which the algorithm in its generalized form (S-15) [= (120)+(119)] predicts for an orthohelium atom simultaneously $S_z = -1$ *and* $|S_z| = 0$.

3.8. Application of the polychotomic algorithm to spin-$\frac{1}{2}$ states

In Section 2.14 we discussed a hidden-variables theory for spin $\frac{1}{2}$ proposed by Bell (1966) in which a parameter λ in the range from $-\frac{1}{2}$ to $+\frac{1}{2}$, together with $\psi(x)$, determined for any direction $\mathbf{n}$ in space whether the spin S_n in that direction, when measured upon the particle at the position $\mathbf{x}_o$, would have the value $+\hbar/2$ or $-\hbar/2$.

We shall now use Wiener–Siegel's polychotomic algorithm for finding $S_n(\mathbf{x}_o)$, and we shall find that Bell's proposal may be regarded as an application of the Wiener–Siegel method to this particular problem, if we identify Bell's λ with an appropriate function of the ψ and ξ determining the particle's microstate according to Wiener and Siegel.

As in Section 2.14, we shall here first consider the problem in the special frame of reference in which the z-direction is chosen parallel to $(\psi^\dagger \sigma \psi)$ at the position $\mathbf{x}_o$ where the spin is measured. We then can easily generalize the result at the end of the calculation to arbitrarily oriented coordinate systems.

We start by expanding both ψ and Ξ in terms of simultaneous eigenfunctions $\phi_i(\mathbf{x}) = \delta_3(\mathbf{x}-\mathbf{x}_o)\chi_\pm$ of the operators $\mathbf{x}_o$ and S_n, by

$$\psi(\mathbf{x}) = \int d^3\mathbf{x}_o \sum_{\eta=\pm} a(\mathbf{x}_o)_\eta\, \delta_3(\mathbf{x}-\mathbf{x}_o)\, \chi_\eta, \tag{133}$$

and similarly for Ξ with coefficients $\xi(\mathbf{x}_o)_\eta$. It follows that

$$\psi(\mathbf{x}_o) = \sum_{\eta=\pm} a(\mathbf{x}_o)_\eta \chi_\eta, \quad \Xi(\mathbf{x}_o) = \sum_{\eta=\pm} \xi(\mathbf{x}_o)_\eta \chi_\eta. \tag{134}$$

Since from here on we do no longer need ψ or Ξ at points $\mathbf{x} \neq \mathbf{x}_o$, we shall in the following simply write ψ, a, Ξ, and ξ without further mentioning $(\mathbf{x}_o)$.

The χ_η are the two-component spin-functions that are eigenfunctions of S_n. They are given by eq. (46) of Section 2.14 if we read $\mathbf{n}$ for $\boldsymbol{\beta}$ and $|\mathbf{n}| = 1$ for β. Inserting them in (134), we can solve for ξ_+ and ξ_- in terms of the top and bottom components of Ξ:

$$\xi_\eta = \sqrt{\tfrac{1}{2}}\left\{\Xi_+(1+\eta n_z)^{\frac{1}{2}} + \eta\,\Xi_-(n_x - i n_y)[1+\eta n_z]^{-\frac{1}{2}}\right\}. \tag{135}$$

We have similar formulas for a_+ and a_- in terms of ψ_+ and ψ_-. If now we temporarily turn at $\mathbf{x}_o$ the z-axis along $\psi^\dagger \sigma \psi$, so that $\psi_- = 0$, we have

$$a_+ = C(1+n_z)^{\frac{1}{2}}, \quad a_- = C(1-n_z)^{\frac{1}{2}}, \tag{136}$$

with $C = \psi_+\sqrt{\frac{1}{2}}$. [Compare eq. (47).] Now S_n will be $+\hbar/2$ if $|a_+/\xi_+|^2 > |a_-/\xi_-|^2$, and S_n will be $-\hbar/2$ otherwise. We therefore obtain S_n positive, when $|a_-|^2|\xi_+|^2 < |a_+|^2|\xi_-|^2$, or, by (136), when $(1-n_z)|\xi_+|^2 < (1+n_z)|\xi_-|^2$, that is, if

$$|\xi_+|^2-|\xi_-|^2 < n_z(|\xi_+|^2+|\xi_-|^2). \tag{137}$$

Inserting here (135), we find for this condition

$$(\Xi^\dagger \mathbf{n}\cdot\sigma\Xi) < n_z(\Xi^\dagger\Xi), \tag{138}$$

where the z-direction was given by

$$\hat{z} = (\psi^\dagger\sigma\psi)/(\psi^\dagger\psi), \tag{139}$$

so that the condition for $S_n = +\hbar/2$ becomes

$$(\Xi^\dagger \mathbf{n}\cdot\sigma\Xi)(\psi^\dagger\psi) < (\Xi^\dagger\Xi)(\psi^\dagger \mathbf{n}\cdot\sigma\psi). \tag{140}$$

Since this is invariant under rotation, (140) is the condition for $S_n = +\hbar/2$ for any orientation of the coordinate system.

Accordingly to Bell, the measured value of S_n would be given by eq. (50) with $\alpha(\mathbf{x}) = 0$ and with $\boldsymbol{\beta}(\mathbf{x}) = (\hbar/2)\mathbf{n}$. That is, he predicts

$$S_n(\lambda) = (\hbar/2)\cdot\text{sign}\,\{2\lambda\,\psi^\dagger\psi+|\psi^\dagger\mathbf{n}\cdot\sigma\psi|\}\cdot\text{sign}\,\{\psi^\dagger\mathbf{n}\cdot\sigma\psi\}, \tag{141}$$

while from (140) we obtain

$$S_n(\lambda) = (\hbar/2)\cdot\text{sign}\,\{(\Xi^\dagger\Xi)(\psi^\dagger\mathbf{n}\cdot\sigma\psi)-(\Xi^\dagger\mathbf{n}\cdot\sigma\Xi)(\psi^\dagger\psi)\}. \tag{142}$$

By comparison, we find the two equations identical when for λ we choose

$$\lambda = -\frac{(\Xi^\dagger\mathbf{n}\cdot\sigma\Xi)(\psi^\dagger\mathbf{n}\cdot\sigma\psi)}{2(\Xi^\dagger\Xi)|(\psi^\dagger\mathbf{n}\cdot\sigma\psi)|}. \tag{143}$$

In the frame of reference given by (139), the $(\Xi^\dagger\Xi)$ and $(\Xi^\dagger\mathbf{n}\cdot\sigma\Xi)$ in (143) are given by the $\xi^\dagger\xi$ and the $\xi^\dagger\sigma_z\xi$ appearing[57] in the right-and-left-hand members of (137), so that, for arbitrary coordinate system orientations, λ is expressed, in terms of the expansion coefficients ξ_η in (134), by[57]

$$\lambda = -\frac{(\xi^\dagger\sigma\xi)\cdot(\psi^\dagger\sigma\psi)}{2(\xi^\dagger\xi)(\psi^\dagger\psi)}\ \text{sign}\,(\psi^\dagger S_n\psi). \tag{144}$$

[57] Here we use for the ξ_+ and ξ_- (coefficients of two eigenfunctions $\chi_\pm$ of S_n) a spin notation as if $\begin{pmatrix}\xi_+\\ \xi_-\end{pmatrix}$ would be a spin function.

3.9. Time dependence in the Wiener–Siegel theory

The Schrödinger picture

As long as the system described is a closed system, the time dependence of its quantum-mechanical state may be described by a Schrödinger equation for its state vector ψ. (This equation may or may not be valid while the system is affected by possibly unknown external influences as during a measurement. See Chapter 4 and the end of Appendix N.) The Schrödinger equation

$$i\hbar\, \partial\psi/\partial t = \mathscr{H}_{\text{op}}\psi(x, t) \tag{145}$$

is solved by

$$\psi(x, t) = U(t)_{\text{op}}\, \psi(x, 0), \tag{146}$$

where $U(t)_{\text{op}}$ is a unitary nonlocal operator formally written as

$$U(t)_{\text{op}} = \exp\,(t\mathscr{H}_{\text{op}}/i\hbar). \tag{147}$$

In practice, this operator acts as follows. If the eigenfunctions of $\mathscr{H}_{\text{op}}$ are $\chi_j(x)$:

$$\mathscr{H}_{\text{op}}\chi_j(x) = \mathscr{E}_j\chi_j(x), \tag{148}$$

we expand $\psi(\mathbf{x}, 0)$ according to

$$\psi(x, 0) = \sum_j c_j(0)\, \chi_j(x), \quad c_j(0) = \int \chi_j(x)^*\, \psi(x, 0)\, dx. \tag{149}$$

Then,

$$\psi(x, t) = U(t)_{\text{op}}\psi(x, 0) = \sum_j c_j(t)\, \chi_j(x) \quad \text{with} \quad c_j(t) = c_j(0)\exp\,(t\mathscr{E}_j/i\hbar). \tag{150}$$

In the Schrödinger picture we use time-independent operators A_{op} for describing observables A. Their eigenfunctions $\phi_i(x)$ are also time independent. When we expand

$$\psi(x, t) = \sum_i a_i(t)\, \phi_i(x), \quad a_i(t) = \int \phi_i^*(x)\, \psi(x, t)\, dx, \tag{151}$$

all time dependence goes into the $a_i(t)$. If we put

$$T_{ij} = \int \phi_i^*(x)\, \chi_j(x)\, dx, \quad \sum_i T_{ij}T_{ik}^* = \delta_{jk}, \quad \sum_k T_{ik}T_{jk}^* = \delta_{ij}, \tag{152}$$

we have

$$\phi_i(x) = \sum_k T_{ik}^*\chi_k(x), \quad \chi_j(x) = \sum_i \phi_i(x)\, T_{ij}, \tag{153}$$

and therefore, by (151), (150), (149),

$$a_i(t) = \sum_j \sum_k T_{ij}\exp\,(t\mathscr{E}_j/i\hbar)\, T_{kj}^*\, a_k(0). \tag{154}$$

Conditions for time dependence of Ξ

We now should define the time dependence of $\Xi(x, t)$. We impose three conditions [Wiener and Siegel (1955)].

(1) For any observable A that commutes with $\mathcal{H}_{op}$, it follows from the above formulas for $\partial\psi/\partial t$ that the quantum-mechanical probability distribution over the eigenvalues of A_{op} is time-independent (that is, $|a_k(t)|^2 = |a_k(0)|^2$, if A_{op} commutes with $\mathcal{H}_{op}$). We now want $\partial\Xi/\partial t$ to provide that, for any system initially in a given microstate, the eigenvalue A_k predicted for the observable A by the polychotomic algorithm will remain the same A_k independent of time, when A_{op} commutes with $\mathcal{H}_{op}$.

(2) Consider a particle whose quantum-mechanical wave packet moves around in a way that can crudely be described classically. Assume that initially we know the particle's microstate so we can predict exactly where in the wave packet the particle can be found. (We would apply for this the polychotomic algorithm, using for $\phi_i(x)$ the eigenfunctions of $\mathbf{x}_{op}$.) Now suppose we repeat this prediction of the exact location at consecutive times. The precise motion $\mathbf{x}(t)$ thus predicted should be smooth and similar to the motion of the wave packet as a whole; that is, $\mathbf{x}(t)$ should not wildly jump around.

(3) If, in an ensemble E_ψ of particles all described quantum mechanically by the same state vector $\psi(x, t)$, the hidden variables are initially distributed canonically according to (118), we want this distribution to prevail at all later times on account of the equation for $\partial\Xi/\partial t$.

If we add to the above three requirements the *principle of simplicity of the theory*, $\partial\Xi/\partial t$ (in absence of external influences) is uniquely determined, as follows.

Satisfying Condition One

Condition (1) means that, when we expand

$$\Xi(x, t) = \sum_i \xi_i(t)\, \phi_i(x), \tag{155}$$

then among the $|a_i(t)/\xi_i(t)|$ the same one $|a_k(t)/\xi_k(t)|$ should remain the largest one for all t, if A is an integral of motion. The latter condition, however, means that in this case the eigenfunctions ϕ_i of A_{op} are at the same time eigenfunctions χ_j of $\mathcal{H}_{op}$, so that the T_{ij} of (152) are Kronecker symbols δ_{ij}, and eq. (154) gives $a_j(t) = a_j(0) \exp(t\mathcal{E}_j/i\hbar)$, or $|a_i(t)| = |a_i(0)|$. Then, condition (1) is easily satisfied by making also $|\xi_i(t)| = |\xi_i(0)|$ for all i.

This condition, however, can be met in more than one simple way. We could make ξ_i and therefore also $\Xi(x)$ time-independent, or we could give $\xi_i(t)$ the same time dependence as $a_i(t)$, and therefore also give $\Xi(x, t)$ the same time dependence as $\psi(x, t)$. The choice between these simple possibilities is to be made on the basis of condition (2) above.[58]

[58] Wiener and Siegel (1953) in their first paper give little discussion of the time dependence of the hidden variables. In some of their equations, even where they discuss the time dependence of $\psi(\mathbf{x}, t)$, they give the impression as if they choose $\Xi(x)$, $\Lambda_i(\alpha)$, $\xi_i(\alpha)$ time-independent. [See also the text underneath eq. (160').] In their Nuovo Cimento paper, Wiener and Siegel (1955) consider the time dependence of $\Xi(x, t)$ and therefore of $\xi_i(t)$ more carefully, and come to the conclusions we shall discuss here.

Satisfying Condition Two

For determining whether or not $\xi_i(t)$ should feature a time dependence like $a_i(t)$ does,[59] we have to consider observables which are *not* integrals of motion, so that there may be interference effects between factors $\exp(t\mathscr{E}_j/i\hbar)$ for different values of j. A simple example is the observable $\mathbf{x}_{op}$, as nonrelativistically $\mathbf{x}_{op}\mathscr{H}_{op}-\mathscr{H}_{op}\mathbf{x}_{op} = i\hbar\mathbf{p}_{op}/m$.

Let us illustrate by a simple example how the time dependence of Ξ can now be obtained from condition (2) above. Suppose the particle is quantum-mechanically described by some wave packet with its center $\langle\mathbf{x}\rangle$ moving approximately according to $\langle\mathbf{x}(t)\rangle = \langle\mathbf{x}(0)\rangle+\mathbf{u}t$. (For instance, the Gaussian packet of Appendix G.) Assume also that the initial $\Xi(\mathbf{x})$ distribution is given by

$$\Xi(\mathbf{x}, 0) = f(\mathbf{x})\,\psi(\mathbf{x}, 0) \quad \text{wherever} \quad \psi(\mathbf{x}, 0) \neq 0,$$
$$\Xi(\mathbf{x}, 0) = \text{any rippled function wherever} \quad \psi(\mathbf{x}, 0) = 0.$$

The "outside ripples" in $\Xi(\mathbf{x}, 0)$ do not affect the polychotomic algorithm at $t = 0$, on account of footnote 56 in Section 3.6. We assume that $f(\mathbf{x})$ is a positive function which has a minimum where $|\psi(\mathbf{x}, 0)|$ has a maximum at the "center" of the wave packet. Then the algorithm (120) will locate the particle at $t = 0$ exactly at the center of the wave packet [at the dip in $f(\mathbf{x})$].[60]

Condition (2) now requires that $\mathbf{x}(t)$ as defined by the algorithm shall approximately move, like the packet, as $\mathbf{x}(t) \approx \mathbf{x}(0)+\mathbf{u}t$. Now if the ψ-wave would move on account of the Schrödinger equation (as in the example of Appendix G), we want to move the maximum of $|a_i(t)/\xi_i(t)| = |\psi(\mathbf{x}_i, t)/\Xi(\mathbf{x}_i, t)|$ along with the wave packet. This will be so if $\Xi(\mathbf{x}, t) \approx f(\mathbf{x}, t)\,\psi(\mathbf{x}, t)$ with $f(\mathbf{x}, t)$ not fluctuating wildly. If, however, we would have $\partial\Xi/\partial t = 0$ [which was the first of the two alternatives mentioned under condition (1)], the ψ-wave would move out while the Ξ-wave packet would stay put. The ψ-wave would move out onto what were the outside ripples of the Ξ-wave, and the maximum of $|\psi/\Xi|$ at later time could be at *any* old place.

The motion $\mathbf{x}(t)$ of the particle as predicted by the polychotomic algorithm in that case would be wild and unpredictable, in disagreement with condition (2).

We therefore must accept the alternative simple assumption, which is that $\xi_i(t)$ varies exactly[59] like $a_i(t)$:

$$\xi_i(t) = \sum_j \sum_k T_{ij} \exp(t\mathscr{E}_j/i\hbar)\, T^*_{jk}\, \xi_k(0). \tag{156}$$

This is achieved by having $\Xi(x, t)$ depend on time in exactly the same way as $\psi(x, t)$[61]:

$$\Xi(x, t) = U(t)_{op}\, \Xi(x, 0), \tag{157}$$

[59] Wiener and Siegel write ξ_i^* for our ξ_i. (See footnote 46 in Section 3.4.) Therefore, *their* ξ_i would feature a time-dependence complex conjugated to the time dependence of *our* ξ_i.

[60] As the $\phi_i(\mathbf{x})$ here are the delta functions $\delta_3(\mathbf{x}-\mathbf{x}_i)$, the $a_i \equiv a(\mathbf{x}_i)$ in $\psi(\mathbf{x}) = \int a(\mathbf{x}_i)\,\phi_i(\mathbf{x})\,d^3\mathbf{x}_i$ are the values of $\psi(\mathbf{x})$ at $\mathbf{x}_i$. Similarly, ξ_i with index i standing for $\mathbf{x}_i$ is equal to the value of $\Xi(\mathbf{x})$ at $\mathbf{x} = \mathbf{x}_i$. Thus, $|a_i/\xi_i| = |\psi(\mathbf{x}_i)/\Xi(\mathbf{x}_i)| = 1/f(\mathbf{x}_i)$.

[61] See Wiener and Siegel (1955), page 999, third paragraph: "The most natural choice..." They write $|P\rangle$ for our Ξ.

with $U(t)_{op}$ given by (150). *Formally*,[62] we may then write

$$i\hbar\, \partial\Xi/\partial t = \mathscr{H}_{op}\Xi(x, t). \tag{158}$$

Thus requirement (G) of Section 1.2 is satisfied.

Satisfying Condition Three

It is easily seen that the transformation from the ξ_i at $t = 0$ to the $\xi_j(t)$ is a unitary transformation. [Wiener and Siegel (1953, 1955).] It follows at once from (156) and (152), which yield

$$\begin{aligned}\sum_j |\xi_j(t)|^2 &= \sum_j\sum_k\sum_l\sum_m\sum_n T_{jk} \exp(t\mathscr{E}_k/i\hbar)\, T^*_{lk}\, \xi_l(0)\, T^*_{jm} \exp(-t\mathscr{E}_m/i\hbar)\, T_{nm}\, \xi^*_n(0)\\ &= \sum_k\sum_l\sum_n \exp(t\mathscr{E}_k/i\hbar)\, T^*_{lk}\, \xi_l(0) \exp(-t\mathscr{E}_k/i\hbar)\, T_{nk}\, \xi^*_n(0)\\ &= \sum_l\sum_n \delta_{ln}\, \xi_l(0)\, \xi^*_n(0) = \sum_n |\xi_n(0)|^2. \end{aligned} \tag{159}$$

As the equilibrium distribution (118) in ξ-space is invariant under unitary transformations, it follows that this distribution, once it has been established, will perpetuate itself. We used this fact early in Section 3.7 for explaining how requirement (C) of Section 1.2 is satisfied in absence of perturbations. See Section 4.4 about external influences upon $\partial\Xi/\partial t$.

The time dependence of $\Psi(\alpha)$

Wiener and Siegel (1953) in their first paper define and often mention the "wavefunction $\Psi(\alpha)$ in the hidden-variables representation" which we defined in eq. (112) at $t = 0$. When we insert in (114) the time dependence of a_i and of ξ_i given by (154) and (156), we find by a derivation almost identical with (159) that

$$\Psi(\alpha, t) = \sum_j a_j(t)\, \xi^*_j(\alpha, t) = \sum_n a_n(0)\, \xi^*_n(\alpha, 0) = \Psi(\alpha, 0). \tag{160}$$

Thus for a single physical system in a given hidden-variables state, the value of Ψ is constant. However, α itself depends on time, as it depends by (79), (83), (76), and (82) on the $\Xi_i(\alpha) \equiv \Xi_\alpha(x_i)$, and $\Xi_\alpha(\mathbf{x})$ was made time-dependent by (157). Thus, the functional dependence of Ψ on $\alpha(t)$ is time-dependent in such a way as to keep Ψ constant. We may indicate this by writing (160) as

$$\Psi(\alpha(t), t) = \Psi(\alpha(0), 0). \tag{160'}$$

In the paper of Wiener and Siegel (1953), a different time dependence of $\Psi(\alpha)$ is found. This is due to the fact that in that paper they did not yet discuss the time dependence of $\Xi(x)$ and the $\Delta_i(\alpha)$, and of our $\xi_i(\alpha)$ [their $\xi^*_i(\alpha)$]. So where they defined $\Psi(\alpha, t)$ they defined it not by our eq. (160), but they combined $a_j(t)$ with $\xi^*_j(\alpha, 0)$. This makes their $\Psi(\alpha, t)$ depend on time, contrary to ours which is constant.

[62] We say "formally" because $\mathscr{H}_{op}$ contains ∇, and for $\Xi(\mathbf{x})$ we may allow nondifferentiable functions. See Wiener and Siegel (1955), page 999, footnote 27.

Since in their later work no important use is made of their $\Psi(\alpha, t)$ [or of their $\Psi(\alpha)$, for that matter], it does not really matter how they defined their $\Psi(\alpha, t)$ [or how we define our $\Psi(\alpha, t)$]. See also Section 3.10.

The Heisenberg picture

In some of their work, Wiener and Siegel use the Heisenberg picture. Basing ourselves upon the correct time dependence of the hidden variables in the Schrödinger picture as derived by Wiener and Siegel (1955), we shall here give the main formulas for the Heisenberg picture.

In the Heisenberg picture in quantum theory,[63] expectation values at a time t are expressed not in terms of a time-independent A_{op} and a time dependent $\psi(x, t)$, but in terms of $\psi(x, t_o)$ at some fixed time t_o, and an operator $A(\Delta t)_{op}$ which, with the help of $\psi(x, t_o)$, predicts $\langle A \rangle$ at the time $t = t_o + \Delta t$. Similarly, in hidden-variables theory, in the Heisenberg picture we would use in the polychotomic algorithm the coefficients of the expansion of $\psi(x, t_o)$ and $\Xi(x, t_o)$ at some fixed time t_o, in terms of the eigenfunctions $\chi_i(x, \Delta t)$ of $A(\Delta t)_{op}$, for determining the precise value of A for a system in a given microstate at the time $t = t_o + \Delta t$.

The formulas follow from Eqs. (146) through (157) above, together with the so-called "group property" of $U(t)_{op}$ given by

$$U(t_1)_{op} U(t_2)_{op} = U(t_1 + t_2)_{op}, \tag{161}$$

and the unitarity property

$$U(t)^{\dagger}_{op} = U(t)^{-1}_{op} \equiv U(-t)_{op}. \tag{162}$$

Then, defining $A(\Delta t)_{op}$ by

$$\int \psi(x, t_o)^* A(\Delta t)_{op} \psi(x, t_o)\, d\mathbf{x} = \langle A(t_o + \Delta t) \rangle_{\psi} = \int \psi(x, t_o + \Delta t)^* A_{op} \psi(x, t_o + \Delta t), \tag{163}$$

we find

$$A(\Delta t)_{op} = U(-\Delta t)_{op} A_{op} U(\Delta t)_{op}. \tag{164}$$

As $\phi_j(x)$ are the eigenfunctions of A_{op}, the eigenfunctions of $A(\Delta t)_{op}$ obviously are given by

$$\phi_j(x, \Delta t) = U(-\Delta t)_{op} \phi_j(x). \tag{165}$$

Therefore, the expansions (151) and (155) for $\psi(\mathbf{x}, t)$ and $\Xi(\mathbf{x}, t)$, by $U(-\Delta t)_{op}$ acting from the left, yield

$$\psi(x, t_o) = \sum_j a_j(t_o + \Delta t) \phi_j(x, \Delta t), \tag{166a}$$

$$\Xi(x, t_o) = \sum_j \xi_j(t_o + \Delta t) \phi_j(x, \Delta t). \tag{166b}$$

This shows that, when in the polychotomic algorithm we use the coefficients of the expansions of ψ and Ξ at the fixed time t_o, in terms of the eigenfunctions of $A(\Delta t)_{op}$, we are actually using $a_j(t_o + \Delta t)$ and $\xi_j(t_o + \Delta t)$, so that we determine the largest among the $|a_j(t_o + \Delta t)/$

[63] For a better-than-average discussion, see ter Haar's 1957 English translation (with comments) of Kramers's 1937 book on quantum theory [listed among our references as Kramers (1957)], sec. 43, pages 158–162.

$\xi_j(t_o+\Delta t)\,|$, and therefore we determine the value of A at the time $(t_o+\Delta t)$, which is just what we want. It shows that *the hidden variables state at time t_o completely determines all later measurements, as long as the Schrödinger equation remains valid in the meantime.*

When we keep Ξ fixed at $t = t_o$, we may want to do the same with $\Delta_i(\alpha) = \Xi(x_i)/N$ and with δ_ι and α_ι and with α. An equation like (101) then becomes

$$\xi_j(t_o+\Delta t) = \int dx\, \phi_j^*(x, \Delta t)\Xi(x, t_o), \tag{167}$$

and (112) and (114) become, with $t = t_o+\Delta t$,

$$\Psi(\alpha(t_o), t_o) = \int \Xi^*_{\alpha(t_o)}(x, t_o)\psi(x, t_o)\, dx \tag{168a}$$

and

$$\Psi(\alpha(t), t) = \sum_j a_j(t)\xi_j^*(\alpha(t), t), \tag{168b}$$

which are equal by (160′).

In the Heisenberg picture, then, usually one takes $t_o = 0$ and omits it where it appears as an argument, and one then writes t for $\Delta t = t-t_o$. Thus the t-dependence moves into the $A(t)_{\text{op}}$ and its eigenfunction $\phi_j(x, t)$, of which the time dependence by $U(-t)_{\text{op}}$ is opposite[64] to what it was in the Schrödinger picture for $\psi(x, t)$. Consequently, the coefficients a_j and ξ_j hold their old t-dependence. All other quantities [including the $\psi, \Xi, \Delta, \delta, \alpha$, our $\Psi(\alpha)$] lose their time dependence if they had any.

3.10. Evaluation of the Wiener–Siegel theory

The meaning of the label α

We mentioned already toward the end of Section 3.2 that α was merely a "label." Its value could hardly have a physical meaning, because, by the "fusion" by which it is obtained from the $2N$ different quantities α_ι [see eq. (78)], its value depends critically on the value of N, which may easily have to be chosen larger than 10^{15} for having a good network of points x_i describing the continuum of the variables x on which ψ depends. (If N is replaced by $N+1$, the value of α for unchanged α_1 through α_N is completely garbled.) Thus there is no such a thing as a limiting value of α for $N \to \infty$.

Moreover, for 1% precision for $N = 10^{15}$, we would need each α_ι in 7 binary digits, and we would need α in 7×10^{15} binary digits, with a precision of 1 part in $10^{2\times10^{15}}$. No physical quantity could be measured that accurately.

For given N and Ξ_i, we find Δ_i by (82), δ_ι by (76), α_ι by (83), and α by fusion. Vice versa, for given α, one can calculate Ξ_i (for given N) by (85). Therefore, there is no real need for mentioning α as an argument for quantities like Ξ_i or Δ_i (or ξ_i, for that matter). This is a good thing, as we just have seen that the value of α is not a useful quantity anyhow.

We thus find that the mathematical formalism of the Wiener–Siegel theory, as developed in Sections 3.1 through 3.4, contains some quantities that are not particularly helpful and that are devoid of a physical meaning. Below, therefore, we will scrutinize this formalism and reduce it to bare necessities.

[64] Compare Wiener and Siegel (1955), page 1000.

Siegel's opinion on what "is" the hidden variable

First, Siegel (like us) is uneasy about calling α the hidden variable. But, even the ξ_i, which appear directly in the polychotomic algorithm, are suspect to him. In Appendix N we show how a canonical distribution of hidden-variables state vectors $\Xi(x)$ (which is a distribution of sets $\{\xi_i\}$) for a composite system may be mutilated as in the "reduction" (N-16) from a distribution (M-14) to a distribution (M-12), and still the mutilated distribution is as good as the original one for predicting quantum-mechanical probabilities for results of future measurements. How could this be possible if the Ξ or $\{\xi_i\}$ would have any physical reality?

Siegel's objection against Ξ or $\{\xi_i\}$ is a different one. He is upset by the fact that the microstate is determined not by Ξ or $\{\xi_i\}$ alone, but by Ξ together with ψ.

Siegel (1966) finally came to the conclusion that probably *none* of the quantities which in Sections 3.1 through 3.4 occasionally were labeled by an α has much of an objective physical meaning.

Still, it might be possible that one could consider the existence of "microstates" for where the potential results of future measurements would be uniquely determined, as long as the system would remain a closed system, not interfered with by external interactions or measurements.

The ψ and Ξ would merely be possible tools for calculating these predictions by expanding these functions "now" (at time t_o) in terms of the eigenfunctions (165) of $A(\Delta t)_{\mathrm{op}}$, and be applying the polychotomic algorithm to the expansion coefficients if we want to predict the potential value of $A(t_o+\Delta t)$.

Thus a "hidden-variable state" would not be defined as a state for which ψ and Ξ are (objectively) given functions, but rather as a "state in which one is able to calculate the precise values of all observables at all future times (as long as the system in the meantime is not disturbed)." Thus the *set of all future values of all observables* might be considered to be the "hidden variable" describing the microstate.

The advantage of such a point of view is that we no longer have to worry about such questions as we discussed in Section 3.7 in connection with requirement (F) of Section 1.2, and in Appendix N, about the reduction of a hidden-variables distribution, nor about such paradoxes as we found in Appendix K, which make us wonder whether there exists a meaningful $\Xi(1, 2)$ for a composite system.

On the other hand, this definition of a hidden-variables state gives us no handle for initiating *calculations* of what the future values of observables will be in a microstate defined somehow indirectly, without starting by telling the result of such calculation beforehand.

Thus Siegel's new definition of a hidden-variables state is conceptionally very nice, but rather useless for practical applications of the theory. Therefore we shall not make use of this definition in the following.

Cutting the deadwood

We shall now answer the question what are the parts of the Wiener–Siegel theory that are essential for its applications.

Obviously α cannot be part of it. Obviously we also should try to avoid replacing x-space by a network of N points x_i, and we should replace all sums over those points by integrals over x, as in (81). Therefore we should work rather with $\Xi(x)$ than with Δ_i.

On the other hand, we need the a_i and the ξ_i that are used in the polychotomic algorithm, which is the backbone of the theory.

We need the a_i and the ξ_i for all possible observables A, that is, for all complete orthonormal sets $\{\phi_i\}$ that are of possible physical interest to us. Therefore, we determine all a_i and ξ_i, as we need them, by

$$a_i = \int \phi_i^*(x)\psi(x)\,dx \quad \text{and} \quad \xi_i = \int \phi_i^*(x)\Xi(x)\,dx, \tag{169}$$

taking as our basic quantities the two "state vectors,"[65] $\psi(x)$ and $\Xi(x)$. Of these, $\psi(x)$ is to be normalized; $\Xi(x)$ is *not* normalized. If only $\psi(x)$ is given, we assume the existence of an *ensemble* of systems with their $\Xi(x)$ distributed in such a way that for *one* orthonormal set $\{\phi_i\}$ the ξ_i are distributed according to (118). Then for *other* sets $\{\phi_j'\}$ the ξ_j' will *automatically* be distributed in the same way, as (118) is invariant under the unitary transformation $\{\phi_i\} \rightarrow \{\phi_j'\}$.

Most everything else said in connection with the theory is deadwood and can be cut out.

We thus end up with a rather simple theory that is easy to comprehend and easy to work with.

Its shortcoming remains that the problems of dealing with composite systems[66] and the problems of the reduction of wavefunctions ψ and Ξ[67] have not yet entirely satisfactorily been solved, and that some paradoxes remain.[66]

[65] See footnote 44 in Section 3.2.
[66] See Appendix K.
[67] See Appendix N and Section 3.7.

CHAPTER 4

THE THEORY OF BOHM AND BUB

4.1. Normalization of Wiener–Siegel's $\Xi(x)$

Bohm and Bub (1966) start with a slight modification of the Wiener–Siegel theory in the simplified form with which we ended up at the end of Section 3.10. The main difference is that in the Bohm–Bub theory $\Xi(x)$ is normalized like $\psi(x)$ is, so that the canonical distribution $\bar{P}$ for Ξ correspondingly differs slightly from the distribution in the Wiener–Siegel theory given by eq. (118).

If there are N eigenfunctions ϕ_i, eq. (118) is a distribution in a $2N$-dimensional real η_λ-space. Suppose we introduce in this space spherical coordinates, so that

$$\prod_{\lambda=1}^{2N} d\eta_\lambda = \mathrm{r}^{2N-1}\, d\mathrm{r}\, d^{2N-1}\Omega, \tag{170}$$

where, as in Section 3.7,

$$\mathrm{r}^2 = \sum_{\lambda=1}^{2N} \eta_\lambda^2 = \sum_{i=1}^{N} |\xi_i|^2 = \int |\Xi(x)|^2\, dx, \tag{171}$$

and where $d^{2N-1}\Omega$ is the $(2N-1)$-dimensional "solid angle" element in Wiener–Siegel's η-space. Then it is easy to give the canonical ξ_i-distribution in the Bohm–Bub theory.

All we have to do is replacing, in the ensemble of hidden-variables states of Wiener–Siegel's canonical distribution, each individual nonnormalized Ξ by its normalized counterpart:

$$\Xi_{\text{Bohm–Bub}} = \Xi_{\text{Wiener–Siegel}}/\mathrm{r}. \tag{172}$$

This normalization of Ξ does not affect at all the workings of the polychotomic algorithm, as it merely changes the ξ_i according to

$$\xi_i^{\text{Bohm–Bub}} = \xi_i^{\text{Wiener–Siegel}}/\mathrm{r}, \tag{173}$$

and, if and only if, $|a_k/\xi_k^{\text{W–S}}| > |a_{\text{NO}(k)}/\xi_{\text{NO}(k)}^{\text{W–S}}|$ by (120), we have

$$|a_k/\xi_k^{\text{B–B}}| > |a_{\text{NO}(k)}/\xi_{\text{NO}(k)}^{\text{B–B}}| \tag{174}$$

on account of (173). Therefore each normalized $\Xi^{\text{B–B}}$ for which $|a_k/\xi_k|$ is larger than all other $|a_i/\xi_i|$ corresponds, in the language we used in Section 3.7, to a red point Ξ_1 in the complex space with coordinates ξ_i, and lies on a radius that is red throughout.

The canonical distribution of the ξ_i in the Bohm–Bub theory

Because of the equivalence of (174) to (120), we can simply take in the Wiener–Siegel real $2N$-dimensional η_λ-space the distribution (118), writing it as

$$d\bar{P} = \pi^{-N} \exp(-\mathrm{r}^2)\, \mathrm{r}^{2N-1}\, d\mathrm{r}\, d^{2N-1}\Omega, \tag{175}$$

and then *collapse it radially in this space onto the unit sphere* $\mathrm{r} = 1$.

This yields the distribution on this sphere given by

$$d\bar{P} = d^{2N-1}\Omega \int_0^\infty \pi^{-N} \exp(-\mathrm{r}^2)\, \mathrm{r}^{2N-1}\, d\mathrm{r}, \tag{176}$$

where the integral is equal to $(N-1)!/(2\cdot\pi^N)$. On the other hand, the area $\int\dots\int d^{2N-1}\Omega$ of the unit sphere imbedded in $2N$-dimensional Euclidian space is $2\cdot\pi^N/(N-1)!$, so that the distribution (176) is properly normalized.

We see from (176) that the Bohm–Bub distribution

$$d\bar{P} = d^{2N-1}\Omega \Big/ \left(\int\dots\int d^{2N-1}\Omega\right) \tag{176'}$$

is a uniform distribution over the unit sphere in Wiener–Siegel's η_λ-space, and that, as long as no functions of r ever enter the theory, Bohm–Bub's Ξ is as good as Wiener–Siegel's. It has the advantage of more symmetry between ψ and Ξ (as also ψ was normalized). The normalization of Ξ is maintained as a function of time, as we know that the Schrödinger equation, which Ξ satisfies according to (158), conserves the integral of the absolute square of the wavefunction.

Proof that in the Bohm–Bub theory the polychotomic algorithm is alright

A disadvantage of the distribution (176′) is that it is much harder to integrate over the $(2N\text{-}1)$-dimensional surface of the unit sphere in $2N$-dimensional space than it is to integrate over the volume of $2N$-dimensional space itself with a weight factor $\exp(-\mathrm{r}^2) = \prod_{\lambda=1}^{2N} \exp(-\eta_\lambda^2)$. Therefore in the proof that the polychotomic algorithm is alright, a direct calculation of $\bar{P}_k$ [by integrating (176′) over the region where in the notation of eqs. (126) we have $X_i > X_k w_i/w_k$ for all $i \neq k$] is much harder than the calculations performed in eqs. (121)–(127), because the X_i in a direct calculation would be subjected to the constraint $\mathrm{r} = 1$, that is, $\sum_{i=1}^{N} X_i = 1$. There is, however, no reason for figuring out the integrals performed under this constraint,[68] because we know already that the application of the algorithm (174) makes the same predictions for the choice of the index k for Bohm–Bub's $\xi_i^{\text{B-B}}$ on the unit sphere, as the algorithm (120) makes for Wiener–Siegel's $\xi_i^{\text{W-S}}$ "radially evaporated" from the unit sphere to a radial distribution given by (175). Therefore, the fact that $\bar{P}^k = w_k$

[68] It can be done! Bohm and Bub (1966a) do it in sec. 5 of their paper, for N as large as 2, but leave the generalization of the proof to $N > 2$ up to the reader.

for Wiener–Siegel's ξ_i^{W-S} distributed by (118) or (175), proves that also $\bar{P}_k = w_k$ for Bohm–Bub's ξ_i^{B-B} distributed by (176′), and we save the work of proving this fact directly.[69]

4.2. Time dependence and reduction of wave packets in the Bohm–Bub theory

The normalization of $\Xi(x)$ alone would make the Bohm–Bub theory only very slightly different from the Wiener–Siegel theory in its simplified form. Therefore most of our conclusions of Chapter 3 would still remain valid.

In Chapter 3 we discussed the time dependence of ψ and Ξ. We concluded:

(1) In absence of unknown external perturbations, ψ would evolve according to a unitary transformation (146)–(150) in Hilbert space, obtainable from a linear Schrödinger equation (145).

(2) *The same must be true for* Ξ,[70] if we want to avoid that a particle would jump around wildly inside the wave packet ψ crudely describing its motion, while the wave packet travels without being influenced by external perturbations. [See Condition Two in Section 3.9, and see eqs. (157)–(158).]

(3) During external influences on the system, the equations for ψ and for Ξ should be modified.

(4) In collisions and the like, the perturbation of Ξ has the effect of making the Ξ distribution in an ensemble of systems canonical. [Here the word "canonical" meant eq. (118) in the Wiener–Siegel theory, and means eq. (176′) in the Bohm–Bub theory.]

(5) The Wiener–Siegel theory tried to approach the problem of "reduction of states after external interactions" by the methods of quantum theory. As explained in Appendix N, this method makes an intrinsically illogical change in the description of a system because such a change is practically efficient, and it justifies this step by showing that for *statistical* purposes the new description is as good as the old one. While this may be satisfactory within a purely statistical theory like quantum theory, it clashes with the (crypto)deterministic philosophy of hidden-variables theory. Comparing the ψ of an observed system before and after an observation, it is clear that the theory somehow should describe the change which quantum theory effectuates by the reduction procedure, or by the shortcut method of von Neumann's projection formalism.[71] As a *rigorous* treatment obviously is impossible without introducing all gory details about the measuring instrument, etc., Bohm and Bub describe this effect *phenomenologically* by adding some extra terms to the Schrödinger equation that have the desired effect. With these terms added, it is to be possible to follow the behavior of the observed object through the measurement without ever introducing a ψ-state that depends upon the apparatus as well as on the object. ***Here is the major deviation of the Bohm–Bub theory from the Wiener–Siegel theory.***

[69] The same may be said differently in the language of Section 3.7. We there proved that w_k was the fraction $\bar{P}_k$ of Ξ points painted red in the distribution (118). Since each radius was either red, or blue, w_k was also the ratio given by the integral of (176′) over the red region in Ξ space, *i.e.* the ratio of the red solid angle, to the sum of the red and blue solid angles. This ratio is equal to the fraction of the area of the unit sphere that is painted red, and therefore is equal to the probability $\bar{P}_k$ in the Bohm–Bub canonical distribution.

[70] Bohm and Bub (1966) seem to have overlooked this fact in their sec. 5.

[71] See the first paragraph of Section 3.6.

Questions about the reduction process left unanswered in the Wiener–Siegel theory (see Appendix N) are thus bypassed. On the other hand, some *paradoxical* predictions which the polychotomic algorithm makes due to Kochen–Specker troubles (see Section 3.7 and Appendix K) remain equally paradoxical in the Bohm–Bub theory, and may require revision or rejection of the general formalism.

The detailed form of the extra terms in the Schrödinger equation for ψ proposed by Bohm and Bub (1966) and the effect of these terms will be discussed in Section 4.3. We will first say something about more general considerations of Tutsch (1969, 1971) about the form which such extra terms should take. This will make the form of the terms adopted by Bohm and Bub more understandable.

We finally may ask whether also to the equation for $\partial\Xi/\partial t$ extra terms should be added. We will discuss this question in Chapter 5 in connection with the discussion of experiments that might confirm or contradict the validity of the polychotomic algorithm.

4.3. Bohm–Bub modification of the Schrödinger equation

Bohm and Bub (1966) assume a Schrödinger equation of the type of

$$\frac{d\psi(x, t)}{dt} = \mathcal{B}(x, t) - \frac{i}{\hbar}\mathcal{H}_{op}\psi(x, t), \tag{177}$$

where the Bohm–Bub terms $\mathcal{B}$ are to take care of what happens during a measurement. In practice, they never discuss the effect of $\mathcal{B}$ and $\mathcal{H}_{op}$ acting simultaneously, considering merely the case where, during the short times that $\mathcal{B}$ does not vanish, $\mathcal{B}$ is far larger than $\mathcal{H}_{op}\psi/\hbar$.

From (177) we obtain, for a measurement described by the set $\{\phi_i(x)\}$ of possible results,

$$\frac{da_i(t)}{dt} = \int dx\; \phi_i^*(x)\mathcal{B}(x, t) - \frac{i}{\hbar}\sum_j \mathcal{H}_{ij}a_j(t). \tag{178}$$

Thence, for $w_i(t) \equiv |a_i(t)|^2$, we obtain

$$\frac{dw_i(t)}{dt} = \mathcal{T}_i(t) - \frac{i}{\hbar}\sum_j \{a_i^*(t)\,\mathcal{H}_{ij}a_j(t) - a_j^*(t)\,\mathcal{H}_{ji}a_i(t)\}, \tag{179}$$

where

$$\mathcal{T}_i(t) = \int dx\,\{a_i^*\phi_i^*(x)\mathcal{B}(x, t) + \mathcal{B}(x, t)^* a_i\phi_i(x)\}. \tag{180}$$

Our task is to find for $\mathcal{B}(x, t)$ an expression which gives $\mathcal{T}_i$ the properties desired.

Instead of just writing down Bohm–Bub's choice for $\mathcal{B}$ and $\mathcal{T}_i$, we shall in an abbreviated form follow the reasoning of Tutsch (1971) leading to a simple form for $\mathcal{T}_i$, and thence we can obtain a corresponding possible form for $\mathcal{B}$ by inspection.

Conditions imposed by Tutsch upon $\mathcal{T}_i$

Obviously, $\mathcal{T}_i(t)$ describing the behavior of an individual system during measurement should depend on the microstate of the individual system [describable by $\psi(x, t)$ and $\Xi(x, t)$] and on the kind of measurement made [describable by the set $\{\phi_i(x)\}$]. As long as the $\phi_n(x)$ appear anyhow as arguments of the functions $\mathcal{T}_n$, we might as well describe the ψ and Ξ by the a_n and ξ_n. Thus we find that the functionals

$$\mathcal{T}_i = \mathcal{T}_i[t, \psi, \Xi, \{\phi_k\}] \tag{181}$$

are functions

$$\mathcal{T}_i = \mathcal{T}_i(t, \{a_n\}, \{\xi_n\}, \{\phi_n\}). \tag{182}$$

The t appears here explicitly because $\mathcal{T}_i$ is to vanish except during the short time while the measurement physically is made. (The *observation* may occur much later, in particular when the apparatus is self-recording.) Of course, if a complete theory of $\mathcal{T}_i$ were possible it should be possible to eliminate explicit t and replace it by such things as spatial overlap integrals of wavefunctions of object and instrument telling when the measurement is made. With the ψ of the object expanded in terms of the $\{\phi_n\}$, such overlap integrals would depend critically upon the *relative phases* of the $\{a_n(t)\}$. However, it obviously is not possible to give a general theory which would give these details which would differ from measurement to measurement. Therefore we take the *phenomenological* approach of slurring over these details, and replacing the explicit time dependence of $\mathcal{T}_i$ by an "on–off" factor $\gamma(t)$, and we postulate that $\mathcal{T}_i$ may be written as a product

$$\mathcal{T}_i = \gamma(t) \cdot G_i. \tag{183}$$

For the sake of simplicity, Tutsch postulates that G_i is a polynomial in the w_n, with coefficients depending upon the u_n, where[72]

$$u_n = |\xi_n|^2. \tag{184}$$

Tutsch claims[73] that the factorizability (183) is explainable when this assumption on G_i being a polynomial is made.

The question may be asked why G_i should not depend on the relative phases of the a_n and the ξ_n. The reason is that G_i is to serve a purpose [see eqs. (186)–(188) below], and this purpose does not depend upon the phases of the a_n and the ξ_n. Therefore for simplicity it is assumed that these phases would appear *only* in the detailed "explanation" of the factor $\gamma(t)$ (which could not be given in a general form).

We thus have found

$$\mathcal{T}_i = \gamma(t) \cdot G_i(\{w_n\}, \{u_n\}). \tag{185}$$

Note that we also have dropped the $\{\phi_n\}$, again for the sake of simplicity. Other people may invent "more general" theories in which the $\mathcal{T}_i$ *might* depend on some of the quantities dropped from (185).

[72] Like $w = |a|^2$ stands for the German word *Wahrscheinlichkeit*, $u = |\xi|^2$ stands for the German word *Unwahrscheinlichkeit* because in eq. (120) it appears in the denominator of $(w_n/u_n)^{1/2}$.

[73] Telephone communication by Tutsch to Belinfante. This point is to be explained in Tutsch (1971a), but did not yet appear in the preprint of it which I saw.

The $\mathcal{T}_i$ now have to serve a purpose, and we shall assume that this purpose is achieved without help from or interference by the terms with $\mathcal{H}_{mn}$ in (179). The purpose of the $\{\mathcal{T}_i\}$ is to effectuate an irreversible transition from initial values

$$a_i(0) = a_i^o, \quad w_i(0) = w_i^o \tag{186}$$

of the coefficients in the expansion of $\psi(x, t)$ and of their absolute squares, to final values

$$|a_i(\infty)| = \delta_{ik}, \quad w_i(\infty) = \delta_{ik}, \tag{187}$$

with

k given by the polychotomic algorithm (120) as the value of i for which (w_i^o/u_i^o) *takes its largest value.* (188)

[As shown in Section 3.7, this guarantees that in ensembles in which the Ξ are canonically distributed there will then be a probability $\bar{P}_n = w_n^o$ that, for $t \to \infty$, $w_n(t)$ will tend to $w_n(\infty) = \delta_{nk}$.]

It now follows at once that the $\mathcal{B}(x, t)$ in (177) cannot be linear in the ψ, so that it must break the validity of the superposition principle for ψ during measurements.[74] The necessity of this nonlinearity follows from the discontinuity of the predictions (187) for the time dependence of the $a_i(t)$ at spots in a_i-space where the algorithm (120) for given $\{\xi_n\}$ changes the value of k. There a *slight* change of initial ψ will cause a *discontinuous* change in ψ at some final large time, in contradiction to the *gradual* change of $\psi(t)$ which a linear Schrödinger equation would suggest, via a slight alteration in the $c_j(0)$ in eq. (150).

There still would be infinitely many ways of choosing $G_j(w_n, u_n)$ in such a way that the goal (186)–(188) is achieved; so Tutsch finally imposes the condition that

the $G_i(w_n, u_n)$ *be a polynomial in* w_n *of the lowest possible degree.* (189)

This will make the G_i unique.

Derivation of the Bohm–Bub formula

Tutsch's derivation of the G_i from the above conditions is algebraic,[75] systematic, and lengthy. We throw out the rigor and proceed intuitively to the final result.

We first establish that, according to (188),

$w_i(t)$ *should be decreasing toward 0 when there is any* $w_j(t)$ *such that* $w_i/u_i < w_j/u_j$, (190)

but that

w_i *should stop decreasing when it reaches the value* $w_i = 0$. (191)

[74] See Tutsch (1971a), below his theorem 1.

[75] For a Hilbert space of $N = 3$ dimensions serving as a model, Tutsch (1969, 1971a) also uses geometrical arguments, considering the trajectory of the point with coordinates $\{w_n(t)\}$ from the point $\{w_n^o\}$ to the corner $\{w_n(\infty) = \delta_{nk}\}$ of the triangle of possible points in the plane $\sum_{i=1}^{N} w_i = 1$. This trajectory would depend upon the initial hidden-variables state $\{u_n^o\}$, which by the algorithm would subdivide this triangle into regions in which the $\{w_n(t)\}$ trajectory would be aiming for any one particular corner $k = 1$ or 2 or 3.

If we introduce the abbreviation

$$R_i = w_i/u_i, \tag{192}$$

the two wishes (190) and (191) are fulfilled if dw_i/dt contains terms like

$$\sum_j Cw_i(R_i - R_j) \tag{193}$$

with positive coefficients C.

Now suppose *all other* w_j have reached the value 0. Then, by $\sum_n w_i = 1$, it follows that w_i has increased to the value 1. Then, w_i should stop increasing.

In the latter case, i must have been the index k of (188), and it follows that all $R_i - R_j = R_k - R_j$ must have been positive. Therefore all terms in (193) were positive, and the only way to stop dw_i/dt from being positive is by making each term in (193) zero for every value of j individually. This suggests putting into every term in (193) an additional factor w_j, and we arrive at

$$\mathcal{T}_i = \gamma \sum_i 2w_i w_j (R_i - R_j), \tag{194}$$

where the $\mathcal{T}_i$, γ, w_n, and R_n all are functions of t, and where, with Tutsch (1971a), we "arbitrarily" threw in a numerical factor 2 in order to comply at the end with the notation of Bohm and Bub (1966).

Equation (194) (apart from the arbitrary numerical factor) is shown by Tutsch (1971a) to be the only polynomial in the w_n that satisfies all conditions listed above. By (192) it is cubic in the $\{w_n\}$, but no term contains any particular factor w_n more than twice.

The Bohm–Bub equation for the a_i and for $\psi(x)$

We see by inspection that eq. (180) with $\mathcal{T}_i$ given by (194) can be solved for $\mathcal{B}(x, t)$ by

$$\mathcal{B}(x, t) = \gamma(t) \sum_i a_i(t)\, \phi_i(t) \sum_j w_j(R_i - R_j). \tag{195}$$

If here we insert eqs. (169) for a_i and ξ_i, after expressing w_j and R_n by

$$w_j = a_j(t)^* a_j(t), \quad R_n = a_n(t)^* a_n(t) \,/\, \xi_n(t)^* \xi_n(t), \tag{196}$$

we get an ugly-looking equation (177) indeed, with integrals in the denominator under a summation sign. According to the above arguments of Tutsch, it is the simplest thing we can do.

The equation looks less ugly when expressed in terms of the $a_n(t)$ and the $w_n(t)$. Therefore Bohm and Bub (1966) give (177) in the form of (178). This gives

$$\frac{da_i}{dt} = \gamma a_i \sum_j w_j(R_i - R_j) - \frac{i}{\hbar} \sum_j \mathcal{H}_{ij} a_j. \tag{197}$$

Conservation of normalization of ψ

We now shall show that the additional terms in the Schrödinger equation do not spoil the conservation of normalization of ψ. That is, we should show

$$\frac{d}{dt}\sum_i |a_i|^2 = 0. \tag{198}$$

For this purpose, proceed from (178) to (179), and sum over i. All that need to be shown is

$$\sum_i \mathcal{T}_i = 0. \tag{199}$$

With $\mathcal{T}_i$ given by (194), the validity of (199), and thence of (198) is trivial.

Nonunitarity of the time dependence of ψ *through a measurement*

Since, by the nonlinearity of $\mathcal{B}(x, t)$ in ψ, $\psi(x, t)$ after a measurement is *no longer linear* in $\psi(x', 0)$, the transformation $\psi(x', 0) \to \psi(x, t)$ obviously is *not a unitary* transformation in this case, even though this transformation is *probability-conserving* on account of (198). [See Tutsch (1969), between his theorem 1 and theorem 2.]

4.4. Application of the Bohm–Bub theory to the collapse of the state vector ψ during a measurement

The effect of the Hamiltonian $\mathcal{H}_{op}$ *during the measurement*

In the following, we shall follow Bohm and Bub in neglecting the effect of the terms with $\mathcal{H}_{ij}$ in the last term of (197).

For an integration of eqs. (197) with (196) from the start till the end of a measurement, we should know also an equation for $d\xi_i/dt$. In Chapter 5 we will investigate a few simple proposals of how ξ_i might be affected by a measurement. The terms in $d\xi_i/dt$ there considered are meant as the analogues of the terms appearing in (197).

One good excuse for neglecting the last term of (197) during a *reproducible* measurement of an observable A is that for such a measurement the Hamiltonian $\mathcal{H}_{op}$ describing the external influence of the apparatus upon the observed object should not contain transition matrix elements between different eigenfunctions ϕ_i of the observed quantity. (The terms with $\mathcal{H}_{ii}$ in first approximation merely affect the energy of the state ϕ_i, if A is an integral of motion. If A is not at least *approximately* an integral of motion, the measurement could at best determine merely an instantaneous value of a rapidly fluctuating quantity, and would not be a very interesting one.)

The hidden variables during the measurement

For an integration of eq. (197) with (196) from the start till the end of a measurement, we should know also an equation for $d\xi_i/dt$. From eq. (158) we gather that $d\xi_i/dt$ should contain at least the terms $-(i/\hbar)\sum_j \mathcal{H}_{ij}\xi_j$, but, if such terms are negligible in (197), they would seem to be negligible for the same reasons in $d\xi_i/dt$ during the measurement. That would leave the hidden variables ξ_i constant during the measurement, and this assumption is made in most of the calculations found in the literature.

It might be conceivable that in the equation for $d\xi_i/dt$ we should add phenomenological terms which are the analogue of the terms with coefficient γ in eq. (197). In Chapter 5 we shall consider this possibility and study the effect which such terms would have.

Is there relaxation of hidden variables during a measurement?

We have postulated in the past that during "collisions" or unsystematic interactions an ensemble of systems originally all in the same hidden-variables state might "relax" to an ensemble in which the hidden variables are distributed canonically. This would mean for each individual system a term in the equation for $d\xi_i/dt$ that would depend upon the happenstances of these collisions, and that for this reason would be unpredictable.

The question may be asked whether such terms (which might well be part of the terms with $\mathcal{H}_{ij}\xi_j$) could be effective during the process of a measurement. If the theory is any good at all, experimental results discussed in Chapter 5 show that relaxation of Ξ occurs indeed in times shorter than it takes to complete a measurement.

If the fluctuating terms in $\partial\Xi/\partial t$ responsible for this are part of $\mathcal{H}_{\text{op}}\Xi$, one would expect that similar terms in $\mathcal{H}_{\text{op}}\psi$ would affect $\partial\psi/\partial t$, and would randomize also ψ, changing pure states into mixtures. The measurement results (Chapter 5) which show the rapid relaxation of Ξ, however, are explainable on the basis of the initial $\psi(x, t)$ resulting by (145) from the system's preparation, and do not show any randomization of ψ.

This suggests that in $\partial\Xi/\partial t$ during collisions some randomizing terms are at work which are absent in $\partial\psi/\partial t$. Our present lack of understanding of the nature of these terms is the reason why so far we cannot reliably estimate how fast the Ξ distribution will relax.

In Section 3.9 we stressed that $\partial\Xi/\partial t$ and $\partial\psi/\partial t$ should be correlated for meeting Condition Two. If, indeed, in collisions this correlation would be lost, Condition Two cannot be met, and particles would emerge from collisions in a way statistically described by ψ, but haphazard in their microstates. Thus *in interactions between particles we would lose the (crypto)-determinism* which Wiener and Siegel postulated in Condition Two and *which was the whole purpose of their hidden-variables theory.*

Collapse of the ψ-function during the measurement

We shall now first investigate how much can be said about the evaluation of $a_i(t)$ according to eq. (197) without specifying the time dependence of the $\xi_i(t)$. Next, we shall consider the $a_i(t)$ under the assumption that the $\xi_i(t)$ are constant during the measurement. In both cases we neglect all $\mathcal{H}_{ij}$-terms, and we neglect any relaxation effects. A case with nonzero $d\xi_i/dt$ will be discussed in Chapter 5.

Case of arbitrary dependence of ξ_i on time

Equation (179) with $\mathcal{T}_i$ from (194) reads in this approximation

$$\frac{dw_i}{dt} = \gamma \sum_j 2w_i w_j \left(\frac{w_i}{u_i} - \frac{w_j}{u_j}\right). \tag{200}$$

Thence

$$\frac{d}{dt} \ln |w_i| = 2\gamma \sum_j w_j(R_i - R_j). \tag{201}$$

Now assume that *initially* R_k is larger than all other R_j:

$$R_k^o > R_j^o \quad \text{for} \quad j \neq k. \tag{202}$$

It follows that $\ln |w_k|$ must be increasing as long as $\gamma \neq 0$ and not all other w_j have decreased to zero. Then, also $|w_k|$ must be increasing. But, since $\Sigma w_i = 1$ is constant, then, some other w_j must be decreasing.

If, at some time t, $R_i > R_n$, we find from

$$\frac{d}{dt}(\ln |w_i| - \ln |w_n|) = 2\gamma \sum_j w_j(R_i - R_n) = 2\gamma(R_i - R_n) \tag{203}$$

that $\ln |w_i|$ must be increasing at that time t more rapidly, or decrease more slowly, than $\ln |w_n|$. Thus the order of the ratios R_i tells us the order in which the $\ln |w_i|$ change with time, and if R_l has the lowest value among all the R_i, $\ln |w_l|$ must be decreasing faster than any other $\ln |w_n|$. If w_k was still increasing, we know that w_l must be decreasing, for, if it were not, all other $\ln |w_n|$ would be increasing even faster than $\ln |w_l|$, and $\sum_i w_i$ could not be conserved.

There are exceptional initial conditions for which two of the w_i may end up with equal R_i.

The most likely final state is one in which $w_i = \delta_{ik}$ with k the index of the largest R_i. It is obvious from (200) that this state, when reached, is stable.

For more definite conclusions, we must assume something about the $\xi_i(t)$.

Case of constant $\xi_i(t)$

When the u_i are simply constants, we can draw more explicit conclusions. Then eq. (203) becomes

$$\frac{d}{dt}(\ln |R_i| - \ln |R_n|) = 2\gamma\,(R_i - R_n) = \frac{d}{dt} \ln \left|\frac{R_i}{R_n}\right|, \tag{204}$$

so that an initial difference between R_i and R_n creates a tendency of $\ln |R_i| - \ln |R_n|$ to increase to $+\infty$ or to decrease to $-\infty$, until either $R_i \to 0$ or $R_n \to 0$. (We cannot have $R_j \to \infty$, as $w_j \leqslant 1$, and $u_j =$ constant.) This means that, if any two R_j are ever different, the one is going to crowd out the other, and, with the u_j constant, also the corresponding w_j is crowded out. This makes all w_j shrink to zero except the one with *initially* the largest R_j, for which then $w_j \to 1$ in order to maintain $\sum_n w_n = 1$. We thus achieve our goal (187)–(188), and not only does this explain, by the reasoning at the end of Section 4.1 together with the calculations in Section 3.7, that the probability $\bar{P}_k$ for ending up with $w_n(\infty) = \delta_{nk}$ for some particular value of k is equal to $w_k^o = |a_k(0)|^2$, but it also shows by $|a_k(\infty)| = \delta_{nk}$ that in this case

$$\psi(x, t \to \text{“}\infty\text{”}) = \sum a_n(\text{“}\infty\text{”})\phi_n(x) = \phi_k(x)\cdot\exp(i\ldots), \quad \text{or} \quad |\psi(x, t \to \text{“}\infty\text{”})| \to |\phi_k(x)|.$$

[Here, "∞" stands for a time T for which $\gamma(t)T \ggg 1$ for most times t between 0 and T, while $\gamma(t) \to 0$ as $t \to T$.]

We thus have shown the collapsing of $\psi(x, t)$ during the measurement.

Exceptional case

There are exceptional cases (of probability zero) in which several R_i would initially be *exactly* equal among each other. As they would differ from other R_j, they either are crowded out by those other R_j, or, if they are not, they crowd out the other R_j, so these equal R_i might survive. The corresponding w_i then could end up having mutual ratios equal to the corresponding mutual ratios of the initial (and assumedly not changing) u_i. Such final states would be metastable during the time that $\gamma(t)$ remains different from zero.

4.5. Application to quantum electrodynamics and to relativistic field theory

We never have stated that $\mathcal{H}_{\text{op}}$ could not include the Hamiltonian of the electromagnetic field as discussed in Appendix H, and that the x on which ψ and Ξ depend could not include the radiation field coordinates q_f. Therefore the Bohm–Bub theory should be applicable to quantum electrodynamics as well as to electron wave mechanics; at least, in the nonrelativistic approximation.

The theory probably can be generalized to a relativistic theory including pair creation and annihilations, etc., by applying its principles to the "second quantized" theory of electrons, or, in general, relativistic quantum field theory, in which case the variables x in $\psi(x)$ and $\Xi(x)$ are taken to be the "occupation numbers" of quantum field theory.

Thus the Bohm–Bub theory seems to be much more general than the old Bohm theory of Chapter 1.

All difficulties that beset quantum field theory due to infinite self interactions and the like, should, of course, be expected to beset also the generalized Bohm–Bub theory.

Of course, the specific mechanism for the collapse of the wavefunction presented by eqs. (197) is not relativistically covariant. However, it is merely meant as a phenomenological equation describing the influence of a measurement, and there exist preferred frames of reference in which the apparatus is at rest, and in which eq. (197) may be as good or as bad an approximation as it was before. Maybe it is for some reasons not quite as good as before, if some parts of the system observed pass the instrument at speeds that are not non-relativistic. In that case, one should try to generalize (197) to a more adequate equation.

There is, however, no *a priori* reason why something similar could not be done. Moreover, the theory is wide open for improvements.

As mentioned before, some improvements are badly needed, for getting rid (if possible) of paradoxes of the type of the ones discussed in Appendix K.

4.6. Possibility of experimental verification

We explained in Section 4.4 that, for conserving the (crypto)deterministic nature of the theory, it is necessary to assume that in meaningful measurements little or no "relaxation of hidden-variables distribution" should take place. That is, we must assume that during the measurement not only ψ, but also Ξ satisfy a deterministic equation. This equation may keep Ξ constant during the measurement, as we assumed in the above discussion of the collapse, or one might suggest some different behavior of Ξ, as we shall try in Chapter 5.

When a measurement has been completed and has given the result A_k, we know not only that we have ended up with $\psi = \phi_k$ but we also know that we have started with $\xi_k/a_k >$ all ξ_i/a_i for $i \neq k$. Assuming any deterministic formula for $d\xi_i/dt$, we know also something in the form of an inequality about the *final* ξ_n, that is, about the final hidden variables state $\Xi(x)$.

If we repeat the same experiment often enough, we work with an ensemble of systems. Knowing its preparation, we would know its initial ψ, and we would assume a canonical initial distribution of its Ξ. *After* the measurement, however, we know some inequalities about Ξ, so, we end up with a noncanonical "biased" Ξ distribution.

If, now, we use this as the initial state for a new experiment, we do no longer have the initial canonical hidden-variables distribution which in Section 3.7 we used for deriving the quantum-mechanical probability distribution from the polychotomic algorithm. That is, the second experiment, performed on all systems in our subensemble for which we found $A = A_k$, should show a probability distribution of results differing from what quantum theory would have predicted on the basis of an initial wave function $\psi = \phi_k$ alone.

Moreover, it can be calculated what the biased Ξ distribution will predict instead. As we will see in Chapter 5, these predictions are insensitive for the special form (197) or (200) of the phenomenological formula used for describing the experiment. Those equations serve merely as a mechanism to reach the goal (187)–(188) of making the polychotomic algorithm valid for the outcome of the experiment. The result of the calculation of what statistics of results will be found by a second measurement rapidly following a first one (or even a third measurement rapidly following a second one) depends merely on the polychotomic algorithm itself and the assumption made about what is the initial state before the first experiment. (Often a "mixed" state.) If we do not assume Ξ to be constant during the measurement, and we write down for $d\xi_i/dt$ an equation which is a variation of eq. (197) for da_i/dt, we have to "integrate" that equation qualitatively like we did for eq. (200) by the reasoning around eqs. (201)–(204), and the statistics predicted for the results of the last measurement will depend on the result of this "integration" of the equation for $d\Xi/dt$.

In Chapter 5 we will execute this program for measurements performed by Papaliolios (1967) and for some similar and slightly simpler experiments that have not yet been performed.

4.7. Evaluation of the Bohm–Bub theory

Among the few hidden-variables theories of the first kind so far in existence, the Bohm–Bub theory seems to be the most promising one. It bypasses difficulties with determinism in the theory of measurements by introducting for the measured object a phenomenological description which remains valid straight through the entire measurement instead of oscillating between descriptions of the object alone and the object as part of a composite system together with the apparatus.

The backbone of the Bohm–Bub theory is the same as of the Wiener–Siegel theory, namely the existence of a (crypto)deterministically behaving Ξ (or set $\{\xi_i\}$) which determines results of measurements by the polychotomic algorithm. As the latter causes a biased Ξ distribution after each measurement, experimental verification or falsification of the theory is possible, but such experiments treat the Wiener–Siegel theory and the Bohm–Bub theory as two variations of one and the same theory. Such an experimental investigation should come first, before we accept or reject any theory based upon the polychotomic algorithm. If results of experiments similar to the one of Papaliolios (1967) or the other one proposed in Chapter 5 remain negative, even with better precautions against relaxation of the Ξ distribution, we may reject all these theories as needless complications of pure quantum theory. (See Part I, Section 1.3.) On the other hand, if we find reasons why we would like *not* to discard these hidden-variables theories of the first kind, we still would have to find a way out of the paradoxes discussed in Appendix K, which are a left-over from the Kochen–Specker troubles (Section 3.7), after the Gleason troubles were removed for ordinary degeneracy cases by Tutsch's rule [eq. (119)] or by the procedure [eq. (S-15)] of Wiener and Siegel (1953).

CHAPTER 5

EXPERIMENTS FOR VERIFYING OR FALSIFYING THE POLYCHOTOMIC ALGORITHM

IN SECTION 4.6 we explained how one could, in principle, verify the polychotomic algorithm if it were true. The possibility of making such an experimental verification of hidden-variables theory has been known for a long time. Wigner (1970) claims in a footnote that already von Neumann was convinced that hidden variables, if they existed and did not quickly change value after measurement, could be partially fixed by successive measurements. Bohm and Bub (1966a) stress the same idea in their section 7, but at the same time warn that relaxation may interfere with such experiments. [See also Siegel (1966), page 154.] We have seen in Section 4.4, however, that we cannot assume that this relaxation takes place while a measurement is made if we do not want to lose completely the cryptodeterministic nature of hidden-variables theory.

The first experimenter who has taken these suggestions serious has been Papaliolios (1967).

5.1. The nature of the experiments of Papaliolios and of a possible variation of it

The hidden-variable experiments proposed by von Neumann [see Wigner (1970)] were based upon spin phenomena of electrons, making use of the fact that an electron, for any direction, can choose between two values of its spin in that direction, and that the hidden variables should determine this choice.

Experiments of this kind are difficult, and therefore it soon was suggested to use, instead, polarization effects of photons. Early experiments of this type[76] were done for check on a hidden-variables theory of the *second* kind.[77] Papaliolios decided to use polarization effects of photons for a verification of the Ξ-bias effects predicted by hidden-variables theories of the *first* kind. In principle his experiments are far simpler than anything suggested before, as they amount to attempts to find deviations from Malus's $\cos^2$ law.

Unpolarized light of weak intensity is incident upon thin polarizers placed immediately behind each other. Each polarizer, from the point of view of classical electromagnetic theory, decomposes the wave into two perpendicular linear polarizations of which the one (polarized in the "preferred" direction) traverses the polarizer and the other one is absorbed.

[76] See Wu (1950).

[77] See Furry (1936), Bohm and Aharonov (1957), and Part III of this book.

We may regard this as a measurement with two possible outcomes where the measurement is reproducible for the one result and irreproducible for the other result. (Or we could say that we make here a reproducible measurement which distinguishes the preferred polarization from the other one followed by an absorption of that other one immediately after the measurement.)

The first filter serves as a polarizer; Papaliolios used a second filter to make a measurement on this polarized light and a third filter as a follow-up measurement.

In the calculations of what is happening according to the polychotomic algorithm, however, we will see that one could do experiments of this kind also with merely two filters. In that case, we follow through both filters a light wave with given Ξ-polarization but originally arbitrary polarization. The first filter will polarize it or reject it, but how many photons with this Ξ go through will depend upon the angle between the Ξ polarization and the filter polarization. Thus the light incident upon the second filter has not only a given ψ-polarization obtained in the first filter, but also has already a biased Ξ-polarization, due to the smaller or larger interval of incident ψ-polarizations that carry different Ξ-polarizations through filter 1. The calculations for this two-filter experiment in Sections 5.3–5.5 and Appendixes P and Q precede the calculations for Papaliolios's three-filter experiment in Section 5.6 and Appendix R. An interesting detail is that the results predicted by the theory for the two-filter experiment depend upon *how* unpolarized the light is that is incident upon filter 1. As explained in Appendix O, we may distinguish between "circular unpolarization," "linear unpolarization," and "elliptical unpolarization."

We will make the calculations first under the assumption of constant hidden variables (compare Section 4.4), and, for the Papaliolios experiment, also for the case that the Ξ distribution relaxed between filters 1 and 2 (which are separated) but not between filters 2 and 3 (which are on top of each other). In Section 5.7 we will also consider other assumptions about the time dependence of Ξ. Finally, in Section 5-8 we shall discuss some experimental details of the Papaliolios experiment and the results.

5.2. Vector addition of polarizations

In the classical theory of light it is well known how we may decompose transverse polarizations along arbitrary orthogonal axes in the plane perpendicular to the propagation of the wave. This, together with the fact that the intensity of the light in the classical theory is proportional to the square of the amplitude, leads at once to Malus's law that the intensity transmitted by a second polarization filter is proportional to the square of the cosine of the angle between the preferred polarization directions of the two filters.

Before we can apply hidden variables theory with its ψ and Ξ to this problem, we first must understand this problem from the point of view of pure quantum electrodynamics, as discussed in Appendix H. The implications of the results of Appendix H for the question of polarizations are explained in Appendix O.

The result is that the wavefunctions ψ are decomposed like the classical lightwave polarizations are. Thus, if ψ_1 and ψ_2 are two photon waves, for photons *traveling in the same direction and with the same wavelength*, but polarized in orthogonal directions, then ψ_1 is wave-mechanically orthogonal to ψ_2. The most general polarization of such a photon wave is then given by

$$\psi = c_1\psi_1 + c_2\psi_2. \tag{205}$$

Then, if ψ_1 corresponds to a polarization $\mathbf{e}_1$ and ψ_2 to a polarization $\mathbf{e}_2$, the wavefunction (205) corresponds to a wave polarized along

$$\mathbf{e} = c_1\mathbf{e}_1 + c_2\mathbf{e}_2. \tag{206}$$

In the Appendix we use (for given wave vector $\mathbf{k}$) the two complex circular-polarization unit vectors $\mathbf{e}_1 = \mathbf{e}_\mathbf{k}^+$ and $\mathbf{e}_2 = \mathbf{e}_\mathbf{k}^-$ defined by eqs. (H-11). By linear combinations similar to (206) we derive from them linear polarizations. For instance, for light traveling in the $+z$-direction, polarization unit vectors along the x- and y-axes are then given by

$$\mathbf{e}^x = \hat{x} = \sqrt{\tfrac{1}{2}}\,(\mathbf{e}^- + \mathbf{e}^+), \quad \mathbf{e}^y = \hat{y} = i\sqrt{\tfrac{1}{2}}\,(\mathbf{e}^- - \mathbf{e}^+). \tag{207}$$

Polarizations along rotated axes

$$\mathbf{e}^{x'} = \hat{x}' = \hat{x}\cos\theta + \hat{y}\sin\theta, \quad \mathbf{e}^{y'} = \hat{y}' = -\hat{x}\sin\theta + \hat{y}\cos\theta \tag{208}$$

are given in terms of the $\mathbf{e}_\mathbf{k}^\pm$ of circular polarization by

$$\mathbf{e}^{x'} = \sqrt{\tfrac{1}{2}}\,(e^{i\theta}\mathbf{e}^- + e^{-i\theta}\mathbf{e}^+), \quad \mathbf{e}^{y'} = i\sqrt{\tfrac{1}{2}}\,(e^{i\theta}\mathbf{e}^- - e^{-i\theta}\mathbf{e}^+). \tag{209}$$

Thus what in Appendix O is proved for linear combinations of circular polarizations we may apply directly to linear polarizations as well, because they, too, are linear combinations of circular polarizations.

The general ψ-wave for a single photon in a lightwave of given wavelength traveling in the z-direction therefore may be written as

$$\psi = a_x\phi_x + a_y\phi_y, \tag{210}$$

if ϕ_x and ϕ_y correspond to one photon in waves polarized in the x- and in the y-direction. For keeping the lightwave linearly polarized and the ψ-wave normalized to one photon, we would choose a_x/a_y real and $|a_x|^2 + |a_y|^2 = 1$, so that

$$a_x = e^{i\alpha}\cos\chi, \quad a_y = e^{i\alpha}\sin\chi \tag{211}$$

with $e^{i\alpha}$ the kind of phase factor which always may occur in any ψ-wave, while we lose no generality by limiting α and χ to the intervals $0 \leq \alpha < 2\pi$, $-\pi/2 < \chi \leq +\pi/2$.

The Ξ-wave was always expanded in terms of the same $\{\phi_i\}$ as the ψ-wave. So we put

$$\Xi = \xi_x\phi_x + \xi_y\phi_y, \tag{212}$$

where in the Bohm–Bub theory we keep Ξ normalized by

$$|\xi_x|^2 + |\xi_y|^2 = 1, \tag{213}$$

but otherwise the ξ_x and ξ_y are arbitrary. We could put

$$\begin{gathered} \xi_x = e^{i\tilde{\theta}_x}\cos\tilde{\chi}, \quad \xi_y = e^{i\tilde{\theta}_y}\sin\tilde{\chi}, \\ 0 \leq \tilde{\theta}_x < 2\pi, \quad 0 \leq \tilde{\theta}_y < 2\pi, \quad \text{and} \quad 0 \leq \tilde{\chi} \leq \pi/2. \end{gathered} \tag{214}$$

The tilde ~ will distinguish phase angles and other quantities for Ξ from similar quantities for ψ.

Totally unpolarized light

The Ξ-wave (212) with the ξ_i of (214) would be a normalized elliptically polarized wave. If light is *totally* unpolarized, also ψ could be of this form, that is, (210) with

$$a_x = e^{i\theta_x}\cos\chi, \quad a_y = e^{i\theta_y}\sin\chi, \qquad 0 \leqslant \theta_x < 2\pi, \quad 0 \leqslant \theta_y < 2\pi, \quad 0 \leqslant \chi \leqslant \pi/2. \tag{215}$$

Light that contains photons described by (215) is called "elliptically unpolarized" in Appendix O.

5.3. Omitting the normalizations

In Section 4.1 verifying the polychotomic algorithm in the Bohm–Bub theory for a canonical Ξ-distribution, we found it expedient to drop the normalization of Ξ for being able to use the calculations of Section 3.7 in the Wiener–Siegel theory for calculating the probability that for a certain result ϕ_k of the measurement the $|a_k/\xi_k|$ could be larger than all other $|a_i/\xi_i|$. The absolute magnitude of Ξ does not affect the probability.

The latter is, of course, also true for ψ. If we multiply all $\{a_i\}$ by a constant factor, this does not affect the fraction of the Ξ's for which the $|\xi_k/a_k|$ is larger than all $|\xi_i/a_i|$ for $i \neq k$ are.

In the calculations for the experiments here considered, this same trick is useful, and here in case of elliptically unpolarized light it is useful for ψ as well as for Ξ. We then would replace (215) and (214) by

$$\left.\begin{aligned} a_x &= \exp(i\theta_x)\, r\cos\chi = a_{xR}+i\,a_{xI}, \\ a_y &= \exp(i\theta_y)\, r\sin\chi = a_{yR}+i\,a_{yI}, \end{aligned}\right\} \tag{216}$$

$$\left.\begin{aligned} \xi_x &= \exp(i\tilde\theta_x)\, \tilde r\cos\tilde\chi = \xi_{xR}+i\xi_{xI}, \\ \xi_y &= \exp(i\tilde\theta_y)\, \tilde r\sin\tilde\chi = \xi_{yR}+i\,\xi_{yI}. \end{aligned}\right\} \tag{217}$$

We then, in the totally unpolarized case, would expect the ψ-state to be a mixture with the same $\{a_i\}$ distribution as the Wiener–Siegel canonical $\{\xi_i\}$ distribution. That is, the probability for a certain Ξ with components ξ_i given by (217) would be

$$d\tilde P = \exp(-|\xi_x|^2-|\xi_y|^2)\, d\xi_{xR}\, d\xi_{xI}\, d\xi_{yR}\, d\xi_{yI}/\pi^2, \tag{218}$$

or, if we write $\tilde x = \tilde r\cos\tilde\chi$, $\tilde y = \tilde r\sin\tilde\chi$,

$$\begin{aligned} d\tilde P &= \exp(-\tilde r^2)\cdot\tilde x\, d\tilde x\, d\tilde\theta_x\cdot\tilde y\, d\tilde y\, d\tilde\theta_y/\pi^2 = \exp(-\tilde r^2)\cdot\tilde x\tilde y\cdot\tilde r\, d\tilde r\, d\tilde\chi\cdot d\tilde\theta_x\, d\tilde\theta_y/\pi^2 \\ &= \exp(-\tilde r^2)\,\tilde r^3\, d\tilde r\cdot\cos\tilde\chi\sin\tilde\chi\, d\tilde\chi\cdot d\tilde\theta_x\, d\tilde\theta_y/\pi^2 \end{aligned} \tag{219}$$

and, for total (elliptical) unpolarization, we similarly would have an ensemble of $\{a_i\}$ with a probability distribution

$$dP = \exp(-r^2)\, r^3\, dr\cdot\cos\chi\sin\chi\, d\chi\cdot d\theta_x\, d\theta_y/\pi^2. \tag{220}$$

If, on the other hand, the incident light is *linearly* unpolarized, the $\{a_i\}$ would be given by

(211) with random α and χ. In this case we will have

$$dP = d\alpha\, d\chi/2\pi^2 \qquad (0 < \alpha < 2\pi, \quad -\pi/2 < \chi < +\pi/2). \tag{221}$$

Finally, if the incident light is *circularly* unpolarized, the incident waves are of the form (210) with

$$a_x = \sqrt{\tfrac{1}{2}}\, e^{i\alpha}, \quad a_y = \pm i\sqrt{\tfrac{1}{2}}\, e^{i\alpha}, \tag{222}$$

with either sign $\pm$ and with arbitrary α, so that

$$dP_{\pm} = d\alpha/4\pi, \quad (0 < \alpha < 2\pi). \tag{223}$$

When we work with (219) or (220) we find it useful to use (*with* as well as *without* tilde)

$$\left.\begin{aligned} \omega &= r^2, \quad v = 1-(1+\omega)\exp(-\omega) \quad (0 < v < 1 \quad \text{for} \quad 0 < r < \infty), \\ dv &= \omega e^{-\omega}\, d\omega = 2r^3 \exp(-r^2)\, dr. \end{aligned}\right\} \tag{224}$$

$$\left.\begin{aligned} u &= \sin^2\chi = \tfrac{1}{2}(1-\cos 2\chi), \quad \lambda = 1-u = \cos^2\chi \quad (0 < u < 1), \\ du &= 2\sin\chi\cos\chi\, d\chi, \end{aligned}\right\} \tag{225}$$

$$dP = du\cdot dv\cdot(d\theta_x/2\pi)\cdot(d\theta_y/2\pi), \quad \text{so} \quad \int dP = 1. \tag{226}$$

5.4. The Ξ bias caused by the first filter

We shall now calculate the Ξ bias caused when unpolarized light is polarized by a filter. We must distinguish here the three cases of unpolarized incident light with distributions (220) or (221) or (223). We assume that the filter transmits light polarized in the x-direction.

For given ξ_x and ξ_y we use the polychotomic algorithm for finding for each possible a_x and a_y of the incident light, whether or not the "measurement" made by the filter will have the result $e^{i\beta}\phi_x$ [that is when $|a_x/\xi_x| > |a_y/\xi_y|$], or whether it will have the result $e^{i\beta}\phi_y$ [that is when $|a_y/\xi_y| > |a_x/\xi_x|$]. (Here β is unimportant.) So, the light with this Ξ transmitted as a wave $\psi = (e^{i\beta}\phi_x$ behind the filter) is the light which had

$$|a_x/a_y| > |\xi_x/\xi_y|. \tag{227}$$

Then, using dP chosen from (220) or (221) or (223), we calculate the probability P_x that this would happen.

Then the canonical distribution $d\tilde{P}$ of (219) incident upon the filter yields a transmitted biased distribution $d\tilde{P}' = P_x\cdot d\tilde{P}$ of the hidden variables.

The calculations are performed in Appendix P. The results are:

For elliptically unpolarized incident light the polarized light emerging from the filter has its Ξ distribution biased by $P_x = \sin^2\tilde{\chi} = \tilde{u}$, so that

$$d\tilde{P}' = \tilde{u}\, d\tilde{u}\, d\tilde{v}\, (d\tilde{\theta}_x/2\pi)\, (d\tilde{\theta}_y/2\pi). \tag{228}$$

For linearly unpolarized incident light we find $P_x = 2\tilde{\chi}/\pi$, and the polarized light emerging from the filter has its Ξ distribution biased according to

$$d\tilde{P}' = (4/\pi)\,\tilde{\chi}\sin\tilde{\chi}\cos\tilde{\chi}\, d\tilde{\chi}\, d\tilde{v}\, (d\tilde{\theta}_x/2\pi)\, (d\tilde{\theta}_y/2\pi). \tag{229}$$

For *circularly unpolarized incident light* we find for P_x the step function $P_x = \Theta(\tilde{\chi} - \frac{1}{4}\pi)$. [See eqs. (P-5)–(P-6).] By (219) and (224)–(225) this gives the biased Ξ distribution

$$d\tilde{P}' = \Theta(\tilde{u} - \tfrac{1}{2}) \cdot d\tilde{u}\, d\tilde{v}\, (d\tilde{\theta}_x/2\pi)(d\tilde{\theta}_y/2\pi). \tag{230}$$

In all cases, but for a possible phase factor, the ψ-wave behind the filter is given by $\psi = \phi_x$ that is,

$$|a_x| = 1, \quad a_y = 0. \tag{231}$$

Intensity transmitted by the first filter

The integrals of (228), of (229) and of (230) over $0 < \tilde{\chi} < \frac{1}{2}\pi$ (that is, $0 < \tilde{u} < 1$), $0 < \tilde{v} < 1$, $0 < \tilde{\theta}_x < 2\pi$, and $0 < \tilde{\theta}_y < 2\pi$, all give the total probability of transmission

$$\mathcal{I}_1 = \tfrac{1}{2}. \tag{232}$$

[For (229) this is seen by taking the average of the integral shown and the integral with the dummy $\tilde{\chi}$ replaced by $(\frac{1}{2}\pi - \tilde{\chi})$.]

5.5. The light transmitted by the second polarization filter

We will see from the results that the difference in angle dependence of the transmitted light, between pure quantum theory or ordinary optics on the one hand, and the theory of the polychotomic algorithm for constant hidden variables on the other hand, is most pronounced when the angle is close to 90°. We therefore will assume that the angle between the polarization axes of the two filters is $\vartheta = \frac{1}{2}\pi - \varepsilon$. More precisely, we shall assume that the primed unit vectors $\hat{x}'$ and $\hat{y}'$ parallel and perpendicular to the polarization direction of the second filter are related to the $\hat{x}$ and $\hat{y}$ of Filter One by

$$\hat{x}' = \hat{x}\sin\varepsilon + \hat{y}\cos\varepsilon, \quad \hat{y}' = -\hat{x}\cos\varepsilon + \hat{y}\sin\varepsilon. \tag{233}$$

The ϕ_x' and ϕ_y' between which the second filter distinguishes are obtained from the ϕ_x and ϕ_y in (210) by the same transformation (233), so that, conversely,

$$\phi_x = \phi_x'\sin\varepsilon - \phi_y'\cos\varepsilon, \quad \phi_y = \phi_x'\cos\varepsilon + \phi_y'\sin\varepsilon. \tag{234}$$

The waves $\psi = \phi_x e^{i\beta}$ and Ξ of (212) incident upon Filter Two must be decomposed along the ϕ_x' and ϕ_y' with coefficients

$$\begin{aligned} a_x' &= e^{i\beta}\sin\varepsilon, & a_y' &= -e^{i\beta}\cos\varepsilon, \\ \xi_x' &= \xi_x\sin\varepsilon + \xi_y\cos\varepsilon, & \xi_y' &= -\xi_x\cos\varepsilon + \xi_y\sin\varepsilon, \end{aligned} \tag{235}$$

before we apply the algorithm. The algorithm then gives as condition for transmission

$$|a_x'/\xi_x'| > |a_y'/\xi_y'|, \tag{236}$$

which by (235) and (217) may be expressed in terms of ε, $\tilde{r}$, $\tilde{\chi}$, $\tilde{\theta}_x$, and $\tilde{\theta}_y$.

We must now integrate the Ξ distribution given by (228) or (229) or (230) under the condition (236) for finding (for fixed ε) what is the intensity of the light transmitted by the second filter, with a new ψ now equal to ϕ_x'.

If there is a third filter behind the second one, it is not sufficient to know what is the total intensity of this transmitted light. We must know how much of it comes with each Ξ-vector. This is just equal to (228) or (229) or (230) whenever condition (236) is satisfied, and is zero otherwise.

The intensities $\mathcal{I}_2$ thus calculated are absolute intensities, counting the intensity before Filter One as unity. If we want the fraction of light transmitted by the second filter, we must divide by the intensity $\mathcal{I}_1$ leaving the first filter.

Results

The calculations are performed in Appendix Q. We find different results depending upon the unpolarization of the incident light. We calculate differently for $\varepsilon < 45°$ and for $\varepsilon > 45°$ ($\vartheta < 45°$). The sum of the two results for two angles ε that add up to 90° is always the total intensity $\mathcal{I}_1 = \frac{1}{2}$ passed through Filter One:

$$\mathcal{I}_2(\varepsilon)+\mathcal{I}_2(\tfrac{1}{2}\pi-\varepsilon) = \mathcal{I}_1. \tag{236a}$$

For $\varepsilon < 45°$, the results are eq. (Q-10) for elliptical unpolarization, eq. (Q-16) for linear unpolarization, and eqs. (Q-21) and (Q-21′) for circular unpolarization. In the latter case, the second filter transmits nothing for $\varepsilon < 22\frac{1}{2}°$ and all of $\mathcal{I}_1$ for $\vartheta < 22\frac{1}{2}°$.

Numerical values for $\mathcal{I}_2/\mathcal{I}_1$ fo each of the three unpolarization cases computed on

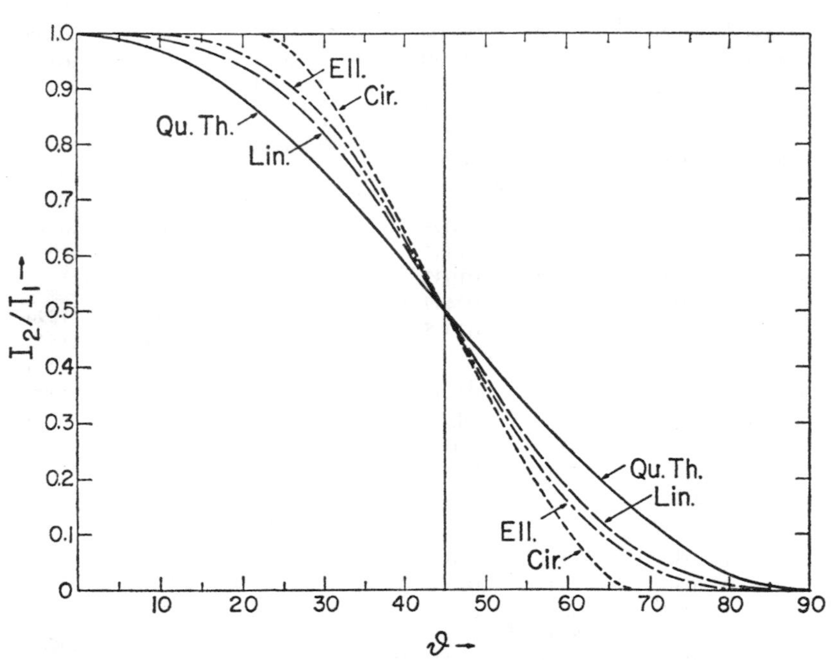

FIG. 5.1. Percentage of light transmitted by a second polarization filter with its polarization direction at an angle ϑ to the polarization of Filter One for constant hidden variables and various kinds of unpolarized light incident upon Filter One, and also for hidden variables relaxed before entering Filter Two (quantum-theoretical prediction).

a Hewlett–Packard desk computer have been listed in Table 5.1 for $\varepsilon < 45°$. For larger ε, use eq. (236a).

The last column in the table gives the $\cos^2\vartheta = \sin^2\varepsilon$ law of Malus, which quantum theory and classical optics predict.

Figure 5.1 shows $\mathcal{J}_2/\mathcal{J}_1$ as a function of ϑ for $0 < \vartheta < 90°$, for the various kinds of unpolarized incident light for constant hidden variables, as well as according to quantum theory. The latter, of course, would follow from the polychotomic algorithm if the hidden variables would completely relax in the back part of Filter One after its front part would have made the measurement.

A case of "known" hidden variables if the latter are constant

It might be interesting to consider the case of circularly unpolarized light incident upon Filter One, and an angle of 67° 20′ between the polarizations of the two filters, so that

TABLE 5.1. INTENSITY $\mathcal{J}_2$ BEHIND A SECOND POLARIZATION FILTER, IN UNITS $\mathcal{J}_1$ = INTENSITY INCIDENT UPON IT FROM FILTER 1, WHEN THE ANGLE BETWEEN THE POLARIZATION DIRECTIONS OF THE FILTERS IS $(90° - \varepsilon)$, FOR $\varepsilon \leq 45°$

(For $\varepsilon \geq 45°$, use $\mathcal{J}_2(\varepsilon)/\mathcal{J}_1 + \mathcal{J}_2(90° - \varepsilon)/\mathcal{J}_1 = 1$.) By ell, lin, cir, we indicate the type of unpolarization of the light incident upon the first filter (see Appendix O).

$\varepsilon =$ (deg.)	$\mathcal{J}_2^{ell}/\mathcal{J}_1 =$	$\mathcal{J}_2^{lin}/\mathcal{J}_1 =$	$\mathcal{J}_2^{cir}/\mathcal{J}_1 =$	$\mathcal{J}_2^{qu}/\mathcal{J}_1 =$	$\varepsilon =$ (deg.)	$\mathcal{J}_2^{ell}/\mathcal{J}_1 =$	$\mathcal{J}_2^{lin}/\mathcal{J}_1 =$	$\mathcal{J}_2^{cir}/\mathcal{J}_1 =$	$\mathcal{J}_2^{qu}/\mathcal{J}_1 =$
0	.00000	.00000	.00000	.00000	23	.0628	.0862	.00195	.1527
1	.00000	.00001	.00000	.00030	24	.0730	.0972	.0107	.1654
2	.00000	.00006	.00000	.00122	25	.0843	.1091	.0215	.1786
3	.00002	.00020	.00000	.00274	26	.0966	.1217	.0354	.1922
4	.00007	.00049	.00000	.00487	27	.1099	.1352	.0513	.2061
5	.00017	.00096	.00000	.00760	28	.1243	.1494	.0688	.2204
6	.00036	.00165	.00000	.01093	29	.1398	.1645	.0879	.2350
7	.00066	.00261	.00000	.01485	30	.1562	.1804	.1082	.2500
8	.00111	.00389	.00000	.01937	31	.1738	.1971	.1296	.2653
9	.00177	.00552	.00000	.02447	32	.1923	.2146	.1522	.2808
10	.00267	.00756	.00000	.03015	33	.2118	.2329	.1756	.2966
11	.0039	.0100	.0000	.0364	34	.2322	.2519	.1999	.3127
12	.0054	.0130	.0000	.0432	35	.2535	.2717	.2249	.3290
13	.0074	.0164	.0000	.0506	36	.2756	.2921	.2506	.3455
14	.0099	.0204	.0000	.0585	37	.2985	.3132	.2768	.3622
15	.0129	.0250	.0000	.0670	38	.3221	.3350	.3036	.3790
16	.0164	.0302	.0000	.0760	39	.3463	.3573	.3309	.3960
17	.0207	.0361	.0000	.0855	40	.3711	.3802	.3585	.4132
18	.0256	.0426	.0000	.0955	41	.3963	.4035	.3864	.4304
19	.0313	.0498	.0000	.1060	42	.4219	.4272	.4146	.4477
20	.0378	.0578	.0000	.1170	43	.4478	.4513	.4430	.4651
21	.0452	.0665	.0000	.1284	44	.4738	.4756	.4714	.4825
22	.0535	.0760	.0000	.1403	45	.5000	.5000	.5000	.5000

$\varepsilon = 22°40'$. In that case, the light transmitted by Filter Two would be very weak. (See Table 5.1.) About the hidden-variables state of the light coming through, we would know from Appendix Q [eq. (Q-1A) and between eqs. (Q-17) and (Q-17')] that

$$\left.\begin{aligned} \tfrac{1}{2} < \tilde{u} = \sin^2\tilde{\chi} < \sin^2 2\varepsilon, \quad &\text{so} \quad 45° < \tilde{\chi} < 45°\,20', \\ \pi - D < \delta < \pi + D, \quad &\text{so} \quad \delta \text{ lies within } 180° \pm D, \end{aligned}\right\} \tag{237}$$

where, by (Q-17),

$$D = \text{arc}\cos\,[\cot 2\varepsilon] = \text{arc}\cos\,(\tan 44°\,40') = 8°\,43'. \tag{238}$$

Thus, on the axes of Filter One, Ξ by (212)–(214) and (Q-2) would be given by

$$\Xi = e^{i\tilde{\theta}}(\phi_x \cdot e^{\frac{1}{2}i\delta}\cos\chi + \phi_y \cdot e^{-\frac{1}{2}i\delta}\sin\tilde{\chi}) \tag{239}$$

with $\tilde{\chi} = 45°10' \pm 10'$ and with $\frac{1}{2}\delta = 90° \pm 4°21'$ as extreme possibilities.

In this (hypothetical) case, two filters would suffice to achieve the fixing of the hidden variables as suggested, for instance, by Wigner (1970).

5.6. The light transmitted by the third polarization filter

Papaliolios (1967) in his experiments placed a third polarization filter behind the second one. Then, for a fixed angle $\vartheta = 90° - \varepsilon$ between filters 1 and 2, he would vary the angle θ between filters 2 and 3. The angle between the polarization directions of filters 1 and 3 was then $(\vartheta - \theta)$. A typical value taken for ε was 10°. We first consider the hypothetical case that Ξ is constant throughout.

No relaxation at all

The photons incident upon Filter Three now have $\psi'' = \phi_x'$, but have $\Xi'' = \Xi$ if we assume constant hidden variables. We must expand ψ'' and Ξ'' in terms of

$$\left.\begin{aligned} \phi_x'' &= \phi_x \sin(\varepsilon+\theta) + \phi_y \cos(\varepsilon+\theta) = \phi_x' \cos\theta - \phi_y' \sin\theta, \\ \phi_y'' &= -\phi_x \cos(\varepsilon+\theta) + \phi_y \sin(\varepsilon+\theta) = \phi_x' \sin\theta + \phi_y' \cos\theta. \end{aligned}\right\} \tag{240}$$

The expansion coefficients are

$$a_x'' = e^{i\beta}\cos\theta, \quad a_y'' = e^{i\beta}\sin\theta, \tag{241}$$

$$\left.\begin{aligned} \xi_x'' &= \xi_x \sin(\varepsilon+\theta) + \xi_y \cos(\varepsilon+\theta), \\ \xi_y'' &= -\xi_x \cos(\varepsilon+\theta) + \xi_y \sin(\varepsilon+\theta). \end{aligned}\right\} \tag{242}$$

Per photon incident upon Filter One, the number of photons incident upon Filter Three with hidden variables (242) is given by (228) or (229) or (230), if $\tilde{u}$, $\tilde{\chi}$, δ were within the intervals used in Appendix Q for calculating the intensity of the light transmitted by Filter Two, and this number was zero when $\tilde{u}$, $\tilde{\chi}$, δ were in one of the forbidden intervals of Appendix Q.

For finding how many of these photons, for given hidden variables, pass Filter Three, we multiply the incident number by 1 or by 0, depending upon whether or not

$$|a_x''/\xi_x''| > |a_y''/\xi_y''|. \tag{243}$$

Integrating the numbers thus obtained over the hidden variables, we obtain the intensity $\mathcal{I}_3$ behind Filter Three. We may divide it by the intensity $\mathcal{I}_2$ behind Filter Two, so $\mathcal{I}_3/\mathcal{I}_2$ is the fraction incident upon Filter Three that is transmitted by it. Pure quantum theory and classical optics predict again $\mathcal{I}_3/\mathcal{I}_2 = \cos^2 \theta$.

Again we will consider the different kinds of unpolarization incident upon Filter One separately. The calculations are discussed in Appendix R.

Results

The numerical results for $\mathcal{I}_3/\mathcal{I}_2$, for the three different kinds of unpolarized incident light on Filter One, are listed in Tables 5.2–5.4 and shown in Figs 5.2–5.4 for a few values of $\varepsilon < 45°$ ($\vartheta > 45°$) as functions of θ in part of the interval $0° < \theta < 90°$. As shown in the second half of Appendix R, we have $\mathcal{I}_3 = \mathcal{I}_2$ for $0° < \theta < 45° - \frac{1}{2}\varepsilon$, and we have $\mathcal{I}_3 = 0$ for $45° + \frac{1}{2}\varepsilon < \theta < 90°$. For the case of circularly unpolarized incident light the θ-interval where $\mathcal{I}_3 = \mathcal{I}_2$ extends to $\theta = 45° + \frac{1}{2}$ arc tan $[(\tan \varepsilon)/(\tan^2 2\varepsilon)]$. [See eq. (R-11).]

It is seen that the drop of $\mathcal{I}_3/\mathcal{I}_2$ from 1 to 0 in the interval between $45° - \frac{1}{2}\varepsilon$ and $45° + \frac{1}{2}\varepsilon$ is not symmetric, and $\mathcal{I}_3/\mathcal{I}_2 \neq \frac{1}{2}$ at $\theta = 45°$. However, the values of $\mathcal{I}_3/\mathcal{I}_2$ at angles θ differing by $\pm 90°$ add up again to 1.

TABLE 5.2. $\mathcal{I}_3^{\text{ell}}/\mathcal{I}_2^{\text{ell}}$ FOR INCIDENT ELLIPTICALLY UNPOLARIZED LIGHT PASSING THROUGH THREE FILTERS, FOR CONSTANT HIDDEN VARIABLES

We tabulate $\mathcal{I}_3^{\text{ell}}/\mathcal{I}_2^{\text{ell}}$ at $\theta = 45° - \frac{1}{2}\varepsilon + x\varepsilon$, as a function of x, for various values of ε. Note that these are also the values of $[1 - (\mathcal{I}_3^{\text{ell}}/\mathcal{I}_2^{\text{ell}})]$ at $\theta = -45° - \frac{1}{2}\varepsilon + x\varepsilon$.

$x =$	$\varepsilon = 5°$	$\varepsilon = 10°$	$\varepsilon = 15°$	$\varepsilon = 20°$	$\varepsilon = 25°$	$\varepsilon = 30°$
0.0	1.0000	1.0000	1.0000	1.0000	1.0000	1.0000
0.1	.9968	.9968	.9968	.9968	.9967	.9966
0.2	.9807	.9806	.9804	.9802	.9799	.9796
0.3	.9436	.9434	.9429	.9423	.9414	.9403
0.4	.8791	.8785	.8777	.8763	.8745	.8722
0.5	.7826	.7818	.7802	.7780	.7749	.7709
0.6	.6525	.6513	.6492	.6461	.6417	.6360
0.7	.4914	.4898	.4875	.4839	.4789	.4721
0.8	.3084	.3071	.3050	.3017	.2970	.2910
0.9	.1249	.1241	.1230	.1213	.1186	.1155
1.0	.0000	.0000	.0000	.0000	.0000	0000

TABLE 5.3. $\mathcal{I}_3^{\text{lin}}/\mathcal{I}_2^{\text{lin}}$ FOR INCIDENT LINEARLY UNPOLARIZED LIGHT PASSING THROUGH THREE FILTERS, FOR CONSTANT HIDDEN VARIABLES

We tabulate $\mathcal{I}_3^{\text{lin}}/\mathcal{I}_2^{\text{lin}}$ at $\theta = 45° - \frac{1}{2}\varepsilon + x\varepsilon$, as a function of x, for various values of ε. Note that these are also the values of $[1-(\mathcal{I}_3^{\text{lin}}/\mathcal{I}_2^{\text{lin}})]$ at $\theta = -45° - \frac{1}{2}\varepsilon + x\varepsilon$.

$x =$	$\varepsilon = 5°$	$\varepsilon = 10°$	$\varepsilon = 15°$	$\varepsilon = 20°$	$\varepsilon = 25°$	$\varepsilon = 30°$
0.0	1.0000	1.0000	1.0000	1.0000	1.0000	1.0000
0.1	.9878	.9878	.9879	.9880	.9881	.9883
0.2	.9485	.9488	.9490	.9494	.9500	.9506
0.3	.8814	.8820	.8824	.8833	.8843	.8856
0.4	.7874	.7879	.7889	.7902	.7918	.7937
0.5	.6691	.6698	.6709	.6726	.6746	.6769
0.6	.5310	.5318	.5331	.5348	.5370	.5395
0.7	.3802	.3807	.3820	.3838	.3858	.3881
0.8	.2268	.2273	.2283	.2295	.2310	.2328
0.9	.0873	.0876	.0881	.0887	.0894	.0904
1.0	.0000	.0000	.0000	.0000	.0000	.0000

TABLE 5.4. $\mathcal{I}_3^{\text{cir}}/\mathcal{I}_2^{\text{cir}}$ FOR INCIDENT CIRCULARLY UNPOLARIZED LIGHT PASSING THROUGH THREE FILTERS, FOR CONSTANT HIDDEN VARIABLES

We tabulate $\mathcal{I}_3^{\text{cir}}/\mathcal{I}_2^{\text{cir}}$ for various values of θ, for $\varepsilon = 25°$ or $30°$. (For $\varepsilon < 22\frac{1}{2}°$, we have $\mathcal{I}_2^{\text{cir}} = 0$.) Note that $1-(\mathcal{I}_3^{\text{cir}}/\mathcal{I}_2^{\text{cir}})$ at the angle $(\theta - 90°)$ has the same value as $\mathcal{I}_3^{\text{cir}}/\mathcal{I}_2^{\text{cir}}$ here has at the angle θ. Here, as in Tables 2 and 3, the angles between the polarization directions are $(90° - \varepsilon)$ between filters 1 and 2, $90° - (\varepsilon + \theta)$ between 1 and 3, and θ between 2 and 3.

$\varepsilon = 25°$:

$\theta =$	54°5′	54°30′	55°0′	55°30′	56°0′	56°30′	57°0′	57°30′
$\mathcal{I}_3^{\text{cir}}/\mathcal{I}_2^{\text{cir}} =$	1.0000	.97909	.8630	.6215	.4062	.2224	.07914	.0000

$\varepsilon = 30°$:

$\theta =$	50°27′	50°30′	51°0′	51°30′	52°0′	52°30′	53°0′	53°30′
$\mathcal{I}_3^{\text{cir}}/\mathcal{I}_2^{\text{cir}} =$	1.0000	.99996	.9953	.9813	.9539	.9021	.8182	.7364

$\theta =$	54°0′	54°30′	55°0′	55°30′	56°0′	56°30′	57°0′	57°30′
$\mathcal{I}_3^{\text{cir}}/\mathcal{I}_2^{\text{cir}} =$	.6568	.5796	.5052	.4337	.3655	.3008	.2399	.1835

$\theta =$	58°0′	58°30′	59°0′	59°30′	60°0′
$\mathcal{I}_3^{\text{cir}}/\mathcal{I}_2^{\text{cir}} =$	.1320	.08607	.04702	.01676	.0000

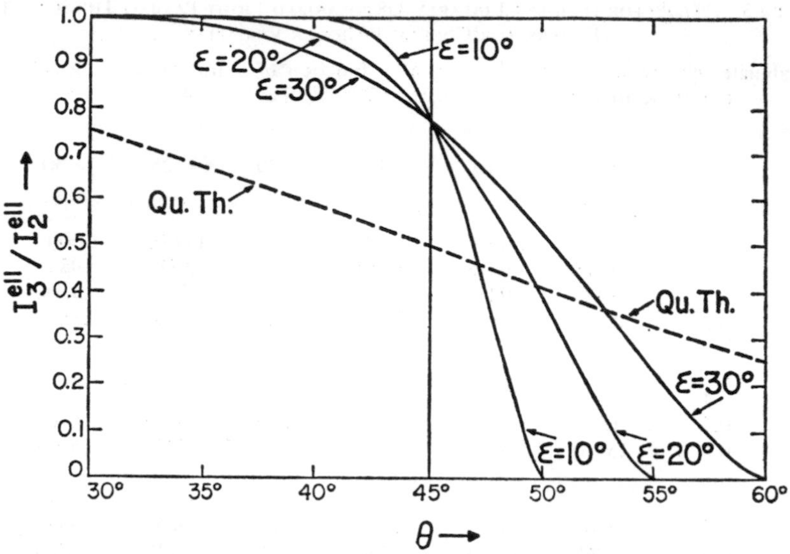

FIG. 5.2. Fraction $\mathcal{I}_3^{ell}/\mathcal{I}_2^{ell}$ transmitted by third filter in absence of relaxation of Ξ, as function of angle θ between filters 2 and 3, for *elliptically* unpolarized light incident upon filter 1, which makes angles $(90° - \varepsilon)$ and $(90° - \varepsilon - \theta)$ with filters 2 and 3. Note $(\mathcal{I}_3/\mathcal{I}_2 \text{ at } \theta) + (\mathcal{I}_3/\mathcal{I}_2 \text{ at } \theta \pm 90°) = 1$.

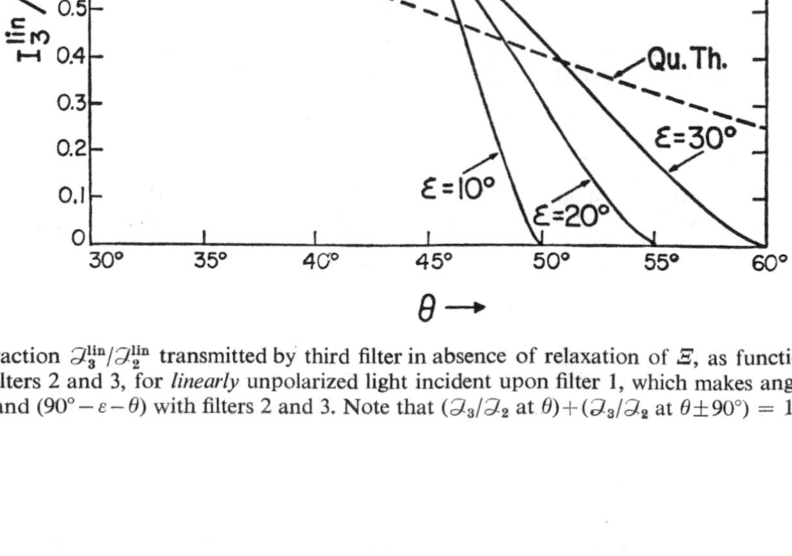

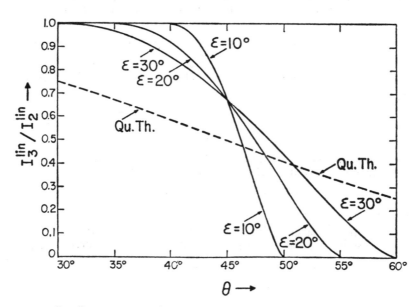

FIG. 5.3. Fraction $\mathcal{I}_3^{lin}/\mathcal{I}_2^{lin}$ transmitted by third filter in absence of relaxation of Ξ, as function of angle θ between filters 2 and 3, for *linearly* unpolarized light incident upon filter 1, which makes angles $(90° - \varepsilon)$ and $(90° - \varepsilon - \theta)$ with filters 2 and 3. Note that $(\mathcal{I}_3/\mathcal{I}_2 \text{ at } \theta) + (\mathcal{I}_3/\mathcal{I}_2 \text{ at } \theta \pm 90°) = 1$.

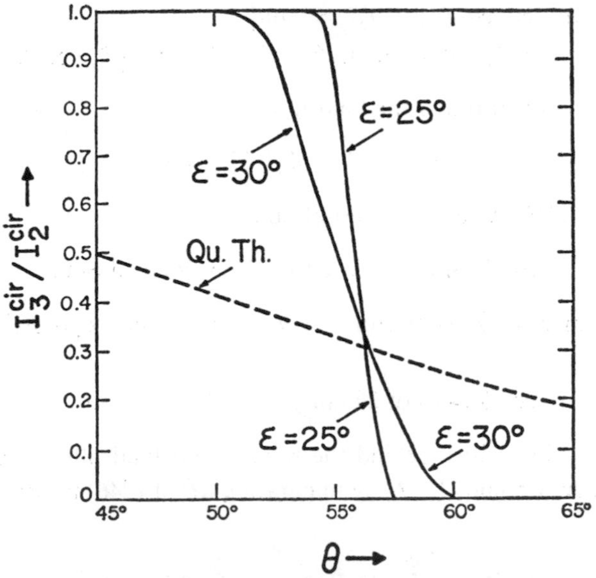

FIG. 5.4. Fraction $\mathcal{J}_3^{cir}/\mathcal{J}_2^{cir}$ transmitted by filter 3 in absence of relaxation of Ξ, as function of angle θ between filters 2 and 3, for *circularly* unpolarized light incident upon filter 1, which makes angle $90°-\varepsilon$ ($< 67\frac{1}{2}°$) with filter 2, and angle $90°-\varepsilon-\theta$ with filter 3. Note that $(\mathcal{J}_3/\mathcal{J}_2$ at $\theta)+(\mathcal{J}_3/\mathcal{J}_2$ at $\theta\pm 90°) = 1$.

Relaxation behind Filter One

In the experimental setup used by Papaliolios (see Section 5.8) it is likely that the hidden variables relax behind Filter One. In that case, his experiment really is a two-filter experiment with linearly polarized instead of unpolarized light incident upon Filter One.

We therefore use the two-filter notation, and assume that upon Filter One (the middle filter in the experiment) the incident light has

$$a_x = e^{i\beta} \sin \varepsilon, \qquad a_y = -e^{i\beta} \cos \varepsilon, \tag{244}$$

$$\xi_x = \tilde{r} \cos \tilde{\chi} \exp\left(i\tilde{\theta}+\tfrac{1}{2}i\delta\right), \quad \xi_y = \tilde{r} \sin \tilde{\chi} \exp\left(i\tilde{\theta}-\tfrac{1}{2}i\delta\right), \tag{245}$$

with a canonical distribution of Ξ unbiased because of the relaxation behind the polarizing front filter, so $d\tilde{P}$ is given by (218)–(220) or (226):

$$d\tilde{P} = d\tilde{v}(d\tilde{\theta}/2\pi)\, d\tilde{u}\, d\delta/2\pi.$$

Integrating over $\tilde{v}$ and $\tilde{\theta}$:

$$d\tilde{P} = d\tilde{u}\, d\delta/2\pi = \pi^{-1} \sin \tilde{\chi} \cos \tilde{\chi}\, d\tilde{\chi}\, d\delta, \quad (0 < \tilde{\chi} < 2\pi, \quad 0 < \delta < 2\pi). \tag{246}$$

Condition for transmission is

$$|a_x \xi_y| > |a_y \xi_x|, \quad \text{or} \quad \sin \varepsilon \sin \tilde{\chi} > \cos \varepsilon \cos \tilde{\chi}. \tag{247}$$

For $0 < \varepsilon < 90°$, this gives

$$\cos(\varepsilon+\tilde{\chi}) < 0, \quad \text{so,} \quad 90°-\varepsilon < \tilde{\chi} < 90°. \tag{248}$$

The light from this Filter One leaves with $|a_x| = 1$, $|a_y| = 0$. We assume that it keeps its Ξ given by (245) until reaching the last filter. So the light is incident upon that filter with

$$|a'_x| = |\cos\theta|, \quad |a'_y| = |\sin\theta|, \tag{249}$$

$$\xi'_x = \xi_x \cos\theta - \xi_y \sin\theta, \quad \xi'_y = \xi_x \sin\theta + \xi_y \cos\theta. \tag{250}$$

The condition for passage through the final filter is

$$|a'_x|^2 |\xi'_y|^2 - |a'_y|^2 |\xi'_x|^2 > 0. \tag{251}$$

We insert here (249)–(250) with (245), and obtain

$$\cos 2\theta \sin^2\tilde{\chi} + \sin 2\theta \sin\tilde{\chi}\cos\tilde{\chi}\cos\delta > 0. \tag{252}$$

We may divide by $\sin\tilde{\chi}\cos\tilde{\chi}$ (>0). For $0 < \theta < 90°$ we may also divide by $\sin 2\theta$ (>0), and we find

$$\cos\delta > -\cot 2\theta \tan\tilde{\chi} \quad \text{if} \quad 0 < \theta < 90°. \tag{253}$$

If $\mathcal{I}_m$ and $\mathcal{I}_f$ are the intensities behind the middle and final filters, and $\mathcal{I}_0$ was incident upon the middle filter, we find $\mathcal{I}_m/\mathcal{I}_0$ by integrating $d\tilde{P}$ of (246) under the condition (248) only:

$$\mathcal{I}_m/\mathcal{I}_0 = \int_{\pi/2-\varepsilon}^{\pi/2} d\tilde{\chi} \sin\tilde{\chi}\cos\tilde{\chi} \int_0^{2\pi} d\delta/\pi$$

$$= \Big[\sin^2\chi\Big]_{\pi/2-\varepsilon}^{\pi/2} = 1-\cos^2\varepsilon = \sin^2\varepsilon. \tag{254}$$

We obtain $\mathcal{I}_f/\mathcal{I}_0$ by performing the same integral while restricting δ, for each value of $\tilde{\chi}$, to the values compatible with (253). This gives:

For $0 < \theta < 45° - \frac{1}{2}\varepsilon$: No restriction on δ for the

$\tilde{\chi}$ interval from $(90°-\varepsilon)$ to $90°$, so $\mathcal{I}_f = \mathcal{I}_m$. (255)

For $45° - \frac{1}{2}\varepsilon < \theta < 45°$:

For $90°-\varepsilon < \chi < 2\theta$ we exclude the interval $\pi - \Delta < \delta < \pi + \Delta$ with $\Delta = \text{arc cos}(\cot 2\theta \tan\tilde{\chi})$, so, $\int d\delta = 2\pi - 2\Delta$. For $2\theta < \tilde{\chi} < 90°$ there is no such restriction, and $\int d\delta = 2\pi$ as in (254). Thus,

$$\mathcal{I}_f = \mathcal{I}_m - \mathcal{I}_0 \int_{\frac{1}{2}\pi-\varepsilon}^{2\theta} d\tilde{\chi} \sin\tilde{\chi}\cos\tilde{\chi} \cdot \frac{2}{\pi} \cdot \text{arc cos}(\cot 2\theta \tan\tilde{\chi}). \tag{256}$$

For $45° < \theta < 45° + \frac{1}{2}\varepsilon$: The δ interval is for $90° - \varepsilon < \tilde{\chi} < 180° - 2\theta$ restricted to the interval $-\Delta' < \delta < +\Delta'$ only, with $\Delta' = \text{arc cos}(\cot[\pi - 2\theta]\tan\tilde{\chi})$. For $180° - 2\theta < \tilde{\chi} < 90°$ no δ at all will satisfy (253). Thence,

$$\mathcal{I}_f = \mathcal{I}_0 \int_{\frac{1}{2}\pi-\varepsilon}^{\pi-2\theta} d\tilde{\chi} \sin\tilde{\chi}\cos\tilde{\chi} \cdot \frac{2}{\pi} \cdot \text{arc cos}(\cot[\pi-2\theta]\tan\tilde{\chi}). \tag{257}$$

For $45° + \frac{1}{2}\varepsilon < \theta < 90°$, no δ at all will satisfy (253) for $\tilde{\chi}$ in the interval (248). Then, $\mathcal{I}_f = 0$. (258)

Results

The integrals (256)–(257) were performed on the Purdue CDC 6500 for six values of ε. Some of the results for $\mathcal{I}_f/\mathcal{I}_m$ are listed in Table 5.5 and shown in Fig. 5.5. Note that, in this case, for $\theta = 45° - \varphi$ and $\theta' = 45° + \varphi$, the two integrals in (256) and (257) are equal, and

TABLE 5.5. $\mathcal{I}_f/\mathcal{I}_m$ FOR THE PAPALIOLIOS EXPERIMENT WITH RELAXATION BETWEEN FILTERS 1 and 2, BUT CONSTANT Ξ BETWEEN FILTERS 2 AND 3

We tabulate $\mathcal{I}_f/\mathcal{I}_m$ at $\theta = 45° - \frac{1}{2}\varepsilon + x\varepsilon$ as a function of x, for various values of ε. Note that here not only $[\mathcal{I}_f/\mathcal{I}_m]$ at θ equals $[1 - \mathcal{I}_f/\mathcal{I}_m]$ at $\theta \pm 90°$, but also equals $[1 - \mathcal{I}_f/\mathcal{I}_m]$ at $90° - \theta$

$x =$	$\varepsilon = 5°$	$\varepsilon = 10°$	$\varepsilon = 15°$	$\varepsilon = 20°$	$\varepsilon = 25°$	$\varepsilon = 30°$
0.0	1.0000	1.0000	1.0000	1.0000	1.0000	1.0000
0.1	.9480	.9480	.9480	.9480	.9480	.9481
0.2	.8576	.8576	.8576	.8577	.8578	.8579
0.3	.7477	.7477	.7477	.7478	.7478	.7480
0.4	.6265	.6265	.6265	.6265	.6266	.6267
0.5	.5000	.5000	.5000	.5000	.5000	.5000
0.6	.3735	.3735	.3735	.3735	.3734	.3733
0.7	.2523	.2523	.2523	.2522	.2522	.2520
0.8	.1424	.1424	.1424	.1423	.1422	.1421
0.9	.0520	.0520	.0520	.0520	.0520	.0519
1.0	.0000	.0000	.0000	.0000	.0000	.0000

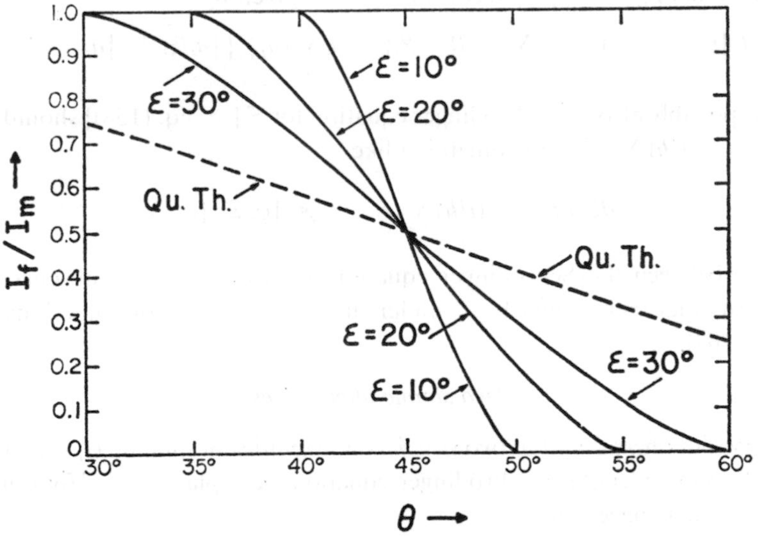

FIG. 5.5. Fraction $\mathcal{I}_f/\mathcal{I}_m$ transmitted by final filter as function of angle θ between final and middle filter in the Papaliolios experiment, for no relaxation of Ξ between these two filters, with light incident upon the middle filter linearly polarized by the front filter (with its polarization at angles $(90° - \varepsilon)$ and $(90° - \varepsilon - \theta)$ with middle and final filter), but with Ξ relaxed to a canonical distribution before reaching the middle filter.

the sum of these two intensities gives

$$\mathcal{I}_f(45^\circ-\varphi)+\mathcal{I}_f(45^\circ+\varphi) = \mathcal{I}_m, \tag{259}$$

a symmetry which was lacking in absence of relaxation between Filters One and Two.

The results (255)–(258) were qualitatively[78] predicted by Papaliolios.

5.7. Other simple possibilities for $d\Xi/dt$

The calculations of Sections 5.3 throught 5.6 were all based upon the assumption that Ξ would be constant during the passage of the light through all three filters. This, of course, is not the case when the filters are thick. In that case, the bias caused in the first atomic layer of each filter where the polarization takes place and thus the "measurement" is made according to the polychotomic algorithm, will be undone (and the Ξ distribution will relax) by the slight interactions of the photons with the bulk of the material in the back, where no new bias is produced because the light here is already polarized to be transmitted for any value of Ξ. Therefore, if the algorithm is correct and hidden variables exist, we could expect to find the values of $\mathcal{I}_2/\mathcal{I}_1$ and of $\mathcal{I}_3/\mathcal{I}_2$ calculated only if each filter is so thin that there is little "bulk" behind the layers that determine the polarization.

Even then, we should consider a possible different reason for a change of Ξ. The change of Ξ might be a systematic change during the measurement itself. As mentioned in 4.4, the fact that the Schrödinger equation for ψ, on the ϕ_i basis, during the measurement is changed from $da_i/dt = -(i/\hbar)\sum_j \mathcal{H}_{ij}a_j$ to

$$da_i/dt = -(i/\hbar)\sum_j \mathcal{H}_{ij}\, a_j+\gamma F_i[\psi, \Xi, \{\phi_n\}] \tag{260}$$

with the functional $F_i[\psi, \Xi, \{\phi_n\}]$ according to (197) given by

$$F_i[\psi, \Xi, \{\phi_n\}] = a_i \sum_j w_j\{R_i-R_j\} = a_i \sum_j |a_j|^2 \{|a_i/\xi_i|^2-|a_j/\xi_j|^2\}, \tag{261}$$

suggests that possibly also the Schrödinger equation for Ξ [see eq. (158)] should be changed from $d\xi_i/dt = -(i/\hbar)\sum_j \mathcal{H}_{ij}\xi_j$ to something like

$$d\xi_i/dt = -(i/\hbar)\sum_j \mathcal{H}_{ij}\xi_j+\gamma G_i[\psi, \Xi, \{\phi_n\}]. \tag{262}$$

The similarity between the Schrödinger equations for ψ and for Ξ in absence of a measurement suggests that there might be a similar similarity between the functionals F_i and G_i in (260) and (262).

Two simple alternatives

We shall consider here only the two simplest possibilities of making G_i similar in form to F_i. Since in the remainder of the Schrödinger equation we replace ψ by Ξ for obtaining (262) from (260), the first suggestion is

$$G_i[\psi, \Xi, \{\phi_n\}] = F_i[\Xi, \Xi, \{\phi_n\}], \tag{263}$$

[78] Papaliolios did not evaluate the integrals in eqs. (256)–(257). [In his paper, read $\frac{1}{4}\pi-\frac{1}{2}\varepsilon$ where he writes $\frac{1}{4}\pi-\varepsilon$.]

replacing ψ by Ξ also in the additional terms. In that case, however,

$$a_i w_j\{R_i - R_j\} = a_i |a_j|^2 \{|a_i/\xi_i|^2 - |a_j/\xi_j|^2\}$$

becomes $\xi_i|\xi_j|^2\{|\xi_i/\xi_i|^2 - |\xi_j/\xi_j|^2\} = 0$, so that (263) is equivalent to leaving the Schrödinger equation (158) for Ξ unchanged during measurements. As explained in the first part of Section 4.4, for a good measurement that does not disturb the observable measured it is plausible to neglect the terms with $\mathcal{H}_{ij}$ during the measurement, and we are back at the case $d\Xi/dt = 0$ considered in the preceding sections.

The alternative to (263) is putting

$$G_i[\psi, \Xi, \{\phi_n\}] = F_i[\Xi, \psi, \{\phi_n\}], \tag{264}$$

which would make the ψ and Ξ appear symmetrically in their equations of motion. This gives (262) the form

$$\begin{aligned} d\xi_i/dt &= -(i/\hbar)\sum_j \mathcal{H}_{ij}\xi_j + \gamma\xi_i \sum_j u_j\{R_i^{-1} - R_j^{-1}\} \\ &= -(i/\hbar)\sum_j \mathcal{H}_{ij}\xi_j + \gamma\xi_i \sum_j |\xi_j|^2 \{|\xi_i/a_i|^2 - |\xi_j/a_j|^2\}. \end{aligned} \tag{265}$$

We shall now briefly consider the consequences which such an equation for $d\Xi/dt$ would have on successive measurements. As before, we neglect the terms with $\mathcal{H}_{ij}$. In particular, we neglect again relaxation of Ξ.

Collapse of hidden variables?

The application of eq. (265) is so similar to the application of (197) in Section 4.4 that we need not give all steps in the reasoning, and will just give the results easily verifiable by the reader.

As the $R^{-1} = |\xi/a|^2$ replace the $R = |a/\xi|^2$ of (197) and therefore of (200), the equation

$$\frac{du_i}{dt} = \gamma \sum_j 2u_i u_j \left(\frac{u_i}{w_i} - \frac{u_j}{w_j}\right), \tag{266}$$

which is the analogue of (200), leads to the conclusion that there is a tendency of the $u_i = |\xi_i|^2$ (normalized by $\sum_j u_j = 1$) to collapse during the measurement upon the one u_k which has its R_k smaller than all other R_i:

$$\left.\begin{aligned} &u_k(t) \to 1 \quad \text{and} \quad u_i(t) \to 0 \quad \text{for} \quad i \neq k, \quad \text{if} \\ &u_k/w_k > u_i/w_i \quad \text{and} \quad R_k < R_i \quad \text{for all} \quad i \neq k. \end{aligned}\right\} \tag{267}$$

Application to one polarization filter

In the case of light passing through a polarization filter, we considered only two states ϕ_i between which the photon could select. Therefore, if a_x and a_y in (210) condense in the filter according to $a_x \to 1$, $a_y \to 0$, the ξ_x and ξ_y contrary to Sections 5.3–5.6 could not remain constant, but would condense according to $\xi_x \to 0$, $\xi_y \to 1$. This strong bias replaces the weak bias calculated in Section 5.4 and Appendix P. Now we need no longer consider three different unpolarization cases.

Application to two polarization filters

If now a second filter under an angle ϑ is placed behind the first one, we will have (but for trivial phase factors $e^{i\alpha}$ and $e^{i\beta}$)

$$a_{x'} = \cos\vartheta, \quad a_{y'} = -\sin\vartheta; \quad \xi_{x'} = \sin\vartheta, \quad \xi_{y'} = \cos\vartheta, \tag{268}$$

where the primed x'-axis is the polarization axis of Filter Two. Then the polychotomic algorithm gives as condition for passage of the photon through Filter Two the condition

$$|a_{x'}/\xi_{x'}| > |a_{y'}/\xi_{y'}|, \quad \text{that is,} \quad |\cot\vartheta| > |\tan\vartheta|.$$

Then,

$$\left.\begin{aligned} \mathcal{J}_2/\mathcal{J}_1 &= 1 \quad \text{if} \quad |\vartheta| < 45° \quad \text{or if} \quad 135° < |\vartheta| < 225°, \\ \mathcal{J}_2/\mathcal{J}_1 &= 0 \quad \text{if} \quad 45° < |\vartheta| < 135°. \end{aligned}\right\} \tag{269}$$

Thus Malus's law for two filters would be changed even more drastically than it would be for (263) or $d\Xi/dt = 0$ with circularly unpolarized light. If $d\Xi/dt$ would be given by the alternative (264), $\mathcal{J}_2/\mathcal{J}_1$ would become a block function (269).

Therefore if hidden variables exist for which the polychotomic algorithm is valid, experimentation with two polarizing filters sufficiently thin to prevent relaxation of Ξ from the first filter to the second one should show a breakdown of Malus's $\cos^2$ law, either in the form of one of the curves of Fig. 5.1 (for incident unpolarized light) or of Fig. 5.5 (for incident linearly polarized light) in case of the alternative (263), or in the form of the block function (269) for any incident light in case of the alternative (264).

If even for the thinnest polarizing filters—barely thick enough to polarize most of the light—we find Malus's law (corrected for incomplete polarization), we should take this as an indication that either hidden variables do not exist or that they do not predict the outcome of experiments by the polychotomic algorithm.

5.8. The experiment of Papaliolios

Papaliolios (1967) used "HN-32 stripable polarizers" supplied by Polaroid Land Corporation. "The 15×10^{-4} cm thick polarizing material was epoxied onto optical flats. The index of refraction of the polarizing sheet is 1.5, resulting in a transit time of about 7.5×10^{-14} sec."

"It can be shown that about 90% of the photons entering the sheet interact in the first 3×10^{-4} cm of the polarizing sheet."

We quoted here literally from the Papaliolios paper.

With the polarizing films of the middle and final polarizer on top of each other, the first polarizer is separated from the middle and final ones by the bulk of mounting material of the middle filter. As the results of the experiment (see below) did show relaxation already in the $15-3 \approx 12$ micron thickness of the film of the middle filter between the layer where most polarizing takes place, and the polarizing front layers of the film of filter 3, it must be assumed that complete relaxation of Ξ takes place in the much thicker material upon which the film of the middle filter was mounted.

Therefore the calculations that apply to the experiment are those of the last part of Section 5.6 (see Fig. 5.5), or those for case (264) in Section 5.7 leading to (269).

The intensities $\mathcal{I}_f$ and $\mathcal{I}_m$ were measured using a photomultiplier. White light was used. The incident light had an intensity of 10^{-10} W cm^{-2}, which is between 10^8 and 10^9 photons cm^{-2} s^{-1}. As the transit time of a photon was less than 10^{-13} s, the photons pass the filters one at a time.

Experimental results

The measurements of $\mathcal{I}_f/\mathcal{I}_m$ as a function of θ at constant ε were taken at 1° steps of θ between 0° and 90°. The results agreed within $1\frac{1}{2}\%$ of $(\mathcal{I}_f/\mathcal{I}_m)_{\max} = 1$ (at $\theta = 0°$) with the quantum-theoretical or optical prediction of Malus's law. Deviations of $1\frac{1}{2}\%$ were understandable from possible systematic errors. There was not the slightest indication of a "plateau" $\mathcal{I}_f/\mathcal{I}_m = 1$ extending to $\theta = 45° - \frac{1}{2}\varepsilon$, or to 45°, or of $\mathcal{I}_f/\mathcal{I}_m = 0$ for $\theta > 45°$ or $\theta > 45° + \frac{1}{2}\varepsilon$.

The experiments should be repeated with thinner polarizing films for avoiding relaxation of Ξ in the back part of the film of the middle filter. If the relaxation would be temperature dependent, cooling of the polarizing films to helium temperatures might reduce any relaxation existing.

If the negative outcome of the experiment was due to relaxation, Papaliolios estimates that in the polarizing film the relaxation time must have been less than 2.4×10^{-14} s. It would be desirable to have some theory of how relaxation takes place for judging whether so fast a relaxation makes theoretically any sense.

The most reasonable explanation of the negative outcome of the Papaliolios experiment, however, would be that no hidden variables Ξ exist that would determine the result of a polarization by the polychotomic algorithm.

The reason why we might as well give up hidden variables if the Papaliolios results are correct was already mentioned in Section 4.4. If hidden variables would exist that determine future happenings by the polychotomic algorithm, only relaxation (randomization) of Ξ inside the middle polarizing film could explain the negative outcome of the experiment. But once a light beam is polarized in the front layer of the film (thereby losing intensity according to Malus's law), the beam keeps its polarization in the remainder of the film without further such intensity losses, showing that the light's state vector ψ is *not* randomized, and that the polarization of the beam as it leaves the film is not due to continuous repolarization of a continuously depolarized (randomized) beam, but shows the polarization obtained at the start and maintained ever since.

In other words, we then would have to conclude that $d\Xi/dt$ contains randomizing terms that do not find their analogue in the expression for $d\psi/dt$. This, however, is contrary to our conclusion (158) that in first approximation Ξ and ψ should depend similarly on time, in order to meet Condition Two of Section 3.9. As explained there, and discussed again in Section 4.4, if Condition Two is not met we cannot expect that a given initial microstate would determine later happenings, as the latter then would be determined mainly by a randomization process for Ξ *for which no deterministic laws are in sight*. Thus the theory practically would lose its cryptodeterministic character. According to Siegel (1966) this would remove the attractiveness of hidden-variables theory, which was invented for making the laws of nature cryptodeterministic rather than governed by unexplainable randomness.

5.9. Conclusions

We have seen in this chapter that if it is possible to make successive measurements fast enough to prevent relaxation of the bias in Ξ distribution during the time between the two measurements, weird effects are predicted by the type of hidden-variables theory of the first kind in which a hidden-variables state vector Ξ determines the results of measurements by the polychotomic algorithm.

Attempts at finding these weird deviations from quantum theory so far have been unsuccessful, but more refined experiments are needed to make the results conclusive. Besides the Papaliolios experiment of Fig. 5.5, the two-filter experiment with two extremely thin polarizing films on top of each other would be worth trying. (See Section 5.5 and Fig. 5.1.)

In the meantime, as these effects have not yet been shown experimentally to exist, ordinary quantum theory suffices to explain the observed facts, without need for hidden variables, however appealing these hidden variables may seem to some people.

Other people, possibly less fascinated by the idea that hidden variables might exist, are not anxious to embrace the latter idea as long as paradoxes of the types discussed in Appendix K keep troubling the theory.

An entirely different category of hidden-variables theories, even more remote from quantum mechanics, will be discussed in Part III.

APPENDICES TO

PART II

APPENDIX A

BOHM'S QUANTUM-MECHANICAL POTENTIAL

WITH π given by (13), and by the Schrödinger equation (7), we obtain

$$\frac{\partial \pi}{\partial t} = Re\left\{\nabla \frac{\hbar}{i}\,\frac{\partial \psi/\partial t}{\psi}\right\} = Re\left\{\nabla\left[\frac{1}{\psi}\left(\frac{\hbar^2}{2m}\nabla^2\psi - V\psi\right)\right]\right\}$$

$$= -\nabla V + \frac{\hbar^2}{2m}\cdot Re\left\{\nabla\left(\frac{1}{\psi}\nabla^2\psi\right)\right\}. \tag{A-1}$$

With $\psi = R\exp(iS/\hbar)$, we find

$$\nabla^2\psi = [\nabla^2 R + (2i/\hbar)(\nabla R)\cdot(\nabla S) + (i/\hbar)R\,\nabla^2 S - (1/\hbar^2)\,R(\nabla S)^2]\exp(iS/\hbar). \tag{A-2}$$

Thence

$$\frac{\partial \pi}{\partial t} = -\nabla V + \frac{\hbar^2}{2m}\nabla\left\{\frac{\nabla^2 R}{R} - \left(\frac{\nabla S}{\hbar}\right)^2\right\}$$

$$= -\nabla V + \frac{\hbar^2}{2m}\nabla\left(\frac{\nabla^2 R}{R}\right) - \nabla\frac{\pi^2}{2m}. \tag{A-3}$$

Therefore (14) gives

$$\mathbf{f} = -\nabla\left[V - \frac{\hbar^2}{2m}\,\frac{\nabla^2 R}{R}\right] - \frac{1}{m}\sum_{i=1}^{3}\pi_i\nabla\pi_i + \frac{1}{m}\sum_{i=1}^{3}\pi_i\nabla_i\boldsymbol{\pi}. \tag{A-4}$$

Since $\pi_i = \nabla_i S$, the j-component of the last two terms in (A-4) is

$$\frac{1}{m}\sum_i \pi_i(\nabla_i\nabla_j S - \nabla_j\nabla_i S) = 0. \tag{A-5}$$

Therefore (A-4) gives

$$\mathbf{f} = -\nabla(V+U), \tag{A-6}$$

with the quantum-mechanical potential U given by

$$U = -\frac{\hbar^2}{2m}\,\frac{\nabla^2 R}{R} = -\frac{\hbar^2}{2m}\,\frac{\nabla^2\left(\varrho^{\frac{1}{2}}\right)}{\varrho^{\frac{1}{2}}}. \tag{A-7}$$

APPENDIX B

HOW IN BOHM'S THEORY $P(x)$ BECOMES $\varrho(x)$

For details, see Bohm and Vigier (1954), where, however, the final conclusion is a plausibility rather than a necessary consequence of the preceding because the symbol $<$ in the formula between their eqs. (11b) and (12) should be a symbol $\leqq$.

The basic idea is the following [Bohm (1953a)]. Let

$$f(\mathbf{x}, t) \equiv P(\mathbf{x}, t)/\varrho(\mathbf{x}, t) \quad \text{with} \quad \varrho \equiv |\psi|^2. \tag{B-1}$$

Since both $P(\mathbf{x}, t)$ and $\varrho(\mathbf{x}, t)$ should satisfy a continuity equation, we find, by the time the velocity $\mathbf{v} = d\boldsymbol{\xi}/dt$ of the former has become equal to the velocity $\nabla S/m$ of the latter,

$$\begin{aligned} 0 &= \left[\frac{\partial P}{\partial t}+\operatorname{div}(P\mathbf{v})\right]-f\left[\frac{\partial \varrho}{\partial t}+\operatorname{div}(\varrho\mathbf{v})\right] \\ &= \varrho\left[\frac{\partial f}{\partial t}+\mathbf{v}\cdot\nabla f\right] = \varrho\frac{df}{dt}, \end{aligned} \tag{B-2}$$

so that df/dt vanishes where ϱ differs from zero. That is, f at the position of some particle at time t is equal to f at a later time at the later position of this particle. (By "particle" we mean here a sample from the ensemble described by ψ.)

Due to outside influences ("collisions"), sample particles that once were close together may be torn far apart, and conversely. This "stirs" the fluid of sample particles so that in a given small volume element $\delta\mathcal{V}$ at a later time one will find a conglomerate of fluid elements that came from all over the original distribution. Though each of the latter fluid elements has kept its own f value, say, $f_i = P_i/\varrho_i$ for fluid element number i, the f value for the whole volume element $\delta\mathcal{V}$ will be a ratio of the kind $\left(\sum_i P_i\, \delta\mathcal{V}_i\right)\Big/\left(\sum_i \varrho_i\, \delta\mathcal{V}_i\right)$, which will be *between* f_{max} and f_{min}. Thus the value of f becomes more and more an average, and thus is likely to become a constant (the same for all volume elements) in the long run. [Bohm and Vigier (1954)].

If both ψ and P are normalized as usually by $\int |\psi|^2\, d\mathcal{V} = \int P\, d\mathcal{V} = 1$, it follows that $f \to 1$.

One necessary condition for the above mechanism is $\varrho \neq 0$ in (B-2), and another necessary condition for successful averaging of f toward a constant value is that volume elements from all over will reach a given $\delta\mathcal{V}$ in the long run. If there are parts in space separated from other parts of space by macroscopic volumes in which $\psi = 0$, $\varrho = 0$, then hopefully there is no transport of particles over these empty spaces,[79] and the above mixing mechanism cannot act across such regions. Therefore, when a ψ-wave breaks up into separate wave packets with $\psi = 0$ spaces in between, and when a particle is trapped in one of these wave packets (so that $P(\mathbf{x})$ is concentrated in one ψ-packet separated by $\varrho = 0$ from other ψ-packets with $P \neq 0$), then the above mechanism does not mix the $f = 0$ packets with the $f > 0$ packet, and cannot free the trapped particle from the wave packet in which it is, so

[79] See Section 2.8.

that, for a description of the further behavior of this particle, we may as well ignore or even omit the empty wave packets (with $P = f = 0$). This is achieved by a reduction of the wave packet. [See under (F) in Section 2.6.]

APPENDIX C

CANCELLATION OF INFINITIES OF U AGAINST THOSE OF V

From $d\boldsymbol{\pi}/dt = -\nabla(V+U)$ [eq. (17)] together with $\boldsymbol{\pi}/m = \mathbf{v} = d\mathbf{x}/dt$ it follows that

$$d(\boldsymbol{\pi}^2/2m) = \mathbf{v}\cdot d\boldsymbol{\pi} = -d\mathbf{x}\cdot\nabla(V+U) = -d(V+U), \tag{C-1}$$

which gives the law of conservation of (imagined or actual) particle energy the form

$$\boldsymbol{\pi}^2/2m+(V+U) = \text{constant}. \tag{C-2}$$

Here, $\boldsymbol{\pi} = \nabla S$, and, unless the wavefunction ψ "curls up infinitely in the complex plane," the gradient of the phase $S/\hbar$ of ψ will remain finite. By (C-2), this means $V+U =$ finite.

From the Schrödinger equation, if we put $\mathcal{E}_{\text{op}} = i\hbar\partial/\partial t$, we obtain

$$V\psi = \mathcal{E}_{\text{op}}\psi+(\hbar^2/2m)\nabla^2\psi, \tag{C-3}$$

so, if we use (A-2) and put div $\{R^2\,\nabla S\} = \hbar\mathcal{Z}$, we find

$$V = \psi^{-1}\mathcal{E}_{\text{op}}\psi+(\hbar^2/2m)\,[R^{-1}\nabla^2R+iR^{-2}\nabla^2\mathcal{Z}-(\nabla S/\hbar)^2]. \tag{C-4}$$

The real part of this gives

$$V = \tfrac{1}{2}\,|\psi|^{-2}(\psi^*\mathcal{E}_{\text{op}}\psi+\psi\mathcal{E}^*_{\text{op}}\psi^*)+(\hbar^2/2m)R^{-1}\nabla^2R-(2m)^{-1}(\nabla S)^2. \tag{C-5}$$

This shows explicitly that V contains a term $(\hbar^2/2m)\,R^{-1}\nabla^2R = -U$, so that all infinities occurring in U are accompanied by opposite infinities in V. In stationary states, eq. (C-4) yields

$$V = \mathcal{E}-U-\boldsymbol{\pi}^2/2m, \tag{C-6}$$

which gives again eq. (C-2).

Equations (C-2) and (C-6) suggest that $(V+U)$ usually will stay finite when $(\nabla^2R)/R$ would become infinite for $|\psi|\to 0$.

For explaining the absence of particles at $|\psi| = 0$, therefore, we should say that small $|\psi|$ means small density ϱ of the ensemble fluid of Section 2.1, so somehow the particles detour around the regions of small $|\psi|$, as we know the continuity equation (10) to hold for this fluid.

EXAMPLE. $\psi = \sqrt{\tfrac{1}{2}}\cdot(e^{ikx}+e^{iky}) = \sqrt{2}\cdot\cos\tfrac{1}{2}\,k(x-y)\exp\tfrac{1}{2}\,ik(x+y)$ gives $\psi = 0$ at $y = x \pm(2n+1)\pi/k$, but gives $(\nabla^2R)/R = -\tfrac{1}{2}\,k^2$ everywhere (also at $R = 0$).

APPENDIX D

MANY-PARTICLE BOHM THEORY

Let $\mathbf{x}$ stand for $\mathbf{x}_1, \mathbf{x}_2, \ldots, \mathbf{x}_N$. Time dependence is understood tacitly. Let

$$\psi(\mathbf{x}) = R(\mathbf{x}) \exp [iS(\mathbf{x})/\hbar] \tag{D-1}$$

and

$$i\hbar \frac{\partial \psi(\mathbf{x})}{\partial t} = \left[-\sum_{j=1}^{N} \frac{\hbar^2}{2m_j} \nabla_j^2 + V(\mathbf{x}) \right] \psi(\mathbf{x}). \tag{D-2}$$

Putting

$$\varrho(\mathbf{x}) \equiv |\psi(\mathbf{x})|^2 = R(\mathbf{x})^2, \tag{D-3}$$

we find

$$\begin{aligned} \frac{\partial \varrho(\mathbf{x})}{\partial t} &= \sum_{j=1}^{N} \frac{i\hbar}{2m_j} \nabla_j \cdot [\psi^* \nabla_j \psi - (\nabla_j \psi^*)\psi] \\ &= -\sum_j \operatorname{div}_j (\varrho \mathbf{v}_j), \end{aligned} \tag{D-4}$$

where $\mathbf{v}_j = \mathbf{v}_i(\mathbf{x})$ is given by

$$\begin{aligned} m_j \mathbf{v}_j = \boldsymbol{\pi}_j(\mathbf{x}) &= \frac{\hbar}{2i} \frac{\psi^* \nabla_j \psi - (\nabla_j \psi^*)\psi}{\psi^* \psi} \\ &= \frac{\hbar}{2i} \nabla_j \ln \left(\frac{\psi}{\psi^*} \right) = \nabla_j S(\mathbf{x}). \end{aligned} \tag{D-5}$$

Now

$$\begin{aligned} \frac{\partial \boldsymbol{\pi}_k}{\partial t} &= Re \left\{ \nabla_k \frac{\hbar}{i} \frac{\partial \psi / \partial t}{\psi} \right\} = Re \left\{ \nabla_k \left[\frac{1}{\psi} \left(\sum_j \frac{\hbar^2}{2m_j} \nabla_j^2 \psi - V\psi \right) \right] \right\} \\ &= -\nabla_k V + \sum_j \frac{\hbar^2}{2m_j} Re \left\{ \nabla_k \left(\frac{1}{\psi} \nabla_j^2 \psi \right) \right\}, \end{aligned} \tag{D-6}$$

where eq. (A-2) holds with all ∇ replaced by ∇_j, so that

$$\begin{aligned} \frac{\partial \boldsymbol{\pi}_k}{\partial t} &= -\nabla_k V + \sum_j \frac{\hbar^2}{2m_j} \nabla_k \left\{ \frac{\nabla_j^2 R}{R} - \left(\frac{\nabla_j S}{\hbar} \right)^2 \right\} \\ &= -\nabla_k V + \sum_j \frac{\hbar^2}{2m_j} \nabla_k \left(\frac{\nabla_j^2 R}{R} \right) - \nabla_k \sum_j \frac{\boldsymbol{\pi}_j^2}{2m_j}. \end{aligned} \tag{D-7}$$

Thence

$$\begin{aligned} \mathbf{f}_k = \frac{d\boldsymbol{\pi}_k}{dt} &= \frac{\partial \boldsymbol{\pi}_k}{\partial t} + \sum_j \frac{\boldsymbol{\pi}_j}{m_j} \cdot \nabla_j \boldsymbol{\pi}_k \\ &= -\nabla_k V + \sum_j \frac{\hbar^2}{2m_j} \nabla_k \left(\frac{\nabla_j^2 R}{R} \right) + \sum_j \left\{ \frac{\boldsymbol{\pi}_j}{m_j} \cdot \nabla_j \boldsymbol{\pi}_k - (\nabla_k \boldsymbol{\pi}_j) \cdot \frac{\boldsymbol{\pi}_j}{m_j} \right\}, \end{aligned} \tag{D-8}$$

so, by $\nabla_j \nabla_k S = \nabla_k \nabla_j S$,

$$\frac{d\boldsymbol{\pi}_k}{dt} = \mathbf{f}_k = -\nabla_k (V + U) \tag{D-9}$$

with

$$U = -\sum_{j=1}^{N} \frac{\hbar^2}{2m_j} \frac{\nabla_j^2 R}{R}, \tag{D-10}$$

with $R = |\psi(\mathbf{x}_1, \mathbf{x}_2, \ldots, \mathbf{x}_N)|$.

Conservation of momentum and energy

From (D-9) we obtain

$$\frac{d}{dt}\sum_k \boldsymbol{\pi}_k = -\sum_k \{\nabla_k(V+U)\}, \tag{D-11}$$

which may be contrasted to the conventional conservation law

$$\frac{d}{dt}\left\langle \sum_k \mathbf{p}_k \right\rangle_\psi = \left\langle -\sum_k \nabla_k V \right\rangle_\psi \tag{D-12}$$

known as Ehrenfest's theorem. [Messiah (1961), page 216.] Here, $\mathbf{p}_{\text{op}} = (\hbar/i)\nabla$. From (D-12) it follows that $\left\langle \sum_k \mathbf{p}_k \right\rangle_\psi$ is conserved when V and ψ depend merely on the *coordinate differences* between the particles. Similarly, $\sum_k \boldsymbol{\pi}_k$ is conserved if $(V+U)$ depends merely on the differences between the particle coordinates.

Conservation of energy takes its simplest form in a *stationary state*, that is, when the conserved energy has a definite value $\mathcal{E}$. Then

$$i\hbar\, \partial\psi/\partial t = \mathcal{E}\psi. \tag{D-13}$$

We combine this with eq. (D-2), and use eqs. (A-2) and (16) both provided with subscripts j (on the ∇ and on the $\boldsymbol{\pi}$). Finally, using (D-1), (D-10), (D-5), and (D-3), we obtain

$$\begin{aligned}(\mathcal{E}-V)\psi &= \sum_j (\hbar^2/2m_j)\, \nabla_j^2\psi \\ &= \Big[U+\sum_j \boldsymbol{\pi}_j^2/2m_j - (i\hbar/2\varrho)\sum_j \nabla_j\boldsymbol{\cdot}(\varrho\mathbf{v}_j)\Big]\psi.\end{aligned} \tag{D-14}$$

Thence

$$\psi^*(\mathcal{E}-V)\psi = \varrho\Big[U+\sum_i \boldsymbol{\pi}_j^2/2m_j\Big] - \tfrac{1}{2}i\hbar\sum_j \nabla_j\boldsymbol{\cdot}(\varrho\mathbf{v}_j). \tag{D-15}$$

Taking the imaginary and the real parts of this equation separately, we find that, for a stationary state,

$$\sum_j \operatorname{div}_j(\varrho\mathbf{v}_j) = 0 \tag{D-16}$$

and

$$\mathcal{E} - V(\mathbf{x}) = U + \sum_j \boldsymbol{\pi}_j^2/2m_j. \tag{D-17}$$

By (D-4), the result (D-16) expresses the well-known fact that in a stationary state the spatial probability distribution does not change:

$$\frac{\partial\varrho(\mathbf{x})}{\partial t} = 0. \tag{D-18}$$

Equation (D-17) may be compared with

$$\langle \mathcal{E} - V(x) \rangle_\psi = \langle \sum_j \mathbf{p}_j^2/2m_j \rangle_\psi. \tag{D-19}$$

As we contrast (D-11) and (D-17) to (D-12) and (D-19), we see that V is used as potential energy when $\mathbf{p} = (\hbar/i)\nabla$ is used as momentum, while $(U+V)$ is used as potential energy when $\boldsymbol{\pi}$ is used as momentum.

APPENDIX E

ANGULAR MOMENTA AND ELECTRON ORBITS IN ATOMIC STATES IN BOHM'S THEORY

(1) A given component of angular momentum

According to eqs. (38)–(39), for a single particle,

$$\mathbf{L}_{\text{op}}\psi(\mathbf{x}) = \boldsymbol{\Lambda}(\mathbf{x})\, \psi(\mathbf{x}) + (\hbar/i)\, \psi(\mathbf{x})\, \mathbf{r} \times \nabla \ln |\psi(\mathbf{x})|. \tag{E-1}$$

If ψ is an eigenfunction of the component of $\mathbf{L}$ in the direction of some unit vector $\mathbf{n}$, it is advantageous to take this direction as the z-axis of a spherical coordinate system. With $L_n \equiv \mathbf{n} \cdot \mathbf{L}$, we obtain in this case

$$L_n\psi = m_l\hbar\, \psi; \quad \psi = F(r, \theta) \exp(im_l\varphi) \equiv R(r, \theta) \exp[i\Phi(r, \theta) + im_l\, \varphi], \tag{E-2}$$

where we put $F \equiv R \exp(i\Phi)$. Since $|\psi| = R$ now is independent of φ, the plane through $\nabla \ln |\psi|$ and $\mathbf{r}$ passes through the z-axis $\mathbf{n}$, so, $\mathbf{L}_{\text{op}}\psi - \boldsymbol{\Lambda}\psi$ by (E-1) is perpendicular to $\mathbf{n}$. Therefore,

$$\Lambda_n\psi = L_n\psi = m_l\hbar\psi. \tag{E-3}$$

In words: "*If ψ is an eigenfunction of some component of $\mathbf{L}_{\text{op}}$ to a certain eigenvalue, the component of $\boldsymbol{\Lambda}$ in that direction has that same value.*"

(2) A one-electron state with quantum numbers l and m_l

Let ψ be given by

$$\psi_{nlm_l}(\mathbf{x}) = \mathcal{R}_{nl}(r) \cdot \Theta_{lm_l}(\theta) \cdot \exp(im_l\varphi) \tag{E-4}$$

with real $\mathcal{R}_{nl}$ and with real Θ_{lm_l}. Then, in (E-2) above, $\Phi(r, \theta) = 0$. We shall write $d_z \equiv (x^2+y^2)^{1/2}$ for the distance from the z-axis. Then, from $S = \hbar m_l\varphi = L_n\varphi = L_z\varphi$ in this case, we obtain

$$\boldsymbol{\pi} = \nabla\left(L_z \arctan \frac{y}{x}\right), \quad \pi_x = -\frac{L_z y}{d_z^2}, \quad \pi_y = +\frac{L_z x}{d_z^2},$$

$$\boldsymbol{\pi} = L_z \mathbf{n} \times \mathbf{r}/d_z^2; \quad |\boldsymbol{\pi}| = L_z/d_z. \tag{E-5}$$

In this case, $\mathbf{v} = \pi/m$ and the imagined electron orbits are in planes perpendicular to the z-direction, as shown in the figure. We have $\Lambda_n = L_n = L_z = d_z|\pi| = mvd_z$, as one would classically expect. The initial hidden variables $\mathbf{x}$ can be arbitrarily chosen, so that the elevation $z = OQ$ of the plane of the orbit, as well as the radius d_z of the orbit, are arbitrary. By (E-5), v is larger when d_z is smaller, so as to provide the correct angular momentum $\Lambda_z = L_z$ according to (E-3). Note, however, that at any point P on the orbit Λ_x and Λ_y do not vanish. It is easily seen that they oscillate in value as P goes around the orbit, and that

$$\mathbf{\Lambda}^2 = L_z^2(1+z^2/d_z^2)\ [\neq \langle \mathbf{L}^2 \rangle_\psi = l(l+1)\hbar^2] \tag{E-6}$$

depends upon m_l and on the arbitrary elevation z and radius d_z rather than upon the quantum number l.

(3) Classical orbital magnetic moment for a Bohm orbit, for ψ an eigenfunction of L_z and of $\mathbf{L}^2$

Let $\varepsilon\ (= -e)$ be the charge of the circulating particle. We then have in Fig. E.1 a circular current $i = \varepsilon v/2\pi d_z = \varepsilon|\pi|/2\pi m d_z = \varepsilon L_z/2\pi d_z^2 m$ around an area $A = \pi d_z^2$. According to Ampère, this is equivalent to a magnetic dipole $\boldsymbol{\mu}$ in the $\mathbf{n}$-direction, with a magnetic moment in Gaussian units given by

$$\mu = \mu_z = Ai/c = (\varepsilon/2mc)\,L_z = (\varepsilon/2mc)\,\Lambda_z. \tag{E-7}$$

We may, *for ψ given by (E-4)*, more generally write

$$\boldsymbol{\mu} = (\varepsilon/2mc)\,\langle \mathbf{L} \rangle_\psi = (\varepsilon/2mc)\,\mathbf{\Lambda}. \tag{E-8}$$

In this special case, Bohm's orbits give zero x- and y-components of $\boldsymbol{\mu}$, even though Λ_x and Λ_y oscillate with $\Lambda_x^2+\Lambda_y^2 = (z/d_z)^2(m_l\hbar)^2$. This is not surprising, as the value assigned

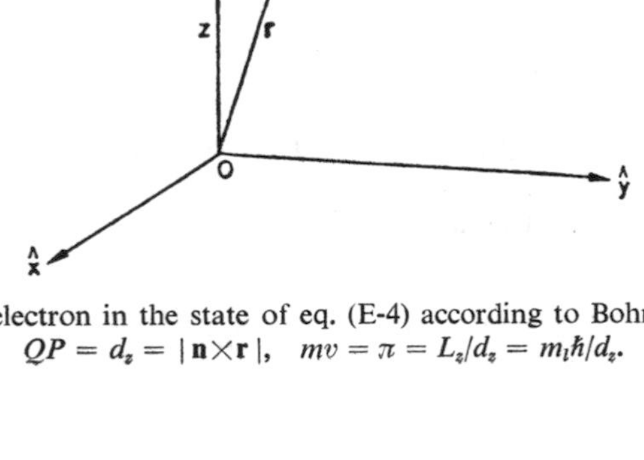

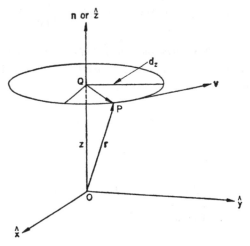

FIG. E.1. The orbit of an electron in the state of eq. (E-4) according to Bohm is a parallel circle. Here, $QP = d_z = |\mathbf{n}\times\mathbf{r}|$, $mv = \pi = L_z/d_z = m_l\hbar/d_z$.

to $\boldsymbol{\mu}$ is based upon the entire closed orbit and not just one line element of it. Thus any linear effect of oscillating components of $\boldsymbol{\Lambda}$ is averaged away in $\boldsymbol{\mu}$. Correspondingly, also in wave mechanics, the expectation value of the vector $(\varepsilon/2mc)\mathbf{L}$ in the state (E-4) is directed along the z-axis, because L_x and L_y have matrix elements only between states which differ ± 1 in m_l, while in (E-4) we have given m_l precisely. Thus *for* ψ *given by* (*E.4*), we find the value

$$\boldsymbol{\mu} = (\varepsilon/2mc)\,\langle\mathbf{L}\rangle_\psi \tag{E-8}$$

for the magnetic moment *for each single Bohm orbit* of the kind shown in the figure.

(4) Orbital magnetic moment averaged over all Bohm orbits, for any wavefunction ψ

When ψ is a superposition of states (E-4) with arbitrary complex coefficients, the phase factor $\exp(iS/\hbar)$ is no longer $\exp(im_l\varphi)$, and the motion of the electron given by $d\mathbf{x}/dt = m^{-1}\nabla S(\mathbf{x})$ may become intricate and certainly quite different from Fig. E.1. It then becomes difficult to calculate $\boldsymbol{\mu}$ for a single orbit, but we can easily calculate the average of $\boldsymbol{\mu}$ over all hidden variable values, which we shall denote by $\langle\boldsymbol{\mu}\rangle$.

We start, this time, with the general formula of Maxwell's theory for the magnetic moment (in Gaussian units) carried by a stationary current density distribution $\mathbf{J}_\varepsilon(\mathbf{x})$, viz.

$$\boldsymbol{\mu} = \int (\mathbf{r}\times\mathbf{J}_\varepsilon(\mathbf{x})/2c)\,d\mathcal{V}. \tag{E-9}$$

[See Jackson (1962), sec. 5.6, or Landau and Lifshitz (1960), sec. 27.] Here it makes no difference whether we take $\mathbf{J}_\varepsilon(\mathbf{x})$ from Bohm's theory or directly from Schrödinger's non-relativistic wave mechanics. In either case, we would write

$$\mathbf{J}_\varepsilon(\mathbf{x}) = \varrho_\varepsilon\mathbf{v}(\mathbf{x}) = \varepsilon\varrho(\mathbf{x})\,\boldsymbol{\pi}(\mathbf{x})/m = \varepsilon Re\{\psi^*\mathbf{p}_{\mathrm{op}}\psi/m\}. \tag{E-10}$$

[Compare eqs. (11)–(12).] Thence, by (38),

$$\langle\boldsymbol{\mu}\rangle = \frac{\varepsilon}{2mc}\int \boldsymbol{\Lambda}(\mathbf{x})\,\varrho(\mathbf{x})\,d\mathcal{V} = \frac{\varepsilon}{2mc}\langle\boldsymbol{\Lambda}\rangle, \tag{E-11}$$

where $\langle\boldsymbol{\Lambda}\rangle$ is the *average of* $\boldsymbol{\Lambda}(\mathbf{x})$ *over the entire hidden-variables distribution* $\bar{P}(\mathbf{x}) = \varrho(\mathbf{x})$, and not just over the $\mathbf{x}$-values around one Bohm orbit as in Fig. E.1.

By integration by parts, it is easily shown that

$$\langle\boldsymbol{\mu}\rangle_\psi \equiv \int \psi^*\,\mathbf{L}_{\mathrm{op}}\psi\,d\mathcal{V} = \int \psi^*\mathbf{r}\times\frac{\hbar}{i}\nabla\psi\,d\mathcal{V} \tag{E-12}$$

is real, and therefore is equal to

$$\int \mathbf{r}\times Re\{\psi^*\mathbf{p}_{\mathrm{op}}\psi\}\,d\mathcal{V} = \int \mathbf{r}\times\varrho(\mathbf{x})\,\boldsymbol{\pi}(\mathbf{x})\,d\mathcal{V} = \langle\boldsymbol{\Lambda}\rangle. \tag{E-13}$$

Therefore (E-11) may be rewritten as

$$\langle\boldsymbol{\mu}\rangle = \frac{\varepsilon}{2mc}\langle\boldsymbol{\Lambda}\rangle = \frac{\varepsilon}{2mc}\langle\mathbf{L}\rangle_\psi. \tag{E-14}$$

The imagined and the conventional angular momentum thus both can be used for finding the average $\langle \mu \rangle$, using the average of the former over all hidden-variables states, or calculating the expectation value of the latter.

APPENDIX F

IMAGINED AND CONVENTIONAL MOMENTUM DISTRIBUTION IN AN *s*-STATE

(1) Electrons standing still in a nonvanishing electrostatic field

For an electron in an *s*-state we can use a real wavefunction ψ, so either $S = 0$ or S is a constant arbitrary phase. In either case, $\boldsymbol{\pi} = \nabla S = 0$, so $\mathbf{v} = 0$. *The imagined or actual particle velocity of an electron in an s-state is zero.*

By (17) we see that in this case the imagined force $\mathbf{f}$ must be zero. That is, the Coulomb force $-\nabla V(\mathbf{x})$ is in this case exactly balanced by the quantum-mechanical force $-\nabla U(\mathbf{x})$.

Electrons may stand still for angular momentum states of arbitrary value of the quantum number l, provided that $m_l = 0$. In that case, the phase factor is absent from eq. (E-4), and $\psi = |\psi|$, which is sufficient for making $\boldsymbol{\pi} = 0$ and $\mathbf{v} = 0$.

Since the z-direction can be chosen in some arbitrary direction $\mathbf{n}$, we may reformulate our result as follows:

> "*If the wavefunction ψ is an eigenfunction of the component of the angular momentum* $\mathbf{L}$ *in some arbitrary direction* $\mathbf{n}$, *and the corresponding eigenvalue of this component is zero, then the imagined velocity of any electron in this ψ-field is zero.*" (F-1)

(2) The momentum distribution in an *s*-state

We consider again an electron in a spherically symmetric state $\psi(\mathbf{x}) = R(r)$. This state $R(r)$ might be the ground state of an atom or it might be the wavefunction of a free electron instantaneously described by a Gaussian wave packet for a particle *on the average* at rest. In the latter case,

$$\psi(\mathbf{x}) = (2\pi\sigma_o^2)^{-\frac{3}{4}} \exp(-\mathbf{x}^2/4\sigma_o^2), \tag{F-2}$$

where σ_o is the initial dispersion of the Gaussian distribution $\varrho(\mathbf{x}) = |\psi(\mathbf{x})|^2$.

In either case, $\boldsymbol{\pi}(\mathbf{x}) = -\nabla S = 0$ everywhere, so that, at least *initially*, the electron is at rest wherever it is located in the wave packet. Due to the dissipation of a Gaussian wave packet on account of the Schrödinger equation, however, the free electron with the initial wavefunction (F-2) would not remain at rest. (See Appendix G.)

Now suppose somehow we make immediately and instantaneously a measurement of the conventional momentum **p**. What Bohm's theory predicts for this measurement, as a function of the hidden variable **x**, is the same as what quantum theory predicts. (See Section 2-11.) We shall here calculate the predicted probability distribution for the measured **p**.

Quantum theory predicts a probability $|A(\mathbf{p})|^2 d^3\mathbf{p}$ that the measured momentum **p** will be within the volume element $d^3\mathbf{p}$ in momentum space, where

$$A(\mathbf{p}) = h^{-\frac{3}{2}} \int d^3\mathbf{x}\, \psi(\mathbf{x}) \exp(i\mathbf{x}\cdot\mathbf{p}/\hbar). \tag{F-3}$$

In the Gaussian wave packet given by (F-2), we find

$$\begin{aligned} |A(\mathbf{p})|^2 &= (2\sigma_o^2/\pi\hbar^2)^{\frac{3}{2}} \exp(-2p^2\sigma_o^2/\hbar^2) \\ &= (2\pi\sigma_p^2)^{-\frac{3}{2}} \exp(-p^2/2\sigma_p^2), \end{aligned} \tag{F-4}$$

which is a Gaussian distribution with dispersion $\sigma_p = \frac{1}{2}\hbar/\sigma_o$ in momentum space. (This value of σ_p is what the Heisenberg "uncertainty" principle predicts.)

In the groundstate of a hydrogen atom, which is given by

$$\psi(\mathbf{x}) = (\pi r_o^3)^{-\frac{1}{2}} \exp(-r/r_o), \tag{F-5}$$

we find from (F-3)

$$|A(\mathbf{p})|^2 = (8r_o^3/\pi^2\hbar^3)\cdot[1+\mathbf{p}^2 r_o^2/\hbar^2]^{-4}, \tag{F-6}$$

so that there is a probability

$$\left\{\frac{2}{\pi} \operatorname{arc\,cot} f - \frac{6f^5+16f^3-6f}{3\pi(1+f^2)^3}\right\} \tag{F-7}$$

that $|\mathbf{p}|$ is found to be *larger* than $f\cdot\hbar/r_o$, where f is any numerical constant.

This shows that the **p** distribution of quantum theory is quite different from the $\boldsymbol{\pi}$ distribution in Bohm's theory, which has *zero* probability of being larger than a fraction f of $\hbar/r_o$.

In Appendix G we elaborate on the case of Eq. (F-2) as a function of time.

APPENDIX G

THE GAUSSIAN WAVE PACKET

Below we shall apply Bohm's theory to a Gaussian wave packet. We write σ_o for the root-mean-square deviation in each coordinate direction, in the ensemble described by ψ, at the instant when this wave packet is a *minimum-uncertainty wave packet*. [See, for instance, Merzbacher (1970), pages 160–161.] Choosing this instant as the zeropoint of time t, and choosing the spatial origin at the center of the wave packet at $t = 0$, we may write ψ at $t = 0$ (except for a complex constant phase factor of absolute value 1 which we arbitrarily choose equal to 1) as

$$\psi_o(\mathbf{x}) = (2\pi\sigma_o^2)^{-\frac{3}{4}} \exp(i\mathbf{k}\cdot\mathbf{x} - \mathbf{x}^2/4\sigma_o^2), \tag{G-1}$$

where $\mathbf{k} = \langle \mathbf{p} \rangle_{\mathrm{Av}}/\hbar$. If the particle is free, solution of $i\hbar\partial\psi/\partial t = -(\hbar^2/2m)\nabla^2\psi$ shows that, at time t, this wave packet is given by

$$\psi_t(\mathbf{x}) = (2\pi s_t^2)^{-\frac{3}{4}} \exp\,[i\mathbf{k}\cdot(\mathbf{x}-\tfrac{1}{2}\,\mathbf{u}t)-(\mathbf{x}-\mathbf{u}t)^2/4\sigma_o s_t], \tag{G-2}$$

where we put

$$s_t = \sigma_o\left(1+\frac{i\hbar t}{2m\sigma_o^2}\right), \quad \mathbf{u} = \frac{\hbar\mathbf{k}}{m} = \frac{\langle\mathbf{p}\rangle}{m}. \tag{G-3}$$

The Fourier transform of ψ_t is defined by

$$A_t(\mathbf{p}) = h^{-\frac{3}{2}}\int d^3\mathbf{x}\,\psi_t(\mathbf{x})\exp\,(-i\mathbf{x}\cdot\mathbf{p}/\hbar), \tag{G-4}$$

so that

$$\psi_t(\mathbf{x}) = h^{-\frac{3}{2}}\int d^3\mathbf{p}\,A_t(\mathbf{p})\exp\,(i\mathbf{x}\cdot\mathbf{p}/\hbar), \tag{G-5}$$

$$1 = \int|\psi_t(\mathbf{x})|^2\,d^3\mathbf{x} = \int|A_t(\mathbf{p})|^2\,d^3\mathbf{p}. \tag{G-6}$$

Using

$$\int_{-\infty}^{+\infty}\exp\,(-zx^2+irx)\,dx = (\pi/z)^{\frac{1}{2}}\exp\,(-r^2/4z) \tag{G-7}$$

for $z = a^2+ib$ with real a, b, and r, we find

$$A_t(\mathbf{p}) = \left(\frac{\sigma_o}{\hbar}\sqrt{\frac{2}{\pi}}\right)^{\frac{3}{2}}\exp\left\{-\sigma_o^2\left(\frac{\mathbf{p}}{\hbar}-\mathbf{k}\right)^2-\frac{i}{\hbar}\left(\frac{\mathbf{p}^2}{2m}\right)t\right\}. \tag{G-8}$$

This gives in momentum space the Gaussian distribution

$$|A_t(\mathbf{p})|^2 = (2\pi\sigma_p^2)^{-\frac{3}{2}}\exp\,[-(\mathbf{p}-\hbar\mathbf{k})^2/2\sigma_p^2] \tag{G-9}$$

with dispersion σ_p given by

$$\sigma_p = \hbar/2\sigma_o, \tag{G-10}$$

while in ordinary space we find from (G-2) the Gaussian distribution

$$\varrho_t(\mathbf{x}) \equiv R_t(\mathbf{x})^2 \equiv |\psi_t(\mathbf{x})|^2 = (2\pi\sigma_t^2)^{-\frac{3}{2}}\exp\,[-(\mathbf{x}-\mathbf{u}t)^2/2\sigma_t^2] \tag{G-11}$$

with dispersion σ_t given by

$$\sigma_t = |s_t| = \sigma_o[1+(\hbar t/2m\sigma_o^2)^2]^{\frac{1}{2}}. \tag{G-12}$$

We define the "*dispersion speed*" of the packet as $d\sigma_t/dt$, and see from (G-12) that for $t \to \infty$ we obtain asymptotically

$$\sigma_t \to \hbar t/2m\sigma_o, \quad d\sigma_t/dt \to \hbar/2m\sigma_o = \sigma_p/m, \tag{G-13}$$

which is for electrons of the order of 5 cm s^{-1} for $\sigma_o \approx 0.1$ cm, but is of the order of 100 km s^{-1} for $\sigma_o \approx 5$ Å.

In general, the uncertainty principle is given by $\sigma_x\cdot\sigma_{p_x} \geqslant \frac{1}{2}\,\hbar$. [See Merzbacher (1970), chap. 8.] For the wave packet here considered, at time t, this takes the form

$$\sigma_t\cdot\sigma_p = \tfrac{1}{2}\,\hbar\,[1+(\hbar t/2m\sigma_o^2)^2]^{\frac{1}{2}}, \tag{G-14}$$

which shows that minimum uncertainty is achieved only instantaneously at $t = 0$.

Application of Bohm's theory

While the wave packet as a whole moves at a uniform average velocity **u**, before $t = 0$ it is shrinking in size, and after $t = 0$ it again is dissipated. Correspondingly, we expec particles in the outer parts of the wave packet (individual particles from the outer parts o the ensemble E_ψ) to be moving inward toward the center of the packet at $t < 0$, and outward at $t > 0$. From a quantum-theoretical point of view, one would expect these free particles to shoot at constant velocity $\mathbf{p}/m$ somewhere through the center of the wave packet $\psi_o(\mathbf{x})$ at $t = 0$, and to emerge at $t > 0$ on the other side of the wave packet. The time independence of the momentum distribution (G-9) agrees with this point of view. In Bohm's theory, however, the velocities of the individual particles are given not by $\mathbf{p}/m$, but by $\boldsymbol{\pi}/m$, which is not the same. (See Section 2.11.) The center of the wave packet provides here by eq. (18) a repulsive quantum-mechanical potential U, which brings the incoming particles to an instantaneous stop at $t = 0$, so that the particles at $t > 0$ will bounce back toward the directions from which they came.

Moreover, the symmetry of the wave packet in Bohm's theory will force all individual particles onto strictly radial trajectories, aiming exactly for the center of the wave packet at $t = 0$ in a frame of reference moving at velocity **u**. This is a marked difference from the lack at correlation at $t = 0$ between **x** and **p** in the distributions $|\psi_o(\mathbf{x})|^2$ and $|A(\mathbf{p})|^2$ expected according to unprejudiced quantum theory.

The qualitative remarks contained in the above are substantiated by the following calcu lations.

Calculation of the Bohm trajectories

We start by writing

$$\psi_t = R_t \exp(iS_t/\hbar) \tag{G-15}$$

with

$$R_t = (2\pi\sigma_t^2)^{-\frac{3}{4}} \exp[-(\mathbf{x}-\mathbf{u}t)^2/4\sigma_t^2], \tag{G-16}$$

$$S_t = \hbar[-3\varphi+\mathbf{k}\cdot(\mathbf{x}-\tfrac{1}{2}\mathbf{u}t)+(\mathbf{x}-\mathbf{u}t)^2\hbar t/8m\sigma_o^2\sigma_t^2], \tag{G-17}$$

where we defined φ by

$$\tan 2\varphi = \hbar t/2m\sigma_o^2, \tag{G-18}$$

so that

$$s_t = \sigma_t \exp(2i\varphi) \quad \text{and} \quad \sigma_o = \sigma_t \cos 2\varphi \tag{G-19}$$

and

$$1/\sigma_o s_t = (\cos 2\varphi - i \sin 2\varphi)/\sigma_o\sigma_t = \sigma_t^{-2}(1-i\hbar t/2m\sigma_o^2). \tag{G-20}$$

From (G-17) we obtain for Bohm's particle motion

$$m\, d\mathbf{x}/dt = \boldsymbol{\pi}_t = \nabla S_t = \hbar\mathbf{k}+(\mathbf{x}-\mathbf{u}t)\hbar^2 t/4m\sigma_o^2\sigma_t^2. \tag{G-21}$$

For the distance

$$\mathbf{X} \equiv \mathbf{x}-\mathbf{u}t \tag{G-22}$$

of a particle from the center of the wave packet, we thus find

$$\frac{d\mathbf{X}}{dt} = \frac{d\mathbf{x}}{dt} - \frac{\hbar\mathbf{k}}{m} = \frac{\mathbf{X}t\hbar^2}{4m^2\sigma_o^4 + \hbar^2t^2}. \tag{G-23}$$

This is integrated by

$$\mathbf{X} = \mathbf{X}_o[1 + (\hbar t/2m\sigma_o^2)^2]^{\frac{1}{2}}, \tag{G-24}$$

where $\mathbf{X}_o = \mathbf{x}_o$. This confirms our above remarks about the motion of individual particles making up the Gaussian wave packet according to Bohm's theory.

The imagined or actual momentum distribution

The $\boldsymbol{\pi}$ distribution of the particles in the ensemble described by ψ_t can be obtained directly from their $\mathbf{x}$ distribution, as $\boldsymbol{\pi}$ and $\mathbf{x}$ at time t according to eq. (G-21) are related by

$$\mathbf{x} - \mathbf{u}t = \frac{4m\sigma_o^2\sigma_t^2}{\hbar^2 t} \cdot (\boldsymbol{\pi}_t - \hbar\mathbf{k}). \tag{G-25}$$

As the probability $\bar{P}_{\pi t}(\boldsymbol{\pi}_t)\, d^3\boldsymbol{\pi}_t$ for an infinitesimal interval of $\boldsymbol{\pi}_t$-values at time t must be equal to the probability $\bar{P}_t(\mathbf{x})\, d^3\mathbf{x} = \varrho_t(\mathbf{x})\, d^3\mathbf{x}$ of the corresponding $\mathbf{x}$-interval, we then have

$$\bar{P}_{\pi t}(\boldsymbol{\pi}_t) = \varrho_t(\mathbf{x}) \cdot \left| \frac{\partial(x, y, z),}{\partial(\pi_{tx}, \pi_{ty}, \pi_{tz})} \right|, \tag{G-26}$$

where the Jacobian, according to (G-25), is equal to $(4m\sigma_o^2\sigma_t^2/\hbar^2 t)^3$. With $\varrho_t(\mathbf{x})$ from (G-11) we thus find the Gaussian distribution

$$\bar{P}_{\pi t}(\boldsymbol{\pi}_t) = (2\pi\sigma_{\pi t}^2)^{-\frac{3}{2}} \exp\left[-(\boldsymbol{\pi}_t - \hbar\mathbf{k})^2/2\sigma_{\pi t}^2\right] \tag{G-27}$$

with

$$\sigma_{\pi t} = \hbar^2 t/4m\sigma_o^2\sigma_t = \sigma_p \cdot [1 + (2m\sigma_o^2/\hbar t)^2]^{-\frac{1}{2}}. \tag{G-28}$$

Notice that, for $t \to \infty$, $\sigma_{\pi t}$ approaches σ_p from below. For $t = 0$ we find $\sigma_{\pi o} = 0$, as, at $t = 0$, all $\boldsymbol{\pi}$ are instantaneously equal to $\hbar\mathbf{k}$ on account of (G-21). This is the instant at which Bohm's electrons have reached their closest approach to the center of the wave packet. Thus $\psi_o(\mathbf{x})$ of eq. (G-1) describes according to Bohm at $t = 0$, a situation where there is indeterminacy of position, but the velocity would exactly be given to be $\hbar\mathbf{k}/m$.

APPENDIX H

BOHM'S QUANTUM ELECTRODYNAMICS

Variation principle and classical equations in the radiation gauge

We shall work in the radiation gauge (Coulomb gauge) in which we take the following equations as definitions:[80]

$$\operatorname{div} \mathbf{A} = 0, \quad \mathbf{B} = \operatorname{curl} \mathbf{A}, \quad \mathbf{E}_\perp = -\partial \mathbf{A}/c\partial t. \tag{H-1}$$

$$\varrho_\varepsilon(\mathbf{x}, t) = \sum_i \varepsilon_i \delta(\mathbf{x}-\mathbf{x}_i(t)), \quad \mathbf{J}(\mathbf{x}, t) = \sum_i \varepsilon_i \mathbf{v}_i \delta(\mathbf{x}-\mathbf{x}_i(t)). \tag{H-2}$$

$$\left.\begin{aligned} &\mathbf{v}_i = d\mathbf{x}_i/dt; \quad r = |\mathbf{x}'-\mathbf{x}|, \quad \mathbf{r}_{ji} = \mathbf{x}_i - \mathbf{x}_j, \quad r_{ij} = |\mathbf{r}_{ij}|. \\ &V_C(\mathbf{x}, t) = \int d^3\mathbf{x}' \varrho_\varepsilon(\mathbf{x}', t)/r, \quad \text{so} \quad V_C(\mathbf{x}_i) = \sum_j \varepsilon_j / r_{ij}. \end{aligned}\right\} \tag{H-3}$$

$$\mathbf{E}_{||} = -\nabla V_C, \quad \text{so} \quad \mathbf{E}_{||}(\mathbf{x}_i) = \sum_j \varepsilon_j \mathbf{r}_{ji}/r_{ij}^3; \quad \mathbf{E} = \mathbf{E}_{||} + \mathbf{E}_\perp. \tag{H-4}$$

A variational principle for classical electrodynamics in the radiation gauge was published by Belinfante (1951b). In it we change the term $\mathbf{A}\cdot\partial\mathbf{E}_\perp/4\pi c\, \partial t$ into $-\mathbf{E}_\perp\cdot\partial\mathbf{A}/4\pi c\, \partial t$, and use the above definitions for expressing everything in terms of the independent variables $\mathbf{x}_i(t)$ and $\mathbf{A}(\mathbf{x}, t)$. We also replace the matter Lagrangian $-\sum_i m_i c^2(1-\mathbf{v}_i^2/c^2)^{1/2}$ by its non-relativistic approximation $\sum_i \frac{1}{2} m_i \mathbf{v}_i^2$. Thus we obtain $\delta \int L\, dt = 0$ with

$$\begin{aligned} L = &\sum_i \tfrac{1}{2} m_i \dot{\mathbf{x}}_i^2 - \sum_i \sum_j \varepsilon_i \varepsilon_j / 2r_{ij} + (8\pi)^{-1} \int d^3\mathbf{x} \{ (\partial \mathbf{A}/c\partial t)^2 - (\operatorname{curl} \mathbf{A})^2 \} \\ &+ \int d^3\mathbf{x} \sum_i \varepsilon_i \delta(\mathbf{x}-\mathbf{x}_i(t))\, \mathbf{A}(\mathbf{x}, t)\cdot d\mathbf{x}_i / c\, dt. \end{aligned} \tag{H-5}$$

Variation of the transverse field $\mathbf{A}(\mathbf{x}, t)$, with $\operatorname{div} \mathbf{A} = 0$, gives

$$(4\pi)^{-1} \{ -\partial^2 \mathbf{A}/c^2 \partial t^2 + \nabla^2 \mathbf{A} + 4\pi \mathbf{J}(\mathbf{x})/c \}_\perp = 0, \tag{H-6}$$

where $_\perp$ means omission of longitudinal waves in a Fourier expansion. Variation of $\mathbf{x}_i(t)$ gives, by (H-1)–(H-4),

$$m_i \ddot{\mathbf{x}}_i = \varepsilon_i [\mathbf{E}(\mathbf{x}_i) + \mathbf{v}_i \times \mathbf{B}(\mathbf{x}_i)/c]. \tag{H-7}$$

From (H-5) we find the canonical momenta and the Hamiltonian:

$$\mathbf{p}_i = \partial L/\partial \dot{\mathbf{x}}_i = m_i \dot{\mathbf{x}}_i + \varepsilon_i \mathbf{A}(\mathbf{x}_i)/c, \tag{H-8}$$

$$\mathbf{P}_A = \delta L/\delta(\partial \mathbf{A}/c\partial t) = \partial \mathbf{A}/4\pi c^2 \partial t = -\mathbf{E}_\perp/4\pi c, \tag{H-9}$$

$$\begin{aligned} \mathscr{H} &= \mathscr{H}(\mathbf{x}_i, \mathbf{A}, \mathbf{p}_i, \mathbf{P}_A) \\ &= \sum_i (2m_i)^{-1} [\mathbf{p}_i - \varepsilon_i \mathbf{A}(\mathbf{x}_i)/c]^2 + \sum_i \sum_j \varepsilon_i \varepsilon_j / 2r_{ij} + \int d^3\mathbf{x}\, [(4\pi c \mathbf{P}_A)^2 + (\operatorname{curl} \mathbf{A})^2] \\ &= \sum_i \tfrac{1}{2} m_i \mathbf{v}_i^2 + \sum_i \sum_j \varepsilon_i \varepsilon_j / 2r_{ij} + (8\pi)^{-1} \int d^3\mathbf{x}\, [\mathbf{E}_\perp^2 + \mathbf{B}^2]. \end{aligned} \tag{H-10}$$

[80] We ignore self-energy problems and will not comment on the $i = j$ case in sums of terms with denominators r_{ij}.

Expansion in polarized waves

We write k for $|\mathbf{k}|$ and $\omega = ck$. We make Fourier expansions in a cubic volume $\mathcal{V} = L^3$, with k_x, k_y, $k_z = \pm$ multiples of $2\pi/L$. We expand vector fields in circularly polarized waves,[81] using complex unit vectors $\mathbf{e}_\mathbf{k}^\eta$ satisfying, for $\eta = \pm$, and with $\mathbf{e}_k^0 \equiv \mathbf{k}/k$,

$$\mathbf{e}_\mathbf{k}^{\eta *} = \mathbf{e}_\mathbf{k}^{-\eta} = -\mathbf{e}_{-\mathbf{k}}^\eta; \quad \mathbf{e}_\mathbf{k}^{\eta *}\cdot\mathbf{e}_\mathbf{k}^{\eta'} = \delta_{\eta\eta'}; \quad \mathbf{e}_\mathbf{k}^0\times\mathbf{e}_\mathbf{k}^\eta = -i\eta\mathbf{e}_\mathbf{k}^\eta; \quad \mathbf{e}_\mathbf{k}^{\eta *}\times\mathbf{e}_\mathbf{k}^\eta = i\eta\mathbf{e}_\mathbf{k}^0. \tag{H-11}$$

Especially, if $\mathbf{k} = \eta' k\hat{z}$ with $\eta' = \pm$, we would satisfy (H-11) by $\mathbf{e}_{\eta' k\hat{z}}^\eta = (\eta'\hat{x} + i\eta\hat{y})\sqrt{\tfrac{1}{2}}$ in terms of unit vectors $\hat{x}$, $\hat{y}$, $\hat{z}$ along the axes.

We then expand

$$\mathbf{A}(\mathbf{x}, t) = c(4\pi/\mathcal{V})^{\frac{1}{2}} \sum_\mathbf{k}\sum_\eta \mathbf{e}_\mathbf{k}^\eta\, q_{\mathbf{k},\eta}(t)\, e^{i\mathbf{k}\cdot\mathbf{x}}, \tag{H-12}$$

so

$$\mathbf{B}(\mathbf{x}, t) = c(4\pi/\mathcal{V})^{\frac{1}{2}} \sum_{\mathbf{k},\eta} \eta k\, \mathbf{e}_\mathbf{k}^\eta\, q_{\mathbf{k},\eta}(t)\, e^{i\mathbf{k}\cdot\mathbf{x}}. \tag{H-13}$$

We expand $\mathbf{E}_\perp$ by

$$\mathbf{E}_\perp(\mathbf{x}, t) = -4\pi c\mathbf{P}_A(\mathbf{x}, t) = -(4\pi/\mathcal{V})^{\frac{1}{2}} \sum_{\mathbf{k},\eta} \mathbf{e}_\mathbf{k}^\eta\, p_{\mathbf{k},\eta}^*(t)\, e^{i\mathbf{k}\cdot\mathbf{x}}, \tag{H-14}$$

so, by (H-1),

$$p_{\mathbf{k},\eta}^*(t) = dq_{\mathbf{k},\eta}(t)/dt. \tag{H-15}$$

The reality of the fields is expressed by the conditions

$$q_{\mathbf{k},\eta}^* = -q_{-\mathbf{k},\eta}, \quad p_{\mathbf{k},\eta}^* = -p_{-\mathbf{k},\eta}. \tag{H-16}$$

We shall also introduce the real and imaginary parts of these coefficients. This, by (H-16), mixes up coefficients for opposite values of $\mathbf{k}$. Then it is preferable to write sums like $\sum_\mathbf{k} F(\mathbf{k})$ as $\sum_{\mathbf{k}/2}\sum_{\eta'} F(\eta'\mathbf{k})$, where $\sum_{\mathbf{k}/2}$ is a sum over $\mathbf{k}$ in a chosen halfspace, and $\eta' = \pm$. Let $\sum_{1/2}$ denote $\sum_{\mathbf{k}/2}\sum_\eta$. For brevity we shall sometimes not write explicitly labels η or $\mathbf{k}$, and by index + or − indicate whether $\mathbf{k}$ or $-\mathbf{k}$ is in the chosen halfspace. Then (H-16) becomes $q_+^* = -q_-$, $q_-^* = -q_+$, and similarly for p. We then put

$$\left.\begin{aligned} q_+ &= (q_R + iq_I)\sqrt{\tfrac{1}{2}}, \quad q_- = (-q_R + iq_I)\sqrt{\tfrac{1}{2}}, \\ p_+ &= (p_R - ip_I)\sqrt{\tfrac{1}{2}}, \quad p_- = (-p_R - ip_I)\sqrt{\tfrac{1}{2}}. \end{aligned}\right\} \tag{H-17}$$

Inserting the expansions in L and $\mathcal{H}$, we find

$$L = \sum_i \left\{\tfrac{1}{2}m_i\dot{\mathbf{x}}_i^2 + \frac{\varepsilon_i}{c}\mathbf{A}(\mathbf{x}_i)\cdot\dot{\mathbf{x}}_i\right\} - \sum_i\sum_j \frac{\varepsilon_i\varepsilon_j}{2r_{ij}} + L_F, \tag{H-18}$$

[81] See Kramers (1957), page 420. His $\mathbf{c}_\lambda$ = our $(k/2\pi)\mathbf{e}_\mathbf{k}^+$. His $\mathbf{c}_{-\lambda}$ = our $-(k/2\pi)\mathbf{e}_{-\mathbf{k}}^+$. We shall here call a wave right-hand circularly polarized when at a fixed point in the beam the polarization vector rotates clockwise *when looking in the direction the beam is propagating*. (In optics this is often called left-hand circularly polarized.)

$$L_F = \tfrac{1}{2}\sum_{\mathbf{k}}\sum_{\eta}\{\dot{q}^*_{\mathbf{k},\eta}\dot{q}_{\mathbf{k},\eta} - c^2k^2 q^*_{\mathbf{k},\eta}q_{\mathbf{k},\eta}\} = \sum_{1/2}(-\dot{q}_+\dot{q}_- + c^2k^2\, q_+q_-)$$
$$= \tfrac{1}{2}\sum_{1/2}[\dot{q}_R^2 + \dot{q}_I^2 - \omega^2(q_R^2+q_I^2)], \tag{H-19}$$

$$\mathscr{H} = \sum_j \frac{1}{2m_j}\left[\mathbf{p}_j - \frac{\varepsilon_j}{c}\mathbf{A}(\mathbf{x}_j)\right]^2 + \sum_i\sum_j \frac{\varepsilon_i\varepsilon_j}{2r_{ij}} + \mathscr{H}_F, \tag{H-20}$$

$$\mathscr{H}_F = \tfrac{1}{2}\sum_{\mathbf{k},\eta}(p_{\mathbf{k},\eta}\, p^*_{\mathbf{k},\eta} + c^2k^2 q^*_{\mathbf{k},\eta}\, q_{\mathbf{k},\eta}) = \sum_{1/2}(-p_+p_- - c^2k^2 q_+q_-)$$
$$= \tfrac{1}{2}\sum_{1/2}[p_R^2 + p_I^2 + \omega^2(q_R^2+q_I^2)]. \tag{H-21}$$

Note that, for all $\mathbf{k}$ and η,

$$p_{R\eta}(t) = dq_{R\eta}(t)/dt, \quad p_{I\eta}(t) = dq_{I\eta}(t)/dt. \tag{H-22}$$

In free-radiation approximation [neglecting the term with $\mathbf{A}$ in (H-18) and (H-20)], we find the time dependence

$$\left.\begin{aligned} q_{\mathbf{k},\eta}(t) &\approx \sum_{\eta''=\pm} q^{\eta''}_{\mathbf{k},\eta}(0)\exp(-i\eta''\omega t),\\ p_{\mathbf{k},\eta}(t) &\approx \sum_{\eta''} i\eta''\omega q^{\eta''*}_{\mathbf{k},\eta}(0)\exp(i\eta''\omega t). \end{aligned}\right\} \tag{H-23}$$

Inserting this in (H-12)–(H-14), we find waves propagating in the direction $\eta''\mathbf{k}$, which are right-hand (left-hand) circularly polarized for $\eta = +$ (for $\eta = -$).

Quantization

We quantize in a representation which diagonalizes the $\mathbf{x}_i$ and the q_R and q_I for all $\mathbf{k}$ and η. We treat the p_i and the p_R and p_I as operators acting on a wavefunction $\psi(\mathbf{x}_i, q_R, q_I)$ according to

$$\mathbf{p}_{j,\,\mathrm{op}} = \frac{\hbar}{i}\nabla_j \equiv \frac{\hbar}{i}\frac{\partial}{\partial \mathbf{x}_j}, \quad p_{f,\,\mathrm{op}} = \frac{\hbar}{i}\nabla_f \equiv \frac{\hbar}{i}\frac{\partial}{\partial q_f}, \tag{H-24}$$

where f means R or I, for both η values and for any $\mathbf{k}$ inside the chosen halfspace. Thus

$$q_f p_{f'} - p_{f'} q_f = i\hbar\,\delta_{f,f'} \tag{H-25}$$

(which includes a factor $\delta_{\mathbf{k},\mathbf{k}'}\delta_{\eta,\eta'}$). Thence it follows that

$$A_m(\mathbf{x})P_{An}(\mathbf{x}') - P_{An}(\mathbf{x}')A_m(\mathbf{x})$$
$$= i\hbar\mathcal{V}^{-1}\sum_{\mathbf{k}} e^{i\mathbf{k}\cdot(\mathbf{x}-\mathbf{x}')}\sum_{\eta}(e^{\eta}_{\mathbf{k}})_m(e^{\eta}_{\mathbf{k}})^*_n = i\hbar\,\delta_{\perp mn}(\mathbf{x}-\mathbf{x}'), \tag{H-26}$$

where $\delta_{\perp mn}(\mathbf{x})$ is the transverse deltafunction. [Belinfante (1946).]

The Schrödinger equation for interacting particles and Maxwell fields now uses a Hamilton operator obtained from (H-20) to (H-21) by (H-24):

$$i\hbar\partial\psi/\partial t = \mathscr{H}_{\mathrm{op}}\psi = \left\{\sum_j \frac{1}{2m_j}\left[\frac{\hbar}{i}\nabla_j - \frac{\varepsilon_j}{c}\mathbf{A}(\mathbf{x}_j)\right]^2 + \sum_i\sum_j\frac{\varepsilon_i\varepsilon_j}{2r_{ij}}\right.$$
$$\left. + \tfrac{1}{2}\sum_f[-\hbar^2\nabla_f^2 + \omega_f^2 q_f^2] - \mathscr{E}_{\mathrm{vac}}\right\}\psi(\mathbf{x}_j, q_f), \tag{H-27}$$

where we subtract an infinite vacuum energy. [See eq. (H-29).[80]]

Perturbation theory

We may now treat

$$\mathcal{H}_M = -\sum_j \frac{\hbar^2}{2m_j} \nabla_j^2 + \sum_i \sum_j \frac{\varepsilon_i \varepsilon_j}{2r_{ij}} \tag{H-28}$$

as the unperturbed matter Hamiltonian, of which the atomic energy levels are eigenvalues, and we may treat

$$\mathcal{H}_F - \mathcal{E}_{\text{vac}} = \tfrac{1}{2} \sum_f [-\hbar^2 \nabla_f^2 + \omega_f^2 q_f^2 - \hbar\omega_f] \tag{H-29}$$

as the Hamiltonian of the free radiation field. The remainder,

$$\mathcal{H}_i = \sum_j [(\hbar\varepsilon_j / i m_j c)\, \mathbf{A}(\mathbf{x}_j) \cdot \nabla_j - (\varepsilon_j^2 / 2m_j c^2)\, \mathbf{A}(\mathbf{x}_j)^2], \tag{H-30}$$

is then the perturbation causing transitions of energy between atoms and radiation field. Here, $\mathbf{A}(\mathbf{x}_j)$ is given by inserting $\mathbf{x}_j$ for $\mathbf{x}$ in

$$\mathbf{A}(\mathbf{x}) = c(2\pi/\mathcal{V})^{\frac{1}{2}} \sum_{\mathbf{k}/2} \sum_{\eta} [q_{R\eta}(\mathbf{e}_{\mathbf{k}}^{\eta} E^+ - \mathbf{e}_{-\mathbf{k}}^{\eta} E^-) + i q_{I\eta}(\mathbf{e}_{\mathbf{k}}^{\eta} E^+ + e_{-\mathbf{k}}^{\eta} E^-)], \tag{H-31}$$

where we wrote $E^{\pm}$ for $\exp(\pm i\mathbf{k} \cdot \mathbf{x})$.

Theoretical photon states

The operator (H-29) is a sum of harmonic oscillator Hamiltonians each with its levels shifted down $\frac{1}{2}\hbar\omega$ so as to make the ground state energy of (H-29) equal to zero. This shift does not affect the ground state eigenfunction, which is[82]

$$\psi_{\text{vac}}^{(F)}(q_f) = \prod_f \left\{ \left(\frac{\omega_f}{\pi\hbar} \right)^{1/4} \exp\left(-\frac{\omega_f}{2\hbar} q_f^2 \right) \right\}. \tag{H-32}$$

The product here is over all q_R and q_I, for both η-values and for all $\mathbf{k}$-values of our $\mathbf{k}$-lattice in the chosen halfspace. Other eigenvalues of the unperturbed radiation field energy lie

$$\mathcal{E}_{\text{phot}} = \sum_f n_f \hbar\omega_f \tag{H-33}$$

above its ground level. The corresponding eigenfunctions are[82]

$$\psi_n^{(F)}(q_f) = \psi_{\text{vac}}^{(F)}(q_f) \cdot \prod_f \left\{ (\sqrt{\tfrac{1}{2}})^n (n!)^{-\frac{1}{2}} H_n\left(q_f \sqrt{\frac{\omega}{\hbar}} \right) \right\}, \tag{H-34}$$

where n is an abbreviation for n_f, and where the H_n are Hermite polynomials.

Standing photons

One might be tempted to call (H-33)–(H-34) a state with n_f photons present of the type f. This, however, would be a misleading name. This is best seen by studying the electromagnetic waves that are excited when a certain q_f and p_f become excited. From (H-31) we find the

[82] See, for instance, Merzbacher (1970), chap. 5, Sec. 3.

A-wave excited at $t = 0$, and thence the **B**-wave; from (H-17) with (H-14) we find the $\mathbf{E}_\perp$-wave. From (H-23) at $t = 0$, we then find the $q_{\mathbf{k}}^{\eta''}(0)$ for the wave excited; thence, by (H-23) for $t \neq 0$, the free-field time dependence of the $q_{\mathbf{k},\eta}$, and thus the time dependence of the wave. We thus find that, if only *one* wave $(q_{R\eta}, p_{R\eta})$ or $(q_{I\eta}, p_{I\eta})$ for one **k** is excited, we have a standing electromagnetic wave!

Running photons

If we want to find photons that correspond to running waves, we must excite at least a $(q_{R\eta}, p_{R\eta})$ wave and the corresponding $(q_{I\eta}, p_{I\eta})$ wave together. If, as a function of time, $q_{R\eta}(t)$ lags one-quarter period behind $q_{I\eta}(t)$, the wave moves in the direction **k**; if the $q_{I\eta}(t)$ wave lags one-quarter period behind the $q_{R\eta}(t)$ wave, the wave moves in the direction $-\mathbf{k}$. If $\eta = +1$, the wave is right-hand circularly polarized; if $\eta = -1$, it is left-hand circularly polarized. For linearly polarized waves, both $\eta = +1$ and $\eta = -1$ should be equally excited. If **k** is in the z-direction, the wave is polarized in the x-direction if the $q_f(t)$ oscillations for both signs of η are in phase, both for $q_{R\eta}$ and for $q_{I\eta}$. To have the wave in the $\pm z$-direction polarized in the y-direction, the $(\eta = +)$-wave and the $(\eta = -)$-wave should be 180° out of phase $(q_{R+} = -q_{R-},\ q_{I+} = -q_{I-})$. Thus for a single linearly polarized wave in one single direction, four of the oscillators labeled by f in (H-29) should be excited.

As the energy levels of all four are degenerate among each other, one might consider eigenfunctions of (H-29) that are linear combinations of the solutions (H-34). Consider, for instance, a case in which most radiation modes are in the ground state, but a number of **k**, η states carry each a photon energy $\hbar\omega$ divided among the $(q_{R\eta}, p_{R\eta})$ and the $(q_{I\eta}, p_{I\eta})$ mode in such a way that the resulting wave is circularly polarized. We make a linear combination of $\sqrt{\tfrac{1}{2}}\, H_1\big(q_{R\eta}\sqrt{\omega/\hbar}\big) = q_{R\eta}\sqrt{2\omega/\hbar}$, and $\sqrt{\tfrac{1}{2}}\, H_1\big(q_{I\eta}\sqrt{\omega/\hbar}\big) = q_{I\eta}\sqrt{(2\omega/\hbar)}$. As the time dependence of the ψ-wave is given by $\exp(-i\omega t)$, the quarter period lag of the q_R oscillation relative to the q_I is naturally indicated by a linear combination $(q_{R\eta} - iq_{I\eta})\sqrt{(\omega/\hbar)}$ of the two Hermite polynomials, so that formally we may write, for each circularly polarized running photon in the $\pm\mathbf{k}$ direction, a factor

$$\big\{(q_{R\eta} \mp iq_{I\eta})\sqrt{(\omega/\hbar)}\big\} = \pm\big\{\tfrac{1}{2} H_1\big(q^*_{\pm\mathbf{k}\eta}\sqrt{(2\omega/\hbar)}\big)\big\} \qquad \text{(H-35)}$$

in (H-34).

Transition probabilities

We can now use time-dependent perturbation theory (the so-called "method of variation of the constants") in the Schrödinger picture in which we treat the p and q as time-independent operators acting on a time-dependent wavefunction ψ. The stationary states of $\mathscr{H}_M + (\mathscr{H}_F - \mathscr{E}_{\text{vac}})$ are not stationary states of $\mathscr{H}$, as $\mathscr{H}_i$ causes transitions. One can also obtain the same results by working in the interaction picture.[83] Either way, one derives the so-called Golden Rule (not discovered, but merely named, by Fermi) for constant rates of transition between the stationary states of $\mathscr{H}_M + (\mathscr{H}_F - \mathscr{E}_{\text{vac}})$, which correspond to transformation of matter energy into photon energy and vice versa.[83]

[83] See Merzbacher (1970), chap. 18, secs. 7–8.

Unrealistic nature of theoretical photon states

The standing photon states (H-34) and the running photon states (H-35) both are theoretical and unrealistic photon states. The waves excited above their ground state energy level in the state (H-35) run through all of space, so that their energy is thinly spread out over the universe. Moreover, they run from spatial infinity on one side to spatial infinity on the other side.

When photons reach our eye from far away, we do not say that the light we see is a solution of Maxwell's homogeneous equations (a wave coming from infinitely far), but we suspect that some astronomical body emitted the light. Thus any light wave has a source, and therefore is a "more or less spherical" wave (apart from polarization effects) before by a reduction of the wavefunction we assign a direction to it after its observation. Even if we would know that an atom placed at the origin O was going to emit its next photon along the $\pm$ z-axis, we would want the wave to run along the $+z$-axis for $z > 0$, and along the $-z$-axis for $z < 0$. The waves (H-35), however, would run in the same directions on both sides of O.

More realistic radiation states

There are various ways of describing the field by more realistic waves. One is by doing away completely with the quantization of the radiation field. Instead, we use the retarded electromagnetic potentials derived from the electric current distribution as a source. We then calculate wave-mechanically the matrix elements of the source between its initial and final state. From the potentials, we calculate **E** and **B**, and thence the Poynting vector. Thus we calculate how much energy is electromagnetically radiated away from the quantized matter field. This is the so-called semi-classical treatment of radiation.[84] The advantage of this method is that it describes the radiated wave realistically immediately after its emission, also near the source, until one starts to make approximations in the calculation.

Sideways limitation of the photon wave packet

If we do not care for the wave near the source and we are willing to describe the radiation by a reduced wavefunction after the observation of the radiation energy, we may do for the radiation field from a source at O something similar to what we did in Section. 2.10 for a matter wave which had been scattered from an atom at O. We remarked that the more-or-less spherical wave leaving O after scattering on a second atom at P would give rise to a spherical wave around P, and, after the fact of this scattering had been observed, we could reduce the final wave to just this second spherical wave. If, however, we then integrate the time-dependent Schrödinger equation back in time, we find that what gave rise to this final secondary wave was not the entire spherical wave around O, but just the part of it that came near the atom at P. We therefore might as well have treated the wave from O to P as a wave packet instead of considering the entire wave in as far as we would be interested merely in the possible interaction of the wave with P.

[84] See H. A. Kramers (1957), chap. 8, sec. 84.

Therefore we could describe a photon that might interact with a particle P, as an electromagnetic wave packet traveling toward P.

This treatment would be justified even more if the source of the light wave were surrounded by a screen with a hole in it, and only photons emerging from the hole could interact with outside particles P. The wave would be sideways limited, and in this case not merely by a trick of imagination, but in a physically real way.

Lengthwise limitation of the photon wave packet

The wave train from the imagined or real hole in a screen surrounding the light source is not an infinitely long wave train either. We should remember that the Golden Rule gives a correct account of the transition probability of an excited atom going to emit a photon, only during the first infinitesimal time duration after the atom was excited. As the probability increases that the photon has been emitted, the probability decreases that the atom is still excited and still is available as a potential source for a photon. If this is taken into account, we find that the probability per second that the atom will emit a photon is not constant but decays exponentially with time, and after a short finite time this transition probability becomes practically zero.[85] After that time, no probability wave for excitation of the radiation field leaves anymore from the atom that was the source of radiation.

This decay of the transition probability is important as it is the cause of the natural line breadth of spectral lines.[85] For us, its importance lies in the fact that it explains that the energy emitted by an atom in a single process (and that is usually what we call a photon) is concentrated in a finite volume, that is, finite not only in width but also in length.

This means, of course, that the wavefunction describing an emitted photon, to start with, does not have a sharply defined momentum. The natural line breadth means that it does not even have a sharp energy. Therefore the state describing a realistic photon is not given by the product of $\psi_{\text{vac}}^{(F)}$ and (H-35), but we first have to smear out (H-35) over a finite region in $\mathbf{k}$-space, writing, for a single photon of momentum nearly equal to $\hbar\mathbf{K}$ and energy nearly $\hbar cK$:

$$\psi_{\mathbf{K}}^{(F)}(q_f) = \psi_{\text{vac}}^{(F)} \cdot \sum_{\mathbf{k}} \sum_{\eta} f_{\eta}^{(\mathbf{K})}(\mathbf{k}) (2ck/\hbar)^{\frac{1}{2}} q_{\mathbf{k},\eta}^{*}, \tag{H-36}$$

where $f_{\eta}^{(\mathbf{K})}(\mathbf{k})$ is a (usually complex) two-component function of $\mathbf{k}$ peaked at $\mathbf{K}$, which, for normalization of $\psi_{\mathbf{K}}^{(F)}$ according to

$$\int_{-\infty}^{+\infty} \cdots \int_{-\infty}^{+\infty} |\psi_{\mathbf{K}}^{(F)}(q_R \ldots, q_I \ldots)|^2 \prod_{\mathbf{k}/2} \prod_{\eta} (dq_{R\eta} dq_{I\eta}) = 1, \tag{H-37}$$

should be normalized according to

$$\sum_{\mathbf{k}} \sum_{\eta} |f_{\eta}^{(\mathbf{K})}(\mathbf{k})|^2 = 1. \tag{H-38}$$

This suggests regarding $\sum_{\eta} |f_{\eta}^{(\mathbf{K})}(\mathbf{k})|^2$ as the probability that the photon described by (H-36) has momentum $\hbar\mathbf{k}$ and energy $c\hbar k$. (We will verify this below.)

[85] See Merzbacher (1970), chap. 18, sec. 9; Heitler (1954), chap. 4, secs. 16 and 18.

For time dependence of ψ, in the free-radiation approximation a factor $\exp(-ickt)$ should be added to the summand in (H-36).

Note that the $f_\mu(\mathbf{k}-\mathbf{k}_0)$ used by Bohm (1952b) differ from our $f_\eta^{(\mathbf{K})}(\mathbf{k})$ by a factor proportional to $\sqrt{k}$.

Energy and momentum distribution in a photon wave packet

The radiation energy density $W^{(F)}$ and momentum density $\mathbf{G}^{(F)}$, as observables, are given by

$$W^{(F)} = (\mathbf{E}_\perp^2 + \mathbf{B}^2)/8\pi, \quad \mathbf{G}^{(F)} = \mathbf{E}_\perp \times \mathbf{B}/4\pi c. \tag{H-39}$$

The corresponding operators are found by inserting here eqs. (H-13), (H-14), (H-17), and (H-24). The expectation values are found by placing the operator between $\psi^{(F)*}$ and $\psi^{(F)}$, and integrating like in (H-37) over all radiation field coordinates. In the vacuum state (H-34), the expectation values of (H-39) are

$$W_{\text{vac}}^{(F)} = \frac{1}{\mathcal{V}} \sum_{\mathbf{k}} \sum_{\eta} \frac{1}{2} \hbar ck, \quad \mathbf{G}_{\text{vac}}^{(F)} = 0. \tag{H-40}$$

The calculation of the expectation values of (H-39) in the state (H-36) is tedious. (Compare Appendix A of Part III.) The results can best be expressed in terms of the following wave-packet functions:

$$\mathbf{g}_n^\eta(\mathbf{x}) \equiv \mathcal{V}^{-\frac{1}{2}} \sum_{\mathbf{k}} f_\eta^{(\mathbf{K})}(\mathbf{k})\, \mathbf{e}_{\mathbf{k}}^\eta\, (\hbar k)^{n/2} \exp(i\mathbf{k}\cdot\mathbf{x}). \tag{H-41}$$

We then find, with some use made of (H-11),

$$\langle W^{(F)} \rangle = W_{\text{vac}}^{(F)} + c \sum_{\eta} \sum_{\eta'} \delta_{\eta'\eta}\, \mathbf{g}_1^{\eta'}(\mathbf{x})^* \cdot \mathbf{g}_1^\eta(\mathbf{x}), \tag{H-42}$$

$$\langle \mathbf{G}_{(F)} \rangle = (2i)^{-1} \sum_{\eta} \sum_{\eta'} \eta [\mathbf{g}_0^{\eta'}(\mathbf{x})^* \times \mathbf{g}_2^\eta(\mathbf{x}) + \mathbf{g}_2^\eta(\mathbf{x})^* \times \mathbf{g}_0^{\eta'}(\mathbf{x})]. \tag{H-43}$$

We interpret $\langle W^{(F)} \rangle - W_{\text{vac}}^{(F)}$ as the average photon energy density $\langle W_{\text{phot}} \rangle$ in the state (H-36) (averaged over the ensemble described by $\psi^{(F)}$).

When we insert (H-41) in the above expressions, we get double sums over $\mathbf{k}$ and $\mathbf{k}'$. In the integrals over space, these reduce to single sums, and we find

$$\begin{aligned} \langle \mathcal{E}_{\text{phot}} \rangle &= \int \langle W_{\text{phot}} \rangle\, d^3\mathbf{x} = \sum_{\mathbf{k}} \sum_{\eta} |f_\eta^{(\mathbf{K})}(\mathbf{k})|^2 c\hbar k, \\ \langle \mathbf{p}_{\text{phot}} \rangle &= \int \langle \mathbf{G}^{(F)} \rangle\, d^3\mathbf{x} = \sum_{\mathbf{k}} \sum_{\eta} |f_\eta^{(\mathbf{K})}(\mathbf{k})|^2\, \hbar\mathbf{k}. \end{aligned} \tag{H-44}$$

This justifies the interpretation of $|f_\eta^{(\mathbf{K})}(\mathbf{k})|^2$ as a probability as mentioned above.

The quantum-mechanical operator for the observable $\mathcal{E}_{\text{phot}}$ is given by (H-29). The operator for $\mathbf{p}_{\text{phot}}$ is given by

$$\mathbf{p}_{\text{phot}} = \mathbf{p}^{(F)} = \int \mathbf{G}^{(F)}\, d^3\mathbf{x} = \sum_{\mathbf{k}/2} \sum_{\eta} i\hbar\mathbf{k}[q_{R,\eta} \nabla_{I,\eta} - q_{I,\eta} \nabla_{R,\eta}]. \tag{H-45}$$

Choice of the radiation field coordinates

So far we have described ordinary quantum electrodynamics, though in a form that during the last decades has lost some of its original familiarity. In Bohm's quantum electrodynamics, the coordinates q_f are taking the role of hidden variables, together with the $\mathbf{x}_i$ which we have found in this role before. In this connection, the reader unfamiliar with the quantum electrodynamics of the 1930s should be warned that the choice of radiation coordinates made above [the q_R and the q_I of (H-17)] is a rather *specific* choice, and *differs*, for instance, from the choice of coordinates of the oscillators representing the photon field, that was made in Pauli's famous *Handbuch* article on wavemechanics, or that was made by Kramers.

While our q_f in (H-17) were defined as the real and imaginary parts of the coefficients $q_{\mathbf{k},\eta}$ appearing in the expansion (H-13) of the magnetic field (and therefore also appearing in the transverse vector potential $\mathbf{A}$), Pauli[86] defined harmonic-oscillator coordinates $\mathbf{q}$ as the imaginary parts of the Fourier components of a field $\mathbf{F}_{\text{Pauli}} = \frac{1}{2}\left\{\mathbf{E}_\perp + i(-\nabla^2)^{-\frac{1}{2}}\operatorname{curl}\mathbf{B}\right\}$. Kramers,[87] instead, used $\mathbf{F}_{\text{Kramers}} = \mathbf{E}_\perp + i\mathbf{B}$ for the same purpose for right-hand polarized waves, and its conjugate $\mathbf{E}_\perp - i\mathbf{B}$ for left-hand polarized waves, but it is easily shown that this is equivalent to using $2\mathbf{F}_{\text{Pauli}}$. Thus both find, for $\mathbf{k}$ in our chosen halfspace, coordinates $\mathbf{q}$ and conjugate momenta $\mathbf{p}$ which in our notation are given by

$$\mathbf{q} = \sqrt{\tfrac{1}{2}}\,[-q_R + p_I/\omega], \quad \mathbf{p} = \sqrt{\tfrac{1}{2}}\,[-p_R - \omega q_I]. \tag{H-46}$$

For $\mathbf{k}$ in the other halfspace, their $\mathbf{q}$ and $\mathbf{p}$ are related to ours by a flip in sign of the terms with p_I and q_I in Kramers's notation, but of those with p_R and q_R in Pauli's notation. The transformation (H-46) between their coordinates and the ones used by Bohm is a canonical transformation. Bohm's choice has the advantage of making the interaction term with $\mathbf{A}$ in $\mathcal{H}_i$ dependent merely on the q_f and not on the p_f.

Bohm's hypothesis

At this point, Bohm deviates from ordinary quantum electrodynamics by making the hypothesis that the $\mathbf{x}_i$ and the q_f (the q_R and the q_I, and therefore also the complex $q_{\mathbf{k},\eta}$) at any time are hidden variables that for a precisely known individual physical system would have unique sharp values instead of having a probability distribution described by $|\psi|^2$. Even if we would not know their exact values when we merely know $\psi(\mathbf{x}_i, q_f)$, we still can *imagine* every system in the ensemble described by ψ to have its private precise values of the $\mathbf{x}_i$ and the q_f. These $\mathbf{x}_i$ and the q_f then would change at imagined velocities given by

$$\mathbf{v}_i = d\mathbf{x}_i/dt \quad \text{and} \quad v_f = dq_f/dt. \tag{H-47}$$

We now want to give equations determining these velocities, for given ψ, as functions of the $\mathbf{x}_i$ and the q_f in such a way that a continuity equation holds for a fluid representing an

[86] Pauli (1933), pages 247–251.
[87] See Kramers (1957), pages 418–422, in particular his eqs. (88)–(89).

ensemble of hidden-variables states with probability density $\varrho = |\psi|^2$ in $\mathbf{x}_i$, q_f-space:

$$\partial(|\psi|^2)/\partial t = \sum_j \nabla_j \cdot (|\psi|^2 \mathbf{v}_j) + \sum_f \nabla_f (|\psi|^2 v_f). \tag{H-48}$$

In the discussion of (H-48) again we factorize ψ according to

$$\psi = R \exp(iS/\hbar) \quad \text{with} \quad R = |\psi|. \tag{H-49}$$

We note that the Hamiltonian $\mathcal{H}_{\text{op}}$ in (H-27) has the form

$$\mathcal{H}_{\text{op}} = \sum_n [-(\hbar^2/2m_n)\nabla_n^2 + i\hbar a_n(x)\nabla_n] + V(x), \tag{H-50}$$

where x stands for all the x_n together (that is, the $\mathbf{x}_j$ and the q_f); we put $m_n = 1$ for the radiation coordinates; V includes the Coulomb term as well as the term with $\mathbf{A}(\mathbf{x}_i)^2$ and the terms $\omega_f^2 q_f^2$. The Hamiltonian (H-50) is slightly more general than the one considered in Appendix D, where we had no terms linear in ∇_n. Following the reasoning of Appendix D step by step for a Schrödinger equation with $\mathcal{H}_{\text{op}}$ given by (H-50), we find that the step (D-4) is possible only if a_n satisfies the condition

$$\sum_n \nabla_n a_n = 0. \tag{H-51}$$

For the a_n of our case (H-27), this becomes

$$\sum_j (\varepsilon_j/m_j c)\nabla_j \cdot \mathbf{A}(\mathbf{x}_j) = 0, \tag{H-52}$$

and this condition is satisfied by our use of the radiation gauge. Then, (D-4) becomes

$$\partial\varrho/\partial t = -\sum_n \nabla_n(\varrho\, v_n) \tag{H-53}$$

with

$$\begin{aligned} v_n &= \frac{\hbar}{2i}\,\frac{\psi^*\nabla_n\psi - (\nabla_n\psi^*)\psi}{m_n\psi^*\psi} - a_n \\ &= (\hbar/2im_n)\nabla_n \ln(\psi/\psi^*) - a_n = m_n^{-1}\nabla_n S - a_n \end{aligned} \tag{H-54}$$

as the counterpart of (D-5). As before, we put

$$\pi_n = \nabla_n S \quad (\text{that is,} \quad \boldsymbol{\pi}_j = \nabla_j S \quad \text{and} \quad \Pi_f = \nabla_f S). \tag{H-55}$$

Then the $\mathbf{v}_j$ and v_f according to (H-54) are determined by

$$m_j \mathbf{v}_j = \boldsymbol{\pi}_j - (\varepsilon_j/c)\mathbf{A}(\mathbf{x}_j), \quad v_f = \Pi_f, \tag{H-56}$$

and (H-53) gives (H-48).

The quantum-mechanical forces on particles in a radiation field

In (D-6) we now find on the right an additional term which is

$$-Re\left\{\nabla_k \sum_n \left(\frac{i\hbar a_n}{\psi}\nabla_n\psi\right)\right\} = \nabla_k \sum_n \left(a_n \frac{\hbar}{2i}\nabla_n \ln\frac{\psi}{\psi^*}\right) = \nabla_k \sum_n (a_n \pi_n). \tag{H-57}$$

Equation (A-2) still can be used to calculate $(1/\psi)\nabla_n^2\psi$, and, if we put

$$U \equiv -\sum_n \frac{\hbar^2}{2m_n}\frac{\nabla_n^2 R}{R} = -\sum_j \frac{\hbar^2}{2m_j}\frac{\nabla_j^2 R}{R} - \sum_f \tfrac{1}{2}\hbar^2 \frac{\nabla_f^2 R}{R} \tag{H-58}$$

with $R = \varrho^{1/2}$, we find in (D-7)

$$\frac{\partial \pi_k}{\partial t} = -\nabla_k V - \nabla_k U - \nabla_k \sum_n \frac{\pi_n^2}{2m_n} + \sum_n a_n \nabla_k \pi_n + \sum_n \pi_n \nabla_k a_n. \tag{H-59}$$

We calculate

$$\frac{d\pi_k}{dt} = \frac{\partial \pi_k}{\partial t} + \sum_n \left(\frac{\pi_n}{m_n} - \alpha_n\right) \nabla_n \pi_k = \frac{\partial \pi_k}{\partial t} + \sum_n \left(\frac{\pi_n}{m_n} - a_n\right) \nabla_k \pi_n, \tag{H-60}$$

by $\nabla_n \nabla_k S = \nabla_k \nabla_n S$. By (H-59), this gives

$$\frac{d\pi_k}{dt} = -\nabla_k U - \nabla_k V + \sum_n \pi_n \nabla_k a_n. \tag{H-61}$$

We now insert

$$V = \sum_i \sum_j \frac{\varepsilon_i \varepsilon_j}{2r_{ij}} + \sum_j \frac{\varepsilon_j^2}{2m_j c^2} \mathbf{A}(\mathbf{x}_j)^2 + \tfrac{1}{2} \sum_f \omega_f^2 q_f^2. \tag{H-62}$$

For $x_k = \mathbf{x}_k$ (particle position), (H-61) gives

$$\frac{d\pi_k}{dt} = -\nabla_k U - \varepsilon_k \nabla_k \sum_j \frac{\varepsilon_j}{r_{kj}} + \frac{\varepsilon_k}{c} \sum_{i=1}^{3} (v_k)_i \nabla_k A_i(\mathbf{x}_k), \tag{H-63}$$

as the terms with $\pm(\varepsilon_k^2/m_k c^2) \sum_i A_i(\mathbf{x}_k) \nabla_k A_i(\mathbf{x}_k)$ cancel. Therefore,

$$\begin{aligned} m_k \frac{d\mathbf{v}_k}{dt} &= \frac{d\pi_k}{dt} - \frac{\varepsilon_k}{c}\left[\frac{\partial \mathbf{A}(\mathbf{x}_k)}{\partial t} + \mathbf{v}_k \cdot \nabla_k \mathbf{A}(\mathbf{x}_k)\right] \\ &= -\nabla_k U - \varepsilon_k \nabla_k V_C(\mathbf{x}_k) + \frac{\varepsilon_k}{c} \mathbf{v}_k \times [\nabla_k \times \mathbf{A}(\mathbf{x}_k)] + \mathbf{E}_\perp(\mathbf{x}_k) \\ &= -\nabla_k U + \varepsilon_k [\mathbf{E}(\mathbf{x}_k) + \mathbf{v}_k \times \mathbf{B}(\mathbf{x}_k)/c]. \end{aligned} \tag{H-64}$$

Thus we find, as before, that Bohm's imagined motion of particles follows an equation of motion differing from the classical one by the quantum-mechanical force $-\nabla U$, this time with U given by (H-58).

Quantum-mechanical influence on the imagined or actual motion of the radiation field coordinates

We shall now apply (H-61) also to the radiation coordinates q_f. We have $a_f = 0$, but the $\mathbf{a}_j$ contain the q_f in $\mathbf{A}(\mathbf{x}_j)$. Thence

$$\begin{aligned} \frac{d^2 q_f}{dt^2} = \frac{dv_f}{dt} = \frac{d\pi_f}{dt} &= -\nabla_f U - \sum_j \frac{\varepsilon_j^2}{m_j c^2} \mathbf{A}(\mathbf{x}_j) \cdot \nabla_f \mathbf{A}(\mathbf{x}_j) - \omega_f^2 q_f \\ &\quad + \sum_j \left[m_j \mathbf{v}_j + \frac{\varepsilon_j}{c} \mathbf{A}(\mathbf{x}_j)\right] \cdot \nabla_f \left[\frac{\varepsilon_j}{m_j c} \mathbf{A}(\mathbf{x}_j)\right], \end{aligned} \tag{H-65}$$

which gives

$$\ddot{q}_f+\omega_f^2 q_f = -\nabla_f U+\sum_j \frac{\varepsilon_j}{c}\, \mathbf{v}_j \cdot \nabla_f \mathbf{A}(\mathbf{x}_j). \tag{H-66}$$

The $\nabla_R \mathbf{A}(\mathbf{x}_j)$ and the $\nabla_I \mathbf{A}(\mathbf{x}_j)$ are easily calculated from (H-31) for **k** in the chosen half-space. We then can use (H-66) to calculate by (H-17) also $\ddot{q}_\pm+\omega^2 q_\pm$, and use this again to calculate $\ddot{\mathbf{A}}/c^2-\nabla^2\mathbf{A}$ at some field point **x**. We then find, by the definition (H-26) of the transverse delta function, that the last terms in (H-66) contribute to $\partial^2\mathbf{A}/c^2\partial t^2-\nabla^2\mathbf{A}$ the transverse part of $4\pi\mathbf{J}(\mathbf{x})/c$. This confirms that the term $-\nabla_f U$ in (H-66) is again the deviation from what one would expect on the basis of the classical theory.

Note that U by (H-58) contains terms calculated from $\nabla_j^2 R$ as well as terms calculated from $\nabla_f^2 R$. In general, after an interaction between matter and radiation, ψ or R will no longer be factorizable into a matter factor and a radiation factor, and then both terms in U will contribute to $-\nabla_f U$ in (H-66). Only in the approximation of a factorizable state vector ψ (as often will be used to describe an initial state), one may neglect the matter term of U in (H-66), like Bohm (1952b) does in eq. (A-8) of his Appendix A.

APPENDIX I

DE BROGLIE'S RELATIVISTIC THEORY OF THE PILOT WAVE

De Broglie (1952b) lists as the two primary problems of a relativistic generalization of his theory of the pilot wave (that is, what we have called Bohm's theory):

(1) We have to decide *which* four-vector field j^μ satisfying a continuity equation $\partial_\mu j^\mu = 0$ shall be interpreted as the density[88] $\varrho = j^0$ and the current density $\mathbf{J} = c\mathbf{j}$ of the ensemble of particles in "hidden-variables states" of given positions;

(2) We have to decide *which* phase factor S in

$$\psi = \phi \exp(iS/\hbar) \tag{I-1}$$

will have its gradients related to the velocity $\mathbf{v} = \mathbf{J}/\varrho$ of the ensemble. These two problems are intimately related to each other.

The choice of density and of velocity

Vigier (1952) and Bohm (1953b) take it for granted that j^μ should be given by

$$j^\mu = \frac{1}{\varepsilon} j_\varepsilon^\mu = i\bar{\psi}\gamma^\mu\psi = \psi^\dagger\alpha^\mu\psi, \tag{I-2}$$

as they want the ensemble density $j^0 = \psi^\dagger\psi$ in hidden-variables space to be identical, at

[88] In view of Section 2.3 it would be more logical to use here the letter $\bar{P}$ rather than ϱ, for the probability density of hidden-variables states **x**.

least for positive-energy electrons, with $1/\varepsilon$ times the charge probability density of quantum theory. De Broglie questions this. Guided by his theory of the double solution [see Section 2.4], he had come to the conclusion that for free electrons in nonrelativistic approximation one should have[88, 89]

$$\mathbf{v} = -c^2[\nabla S - \varepsilon \mathbf{A}/c]/(\partial S/\partial t + \varepsilon\Phi), \tag{I-3}$$

$$\varrho = (-\partial S/\partial t - \varepsilon\Phi)K|\psi|^2, \tag{I-4}$$

where Φ, $\mathbf{A}$ form the potential four-vector A^μ in the Lorentz gauge, and where K is a constant. Now it is easy to find a four-vector different from (I-2) which satisfies a continuity equation. For finding it, we first calculate the four-vector (I-2) from the Dirac equation as follows.

We write the Dirac equation in the form

$$mc^2\psi + c\hbar\gamma^\mu\, \partial_\mu\psi - i\varepsilon A_\mu\gamma^\mu\psi = 0, \tag{I-5a}$$

so that $\bar{\psi} = \psi^\dagger\beta$ satisfies

$$mc^2\bar{\psi} - c\hbar\, \partial_\mu\bar{\psi}\gamma^\mu - i\varepsilon\bar{\psi}\gamma^\mu A_\mu = 0. \tag{I-5b}$$

We multiply (I-5a) by $\bar{\psi}\gamma^\lambda$ in front, and (I-5b) by $\gamma^\lambda\psi$ in back, and add the results. Using

$$\gamma^\lambda\gamma^\mu = g^{\lambda\mu} + \tfrac{1}{2}I^{\lambda\mu} \quad \text{with} \quad I^{\lambda\mu} \equiv \gamma^\lambda\gamma^\mu - \gamma^\mu\gamma^\lambda, \tag{I-6}$$

we obtain

$$2mc^2\bar{\psi}\gamma^\lambda\psi + c\hbar[\bar{\psi}\, \partial^\lambda\psi - (\partial^\lambda\bar{\psi})\psi] + \tfrac{1}{2}c\hbar\, \partial_\mu(\bar{\psi}I^{\lambda\mu}\psi) - 2i\varepsilon A^\lambda\bar{\psi}\psi = 0. \tag{I-7}$$

Thence, by (I-2),

$$j^\lambda = j^{(1)\lambda} + j^{(2)\lambda} \tag{I-8}$$

with

$$j^{(1)\lambda} = \frac{1}{mc}\left\{\frac{\hbar}{2i}[\bar{\psi}\, \partial^\lambda\psi - (\partial^\lambda\bar{\psi})\psi] - \frac{\varepsilon}{c}A^\lambda\bar{\psi}\psi\right\}. \tag{I-9}$$

$$j^{(2)\lambda} = \frac{\hbar}{4imc}\partial_\mu(\bar{\psi}I^{\lambda\mu}\psi). \tag{I-10}$$

Since, however,

$$\partial_\lambda j^{(2)\lambda} \equiv 0 \tag{I-11}$$

by the antisymmetry of $I^{\mu\lambda}$, it follows from $\partial_\lambda j^\lambda = 0$ that also

$$\partial_\lambda j^{(1)\lambda} = 0, \tag{I-12}$$

so that De Broglie (1952b) asks himself whether he possibly could identify ϱ with $j^{(1)0}$. Before this question can be answered, first S in (I-4) has to be defined.

The choice of the scalar wavefunction phase S

This brings us to problem (2). In the Dirac theory, if each component ψ_A of the wavefunction were written like eq. (15a) as

$$\psi_A = R_A \exp(iS_A/\hbar), \tag{I-13}$$

[89] See De Broglie (1927b).

the R_A would no longer form an undor[90] and would transform in a messy way under coordinate transformations. Moreover, it would not be clear which of the four S_A to use in (I-3) or (I-4). Therefore, we replace (I-13) by (I-1), where we postulate S to be scalar, and where the four components ϕ_A of the undor ϕ are complex. If the latter are written as

$$\phi_A = R_A \exp (i\Theta_A/\hbar), \tag{I-14}$$

we can relate (I-1) to (I-13) by

$$S_A = S+\Theta_A. \tag{I-15}$$

The problem now is to define S in a unique way. If we insert (I-13) in (I-9), we find

$$mcj_\mu^{(1)} = \sum_A \bar{R}_A R_A [\partial_\mu S_A-(\varepsilon/c)A_\mu], \tag{I-17}$$

so, by (I-15),

$$mcj_\mu^{(1)} = (\bar{\psi}\psi)[\partial_\mu S-(\varepsilon/c)A_\mu]+\sum_{A=1}^{4} \bar{R}_A R_A \partial_\mu \Theta_A. \tag{I-18}$$

De Broglie (1952b) suggests imposing the four conditions that the last term in (I-18) for $\mu = 0, 1, 2, 3$ shall vanish:

$$\sum_{A=1}^{4} \bar{R}_A R_A \partial_\mu \Theta_A = 0. \tag{I-19}$$

In that case, $\partial_\mu S$ can be calculated directly from ψ and A_μ by

$$\partial_\mu S = (\varepsilon/c)A_\mu+mcj_\mu^{(1)}/(\bar{\psi}\psi), \tag{I-20}$$

that is, by (I-9),

$$\partial_\mu S = \frac{\hbar}{2i}[\bar{\psi}\partial_\mu\psi-(\partial_\mu\bar{\psi})\psi]/(\bar{\psi}\psi). \tag{I-21}$$

Moreover, (I-18) then would give

$$j^{(1)\mu} = (mc^2)^{-1}[c\partial^\mu S-\varepsilon A^\mu](\bar{\psi}\psi), \tag{I-22}$$

of which the $\mu = 0$ component would yield (I-4) with $K = (mc^2)^{-1}$ and with the nonrelativistic $|\psi|^2$ replaced by the relativistic scalar expression $(\bar{\psi}\psi)$. This, to De Broglie, suggested that

$$\bar{P} = j^{(1)0} = (mc^2)^{-1}\left[-\frac{\partial S}{\partial t}-\varepsilon\Phi\right](\psi^\dagger\beta\psi) \tag{I-23}$$

would be preferable to $\bar{P} = j^0$ for the equilibrium distribution over hidden-variables states $\mathbf{x}$.

However, the integrability equation for the differential equations (I-21) for S are

$$\partial_\lambda\{[\bar{\psi}\partial_\mu\psi-(\partial_\mu\bar{\psi})\psi]/(\bar{\psi}\psi)\} = \text{symmetric in } \lambda \text{ and } \mu. \tag{I-24}$$

This is equivalent to

$$\sum_A\sum_B(\partial_\mu\bar{\psi}_A)\bar{\psi}_B[\psi_A\partial_\lambda\psi_B-(\partial_\lambda\psi_A)\psi_B] = 0, \tag{I-25}$$

[90] See Belinfante (1939a) for this name for a four-component Dirac wavefunction.

and De Broglie does not explain why these conditions would be satisfied. Therefore it is not clear whether it is possible to define a unique S by the conditions (I-19), and whether (I-22) is feasible. (Also, it is not clear how $\bar{P} = j^{(1)0}$ can explain $\langle \varrho_\varepsilon \rangle = \varepsilon j^0 = i\varepsilon \bar{\psi}\gamma^0\psi$.)

APPENDIX J

LACK OF UNIQUENESS OF THE POLYCHOTOMIC ALGORITHM WHEN DEGENERACY MAKES THE CHOICE OF THE SET $\{\phi_i\}$ OF POSSIBLE RESULTS NONUNIQUE

In the appendix of their 1955 paper (before Gleason's work), Wiener and Siegel showed that already *two*-dimensional degeneracy in the set $\{\phi_i\}$ suffices to make the predictions of their polychotomic algorithm ambiguous. We shall make their example more easily understandable by making it more specific, making it completely numerical.

We shall also use the same example for showing how the ambiguity is removed when we add to the scheme determining the set $\{\phi_i\}$ Tutsch's additional rule (119) given in Section 3.6.

Consider the measurement of an observable A which possesses a degenerate eigenvalue A_k to which two orthogonal eigenfunctions ϕ_{k1} and ϕ_{k2} belong. Let A_l be some other (nondegenerate) eigenvalue, and ϕ_l the corresponding eigenfunction. Let ψ and Ξ be given by[91]

$$\psi = \sum a_i \phi_i = \frac{1}{\sqrt{3}} \cdot (\phi_{k1} + \phi_{k2} + \phi_l), \tag{J-1}$$

$$\Xi = \sum \xi_i \phi_i = \tfrac{3}{13}\phi_{k1} + \tfrac{12}{13}\phi_{k2} + \tfrac{4}{13}\phi_l. \tag{J-2}$$

Since, with this choice of the $\{\phi_i\}$, we have[92]

$$\left|\frac{a_{k1}}{\xi_{k1}}\right| = \frac{13}{3\sqrt{3}} > \frac{a_l}{\xi_l} = \frac{13}{4\sqrt{3}} > \left|\frac{a_{k2}}{\xi_{k2}}\right| = \frac{13}{12\sqrt{3}} > \text{other} \left|\frac{a_i}{\xi_i}\right| = \frac{0}{0} = 0, \tag{J-3}$$

the algorithm would predict that the system upon measurement of A is found in the state ϕ_{k1}, with A_k the measured value of A, provided that our pick of $\{\phi_i\}$ was correct. This choice of $\{\phi_i\}$ might have been suggested by some peculiarity of the apparatus that would make ϕ_{k1} and ϕ_{k2} rather than any linear combinations of them the only reasonable possible results of the measurement.

If, however, the apparatus does not possess such peculiarities, the measurement apparently could just as well have been described using as basic Hilbert vectors $\{\phi_i\}$ the vector ϕ_l

[91] See eq. (113) and the generalization of (105) from $\Xi_j(\alpha) \equiv \Xi_\alpha(x_j)$, to $\Xi_\alpha(x)$ at some arbitrary point x. There is in the Wiener–Siegel theory no reason why Ξ should be normalized; so, we could omit from (J-2) the denominators 13 or multiply Ξ by any number. There is, however, no reason either why Ξ could not happen to be normalized, and, with a normalized Ξ in our example, the example applies to the Bohm–Bub theory as well as to the Wiener–Siegel theory.

[92] For 0/0 see footnote 56, p.135.

together with

$$\phi_{k1'} = \sqrt{\tfrac{1}{2}}\cdot(\phi_{k2}+\phi_{k1}) \quad \text{and} \quad \phi_{k2'} = \sqrt{\tfrac{1}{2}}\cdot(\phi_{k2}-\phi_{k1}) \tag{J-4}$$

instead of ϕ_{k1} and ϕ_{k2}. Then,

$$\psi = \sum a_{i'}\phi_{i'} \quad \text{and} \quad \Xi = \sum \xi_{i'}\phi_{i'}$$

with

$$\left.\begin{aligned} a_{k1'} &= \sqrt{\tfrac{1}{2}}\cdot(a_{k2}+a_{k1}) = \sqrt{\tfrac{2}{3}}, & a_{k2'} &= \sqrt{\tfrac{1}{2}}\cdot(a_{k2}-a_{k1}) = 0, & a_l &= \sqrt{\tfrac{1}{3}}, \\ \xi_{k1'} &= \sqrt{\frac{1}{2}}\cdot(\xi_{k2}+\xi_{k1}) = \frac{15}{13\sqrt{2}}, & \xi_{k2'} &= \sqrt{\frac{1}{2}}\cdot(\xi_{k2}-\xi_{k1}) = \frac{9}{13\sqrt{2}}, & \xi_l &= \frac{4}{13}, \end{aligned}\right\} \tag{J-5}$$

and

$$\left|\frac{a_l}{\xi_l}\right| = \frac{13}{4\sqrt{3}} > \left|\frac{a_{k1'}}{\xi_{k1'}}\right| = \frac{26}{15\sqrt{3}} > \left|\frac{a_{k2'}}{\xi_{k2'}}\right| = 0 = \text{other} \left|\frac{a_{i'}}{\xi_{i'}}\right| = \frac{0}{0} = 0, \tag{J-6}$$

so, this time, the algorithm would tell us that A, upon measurement, would be found to have the value A_l instead of A_k.

Therefore Wiener and Siegel (1955) concluded that the polychotomic algorithm woulp not uniquely determine whether the result of the measurement is A_k or A_l, *unless we can detect some peculiarities in the apparatus which makes some particular choice of linear combinations of* ϕ_{k1} *and* ϕ_{k2} *preferable as member of the set* $\{\phi_i\}$, *over any other linear combinations.*

Uniqueness restored by Tutsch's Rule[93]

We can restore uniqueness by postulating that, *in absence of other criteria like observables measured simultaneously*, we should follow Tutsch's rule (119) discussed in Section 3.6.[93]

As in the above example we found already $a_{k2'} = 0$, we need no longer search in this case for a third set of basic vectors $\phi_{k1''}$ and $\phi_{k2''}$ that would have the desired property. Our second set $\phi_{k1'}$, $\phi_{k2'}$, with ϕ_l meets the requirements, with $\phi_s = \phi_{k1'}$, and with ϕ_{k2} normal to ψ. Thus $\phi_{k2'}$ is *out* and only $\phi_{k1'}$ and ϕ_l need to be considered as possible results of the measurement. Since $|a_l/\xi_l| > |a_{k1'}/\xi_{k1'}|$ in this case, the result of the measurement now uniquely is A_l and not A_k.

Conventional alternative to Tutsch's choice of $\{\phi_i\}$

Without Tutsch's specification (119) for making the choice of the set $\{\phi_i\}$ unique for a measurement that seems to suffer from "degeneracies", one can take the conventional excuse out of the dilemma by assuming that slight perturbations of the measuring apparatus unknown to us break up all degeneracies, and determine what is the correct set $\{\phi_i\}$ to be

[93] See Tutsch (1969), page 1117, top half of second column.

used in the polychotomic algorithm. This makes the result of the measurement extremely sensitive to unknown small perturbations, like it was also in Bohm's 1951 theory (though for other reasons in that case). We discussed this oversensitivity of the result of a measurement for trifling circumstances in Section 3.7 of Part I, and explained why a theory of this type would remain *crypto*deterministic even if the hidden variables were known to us.

While Tutsch's proposal (119) sounds reasonable, and leads in many cases to satisfactory results, we shall find in Appendix K that it leads to paradoxes when applied to composite systems that could be the subject of the Einstein–Podolsky–Rosen paradox. We will also find there that paradoxes may occur already when Tutsch's rule is applied to *single* systems.

See Appendix S for a method of applying Tutsch's rule (119) without explicitly calculating the normalized orthogonal projections ϕ_s of ψ upon the Hilbert subspaces S of degeneracy. This method was already known to Wiener and Siegel (1953), though they did not relate it to projections of ψ upon Hilbert subspaces of degeneracy.

APPENDIX K

PROBLEMS ARISING WHEN THE POLYCHOTOMIC ALGORITHM IS APPLIED TO COMPOSITE SYSTEMS OR TO SINGLE SYSTEMS

In this appendix we shall first show how the polychotomic algorithm sometimes makes paradoxical predictions when applied to a given hidden-variables state of a composite system.[93a] We shall consider a system consisting of two parts 1 and 2, which may or may not be separated spatially, and we shall assume that somehow we know the initial microstate [$\psi(1, 2)$ with $\Xi(1, 2)$] of this system. We then shall several times apply the polychotomic algorithm and see what it predicts for various possible measurements on this system, of which some are measurements on one part of the system only, and other measurements are simultaneous measurements on both parts, which may be realized by coincidence and anti-coincidence techniques.

Description of the system considered

We shall consider here the system already discussed in Section 3.8 and Appendix G, both of Part I, where we had a system in a zero-angular-momentum state splitting up into two separate systems each with angular momentum $\hbar$. We shall use the notation explained in Part I, in which p, z, n are eigenfunctions of J_z belonging to *p*ositive, *z*ero, or *n*egative eigenvalues of J_z ($= \hbar, 0, -\hbar$), and in which r, m, l (standing for *r*ight, *m*iddle, *l*eft) are the eigenfunctions of J_x to the eigenvalues $\hbar, 0, -\hbar$.

[93a] Belinfante (1972).

The quantum state of the composite system

As before [see eqs. (I:82) and I:84)], the initial quantum state will be assumed to be

$$\psi(1,2)=\sqrt{\tfrac{1}{3}}\cdot[p_1n_2-z_1z_2+n_1p_2], \tag{K-1a}$$

$$\psi(1,2)=\sqrt{\tfrac{1}{12}}\cdot(p_1+n_1)(r_2-l_2)+\sqrt{\tfrac{1}{6}}\cdot[i(n_1-p_1)m_2-z_1(l_2+r_2)], \tag{K-1b}$$

$$\psi(1,2)=\sqrt{\tfrac{1}{3}}\cdot[r_1l_2-m_1m_2+l_1r_2]. \tag{K-1c}$$

These three forms of $\psi(1, 2)$ are obtainable from each other by the transformations I:(G-1).

Choice of the microstate

For the sake of our illustration, we shall assume that the initial microstate of the composite system is given by

$$\Xi(1,2)=\sqrt{\tfrac{1}{2}}\cdot\{0.6p_1p_2-\sqrt{\tfrac{1}{2}}\cdot p_1z_2+0.4p_1n_2+0.7z_1p_2+0.3z_1n_2+0.6n_1p_2+0.2n_1n_2\}. \tag{K-2a}$$

(The normalizing factor $\sqrt{\frac{1}{2}}$ in front is not needed in the Wiener–Siegel theory, where most Ξ are not normalized, but it would be needed if we want to apply the following to the Bohm–Bub theory.) We expressed Ξ like ψ in (K-1a) as an expansion in the simultaneous eigenfunctions of J_{z1} and J_{z2} used as the set $\{\phi_i\}$. If, instead, we use for $\{\phi_i\}$ the simultaneous eigenfunctions of J_{z1} or J_{x1} and of J_{x2}, as in (K-1b) and (K-1c), we find

$$\Xi(1,2)=0.1i(p_1+2z_1+2n_1)m_2+\sqrt{\tfrac{1}{2}}\cdot\{-p_1l_2+(0.5z_1+0.4n_1)(r_2-l_2)\}, \tag{K-2b}$$

$$\begin{aligned}\Xi(1,2)=\;&0.3914r_1r_2+0.2914ir_1m_2-0.7450r_1l_2-0.2im_1r_2+0.0707m_1m_2-0.3im_1l_2\\&+0.1086l_1r_2-0.00858il_1m_2+0.2450l_1l_2.\end{aligned} \tag{K-2c}$$

Predictions made in this microstate for various measurements

We can now answer some of the questions we asked in Section 3.8 of Part I if we remove the ambiguity of the set $\{\phi_i\}$ for measurements on one of the two particles alone, by adopting Tutsch's additional requirement (119) of Section 3.6. Thus we find, by means of the polychotomic algorithm applied to the composite system (1, 2):

(A) If we measure J_{z1} and J_{z2} simultaneously, we should use for $\{\phi_i\}$ the set used in (K-1a) and (K-2a). With a_i and ξ_i the coefficients of ϕ_i in these expansions of ψ and of Ξ, we find that $\phi_k = z_1z_2$ has the largest $|a_k/\phi_k| = \infty$, as there is no z_1z_2 term in (K-2a), while there is a z_1z_2 term in (K-1a). Therefore, the result of the measurement is $J_{z1} = J_{z2} = 0$.

(B) The same result is obtained if we first measure J_{z1} alone and later J_{z2}, or vice versa, as (119) will still require in this case the use of the expansions (K-1a) and (K-2a).

(C) If we measure simultaneously J_{z1} and J_{x2}, we must use the expansions of eqs. (K-1b) and (K-2b). Since $\phi_k = p_1r_2$ occurs in (K-1b), but not in (K-2b), this ϕ_k has the largest

$|a_k/\xi_k| = \infty$. So, the result of the measurement is now $J_{z1} = \hbar$, $J_{x2} = \hbar$. Comparison with the result obtained in (A) shows that, indeed, the result of the measurement of J_{z1} may depend upon whether we measure simultaneously J_{x2} or J_{z2}, as we predicted in Section 3.8 of Part I.

(D) If, however, we first measure just J_{z1} alone, so that in the two-particle Hilbert space each eigenvalue of J_{z1} is degenerate as the corresponding eigenfunction p_1 or z_1 or n_1 may be multiplied by an arbitrary function of the second particle, Tutsch's rule specifies that the two-particle eigenfunctions $\{\phi_i\}$ to be used shall be such that each eigenvalue of the measured observable J_{z1} shall be represented in ψ by one single term only. This is not so if we use for $\{\phi_i\}$ the simultaneous eigenfunctions of J_{z1} and of J_{x2} as in (K-1b), in which case, for instance, the eigenvalue $J_{z1} = +\hbar$ is represented in ψ by three terms containing p_1r_2 and p_1l_2 and p_1m_2. We can satisfy Tutsch's rule, however, by using for $\{\phi_i\}$ the simultaneous eigenfunctions of J_{z1} and J_{z2}, as in (K-1a). Then this set must also be used for expanding Ξ, as in (K-2a). As, then, $\phi_k = z_1z_2$ has a nonzero coefficient $a_k = -\sqrt{\frac{1}{3}}$ in ψ, but a zero coefficient $\xi_k = 0$ in Ξ, this $|a_k/\xi_k| = \infty$ is the largest among the $|a_i/\xi_i|$, and the result of measuring J_{z1} alone will be $J_{z1} = 0$, and the reduced wavefunction $\psi(1, 2)$ of the two-particle system should be z_1z_2 after the measurement, which is equal to $\sqrt{\frac{1}{2}} \cdot z_1(r_2+l_2)$.

(E) If we now follow up by a measurement of J_{x2}, we see that Tutsch's requirement postulates using r_2z_1 and l_2z_1 as some of the $\{\phi_i\}$, so that now (K-2b) should be used for Ξ. If we assume that $\Xi(1, 2)$ was not affected by the first measurement and did not "relax", we meet here the unlikely case not discussed in Section 3.6 that both $\phi_k = r_2z_1$ and $\phi_l = l_2z_1$ feature the same maximal value of the ratio $|a_k/\xi_k| = |a_l/\xi_l|$, so that it is not clear from the algorithm in this case whether we will find $J_{x2} = +\hbar$ of $J_{x2} = -\hbar$.

(F) If we had started with measuring J_{x2} alone, removal of the arbitrariness in the $\{\phi_i(1, 2)\}$ by requirement (119) forces us to use for the $\{\phi_i\}$ the simultaneous eigenfunctions of J_{x1} and J_{x2} as in eq. (K-1c). Then, only $\phi_i = r_1l_2$ and m_1m_2 and l_1r_2 have nonvanishing $|a_i|$, and these $|a_i|$ are equal; but, from these three ϕ_i, $\phi_k = m_1m_2$ has in (K-2c) the smallest $|\xi_k|$ $(= 0.07071)$, and thus has the largest $|a_k/\xi_k|$. Therefore, the result of measuring only J_{x2} is $J_{x2} = 0$, with $\psi(1, 2) = m_1m_2$ as the wavefunction after this measurements. That is, measurement of J_{x2} on particle 2 makes the wavefunction after the measurement an eigenfunction not only of J_{z2} but also of J_{x1}, for our initial state (K-1c).

Discrepancies between the results

We could consider many other special cases, but the above cases suffice for showing where the trouble lies. When system 1 and 2 are spatially separated when the measurements are made, or do no longer interact, like an observed object 1 and a previous measuring apparatus 2, then upon measurement of J_{z1} on 1 and J_{x2} on 2 it cannot really be told by the apparatus whether we measure these two simultaneously, or first J_{z1} and later J_{x2}, or first J_{x2} and later J_{z1}. When, however, we compare the results $J_{z1} = J_{x2} = \hbar$ of the simultaneous measurement in (C) with the results $J_{z1} = 0$ in (D) and $J_{x2} = 0$ in (F) when one measurement precedes the other, we find a discrepancy that does not seem to be acceptable.[93a]

What should a "reduced" ψ and Ξ for "part 1 alone" predict?

If, now, we want to consider a complete reduction of the state of system 1, for a given "reading" of system 2, we should try to define a state $\Xi(1)$ and $\psi(1)$ as a function of $\Xi(1, 2)$ and $\psi(1, 2)$ in such a way that $\Xi(1)$ with $\psi(1)$ would predict for system 1 the same as $\Xi(1, 2)$ with $\psi(1, 2)$. Since, however, the predictions by $\Xi(1, 2)$ and $\psi(1, 2)$ themselves for spacelike distances between 1 and 2 contradict common sense [as between (C), and (D) or (F), in the above examples], it is not clear what we would want $\Xi(1)$ and $\psi(1)$ to predict for system 1 alone.

If we try to avoid the above discrepancies by rejecting Tutsch's additional requirement (119), we do not really improve the situation. We worsen it, as then the questions asked in Section 3.8 of Part I will have no unique answer at all, and the microstate of the composite system will not uniquely predict at all what will be the outcome of a measurement on part 1 alone: There would no longer be any reason why the measurement (D) of J_{z1} alone should yield the unambiguous result of (A) and (B), and why it should not yield the result of (C) for J_{z1}.

Paradoxes for single systems

The occurrence of paradoxes when Tutsch's rule is applied is not confined to composite systems. Tutsch (1969) uses the orthohelium spin (Part I, Chapter 3) as an example.[94] If one measurement distinguishes between the values 1, 0, and -1 of S_z, but another measurement distinguishes merely between the values 1 and 0 of $|S_z|$, it is easy to construct examples[94] where $S_z = -1$ by the first measurement, but $|S_z| = 0$ by the second measurement, if we apply the polychotomic algorithm augmented either by Tutsch's rule (119), or by the generalization of the algorithm proposed by Wiener and Siegel (1953) discussed in our Appendix S.

[94] Tutsch seems to use in his example $\left(\sum_m |a_{im}|^2\right) \Big/ \left(\sum_m |\xi_{im}|^2\right)$ as the polychotomic criterion for deciding which result A_i a measurement has, when several eigenfunctions ϕ_{im} belong to A_i. However, as shown in our Appendix S, the correct formula to use according to Tutsch's rule is the formula proposed for the criterion already by Wiener and Siegel (1953), which is $\left(\sum_m |a_{im}|^2\right) \Big/ \left|\sum_m \xi^*_{im} a_{im}\right|$. Using the latter criterion, we find that Tutsch's numerical example is not an example of the paradox he mentions. A correct example would be given by

$$\xi_1 = \xi = \xi_{-1} = \sqrt{(1/3)}, \quad a_1 = \sqrt{0.10}, \quad a_0 = \sqrt{0.40}, \quad a_{-1} = \sqrt{0.50},$$

which makes $|a_{-1}/\xi_{-1}| > |a_0/\xi_0| > |a_1/\xi_1|$, but which makes

$$|a_0/\xi_0| > (|a_1|^2 + |a_{-1}|^2)/|a_1\xi_1^* + a_{-1}\xi_{-1}^*|.$$

APPENDIX L

APPLICATION OF THE POLYCHOTOMIC ALGORITHM FOR A MEASUREMENT OF AN OBSERVABLE WHEN THE STATE DEPENDS ALSO ON A COORDINATE OF A DIFFERENT SYSTEM

Suppose we make a measurement of A on a particle with state vectors depending on x, but the state vectors depend also on the coordinate y of some second particle. In that case, $\psi(x, y)$ is normalized by

$$\int |\psi(x, y)|^2 \, dx \, dy = 1. \tag{L-1}$$

We shall assume here that the eigenvalues of A_{op} are all nondegenerate, so that A_{op} determines a unique set $\{\phi_i(x)\}$ of eigenfunctions (but for constant phase factors). If we then expand

$$\psi(x, y) = \sum_i a_i(y)\phi_i(x), \tag{L-2}$$

we will have

$$\sum_i \int dy \, |a_i(y)|^2 = 1. \tag{L-3}$$

This suggests identifying the probability w_i of measuring A_i with

$$w_i = \int dy \, |a_i(y)|^2. \tag{L-4}$$

We shall also assume that the second state vector $\Xi(x, y)$ has been expanded according to

$$\Xi(x, y) = \sum_i \xi_i(y)\phi_i(x). \tag{L-5}$$

In order to apply the polychotomic algorithm in this case to the measurement of A, we must remember that in this case in (x, y)-space every A_i is degenerate, as $\phi_i(x)$ times any function of y is an eigenfunction of A_{op}. Therefore, according to (119), for every value of i, we should determine an orthonormal set of functions $\chi_{im}(y)$, so that all except $\chi_{i1}(y)$ are orthogonal to $\psi(x, y)$. We then may expand $\psi(x, y)$ according to

$$\psi(x, y) = \sum_i b_i \phi_i(x)\chi_{i1}(y). \tag{L-6}$$

Thence, by (L-2),

$$b_i \chi_{i1}(y) = \int \phi_i(x)^* \psi(x, y) \, dx = a_i(y). \tag{L-7}$$

Normalization of $\chi_{i1}(y)$ gives

$$|b_i|^2 = \int |a_i(y)|^2 \, dy. \tag{L-8}$$

We might as well choose $b_i = |b_i|$ and define $\chi_{i1}(y)$ by

$$\chi_{i1}(y) = a_i(y)/|b_i|, \tag{L-9}$$

with $|b_i|$ the square root of (L-8).

We now complete the orthonormal set $\chi_{im}(y)$ so that

$$\int \chi^*_{im}(y)\chi_{in}(y)\,dy = \delta_{mn}, \tag{L-10}$$

ut we should remember that $\chi_{im}(y)$ and $\chi_{jn}(y)$ for $i \neq j$ need not be orthogonal.

We then use the $\phi_i(x)\chi_{im}(y)$ as the complete orthonormal set in which to expand $\Xi(x, y)$ for the application of the polychotomic algorithm for determining which eigenvalue A_k of A_{op} will be realized. So we put

$$\Xi(x, y) = \sum_i \phi_i(x) \sum_m \zeta_{im}\chi_{im}(y), \tag{L-11}$$

and split up the $\zeta_{im} = \zeta_k$ according to

$$\zeta_k = \mu_{2k-1} + i\mu_{2k}. \tag{L-12}$$

The canonical distribution of the $\Xi(x, y)$ then is given by

$$d\bar{P} = \prod_{\lambda=1}^{2MN} \left\{\pi^{-\frac{1}{2}} \exp(-\mu_\lambda^2)\,d\mu_\lambda\right\}, \tag{L-13}$$

if $1 \leqslant i \leqslant N$ and $1 \leqslant m \leqslant M$ in (L-11). The polychotomic algorithm then asks for the largest among the $|b_{im}/\zeta_{im}|$, if

$$\psi(x, y) = \sum_i \phi_i(x) \sum_m b_{im}\chi_{im}(y), \tag{L-14}$$

but from (L-6) we see that $(M-1)N$ of the MN coefficients b_{im} vanish (for $m \neq 1$), and therefore, according to footnote 56 (p. 135), the corresponding ζ_{im} may have any value whatsoever.

We now introduce the notation of eqs. (121)–(124), integrate over complex phase angles, and, as in (127), calculate the integral of $d\bar{P}$ under the condition that some particular $|b_{k1}/\zeta_{k1}|$ is larger than any other $|b_{im}/\zeta_{im}|$. As this by $b_{im} = 0$ imposes no conditions upon ζ_{im} for $m \neq 1$, the corresponding $(M-1)N$ integrals $\int_0^\infty \exp(-X)\,dX$ give factors 1, and only N integrals are left, giving the expression (127), if we interpret here X_i as $|\zeta_{i1}|^2$ and w_i as $|b_{i1}|^2 = |b_i|^2$. The latter interpretation agrees with (L-8) and (L-4). Then, (127) shows that we obtain the desired probability distribution of the A_k, if we use the ratios $|b_i/\zeta_{i1}|$ as the criterion in the polychotomic algorithm.

For applying this algorithm in practice, we should express $|b_i/\zeta_{i1}|$ directly in terms of $\psi(x, y)$ and $\Xi(x, y)$ or in terms of the $a_i(y)$ and $\xi_i(y)$ in the expansions (L-2) and (L-5) From (L-5) and (L-11), we obtain

$$\int \phi_i^*(x)\Xi(x, y)\,dx = \xi_i(y) = \sum_m \zeta_{im}\chi_{im}(y). \tag{L-15}$$

Thence, by (L-7) and (L-10),

$$\int \xi_i^*(y)a_i(y)\,dy = \sum_m \zeta^*_{im} \int \chi^*_{im}(y) b_i \chi_{i1}(y)\,dy = \zeta^*_{i1} b_i. \tag{L-16}$$

We conclude that

$$|b_i/\zeta_{i1}| = |b^* b_i/\zeta^*_{i1} b_i| = \int |a_i(y)|^2\,dy \Big/ \left|\int \xi_i^*(y)a_i(y)\,dy\right|. \tag{L-17}$$

Thus the undesired coordinate y becomes a dummy integration variable, and the algorithm can be applied directly without questions being asked about the value taken by y.

We use the above result in Appendix M in the discussion of the measurement problem.

Degenerate operator and external coordinate

If in the above we drop the condition that the set $\{\phi_i(x)\}$ is determined uniquely by A_{op}, complications arise with the determination of a complete set of functions of x and y that are eigenfunctions of A_{op}, and of which for each eigenvalue of A_{op} at most one is not normal to $\psi(x, y)$. We have not tried to generalize the above method to that case. For any choice of a set $\{\phi_i\}$ of eigenfunctions of A_{op}, one still can follow the above reasoning, and for a canonical distribution one will obtain agreement with quantum-mechanical probabilities. However, for a given $\Xi(x, y)$ instead of for a distribution, it has not been verified whether it is possible to make the prediction of k by the polychotomic algorithm independent of the choice of $\{\phi_i(x)\}$ or whether a unique choice of an orthonormal set of functions of x and y can be defined on the basis of Tutsch's condition (119).

APPENDIX M

VARIOUS CANONICAL HIDDEN-VARIABLES DISTRIBUTIONS FOR A COMPOSITE SYSTEM WITH FACTORIZABLE STATE VECTOR ψ

We shall start by showing that, if parts 1 and 2 of a composite system (1, 2) never interacted at all, so that $\psi(1, 2)$ if factorizable as

$$\psi(1, 2) = \psi_1(1)\cdot\psi_2(2), \tag{M-1}$$

we may well assume that also the hidden-variables state vector Ξ is factorizable as

$$\Xi(1, 2) = \Xi_1(1)\cdot\Xi_2(2). \tag{M-2}$$

Predicting the results of measurements

Suppose we make a simultaneous measurement of A on system 1, and B on system 2. Let $\{\phi_{Ai}(x_1)\}$ and $\{\phi_{Bj}(x_2)\}$ be the eigenfunctions of A and B, and let (with $x \equiv x_1, x_2$)

$$\phi_k(x) \equiv \phi_{i,j}(x_1, x_2) = \phi_{Ai}(x_1)\cdot\phi_{Bj}(x_2). \tag{M-3}$$

Then the $\{\phi_k(x)\}$ are the possible outcomes of the simultaneous measurement and should be used in the application of the polychotomic algorithm for the composite system.

If we have

$$\psi_1(x_1) = \sum_i a_{1i}\phi_{Ai}(x_1), \quad \psi_2(x_2) = \sum_j a_{2j}\phi_{Bj}(x_2), \tag{M-4}$$

then (M-1) gives us

$$\psi(x) \equiv \psi(x_1, x_2) = \sum_k a_k \phi_k(x) \quad \text{with} \quad a_k \equiv a_{i,j} = a_{1i}\, a_{2j}. \tag{M-5}$$

Similarly,

$$\Xi_1(x_1) = \sum_i \xi_{1i} \phi_{Ai}(x_1) \quad \text{and} \quad \Xi_2(x_2) = \sum_j \xi_{2j} \phi_{Bj}(x_2) \tag{M-6}$$

by (M-2) will provide

$$\Xi(x) = \sum_k \xi_k \phi_k(x) \quad \text{with} \quad \xi_k \equiv \xi_{i,j} = \xi_{1i}\xi_{2j}. \tag{M-7}$$

Let $|a_{1I}/\xi_{1I}|$ be the largest among all $|a_{1i}/\xi_{1i}|$, and let $|a_{2J}/\xi_{2J}|$ be the largest among all $|a_{2j}/\xi_{2j}|$. Then, clearly, $|a_K/\xi_K|$ is the largest among all $|a_k/\xi_k|$, with $K \equiv I, J$. So, the outcome of the combined measurement will be

$$\phi_K(x) = \phi_{1I}(x_1)\phi_{2J}(x_2). \tag{M-8}$$

Thus the polychotomic algorithm applied to the *composite* system correctly predicts in this case the measurements A_I and B_J on the *two parts separately*. Our success here is due to the factorizability of both ψ and Ξ. The troubles we had in Appendix K were due to nonfactorizability in the example discussed there.

Canonical distribution of hidden variables for the two-part system in the Wiener–Siegel theory

Consider now the case that $\Xi(1, 2)$ is unknown, but we do know $\psi(1, 2) = \psi_1(1)\ \psi_2(2)$. If parts 1 and 2 never interacted in the past, we will assume that, in an ensemble E_ψ of systems all described by this $\psi(1, 2)$, each part by itself by perturbations will have arrived at a canonical distribution of its Ξ vectors in the ensemble. If, as in Section 3.7, we introduce the notation

$$X_k \equiv X_{i,j} = |\xi_k|^2, \tag{M-9}$$

and, similarly, for the two parts,

$$Y_i = |\xi_{1i}|^2, \quad Z_j = |\xi_{2j}|^2, \tag{M-10}$$

then the factorizability assumption (M-2) for parts that never interacted leads for any sets $\{\phi_{1i}\}$ and $\{\phi_{2j}\}$, by (M-7), to

$$X_{i,j} = Y_i Z_j \tag{M-11}$$

for the composite system as a whole. The canonical distributions for the separate parts are given according to eq. (124), by

$$d\bar{P}_1 = \prod_{i=1}^{N_1} \{\exp(-Y_i)\, dY_i\}, \quad d\bar{P}_2 = \prod_{j=1}^{N_2} \{\exp(-Z_j)\, dZ_j\}. \tag{M-12}$$

They combine to a distribution for the products $\xi_k = \xi_{i,j} = \xi_{1i}\xi_{2j}$ in the ensemble E_ψ given by

$$d\bar{\bar{P}}_{1,2} = d\bar{P}_1\, d\bar{P}_2 = \exp\left\{-\sum_{i=1}^{N_1} Y_i - \sum_{j=1}^{N_2} Z_j\right\} \cdot \prod_{i=1}^{N_1} dY_i \cdot \prod_{j=1}^{N_2} dZ_j. \tag{M-13}$$

However, if the two parts ever in the past interacted, we would expect that during this interaction, in an ensemble E_ψ of these systems, the Ξ distribution would have relaxed to

$$d\bar{P}_{(1,2)} = \prod_{k=1}^{N_1 N_2} \{\exp(-X_k)\, dX_k\} = \exp\left(-\sum_{i=1}^{N_1}\sum_{j=1}^{N_2} Y_i Z_j\right)\cdot \prod_{k=1}^{N_1 N_2} dX_k, \qquad \text{(M-14)}$$

which is quite a different thing.

Deriving the quantum-mechanical probability distributions

According to eqs. (121)–(127) of Section 3.7, the distributions (M-12) give, by the polychotomic algorithm applied to each part separately, probabilities $w_{1I} = |a_{1I}|^2$ and $w_{2J} = |a_{2J}|^2$ for finding part 1 by an appropriate reproducible measurement in state ϕ_{1I}, and finding part 2 in a state ϕ_{2J}. According to the reasoning between (M-7) and (M-8), then also the algorithm applied to the $N_1 N_2$ states $\phi_{1I}(x_1)\, \phi_{2J}(x_2)$ of the composite system will give a probability $w_{1I} w_{2J}$ for measuring simultaneously $A_1 = A_{1I}$ in part 1, and $A_2 = A_{2J}$ in part 2, assuming the distribution $d\bar{P}_{1,2}$ (M-13) to be valid.

The result $w_{ij} = |a_{1i} a_{2j}|^2 \equiv |a_k|^2$, however, is equally well obtained by eqs. (121)–(127) if for the composite system we assume the distribution $d\bar{P}_{(1,2)}$ (M-14) for $N_1 N_2$ independent variables ξ_k. If the ξ_k are independent, of course, this means giving up the factorizability (M-2), because (M-2), through (M-7), yields

$$\xi_k \xi_{k'} \equiv \xi_{i,j}\, \xi_{i',j'} = \xi_{1i}\xi_{1i'}\xi_{2j}\xi_{2j'} = \xi_{i,j'}\xi_{i',j} \equiv \xi_{k''}\xi_{k'''}, \qquad \text{(M-15)}$$

which defeats independence of the ξ_k. That is, in the distribution (M-14) there occur many states $\Xi(1, 2)$ for which the

$$\xi_k = \xi_{i,j} = \iint \phi_k^*(1,2)\, \Xi(1,2)\, dx_1\, dx_2 = \iint \phi_{1i}^*(x_1)\, \Xi(x_1, x_2)\, \phi_{2j}^*(x_2)\, dx_1\, dx_2 \qquad \text{(M-16)}$$

do *not* satisfy (M-15). On the other hand, the distribution $d\bar{P}_{1,2}$ (M-13) contained merely factorizable $\Xi(1, 2)$. We thus see that the same quantum-mechanical results for a factorizable $\psi(1, 2)$ are obtainable from two entirely different distributions of hidden variables $\Xi(1, 2)$.

If ψ is *not* factorizable, only the distribution $d\bar{P}_{(1,2)}$ of (M-14) can be expected to yield the quantum-mechanical probability predictions $w_k \equiv w_{ij} = |a_{i,j}|^2$, with

$$a_k \equiv a_{i,j} = \iint \phi_{1i}^*(x_1)\, \phi_{2j}^*(x_2)\, \psi(x_1, x_2)\, dx_1\, dx_2, \qquad \text{(M-17)}$$

and with $w_k = |a_k|^2$ in general not factorizable into a product $w_{1i} w_{2j}$. [That is, $a_k a_{k'} \neq a_{k''} a_{k'''}$ in the notation of (M-15).]

The fact that $d\bar{P}_{1,2}$ does not give the desired results when ψ is not factorizable suggests that, in a composite system in which the parts interact or have interacted, $d\bar{P}_{(1,2)}$ will be the established canonical distribution. On the other hand, $d\bar{P}_{(1,2)}$ would not be understandable for a two-part system in which the two parts have never met before. That is why we started out with proposing $d\bar{P}_{1,2}$ as the canonical distribution in a system that only formally is regarded as composite, but that in reality consists of two parts which have never interacted in the past.

Complete incompatibility of $d\bar{\bar{P}}_{1,2}$ and $d\bar{P}_{(1,2)}$

Returning to the case where $\psi(1, 2)$ is factorizable, and where both distributions (M-13) and (M-14) would yield the quantum-mechanical probability distributions over measurement results by means of the polychotomic algorithm, we want to point out the complete incompatibility of these distributions with each other. In fact, they are defined in spaces of different dimensionality. The distribution $d\bar{\bar{P}}_{1,2}$ has its "volume element" defined in an (N_1+N_2)-dimensional complex $\{\xi_{1i}, \xi_{2j}\}$-space, or in a real $(2N_1+2N_2)$-dimensional $\{\eta_{1\lambda}, \eta_{2\mu}\}$-space. After integration over the phases in the complex planes, we have left in (M-13) an (N_1+N_2)-dimensional real volume element in $\{Y_i, Z_j\}$-space.

On the other hand, the distribution $d\bar{P}_{(1,2)}$ is defined in an N_1N_2-dimensional complex $\{\xi_k\}$-space, or a $2N_1N_2$-dimensional real $\{\eta_\nu\}$-space, or, in (M-14), in an N_1N_2-dimensional real $\{X_k\}$-space.

The latter spaces for $d\bar{P}_{(1,2)}$ have far greater dimensionality than the former ones for $d\bar{\bar{P}}_{1,2}$, which are merely strangely shaped hypersurfaces in the spaces for $d\bar{P}_{(1,2)}$. Most individual hidden-variables states $\Xi(1, 2)$ occurring in the ensemble $d\bar{P}_{(1,2)}$, therefore, do not occur at all in the ensemble $d\bar{\bar{P}}_{1,2}$. This is not surprising: most of these $\Xi(1, 2)$ in the ensemble $d\bar{P}_{(1,2)}$ are not factorizable, and therefore could not possibly occur in $d\bar{\bar{P}}_{1,2}$.

Corresponding canonical ensembles in the Bohm–Bub theory

The Ξ of the Bohm–Bub theory are obtainable from those in the Wiener–Siegel theory by normalizing them. [See eq. (172) of Section 4.1.] Thus the distributions in $2N$-dimensional $\{\eta\}$-space are radially condensed onto the $(2N-1)$-dimensional surface of a real sphere in that η-space. They then become uniform distributions. Putting here N equal to N_1+N_2 for $d\bar{\bar{P}}^{\text{B-B}}_{1,2}$ and to N_1N_2 for $d\bar{P}^{\text{B-B}}_{(1,2)}$, we can easily see what happens to the Wiener–Siegel ensembles in the Bohm–Bub theory.

We will not try to write down explicit formulas for $d\bar{\bar{P}}^{\text{B-B}}_{1,2}$ or for $d\bar{P}^{\text{B-B}}_{(1,2)}$, as we see from Section 4.1 that use of these explicit formulas can be avoided because of the correspondence existing between the Bohm–Bub distributions and the Wiener–Siegel distributions.

It remains true that $d\bar{P}^{\text{B-B}}_{(1,2)}$ and $d\bar{\bar{P}}^{\text{B-B}}_{1,2}$ are utterly different.

APPENDIX N

REDUCTIONS OF STATE VECTORS IN THE THEORY OF WIENER AND SIEGEL

The "measurement problem" in the Wiener–Siegel theory

In the Wiener–Siegel theory the description of a measurement of an observable A, taking into account the measuring apparatus, requires consideration of at least the following two different cases:

(1) Somehow the initial hidden-variables states are known, and somehow no relaxation of these states occurs.

(2) In the initial state, hidden variables are, separately for object 1 and for instrument 2, distributed canonically over an ensemble. Thus, the initial distribution is given by (M-13). While the measurement takes place, the hidden variables of the composite system of apparatus and object completely relax to the distribution (M-14).

In each case, we end up with predictions for the final state of the composite system. We could use these for predicting the results of a subsequent measurement of a different observable B.

However, we also want to get rid of the appearance of the first measurement apparatus in the description of the second measurement. For this we want to find formulas describing a "reduction" of the state of the composite system of object and first apparatus to a state of the object by itself.

Then, in each of the above cases, two questions arise. (1) Can we write down reduction formulas such that the reduced formalism yields for the second measurement the same results as the original "full" description did? (2) What is the nature of these reduction formulas, how do they fit into a supposedly (pseudo)deterministic theory, and what are the implications of these formulas for the "objective reality" of the quantities appearing in these formulas?

A general and satisfactory discussion of all of these questions, to my knowledge, has not been given in the literature. Siegel and Wiener (1956) give only a partial discussion. In the following, therefore, we will consider only an oversimplified model of the theory for arriving at some tentative conclusions.

Idealized measurements

We will make the following simplifying assumptions about the interaction between the object (1) and the measuring apparatus (2):

(a) When an observable A is measured, the object will not make transitions between different eigenstates of A_{op}.

(b) The different "dial readings" on the apparatus may be described as eigenstates of a corresponding operator A'_{op}.

(c) There will be a strict one-to-one correspondence between eigenvalues A_i of A_{op} and eigenvalues A'_i of A'_{op}, so that the dial reading will tell us uniquely the final state of the object.

(d) The same dial-reading state will be obtained irrespective of the initial state of the instrument.

(e) The lack of dependence of the final state upon the initial instrument state is for simplicity assumed to be so general, that the equivalence between the equations (146) and (157), for the evolutions of the ψ-state and of the Ξ-state, will also make the final hidden-variables state Ξ independent of the initial instrument hidden-variables state. (This may be too much of an idealization.)

(f) We assume A_{op} to have no degenerate eigenvalues, so that the formalism of Appendix L is applicable.

(g) Dial reading states are assumed to be permanent, so that the dial reading for measurement A is not affected by the later measurement of B.

Case One: Given hidden-variables state and no relaxation

We start with an object in an initial state

$$\psi_1^{(o)}(x_1) = \sum_i a_{1i}^{(o)} \phi_{1i}(x_1), \quad \Xi_1^{(o)}(x_1) = \sum_i \xi_{1i}^{(o)} \phi_{1i}(x_1), \tag{N-1}$$

and with two instruments, systems 2 and 3, of which first 2 will be used as apparatus for measuring A, and later 3 for measuring B. Let the initial state of system n be given by $\psi_n^{(o)}(x_n)$ and $\Xi_n^{(o)}(x_n)$. Thus, the combined initial state is

$$\left.\begin{aligned} \psi^{(o)} &= \sum_i a_{1i}^{(o)} \phi_{1i}(x_1)\, \psi_2^{(o)}(x_2)\, \psi_3^{(o)}(x_3), \\ \Xi^{(o)} &= \sum_i \xi_{1i}^{(o)} \phi_{1i}(x_1)\, \Xi_2^{(o)}(x_2)\, \Xi_3^{(o)}(x_3). \end{aligned}\right\} \tag{N-2}$$

After systems 1 and 2 have interacted, the state (M) after the measurement of A, according to the various assumptions (a)–(g), will be

$$\left.\begin{aligned} \psi^{(M)} &= \sum_i a_{1i}^{(o)} \phi_{1i}(x_1)\, \phi_{2i}(x_2)\, \psi_3^{(o)}(x_3), \\ \Xi^{(M)} &= \sum_i \xi_{1i}^{(o)} \phi_{1i}(x_1)\, \phi_{2i}(x_2)\, \Xi_3^{(o)}(x_3). \end{aligned}\right\} \tag{N-3}$$

This specific form for $\Xi^{(M)}$ depends heavily upon the idealization (e).

We expand the eigenstates ϕ_{1i} of $A_{1,\text{op}}$ in term of the eigenstates χ_{1i} of $B_{1,\text{op}}$, by

$$\phi_{1i}(x_1) = \sum_j T_{ij}\, \chi_{1j}(x_1), \quad T_{ij} = \int \chi_{1j}^*(x_1)\, \phi_{1i}(x_1)\, dx_1, \tag{N-4}$$

so that

$$\left.\begin{aligned} \psi^{(M)} &= \sum_j \chi_{1j}(x_1) \sum_i a_{1i}^{(o)} T_{ij}\, \phi_{2i}(x_2)\, \psi_3^{(o)}(x_3), \\ \Xi^{(M)} &= \sum_j \chi_{1j}(x_1) \sum_i \xi_{1i}^{(o)} T_{ij}\, \phi_{2i}(x_2)\, \Xi_3^{(o)}(x_3). \end{aligned}\right\} \tag{N-5}$$

After interaction of object 1 with B-apparatus 3, we obtain a final state (f) given by

$$\left.\begin{aligned} \psi^{(f)} &= \sum_j \chi_{1j}(x_1) \sum_i a_{1i}^{(o)} T_{ij}\, \phi_{2i}(x_2)\, \chi_{3j}(x_3), \\ \Xi^{(f)} &= \sum_j \chi_{1j}(x_1) \sum_i \xi_{1i}^{(o)} T_{ij}\, \phi_{2i}(x_2)\, \chi_{3j}(x_3). \end{aligned}\right\} \tag{N-6}$$

Prediction of results of measurements by the polychotomic algorithm

For determining which instrument reading $\phi_{2i}(x_2)$ is the result of the first measurement, we must apply the polychotomic algorithm to the statevectors (N-3), using according to (L-17) as the deciding ratio the expression

$$\left\{ \int | a_{1i}^{(o)}\, \phi_{1i}(x_1)\, \psi_3^{(o)}(x_3) |^2\, dx_1\, dx_3 \right\} / \left| \int \xi_{1i}^{(o)*}\, \phi_{1i}^*(x_1)\, \Xi_3^{(o)*}(x_3) \cdot a_{1i}^{(o)}\, \phi_{1i}(x_1)\, \psi_3^{(o)}(x_3)\, dx_1\, dx_3 \right| . \tag{N-7}$$

The integrals over x_1 here cancel out, and, if we put

$$K = \left| \int \Xi_3^{(o)*}(x_3)\, \psi_3^{(o)}(x_3)\, dx_3 \right|, \tag{N-8}$$

the above ratio comes out to be

$$K^{-1} \cdot \left| a_{1i}^{(o)} / \xi_{1i}^{(o)} \right|. \tag{N-9}$$

If $i = k$ for the largest among the quantities (N-9), the same is true for the largest among the ratios $| a_{1i}^{(o)}/\xi_{1i}^{(o)} |$ occurring in (120), so that we find that the complete theory of measurement (taking into account the instrument) makes in a given hidden-variables state the same prediction for the result A_i, as made by the algorithm applied in its primitive form (120) directly to the initial state vectors (N-1) of the object alone.

If, now, we use (N-6) with (L-17) for finding the ratio that is to determine by the polychotomic algorithm which combination of values A_i and B_j the double measurement is to yield, we should use $\phi_{2i}(x_2) \cdot \chi_{3j}(x_3)$ like we used above $\phi_{2i}(x_2)$. We thus find the ratio

$$\left\{ \int | \chi_{1j}(x_1)\, a_{1i}^{(o)} T_{ij} |^2\, dx_1 \right\} \Big/ \left| \int \chi_{1j}^*(x_1)\, \xi_{1i}^{(o)*} T_{ij}^* \chi_{1j}(x_1)\, a_{1i}^{(o)} T_{ij}\, dx_1 \right| = | a_{1i}^{(o)}/\xi_{1i}^{(o)} |. \tag{N-10}$$

Thus the complete theory in this idealized case predicts again the same result A_i as before, but leaves B_j completely undetermined, as we meet here the exceptional case that the maximum of the ratios compared in the algorithm is shared by all values of j, probably because of the oversimplification (e) of our model.

Since the "rigorous theory" here gives no result, it is impossible for our model to compare the "reduced" theory of Siegel and Wiener (1956) with the "rigorous" theory. In this "reduced" treatment, if A_l was observed, $\psi^{(M)}$ and $\Xi^{(M)}$ of (N-3) would be replaced by the reduced wave functions $\psi^{(r)} = \phi_{1l}(x_1)\, \psi_3^{(o)}(x_3)$ and $\Xi^{(r)} = \phi_{1l}(x_1)\, \Xi_3^{(o)}(x_3)$ before the measurement of B, without an explanation why the hidden-variables state should change in this way by the measurement of an observable.

Case Two: Relaxing hidden-variables states

Here we start out with an ensemble with the same ψ as in Case One, and with initially separate canonical distributions for $\Xi_1^{(o)}(x_1)$, for $\Xi_2^{(o)}(x_2)$, and for $\Xi_3^{(o)}(x_3)$.

After the first measurement, $\Xi_3^{(o)}$ has maintained its separate canonical distribution, but there is a single canonical distribution of the kind of eq. (M-14) for the hidden-variables state $\Xi_{(1,2)}(x_1, x_2)$ of the composite system of 1 and 2 together.

After the second measurement, we shall assume for simplicity that there is a canonical distribution for the composite system of all three parts, although this is a dubious assumption, as parts 2 and 3 never interact directly and never interact both with part 1 at the same time.

"Rigorous" treatment

We now use the formalism of Appendix L for deriving in each case the ensemble probabilities for the various possible results of measurement from the formulas for ψ. They are given by eqs. (L-4) with (L-2).

After the first measurement, we use $\psi^{(M)}$ from (N-3) for calculating the probability of $\phi_{2i}(x_2)$. We must integrate over x_1 and x_3 acting as the y of eq. (L-4). Normalization of $\phi_{1i}(x_1)$ and of $\psi_3^{(o)}(x_3)$ gives

$$w(A_i) = |a_{1i}^{(o)}|^2. \tag{N-11}$$

After the second measurement, we use $\psi^{(f)}$ from (N-6) for calculating the probability of $\phi_{2i}(x_2)\chi_{3j}(x_3)$. Here, x_1 acts as y. We find

$$w(A_i, B_j) = |a_{1i}^{(o)}T_{ij}|^2. \tag{N-12}$$

We can also use $\psi^{(f)}$ for calculating the probability of $\chi_{3j}(x_3)$ by itself. In this case, not only x_1 but also x_2 and i take the role of y. An "integration" over i is, of course, just a sum. We find

$$w(B_j) = \sum_i |a_{1i}^{(o)}T_{ij}|^2. \tag{N-13}$$

The probability with which B_j follows a *given* result A_i is found by

$$w(B_j, \text{ if } A_i) = \frac{w(A_i, B_j)}{w(A_i)} = |T_{ij}|^2. \tag{N-14}$$

Thus the "rigorous" method without reductions yields the familiar results.

"Reduced" treatment

In the "reduced" treatment, we return after each interaction to a "reduced" wavefunction $\psi_1^{(r)}(x_1)$ of the object alone, using the result of the measurement just completed.

We know how to do this quantum mechanically. The question merely is: If we want to explain the same probabilities using the polychotomic algorithm, then, what information on some $\Xi_1^{(r)}(x_1)$ should accompany $\psi_1^{(r)}(x_1)$? Can we use this $\psi_1^{(r)}$ and $\Xi_1^{(r)}$ information for the discussion of the measurement of B, like we used $\psi_1^{(o)}$ and $\Xi_1^{(o)}$-information for the discussion of measurement A?

If A_I is the result of the measurement of A, we reduced $\psi^{(M)}$ to

$$\psi_1^{(r)}(x_1) = \phi_{1I}(x_1). \tag{N-15}$$

We here omitted the instrument factors in the term of the sum corresponding to the result $i = I$ of the measurement, and then renormalized $\psi^{(r)}(x_1)$.

We had a canonical distribution $d\bar{P}_{(1,2)}$ of $\Xi_{(1,2)}(x_1, x_2)$ at the end of the measurement, but any reduced $\Xi_1^{(r)}(x_1)$ distribution should not depend on x_2. Therefore

> we reduce this $d\bar{P}_{(1,2)}$ distribution to the $d\bar{P}_1$ distribution of (M-12), *even though the two distributions are totally unrelated, and $d\bar{P}_1$ is not even a factor of $d\bar{P}_{(1,2)}$*. (N-16)

The only excuse which we have for such a reduction is that *it explains the things further happening to the object*. Like we do not ask in quantum theory how a pure state of a composite system can change, via a mixed state of the object alone, to a pure state of the object alone, we do not ask questions here. The only excuses are the effectiveness of the change and the strong desire to rid ourselves of things reminiscent of instruments used in the past.

We now start with $\psi_1^{(r)}(x_1)$ toward the second measurement, as we started from (N-1) toward the first measurement.

For the composite system of 1 and 3, we write

$$\psi^{(o)}(1, 3) = \psi_1^{(r)}(x_1)\psi_3^{(o)}(x_3) = \sum_j T_{Ij}\chi_{1j}(x_1)\psi_3^{(o)}(x_3), \tag{N-17}$$

$$\psi^{(f)}(1, 3) = \sum_j T_{Ij}\chi_{1j}(x_1)\chi_{3j}(x_3), \tag{N-18}$$

and the $\Xi^{(f)}(1, 3)$ states in the ensemble will be distributed according to a composite canonical distribution $d\bar{P}_{(1,3)}$ similar to (M-14), with 3 replacing 2.

Now (L-4) predicts for (N-17) the probability for the dial reading $\chi_{3j}(x_3)$ according to the polychotomic algorithm. We use again x_1 like the y of (L-4). We find

$$W(B_j \text{ after } A_I) = |T_{Ij}|^2, \tag{N-19}$$

in agreement with (N-14). It shows that the reduction procedure (N-15) with (N-16) "works" for statistical predictions if continuous relaxation during interaction is *assumed* and if some of this relaxation is "undone by postulate" in reductions as in (N-16).

Evaluation of the procedure

The above reduction procedures are *postulated* and *not explained.* As hidden-variables theory is supposedly a (crypto)deterministic theory, this authoritarian treatment of the state vector is *contradictory to the spirit of the theory.*

The Bohm–Bub approach

In order to avoid messing up the description of the object by coordinates describing previously used instruments, Bohm and Bub never introduce these coordinates. To some extent this may be considered a phenomenological treatment, as it is a physical fact that object and instrument form a composite and interacting system during the measurement. If we leave out the instrument, replacing it by some kind of an external influence upon the single object system 1, we cannot treat the object during the measurement as a closed system, and the Schrödinger equation does not apply during that time without modifications.

As discussed in Section 4.1, a phenomenological description of what for all practical purposes seems to happen during a measurement requires additional terms in the Schrödinger equation for the object that are not linear in the state vector of the object.

See Chapter 4 for a discussion of what such terms might look like.

APPENDIX O

"POLARIZATION" OF A PHOTON IN QUANTUM THEORY

In electrodynamics a polarized wave can be decomposed into different polarizations. A linearly polarized wave can be decomposed into two oppositely circularly polarized waves with equal amplitudes and with a phase difference. Two perpendicularly linearly or oppositely circularly polarized waves traveling in the same direction, with the same wavelength, with unequal amplitudes and phases, can be superposed to an *elliptically* polarized wave.

In quantum electrodynamics we describe a photon corresponding to waves of a certain wavelength and polarization by a state vector ψ which is a traveling wave packet [see eq. (H-36)]. The question we want to answer is: How do we superpose such ψ wave packets for describing the superposition of polarization of the electromagnetic waves corresponding to the photons described by the ψ-waves?

In Appendix H we used complex Fourier-component waves for describing real electromagnetic waves. We used complex polarization vectors $\mathbf{e}_{\mathbf{k}}^{\eta}$ for describing such waves.

We first must answer the question: What electromagnetic waves do correspond to the presence of a photon? It can be shown that the expectation values of the fields $\mathbf{E}_{\perp}$ and $\mathbf{B}$ *vanish* if it is certain that only one photon is present. (In that case, $\psi^{*}\mathbf{E}_{\perp}\,\psi$ and $\psi^{*}\,\mathbf{B}\,\psi$ are odd in the q_f, so their integrals $\langle \mathbf{E}_{\perp}\rangle = \int \psi^{*}\mathbf{E}_{\perp}\psi\, dq_f \ldots$, etc., vanish.) However, we have seen in eqs. (H-42)–(H-43) that the electromagnetic energy and momentum distributions in the presence of one photon are described by wave packets $\mathbf{g}_n^{\eta}(\mathbf{x})$. These waves, given by (H-41), differ among each other by the power with which k occurs. For a photon which corresponds pretty strictly to a wave vector $\mathbf{k} \approx \mathbf{K}$ (and which therefore is not well localized), $f_{\eta}^{(\mathbf{K})}(\mathbf{k})$ depends on $\mathbf{k}$ nearly by a delta-function factor $\delta_3(\mathbf{k}-\mathbf{K})$, and then the differences between $\mathbf{g}_0^{\eta}(\mathbf{x})$, $\mathbf{g}_1^{\eta}(\mathbf{x})$, and $\mathbf{g}_2^{\eta}(\mathbf{x})$ can be factorized out as powers of $(\hbar K)^{\frac{1}{2}}$, that is, merely normalization factors, and these three waves otherwise are then identical among each other.

We therefore regard these $\mathbf{g}_n^{\eta}(\mathbf{x})$ waves as the waves that come most closely to describing the one-photon-phenomenon electromagnetically. The vector $\mathbf{e}_{\mathbf{k}}^{\eta}$ occurring in $\mathbf{g}_n^{\eta}(\mathbf{x})$ then tells as the polarization of the photon described by the scalar ψ wave packet (H-36).

Now suppose we replace the latter scalar ψ wave by a "superposition of polarizations"

$$\psi^{\text{new}} = \sum_{\eta} c_{\eta}\psi_{\mathbf{K},\eta}^{(F)}(q_f) \quad \text{with} \quad \sum_{\eta=\pm 1} |\,c_{\eta}\,|^2 = 1. \tag{O-1}$$

Repeating the calculations that led to eqs. (H-42) and (H-43), the effect is merely to replace each $\mathbf{g}_n^{\eta}(\mathbf{x})$ in the result by $\mathbf{g}^{\text{new}}(\mathbf{x}) = \sum_{\eta} c_{\eta}\mathbf{g}_n^{\eta}(x)$, where the new waves $\mathbf{g}_n^{\text{new}}$ according to (H-41) are obtained from the old ones merely by changing the polarization factors $\mathbf{e}_{\mathbf{k}}^{\eta}$ in them by

$$\mathbf{e}_{\mathbf{k}}^{\text{new}} = \sum_{\eta} c_{\eta}\,\mathbf{e}_{\mathbf{k}}^{\eta}. \tag{O-2}$$

Comparing (O-1) with (O-2) we see that the wavefunctions ψ descriptive of the photons are superposed with the same coefficients as with which we superpose in (O-2) (and thence in the electromagnetic waves) the polarization vectors.[95]

Is the photon polarized?

We note that the distinctions between photons polarized one way or another are *in the wavefunctions*, be those **g**(**x**) or ψ. The polarization of a photon, in this regard, is like the spin of an electron. The latter, too, is a property of the wave, which, however, in that case is a "spinor" wave instead of a "vector" wave.

A photon in unpolarized light

Suppose very weak unpolarized light is incident, and after a long wait finally the first photon is on its way. How is this described quantum theoretically?

We should describe a single photon by a wave like (H-36) if the light is circularly polarized, or by a linear combination (O-1) if the polarization is different according to (O-2). As in the present case the polarization direction is unknown, the photon is described by a *mixed state*, that is, there are equal probabilities for "all" waves (O-1) corresponding to "all" polarizations (O-2).

The question remains: What means "all"?

This may well depend upon the source of the unpolarized light.

Circularly unpolarized light

It may be that this source will emit merely circularly polarized waves half left-hand and half right-hand, and each with an arbitrary phase. We can describe this as a mixed state in which half of the state vectors ψ are given by (O-2) with $c_+ = \exp(i\theta)$ and $c_- = 0$, and the other half by $c_+ = 0$ and $c_- = \exp(i\theta)$. In each case there is equal probability of any θ between 0 and 2π.

Linearly unpolarized light

We may have a source which emits linearly polarized light, but the linear polarization has *any* direction perpendicular to **K**, and the phases are arbitrary. We have again a mixed state, but, this time, the state vectors that occur with equal probability are of the form (O-1) with $c_\eta = \sqrt{\tfrac{1}{2}}\cdot\exp(i\theta_\eta)$ where both θ_+ and θ_- each independently have equal probability within the interval from 0 to 2π.

[95] Therefore the ψ waves are often called "vector" waves and the photons vector particles. Actually, each $\psi^{(F)}$ is a *scalar* function of q_f, but the index η on the $\psi^{(F)}_{\mathbf{K},\eta}(q_f)$ makes each pair $\psi^{(F)}_{\mathbf{K},+}(q_f)$ and $\psi^{(F)}_{\mathbf{K},-}(q_f)$ look somewhat like two components of a vector perpendicular to **K**.

Elliptically unpolarized light

This most arbitrary polarization is described by a mixed state in which the state vectors (O-1) have their c_η limited merely by the condition $|c_+|^2+|c_-|^2 = 1$. That is, we have here

$$c_+ = \exp(i\theta_+) \cos\chi, \quad c_- = \exp(i\theta_-) \sin\chi, \tag{O-3}$$

with χ, θ_+, and θ_- independent of each other and stochastically distributed over the intervals $0 < \chi < \pi/2$ and $0 < \theta_\eta < 2\pi$.

APPENDIX P

HIDDEN-VARIABLES BIAS BEHIND FIRST POLARIZATION FILTER

For hidden variables given by (217), the incident ψ wave will be transmitted by the first filter with its polarization direction $\hat{x}$ if

$$|a_x/a_y| > |\xi_x/\xi_y| = \cot\tilde{\chi}. \tag{P-1}$$

For *incident elliptically unpolarized light* (216), this requirement means

$$\cot\chi > \cot\tilde{\chi}, \quad \text{so,} \quad |\chi| < \tilde{\chi}. \tag{P-2}$$

From an ensemble of incident photons with all possible χ-polarizations distributed by (220) or (226), but with the given Ξ-polarization (217), the fraction transmitted will then be

$$P_x = \int_{\chi=0}^{\chi=\tilde{\chi}} \int dP = \int_{u=0}^{u=\sin^2\tilde{\chi}} du = \sin^2\tilde{\chi} \equiv \tilde{u}. \tag{P-3}$$

For *incident linearly unpolarized light* (211), the condition (P-1) gives again (P-2); but, with an incident distribution (221), the fraction transmitted is

$$P_x = \int_{-\tilde{\chi}}^{\tilde{\chi}} d\chi \int_0^{2\pi} d\alpha/2\pi^2 = 2\tilde{\chi}/\pi. \tag{P-4}$$

For *incident circularly unpolarized light* (222), we have $|a_x/a_y| = 1$, so that the condition (P-1) gives full transmission if $1 > \cot\tilde{\chi}$ and $\frac{1}{4}\pi < \tilde{\chi} < \frac{1}{2}\pi$, and no transmission if $1 < \cot\tilde{\chi}$ and $0 < \tilde{\chi} < \frac{1}{4}\pi$. Let

$$\Theta(x) = 1 \quad \text{for} \quad x > 0, \qquad \Theta(x) = 0 \quad \text{for} \quad x < 0. \tag{P-5}$$

Then, in the interval $0 < \tilde{\chi} < \frac{1}{2}\pi$, we find

$$P_x = \Theta(\tilde{\chi} - \tfrac{1}{4}\pi). \tag{P-6}$$

APPENDIX Q

THE TRANSMISSION OF LIGHT BY THE SECOND POLARIZATION FILTER

The condition (236), or $|a'_x\xi'_y|^2 - |a'_y\xi'_x|^2 > 0$, by (235) and (217) becomes

$$-\tilde{r}^2 \sin^2 \tilde{\chi} \cos 2\varepsilon - \tilde{r}^2 \sin \tilde{\chi} \cos \tilde{\chi} \sin 2\varepsilon \cos (\tilde{\theta}_x - \tilde{\theta}_y) > 0,$$

which is equivalent, for $0 < \varepsilon < 90°$ and $0 < \tilde{\chi} < 90°$, to

$$\cos \delta < -(\tan \tilde{\chi})/(\tan 2\varepsilon) = (\tan \tilde{\chi})/(\tan 2\vartheta), \tag{Q-1}$$

where we put

$$\delta = \tilde{\theta}_x - \tilde{\theta}_y, \quad \tilde{\theta} = \tfrac{1}{2}(\tilde{\theta}_x + \tilde{\theta}_y), \quad \vartheta = \pi/2 - \varepsilon. \tag{Q-2}$$

The biased Ξ distribution behind Filter Two may be written as (228) or (229) or (230) with $d\tilde{\theta}_x\, d\tilde{\theta}_y$ replaced by $d\tilde{\theta}\, d\delta$, if the inequality (Q-1) is satisfied, while it is zero otherwise. The intensity $\mathcal{J}_2$ through Filter Two (as fraction of what is incident upon Filter One) is the integral of this. Here, $\int_0^1 d\tilde{v} \int_0^{2\pi} d\tilde{\theta}/2\pi$ gives a factor 1 in all cases.

Note that, by (225),

$$\tilde{u} = 1 - [1 + \tan^2 \tilde{\chi}]^{-1}. \tag{Q-3}$$

We will distinguish *Case A* with $\varepsilon < 45°$, $\tan 2\varepsilon > 0$, from *Case B* with $45° < \varepsilon < 90°$, $\tan 2\varepsilon < 0$, $\vartheta < 45°$. For these two cases, (Q-1) limits $\tan \tilde{\chi}$ differently:

CASE A: $$\tan \tilde{\chi} < -\cos \delta \tan 2\varepsilon. \tag{Q-1A}$$

CASE B: $$\tan \tilde{\chi} > \cos \delta \tan 2\vartheta. \tag{Q-1B}$$

Therefore, Case A allows no χ at all when $\cos \delta > 0$, and limits $\tilde{\chi}$ and $\tilde{u}$ at the top for $\cos \delta < 0$, while Case B allows any $\tilde{\chi}$ in the original interval $0 < \tilde{\chi} < \frac{1}{2}\pi$ for $\cos \delta < 0$, and limits $\tilde{\chi}$ at the bottom for $\cos \delta > 0$.

We will use this while integrating (228)–(230) under the conditions (Q-1A, B), integrating first over $\tilde{\chi}$ or $\tilde{u}$, and then over δ. It is convenient to integrate over δ from $-\frac{1}{2}\pi$ to $+3\pi/2$.

Case of elliptically unpolarized light incident upon Filter One

We integrate (228) under the above conditions.

CASE A gives

$$\mathcal{J}_{2A}^{\mathrm{ell}}(\varepsilon) = \int_{\pi/2}^{3\pi/2} d\delta \int_0^{\tilde{u}_m} \tilde{u}\, d\tilde{u}/2\pi = (8\pi)^{-1} \oint d\delta \cdot \tilde{u}_m^2, \tag{Q-4}$$

where (Q-1A) and (Q-3) give

$$\tilde{u}_m = 1 - [1 + \alpha \cos^2 \delta]^{-1} \tag{Q-5}$$

with

$$\alpha = \tan^2 2\varepsilon = \tan^2 2\vartheta. \tag{Q-6}$$

CASE B gives

$$\begin{aligned}\mathcal{J}_{2B}^{\text{ell}}(\vartheta) &= \int_{-\pi/2}^{+\pi/2} d\delta \int_{\tilde{u}m}^{1} \tilde{u}\, d\tilde{u}/2\pi + \int_{\pi/2}^{3\pi/2} d\delta \int_{0}^{1} \tilde{u}\, d\tilde{u}/2\pi \\ &= \oint d\delta \int_{0}^{1} \tilde{u}\, d\tilde{u}/2\pi - \int_{-\pi/2}^{+\pi/2} d\delta \int_{0}^{\tilde{u}_m} \tilde{u}\, d\tilde{u}/2\pi \\ &= \tfrac{1}{2} - \mathcal{J}_{2A}^{\text{ell}}(\varepsilon).\end{aligned} \tag{Q-7}$$

That is, $\mathcal{J}_2^{\text{ell}}$ at a value of $\varepsilon < 45°$ and $\mathcal{J}_2^{\text{ell}}$ for the same value of $\vartheta < 45°$ (for the complementary value of ε), added together, give $\frac{1}{2} = \mathcal{J}_1$.

Considering again *Case A*, we put
$t = \alpha \cos^2 \delta$, so that

$$u_m^2 = \left(\frac{t}{1+t}\right)^2 = t^2(1-2t+3t^2-\ldots) = \sum_{n=2}^{\infty} (n-1)(-t)^n, \tag{Q-8}$$

$$\begin{aligned}\mathcal{J}_{2A}^{\text{ell}} &= (8\pi)^{-1} \sum_{n=2}^{\infty} (n-1)(-\alpha)^n \oint \cos^{2n} \delta\, d\delta \\ &= \tfrac{1}{4} \sum_{n=2}^{\infty} (n-1) \binom{-\frac{1}{2}}{n} \alpha^n = \tfrac{1}{4}\alpha^2 \cdot \frac{d}{d\alpha} \left[\frac{1}{\alpha} \cdot \sum_{n=2}^{\infty} \binom{-\frac{1}{2}}{n} \alpha^n\right] \\ &= \frac{\alpha^2}{4} \frac{d}{d\alpha} \left\{\frac{1}{\alpha(1+\alpha)^{\frac{1}{2}}} - \frac{1}{\alpha} + \frac{1}{2}\right\} \\ &= \tfrac{1}{4} \left\{1-(1+\alpha)^{-\frac{1}{2}} - \tfrac{1}{2}(1+\alpha)^{-\frac{3}{2}}\alpha\right\}.\end{aligned} \tag{Q-9}$$

Inserting here (Q-6), we find

$$\begin{aligned}\mathcal{J}_{2A}^{\text{ell}} &= \tfrac{1}{4}\{1-\cos(2\varepsilon) - \tfrac{1}{2}\cos(2\varepsilon)\sin^2(2\varepsilon)\} \\ &= \tfrac{1}{4} - \tfrac{3}{8}\cos(2\varepsilon) + \tfrac{1}{8}\cos^3(2\varepsilon) \\ &= \tfrac{1}{2}\sin^2\varepsilon - \tfrac{1}{8}\sin^2(2\varepsilon)\cos(2\varepsilon),\end{aligned} \tag{Q-10}$$

so that, for $\varepsilon > 45°$, $\vartheta < 45°$, we find from (Q-7)

$$\mathcal{J}_{2B}^{\text{ell}} = \tfrac{1}{4} + \tfrac{3}{8}\cos(2\vartheta) - \tfrac{1}{8}\cos^3(2\vartheta). \tag{Q-11}$$

Both (Q-10) and (Q-11) may be written as

$$\mathcal{J}_2^{\text{ell}} = \tfrac{1}{2}\cos^2\vartheta + \tfrac{1}{8}\sin^3(2\vartheta)\cos(2\vartheta), \tag{Q-12}$$

where the last term is the effect of the bias in Ξ upon the intensity transmitted by two filters, as the first term (with $\frac{1}{2} = \mathcal{J}_1$) is the intensity predicted by Malus's law.

Case of linearly unpolarized light incident upon Filter One

This time we integrate (229) instead of (228) over the same $\tilde{u}$, δ intervals as above: The integrand (229) differs from the one in (228) by featuring a factor $(2\chi/\pi)$ instead of a factor $\tilde{u}$.

We obtain

$$\mathcal{J}_{2A}^{\text{lin}}(\varepsilon) = \int_{\pi/2}^{3\pi/2} d\delta \int_0^{\tilde{u}_m} (2\tilde{\chi}/\pi)\, d\tilde{u}/2\pi, \tag{Q-13}$$

$$\mathcal{J}_{2B}^{\text{lin}}(\vartheta) = \int_{-\pi/2}^{+\pi/2} d\delta \int_{\tilde{u}_m}^{1} (2\tilde{\chi}/\pi)\, d\tilde{u}/2\pi + \int_{\pi/2}^{3\pi/2} d\delta \int_0^1 (2\tilde{\chi}/\pi)\, d\tilde{u}/2\pi$$

$$= \oint d\delta \int_0^1 (2\chi/\pi)\, d\tilde{u}/2\pi - \int_{-\pi/2}^{+\pi/2} d\delta \int_0^{\tilde{u}_m} (2^{\sim}/\pi)\, d\tilde{u}/2\pi, \tag{Q-14}$$

while (232) gives

$$\mathcal{J}_1 = \oint d\delta \int_0^1 (2\tilde{\chi}/\pi)\, d\tilde{u}/2\pi = \tfrac{1}{2}, \tag{Q-15}$$

so again we find that the $\mathcal{J}_2$ for complementary angles ε add up to $\mathcal{J}_1 = \frac{1}{2}$.

For the computation of $\mathcal{J}_{2A}^{\text{lin}}$ (ε), we may interchange the order of integration of δ and of $\tilde{\chi}$ or $\tilde{u}$. In this case of $\varepsilon < 45°$, the condition (Q-1) gives no possible δ for $\chi > 2\varepsilon$, while for $\tilde{\chi} < 2\varepsilon$ we integrate δ over an interval from $(\pi—\Delta)$ to $(\pi+\Delta)$, where $\Delta = \text{arc cos}$ $[(\tan \tilde{\chi})/(\tan 2\varepsilon)]$. Thus we get

$$\mathcal{J}_{2A}^{\text{lin}} = \int_0^{2\varepsilon} (2/\pi^2)\, \tilde{\chi} \sin \chi \cos \chi \, d\chi \int_{\pi-\Delta}^{\pi+\Delta} d\delta$$

$$= (4/\pi^2) \int_0^{2\varepsilon} \tilde{\chi} \sin \tilde{\chi} \cos \tilde{\chi} \text{ arc cos } [(\tan \chi)/(\tan 2\varepsilon)]\, d\tilde{\chi}. \tag{Q-16}$$

Twice this integral, computed numerically on a Hewlett–Packard desk computer, is tabulated as $\mathcal{J}_2^{\text{lin}}/\mathcal{J}_1$ in Table 5.1 (p. 172) as a function of $\varepsilon = 90° - \vartheta$ for $\varepsilon < 45°$.

Case of circularly unpolarized light incident upon Filter One

This time we integrate (230) over the same $\tilde{u}$, δ intervals as before. Because of the factor $\Theta(\tilde{u}-\frac{1}{2})$ in (230), this amounts to integrating $d\tilde{u}\, d\delta/2\pi$ under the condition $\frac{1}{2} < \tilde{u} < 1$ (or $\tan \tilde{\chi} > 1$) as well as the condition (Q-1).

In *Case A*, we consider separately $\tan 2\varepsilon > 1$ and $0 < \tan 2\varepsilon < 1$. We put

$$D = \text{arc cos}\,[1/(\tan 2\varepsilon)], \quad \text{if} \quad \tan 2\varepsilon > 1, \quad 22°30' < \varepsilon < 45°. \tag{Q-17}$$

In that case, for $\pi - D < \delta < \pi + D$, $\tilde{u}$ runs from $\frac{1}{2}$ to $\tilde{u}_m$ of eq. (Q-5). For $|\delta - \pi| > D$, no possible values of $\tan \tilde{\chi}$ exist. We put

$$D = 0, \quad \text{if} \quad 0 < \tan 2\varepsilon < 1, \quad 0 < \varepsilon < 22°30'. \tag{Q-17'}$$

In that case, no $\tilde{\chi}$ will ever satisfy both conditions imposed, and we find $\mathcal{J}_{2A}^{\text{cir}} = 0$ for $0 < \varepsilon < 22°\,30'$.

In either case we get

$$\mathcal{J}_{2A}^{\text{cir}}(\varepsilon) = \int_{\pi-D}^{\pi+D} d\delta \int_{\frac{1}{2}}^{u_m} d\tilde{u}/2\pi = \int_{\pi-D}^{\pi+D} \frac{\tilde{u}_m\, d\delta}{2\pi} - \frac{D}{2\pi}. \tag{Q-18}$$

In *Case B*, we define D by

$$D = \text{arc cos}\,[1/(\tan 2\vartheta],\quad \text{if}\quad \tan 2\vartheta > 1, \tag{Q-19}$$
$$D = 0,\quad \text{if}\quad \tan 2\vartheta < 1,\quad \text{when}\quad \vartheta < 22°30'. \tag{Q-19'}$$

Then it is useful to let δ run from $-D$ to $(2\pi - D)$, and, for $-D < \delta < +D$, $\tilde{u}$ will run from $\tilde{u}_m$ to 1, while, for $D < \delta < (2\pi - D)$, $\tilde{u}$ will run from $\frac{1}{2}$ to 1.

Thus we get

$$\mathcal{J}^{\text{cir}}_{2B}(\vartheta) = \int_{-D}^{+D} d\delta \int_{\tilde{u}_m}^{1} \frac{d\tilde{u}}{2\pi} + \int_{D}^{2\pi-D} d\delta \int_{\frac{1}{2}}^{1} \frac{d\tilde{u}}{2\pi} = \frac{\pi + D}{2\pi} - \int_{-D}^{+D} \frac{\tilde{u}_m\, d\delta}{2\pi}. \tag{Q-20}$$

As before, take ϑ in $\mathcal{J}_{2B}$ equal to ε in $\mathcal{J}_{2A}$ (so that both have the same value for D), and add. Since $\tilde{u}_m$, by (Q-5), as a function of δ is the same at δ and at $\delta + \pi$, we find the sum again to be $\frac{1}{2} = \mathcal{J}_1$. Since $\mathcal{J}^{\text{cir}}_{2A}(\varepsilon) = 0$ for $\varepsilon < 22°\,30'$, we have $\mathcal{J}^{\text{cir}}_{2B}(\vartheta) = \frac{1}{2} = \mathcal{J}_1$ for $\vartheta < 22°\,30'$.

Returning to Case A, with $\alpha = \tan^2 2\varepsilon$ as before, for $22°\,30' < \varepsilon < 45°$ we obtain

$$\mathcal{J}^{\text{cir}}_{2A}(\varepsilon) = \int_{-D}^{+D} \left[1 - \frac{1}{1 + \alpha \cos^2 \delta}\right] \frac{d\delta}{2\pi} - \frac{D}{2\pi} = \frac{1}{2\pi}\left\{D - \frac{2}{(1+\alpha)^{\frac{1}{2}}}\,\text{arc tan}\left[\frac{\tan D}{(1+\alpha)^{\frac{1}{2}}}\right]\right\}$$
$$= \frac{1}{2\pi}\left\{\text{arc cos}\,[\cot 2\varepsilon] - 2(\cos 2\varepsilon)\,\text{arc tan}\left[(-\cos 4\varepsilon)^{\frac{1}{2}}\right]\right\}, \tag{Q-21}$$

while we found already

$$\mathcal{J}^{\text{cir}}_{2A}(\varepsilon) = 0 \quad \text{for} \quad \varepsilon < 22°\,30'. \tag{Q-21'}$$

APPENDIX R

THE TRANSMISSION OF LIGHT BY THE THIRD POLARIZATION FILTER

By inserting (241)–(242) in the polychotomic condition (243) for Filter Three, we obtain

$$0 < |a''_x \xi''_y|^2 - |a''_y \xi''_x|^2$$
$$= |\xi_x|^2 \cos\varepsilon \cos(\varepsilon + 2\theta) - \tfrac{1}{2}(\xi^*_x \xi_y + \xi^*_y \xi_x)\sin(2\varepsilon + 2\theta) + |\xi_y|^2 \sin\varepsilon \sin(\varepsilon + 2\theta), \tag{R-1}$$

so, by (217) and (Q-2), and putting

$$\left.\begin{aligned} A &= \cos\varepsilon\cos(\varepsilon + 2\theta), \quad B = \cos(2\varepsilon + 2\theta), \quad C = \sin(2\varepsilon + 2\theta), \\ D &= B\sin^2\tilde{\chi} - A, \qquad\qquad E = C\sin\tilde{\chi}\cos\tilde{\chi}, \end{aligned}\right\} \tag{R-2}$$

we find

$$E\cos\delta \leqslant -D. \tag{R-3}$$

This condition, for given $\tilde{\chi}$, places an additional condition upon δ. [From Filter Two, condition (Q-1) was already imposed upon δ.] We can now calculate $\mathcal{J}_3$ by a method similar to the one used in Eq. (Q-16). We consider the case $\varepsilon < 45°$ only. We find

$$\mathcal{J}_3 = 2\mathcal{J}_3/\mathcal{J}_1 = (2/\pi) \int_{\text{bottom}}^{\text{top}} P_x(\tilde{\chi}) \sin\tilde{\chi} \cos\tilde{\chi}\, h(\chi)\, d\tilde{\chi}, \tag{R-4}$$

where $P_x(\tilde{\chi})$ is given by one of the equations (P-3), (P-4), or (P-6), and where $h(\tilde{\chi})$ is half the δ-interval admitted by conditions (Q-1) and (R-3), while bottom $= 0$ and top $= 2\varepsilon$. For the case (P-6) of circular unpolarization, we may replace $P_x(\tilde{\chi})$ by 1, if we replace bottom $= 0$ by bottom $=$ minimum $(\frac{1}{4}\pi$, top). (Thus there is no integral if $\varepsilon < \frac{1}{8}\pi = 22\frac{1}{2}°$.)

The value of h can never be more than

$$\Delta = \arc\cos\,[(\tan\tilde{\chi})\,/(\tan 2\varepsilon)] \tag{R-5}$$

allowed by condition (Q-1). [See above eq. (Q-16).] Also, if $0 < D < E$, the interval can be at most the interval from $(\pi - \Delta')$ to $(\pi + \Delta')$ with

$$\Delta' = \arc\cos\,(D/E), \tag{R-6}$$

so that in this case $h =$ minimum (Δ', Δ). If $D < E$ *and* $D < 0$, we find $h = \Delta$. If $0 > D > E$, the allowable δ-interval is the intersection of the intervals *inside* $(\pi - \Delta, \pi + \Delta)$ and *outside* $(\pi - \Delta', \pi + \Delta')$, so that in this case $h =$ minimum $(\Delta - \Delta', 0)$. If $D > E$ *and* $D > 0$, condition (R-3) alone makes $h = 0$.

Defining the function h by these conditions, we can calculate (R-4) on the computer as a single integral over $\tilde{\chi}$. Changing the integration variable from $\tilde{\chi}$ to $\tilde{u} = \sin^2\tilde{\chi}$, this calculation was performed on the CDC 6500 at Purdue University for $\varepsilon = 5°, 10°, 15°, 20°, 25°$, and $30°$ for θ at $\frac{1}{2}°$ intervals from $\theta = 45° - \frac{1}{2}\varepsilon$ to $\theta = 45 + \frac{1}{2}\varepsilon$, and (first) at $\theta = 0$ for a check on $\mathcal{J}_2/\mathcal{J}_1$, enabling us to list also $\mathcal{J}_3/\mathcal{J}_2$. See Section 5.6 for the results.

Conditions for $\mathcal{J}_3/\mathcal{J}_2 = 0$ and for $\mathcal{J}_3/\mathcal{J}_2 = 1$

For $|\theta| < 45° - \frac{1}{2}\varepsilon$, one finds from the computations $\mathcal{J}_3/\mathcal{J}_2 = 1$; for $45° + \frac{1}{2}\varepsilon < |\theta| < 135° - \frac{1}{2}\varepsilon$, one finds $\mathcal{J}_3/\mathcal{J}_2 = 0$; for $135° + \frac{1}{2}\varepsilon < |\theta| < 225° - \frac{1}{2}\varepsilon$, again $\mathcal{J}_3/\mathcal{J}_2 = 1$. This is best understood by repeating all calculations on the basis $\phi_{x'}$ and $\phi_{y'}$ instead of ϕ_x and ϕ_y. This gives somewhat messy expressions for the bias factors P'_x in the first filter in terms of the $\tilde{\chi}'$ and δ' which express ξ'_x and ξ'_y by

$$\xi'_x = \tilde{r}\cos\chi' \exp(i\tilde{\theta}' + \tfrac{1}{2}i\delta'), \quad \xi'_y = \tilde{r}\sin\tilde{\chi}' \exp(i\tilde{\theta}' - \tfrac{1}{2}i\delta'). \tag{R-7}$$

The condition for passage through Filter Two now becomes simply

$$\tan^2\tilde{\chi}' > \cot^2\varepsilon, \quad \text{that is,} \quad 90° - \varepsilon < \tilde{\chi}' < 90°. \tag{R-8}$$

The condition for passage throught Filter Three becomes

$$\cos 2\theta \sin\tilde{\chi}' + \sin 2\theta \cos\tilde{\chi}' \cos\delta' > 0. \tag{R-9}$$

In the case of circular unpolarization before Filter One, there is the additional condition

$$\cos 2\varepsilon \cos 2\tilde{\chi}' + \sin 2\varepsilon \sin 2\tilde{\chi} \quad \text{os}\, \delta' > 0 \tag{R-10}$$

for passage throught Filter One. [This replaces (P-5).]

For $\varepsilon < 45°$ we now must consider separately each quadrant for θ. For instance, for $0 < \theta < 90°$ we see that (R-9) cannot be satisfied for any δ', if it cannot be satisfied for $\cos \delta' = 1$, which gives us $\sin (2\theta + \tilde{\chi}') > 0$, but by the restriction (R-8) on the values of $\tilde{\chi}'$ this is impossible if $2\theta + (90° - \varepsilon) > 180°$, that is, for $\theta > 45° + \frac{1}{2}\varepsilon$. In that case, $\mathcal{J}_3/\mathcal{J}_2 = 0$.

On the other hand, for $0 < \theta < 90°$, condition (R-9) is *automatically* satisfied for all δ', if for all $\tilde{\chi}'$ of condition (R-8) it is valid for $\cos \delta' = -1$, that is, if $\sin (2\theta - \tilde{\chi}') < 0$ even for $\tilde{\chi}' = 90° - \varepsilon$, so that $2\theta - (90° - \varepsilon) < 0$, or $\theta < 45° - \frac{1}{2}\varepsilon$. In that case, $\mathcal{J}_3/\mathcal{J}_2 = 1$.

The case of circularly unpolarized light

From Fig. 5.4 (p. 177) we see that in the case of circularly unpolarized light on Filter One we achieve $\mathcal{J}_3/\mathcal{J}_2 = 1$ for angles θ much larger than $45° - \frac{1}{2}\varepsilon$. This is due to the additional condition (R-10) imposed already upon the light from Filter One. In this case, (R-9) for $0 < \theta < 90°$ is automatically satisfied by all $\tilde{\chi}'$ and δ' allowed by (R-8) and (R-10) combined, if

$$\cos \delta' > -\cot 2\theta \tan \tilde{\chi}' \tag{R-9'}$$

follows automatically for $90° - \varepsilon < \tilde{\chi}' < 90°$ from

$$\cos \delta' > -\cot 2\varepsilon \cot 2\tilde{\chi}'. \tag{R-10'}$$

This is so, if $-\cot 2\theta \tan \tilde{\chi}' < -\cot 2\varepsilon \cot 2\tilde{\chi}'$ for these $\tilde{\chi}'$, that is, if

$$\begin{aligned}\cot (180° - 2\theta) &< \cot 2\varepsilon\, [-\cot 2\tilde{\chi}'/\tan \tilde{\chi}']_{\min} \\ &= \tfrac{1}{2} \cot 2\varepsilon \cdot (1 - \cot^2 \tilde{\chi}'_{\min}) = \tfrac{1}{2} \cot 2\varepsilon \cdot (1 - \tan^2 \varepsilon),\end{aligned}$$

which gives

$$\theta < 45° + \tfrac{1}{2} \operatorname{arc\,tan} [(\tan \varepsilon)/(\tan^2 2\varepsilon)] \tag{R-11}$$

as the condition for $\mathcal{J}_3^{\text{cir}} = \mathcal{J}_2^{\text{cir}}$ for $22\frac{1}{2}° < \varepsilon < 45°$, $0 < \theta < 90°$ in absence of relaxation of Ξ.

APPENDIX S

WIENER-SIEGEL'S APPLICATION OF THE POLYCHOTOMIC ALGORITHM TO EIGENVALUES INSTEAD OF TO EIGENFUNCTIONS

The following method [see eq. (S-15) below] was suggested by Wiener and Siegel (1953) [see also Siegel and Wiener (1956)] for determining by the polychotomic algorithm the eigenvalue measured of an observable, if the set $\{\phi_{im}(x)\}$ of eigenfunctions is not for all eigenvalues A_i uniquely given. It is similar to the method explained in Appendix L for the case of external coordinates. The y there is replaced by the m here. The main difference is that in Appendix L the ψ and Ξ depended on the y, and the $\phi_i(x)$ did not. Here, the $\phi_{im}(x)$

do depend on the m, and the $\psi(x)$ and $\Xi(x)$ are sums over a dummy index m. So this time we have, instead of eqs. (L-1) through (L-17), the equations

$$\int |\psi(x)|^2 \, dx = 1, \tag{S-1}$$

$$\psi(x) = \sum_i \sum_m a_{im} \, \phi_{im}(x), \tag{S-2}$$

$$\sum_i \sum_m |a_{im}|^2 = 1, \tag{S-3}$$

$$w_i = \sum_m |a_{im}|^2, \tag{S-4}$$

$$\Xi(x) = \sum_i \sum_m \xi_{im} \, \phi_{im}(x). \tag{S-5}$$

We introduce a new orthonormal set of eigenfunctions

$$\phi'_{in}(x) = \sum_m c^i_{nm} \, \phi_{im}(x) \tag{S-6a}$$

in such a way that $\phi'_{i1}(x)$ is the normalized orthogonal projection of ψ onto the Hilbert subspace S_i of the $\phi_{im}(x)$ for each given eigenvalue A_i. The other $\phi'_{in}(x)$ ($n \neq 1$) are to be orthogonal to ψ, in agreement with Tutsch's rule (119). Then,

$$\psi(x) = \sum_i b_i \phi'_{i1}(x) = \sum_i b_i \sum_m c^i_{1m} \, \phi_{im}(x). \tag{S-6b}$$

Since in S_i we must have

$$\sum_m a_{im} \, \phi_{im}(x) = \sum_n b_{in} \, \phi'_{in}(x) = b_i \, \phi'_{i1}(x), \tag{S-7}$$

it follows from $\int \psi^*(x) \, dx \sum_m a_{im} \, \phi_{im}(x) = \ldots$ that

$$\sum_m |a_{im}|^2 = \sum_n |b_{in}|^2 = |b_i|^2. \tag{S-8}$$

From the orthonormality of the eigenfunctions, together with (S-7) and (S-6a), we obtain

$$a_{im} = \int \phi^*_{im}(x) \, b_i \, \phi'_{i1}(x) \, dx = b_i \, c^i_{1m}, \tag{S-9}$$

so we have

$$c^i_{1m} = a_{im}/b_i. \tag{S-10}$$

According to Tutsch's rule (119), we should use the $\phi'_{in}(x)$ and not the $\phi_{im}(x)$ in the application of the polychotomic algorithm. (This restores the uniqueless lost according to the first part of Appendix J, though at the cost of the paradoxes noted in Appendix K.) So we put

$$\Xi(x) = \sum_i \sum_n \zeta_{in} \, \phi'_{in}(x) = \sum_i \sum_n \sum_m \zeta_{in} \, c^i_{nm} \, \phi_{im}(x). \tag{S-11}$$

The canonical distribution is, with

$$\zeta_{in} \equiv \zeta_k = \mu_{2\lambda-1} + i\mu_{2\lambda}, \tag{S-12}$$

given by

$$d\bar{P} = \prod_\lambda \left\{ \pi^{-\frac{1}{2}} \exp(-\mu^2_\lambda) \, d\mu_\lambda \right\}. \tag{S-13}$$

The polychotomic algorithm then asks for the largest among the $|b_{in}/\zeta_{in}|$, but these ratios all are zero for $n \neq 1$, so we have reduced our problem to finding the largest among the $|b_i/\zeta_{i1}|$, and the values of the ζ_{in} for $n \neq 1$ are irrelevant.

We now got

$$\zeta_{i1} = \int \phi_{i1}'^{*}(x)\, \Xi(x)\, dx = \sum_m c_{1m}^{i*} \int \phi_{im}^{*}(x)\, \Xi(x)\, dx$$

$$= \sum_m c_{1m}^{i*}\, \xi_{im} = \frac{1}{b_1^*} \sum_m a_{im}^{*}\, \xi_{im}. \tag{S-14}$$

Since in the algorithm we need only the absolute values $|b_i/\zeta_{i1}|$, the quantities among which we must search for the biggest one are

$$|b_i/\zeta_{i1}| = |b_i|^2 \Big/ \Big|\sum_m a_{im}^{*}\, \xi_{im}\Big| = \Big(\sum_m |a_{im}|^2\Big) \Big/ \Big|\sum_m \xi_{im}^{*}\, a_{im}\Big|, \tag{S-15}$$

which enables us to calculate the unique set $\{|b_i/\zeta_{i1}|\}$ on the basis of any set $\{\phi_{im}\}$ of eigenfunctions of A_{op}, without first calculating explicitly the set $\{\phi_{in}'\}$.

The prescription (S-15) was given by Wiener and Siegel (1953) without mentioning that it is equivalent to basing the algorithm exclusively upon the use of Tutsch's set $\{\phi_{in}'\}$ of the normalized orthogonal projections of ψ upon the subspaces S_i.

PART III

THEORIES OF THE SECOND KIND

FOREWORD TO PART III

WHEN the wavefunction of a composite system is not factorizable, the behaviors of its parts are correlated. In quantum theory this leads to results that seem paradoxical when considered from a point of view that assumes local causal behavior of elementary particles and quanta, if among the quantities correlated there are some that are not commutative. The resulting "nonlocality paradox" (which is a side aspect of the Einstein–Podolsky–Rosen paradox) could be solved by assuming the existence of hidden variables "of the second kind" only if one is willing to accept deviations from quantum theory already at the quasistatic level at which theories "of the first kind" would agree with quantum theory.

Most applications of theories of the second kind considered in Part III deal with a pair of photons emitted from a single source. In the quantum-theoretical treatment of such problems, considerations of angular momentum and of photon spin are useful. For photons passing more than one polarization filter, quantum theory and reasonable hidden-variables theories should predict Malus's $\cos^2$ law. For a pair of photons analyzed by separate filters, both types of theory lead to polarization correlations, but these correlations are for hidden-variables theories of the second kind different from what they are for quantum theory. This is demonstrated in Chapter 3 by some simple model theories, and is generally proved in Chapter 4, in which we also discuss the crucial experiments of Clauser and of Holt which are to decide which type of theory is the correct one.

In Chapter 5 we consider experiments in which the photon pair arises from positon–negaton annihilation. For photons that hard, no polarization filters are known. Therefore the polarization dependence of Compton scattering is used for an indirect statistical observation of polarization. The experimental results suffice for disproving an idea of Furry that nonfactorizable wavefunctions of composite systems by an alteration of quantum theory would go over into mixtures of factorizable wavefunctions when the parts of the system become macroscopically separated from each other. Also, the experiments disprove a certain type of hidden-variables theories of the second kind, but they cannot disprove *arbitrary* hidden-variables theories of the second kind.

In Chapter 6 we review briefly some results of all three parts of this survey and conclude that hidden-variables theory at this stage appears to be unacceptable. A number of attitudes one can take in view of this fact are enumerated.

CHAPTER 1

INTRODUCTION

IN PART III we shall discuss hidden-variables theories for the second kind and experimental evidence that may help us decide whether or not such theories are in agreement with the facts.

We discussed already in Section 1.5 of Part I the purpose of theories of the second kind. Quantum theory predicts, for composite systems that interacted in the past, that measurements on one part may provide us information on the other part even when both parts are spatially separated. This becomes paradoxical when one may thus get information on quantities that do not commute. For instance, consider a system of two spin-1/2 particles in the singlet state with particle 1 moving away in the $+z$-direction, and particle 2 moving away in the $-z$-direction. Let $\psi^{+}_{x'1}$ be a wavefunction for particle 1 with its spin directed in the x'-direction of some primed frame, and $\psi^{-}_{x'2}$ a wavefunction for particle 2 with its spin in the $-x'$-direction. The singlet state is then given by

$$\psi = \psi(1, 2) = \sqrt{\tfrac{1}{2}}[\psi^{+}_{x'1}\psi^{-}_{x'2} - \psi^{-}_{x'1}\psi^{+}_{x'2}] \tag{1}$$

if the particles are distinguishable. (If they both are electrons, there should be additional antisymmetrization between either particle traveling either way. We here shall omit this complication.) Equation (1) is invariant when the one-particle wavefunctions transform like spinors under rotation of the coordinate system.[1]

Now introduce an arbitrary double-primed coordinate system, and, for given momenta of the two particles, expand ψ in terms of the four two-particle functions $\psi^{\varepsilon'}_{x'1}\psi^{\varepsilon''}_{x''2}$, where $\varepsilon' = \pm$ and $\varepsilon'' = \pm$:

$$\psi(1, 2) = \sum_{\varepsilon'} \sum_{\varepsilon''} c_{\varepsilon'\varepsilon''}\psi^{\varepsilon'}_{x'1}\psi^{\varepsilon''}_{x''2}. \tag{2}$$

If we would simultaneously perform two Stern–Gerlach-type experiments on the two particles (which might be atoms), using x' and x'' as the directions into which the particle beams[2] in the $+z$- and $-z$-directions are split up by inhomogeneous magnetic fields, we would interpret

$|c_{\varepsilon'\varepsilon''}|^2$ = probability that particle 1 has its spin in the $\varepsilon'x'$ direction while particle 2 has its spin in the $\varepsilon''x''$ direction. (3)

[1] For instance, if $x' \perp z$ and we rotate 180° around the z-axis to a double-primed frame so that $x'' = -x'$, spinor transformation $(\psi^{\pm}_{x'1})'' = i\psi^{\mp}_{x'1}$ gives $\psi'' = i^2[\psi^{-}_{x'1}\psi^{+}_{x'2} - \psi^{+}_{x'1}\psi^{-}_{x'2}] = \psi$.

[2] We speak of "beams" as we think of repeating the same experiment over and over again.

In particular, we see by comparison of (2) with (1) for the special case $x'' = x'$ that in the singlet state (1) there is a 50% probability for having the first spin S_1 in the $+x'$- and the second spin S_2 in the $-x'$-direction; a 50% probability that these spins are in the $-x'$- and $+x'$-direction; and a zero probablity that both spins are in the $+x'$-direction or both in the $-x'$-direction. Therefore, if we measure $S_1 || x'$, we know from the quantum-mechanical description discussed above that $S_2 || -x'$, and this could be verified statistically by measuring $S_{2x'}$ with a Stern–Gerlach-type experiment set up in the x'-direction on the $-z$-axis. Thus the way the second particle is deflected in the apparatus on the $-z$-axis according to quantum theory depends on how the first particle is deflected by the apparatus on the $+z$-axis, even though the instruments are spatially separated and there may be NO time for a light signal between the two apparatus to let the second one know what deflection the first one caused before the second particle reaches the second apparatus.

The question then arises how particle 2 in apparatus 2 "knows" how it should be deflected so as to satisfy the requirement of quantum theory that the deflection of the two particles be opposite. This suggests that the particles would "carry along," from the time they were put into the singlet state, information on how they should be deflected upon reaching some apparatus.

It then may be asked whether this information can be described quantum mechanically. It is easily seen that this is impossible. In order to see this, we perform similar experiments with the x'-direction replaced by a different x''-direction, so that $S_{x'}(2)$ and $S_{x''}(2)$ are not commutative. As the particles upon leaving the source that put them into the singlet state, do not know whether particle 1 is going to meet a Stern–Gerlach setup in the x'- or in the x''-direction, particle 2 should carry information on how to match *in either case* the deflection of particle 1. This would mean that particle 2 would have to carry simultaneously information about $S_{x'}(2)$ and about $S_{x''}(2)$, while it is known that quantum theory does not provide any possibility of simultaneously determining $S_{x'}(2)$ and $S_{x''}(2)$.

We thus meet the paradoxical result that the principles of quantum mechanics *prescribe correlations between measurement results* at spatially separated positions on separate particles that previously by interaction formed a composite system which was put into some quantum-mechanical state $\psi(1, 2)$ which, like (1), is not factorizable into a product $\psi_1(1)\, \psi_2(2)$; but, on the other hand, quantum mechanics does *not* provide any formalism by which one could understand *how* the separate particles upon reaching the separate apparatus would acquire the information that seems necessary to "make" the two apparatus find results that show the quantum-mechanically predicted correlations.

That these correlations actually exist, at least qualitatively in agreement with quantum-mechanical predictions, is known from a number of experiments of which some will be discussed in this Part III (see Chapters 4 and 5). As the above example shows, such correlations can well be calculated theoretically, so that there is no mystery in what quantum theory predicts. The "mystery" lies in our lack of understanding why in no individual pair of measurements on the two particles ever a contradiction would occur in those cases where the correlations make definite predictions (as above with the two Stern–Gerlach setups parallel to each other). That is, we are mystified by an *absence of a deterministic mechanism* in quantum theory for enforcing its own predictions.

This paradox (called the "nonlocality aspect of the Einstein–Podolsky–Rosen paradox") would suggest that quantum theory by itself would be incomplete, and that we need additional quantities (so-called hidden variables) which might be considered as properties of the particles, which these particles from the location of their original interaction (or common creation) could carry along and which would tell them how to behave when reaching measuring apparatus. The hidden variables of the two particles could be correlated, and the observed correlation between the behavior of the two particles in the apparatus could be a consequence of this correlation at the source.

In the following, however, we shall find that there exist cases in which any correlations between measurements that could be described by any hidden variables that are carried along by the particles, will always be different from the correlations predicted by quantum mechanics. As the differences are not large, it is thinkable that they could have escaped detection in the past. While this is being written, experiments are underway for verifying whether the correlations actually existing in nature are the ones predicted by quantum theory or whether they actually contradict quantum theory and could be explained by these information-carrying "hidden variables of the second kind" (see Section 4.4).

The deviations from quantum theory about which we are talking here have nothing to do with the instability effects of hidden variables of the first kind which Papaliolios (1967) tried to measure but failed to find (see Part II, Chapter 5). In the theories which here are considered, the hidden variables are quantities that are like "instruction manuals" carried along by the particles until they reach apparatus, and that do not change by any "relaxation" along the particles' path. We do not attempt here to construct elaborate theories that try to explain the results found in arbitrary experiments, and we *certainly* do not try to prove that the hidden variables considered in this Part III would explain all results of quantum mechanics, because we shall see that they *cannot* do this. We will simply *postulate some properties* which these "hidden variables of the second kind" ought to have if they are to "explain" the existence of correlations between measurement results on parts of a composite system separated spatially, and we will find that *this suffices for proving contradictions with quantum theory of a nature that is experimentally verifiable.*

Where some of the discussion is kept rather general and vague, "examples" of this kind of hidden-variables theory will be given merely for demonstrating that our general conclusions really are valid for such specific cases.[3] We will not try to determine the relative merits of these models nor do we investigate whether some of these models are unacceptable for reasons different from the particular applications we are making of them.

We then will discuss the experiments of Clauser and of Holt which are in progress and hopefully will settle the question whether or not hidden variables of the second kind govern the happenings at the apparatus.

We, finally, will briefly discuss the experiments of Langhoff (1960) and others which were done mainly for verifying the validity of quantum theory and which have some bearing

[3] To mathematicians this will seem to be a wasted effort, as, what is true in general, *obviously* must be true in specific cases. This effort is merely made here for reassurance of those less mathematically minded, for having some models of hidden-variables theories of the second kind, and for showing that the same experiments that are to decide between quantum theory on the one hand and hidden-variables theories of the second kind on the other hand, could also be used for a distinction between various models of hidden-variables theory in case quantum theory by these experiments would be disproved.

upon the question of hidden variables of the second kind, although they are less conclusive as proofs of the incorrectness of such theories. These experiments, however, may serve as a proof of the invalidity of a *different* method suggested earlier by Furry (1936) and by Bohm and Aharonov (1957) for explaining away the nonlocality aspect of the Einstein–Podolsky–Rosen paradox, without introducing hidden variables, but altering quantum theory as applied to spatially widely separated systems (see Section 5.5).

CHAPTER 2

QUANTUM THEORY OF POLARIZATION CORRELATION IN TWO-PHOTON EMISSION

WE SHALL start here by explaining the predictions of quantum theory for the polarization correlations between two photons consecutively emitted by an atom cascading toward its ground state. These polarizations are related to the values of the quantum number j of the atom in the states between which the transitions take place.

2.1. The 0–1–0 case and the 1–1–0 case

In practice we will apply the theory mainly to two special cases. In the one case, called the *0–1–0 case*, a calcium atom excited from the 4^1S_0 ground state to the 6^1P_1 state by absorption of a photon of 2275 Å light, cascades first to the 6^1S_0 state ($j = 0$) (we do not care for the photon here emitted), then under emission of "photon 1" (5513 Å) to the 4^1P_1 state ($j = 1$) and, finally (with a mean decay time of around 4.5×10^{-9} second), to the 4^1S_0 ground state ($j = 0$) under emission of "photon 2" (4227 Å). The 0–1–0 case is named after the j-values mentioned.

A number of other methods of cascading of the atom are possible. They do not interest us directly, as in Clauser's experiments (Chapter 4) photons of wavelengths different from 5513 Å and 4227 Å are eliminated by color filters. They merely cause background difficulties in these experiments, as not all "second photons" have a 5513 Å "first photon" accompanying it. (See Fig. 4.1, p. 285.)

The other case is the *1–1–0 case*, used by Holt (Chapter 4). In it, a mercury atom excited by electron bombardment to one of the $9P_1$ states ($j = 1$) cascades via the 7^3S_1 state ($j = 1$) to the 6^3P_0 triplet ground state ($j = 0$) under emission of a first photon of 5677 Å and a second photon of 4048 Å. The mean decay time of the 7^3S_1 state between the two photons is here of the order of 8.3×10^{-9} s. The initial $9P_1$ state is a superposition of over 50% singlet state and less than 50% triplet state. Its triplet part makes the transition to the 7^3S_1 state possible. (Its singlet part facilitates excitation from the singlet ground state.)

2.2. Abbreviated notation for polarized photon states

In Section 5.2 of Part II we have discussed how normalized linear superpositions of wavefunctions ψ describing polarized one-photon states [such as given by eq. (II:H-36)] give rise to wave packet functions **g** [of the type occurring in eqs. (II:H-41/43)] which are similar linear superpositions of the wave-packet functions **g** corresponding to the original polarizations. Thence we concluded that combining and decomposing different polarizations in classical optics finds its analogue in similar operations on quantum-mechanical photon wave functions. [Compare eqs. (II:205–206).]

We shall make use of this similarity by using polarization directions as labels indicating the one-photon wavefunctions, changing the boldface vector notation **e** and the unit vector notation $\hat{x}$ of polarization directions into italic notation with a tilde ($\tilde{e}$, $\tilde{x}$). Thus we will write $\tilde{e}^{\eta}$ for the wavefunction (II:H-36) of a photon with wave-packet functions $\mathbf{g}_n^{\eta}(\mathbf{x})$ containing the polarization unit vector $\mathbf{e}_{\mathbf{k}}^{\eta}$ as in eq. (II:H-41). We need not give **k**, when it is understood that we talk about one of the two photons of the 0–1–0 or the 1–1–0 case described above. We then will simply write an index 1 for photon 1 with $\mathbf{k}_1$ in the $+\hat{z}$-direction, and an index 2 for photon 2 with its $\mathbf{k}_2$ in the $-\hat{z}$-direction. We will later consider various coordinate systems with singly and doubly primed axes, but with $\hat{z}'' = \hat{z}' = \hat{z}$ all coinciding, so that $\{\tilde{x}_i, \tilde{y}_i\}, \{\tilde{x}_i', \tilde{y}_i'\}, \{\tilde{x}_i'', \tilde{y}_i''\}$ will describe three pairs of perpendicularly polarized one-photon wavefunctions all for photons traveling in the $+\hat{z}$-direction if $i = 1$, or in the $-\hat{z}$-direction if $i = 2$.

We will write $\tilde{R}$ and $\tilde{L}$ for the wavefunctions, and $\hat{R}$ and $\hat{L}$ for the complex polarization vectors, of right and left-hand circularly polarized waves. As explained below eq. (II:H-23), they correspond to taking $\eta = +$ and $\eta = -$ in $\mathbf{e}_{\mathbf{k}}^{\eta}$. Below eq. (II:H-11) we defined, for photons 1 and 2,

$$\mathbf{e}^{\eta}_{(\eta' k\hat{z})} = (\eta'\hat{x} + i\eta\hat{y})\sqrt{\tfrac{1}{2}}, \tag{4a}$$

where $\eta' = +$ for photon 1 and $\eta' = -$ for photon 2. Thus we obtain the relations

$$\left.\begin{aligned} \tilde{R}_1 &= \sqrt{\tfrac{1}{2}}\cdot(\tilde{x}_1 + i\tilde{y}_1), & \tilde{L}_1 &= \sqrt{\tfrac{1}{2}}\cdot(\tilde{x}_1 - i\tilde{y}_1), \\ \tilde{R}_2 &= \sqrt{\tfrac{1}{2}}\cdot(-\tilde{x}_2 + i\tilde{y}_2), & \tilde{L}_2 &= \sqrt{\tfrac{1}{2}}\cdot(-\tilde{x}_2 - i\tilde{y}_2), \end{aligned}\right\} \tag{4b}$$

between linearly and circularly polarized one-photon wavefunctions.

Photon angular momentum and justification of use of photon wavefunctions and of abbreviated notation

Sections 2.3 through 2.6 contain much that is not directly used later in the discussion of the experiments of Clauser and of Holt and of Langhoff and others, as the simple notation introduced above suffices for that purpose. However, angular-momentum considerations throw more light on some of our results, and those who want to understand why and how the wave-packet functions of Appendix H of Part II are applicable as photon wavefunctions, will find this explanation in what follows. Moreover, these details are needed for an understanding why in the Langhoff experiment the two-photon positronium decay starts from the $j = 0$ singlet state, and not from the $j = 1$ triplet state of positronium. Those who find some of the material in Sections 2.3–2.6 too heavy, however, may, after a glance at what is being offered, *proceed to Section 2.7 and Appendix E* without too much of a loss in understanding what follows.

2.3. Photon angular momentum

According to field theory, the momentum $\mathbf{p}$, orbital angular momentum $\mathbf{L}$, and spin angular momentum $\mathbf{S}$ of a field $q_n(x)$ with canonically conjugated field components $p_n(x) = \partial\mathcal{L}/\partial\dot{q}_n(x)$ (where $\mathcal{L}$ is the Lagrangian density) are given by[4]

$$\mathbf{p} = -\int d^3\mathbf{x} \sum_n p_n(\mathbf{x})\, \nabla q_n(\mathbf{x}), \tag{5a}$$

$$\mathbf{L} = -\int d^3\mathbf{x} \sum_n p_n(\mathbf{x})\, \mathbf{r}\times \nabla q_n(\mathbf{x}), \tag{5b}$$

$$S_x = \int d^3\mathbf{x} \sum_{n,m} p_n(\mathbf{x})\, [(S_{zy})_{nm}-(S_{yz})_{nm}]\, q_m(\mathbf{x}), \text{ and cyclic,} \tag{5c}$$

if under the infinitesimal rotation

$$x_i' = x_i+\delta x_i = x_i+\sum_j \omega_{ij}x_j \quad (\text{with} \quad \omega_{ij} = -\omega_{ji}) \tag{6a}$$

the field components q_n transform according to

$$q_n' = q_n+\delta q_n = q_n+\sum_i\sum_j\sum_m \omega_{ij}(S_{ij})_{nm}q_m\,. \tag{6b}$$

For electrons, with $\mathcal{L} = \psi^\dagger(i\hbar\, \partial\psi/\partial t-\mathcal{H}_{\text{op}}\psi)$, and with the spinor ψ transforming under the infinitesimal rotation $\delta y = z\, \delta\theta$, $\delta z = -y\, \delta\theta$, $\omega_{yz} = -\omega_{zy} = \delta\theta$ according to $\delta\psi = \frac{1}{2}i\, \delta\theta\cdot\sigma_x\psi$, the above formulas give the familiar results $\mathbf{p} = \int d^3\mathbf{x}\psi^\dagger(\hbar/i)\, \nabla\psi$, $\mathbf{L} = \int d^3\mathbf{x}\psi^\dagger(\hbar/i)\mathbf{r}\times\nabla\psi$, and $\mathbf{S} = \int d^3\mathbf{x}\psi^\dagger(\hbar/2)\sigma\psi$. For photons, that is, for the electromagnetic radiation field, with[5]

$$\mathcal{L} = (8\pi)^{-1}\int \{(\dot{\mathbf{A}}/c)^2-(\text{curl } \mathbf{A})^2\}\, d^3\mathbf{x},$$

the vector field $\mathbf{A}$ transforms like the coordinates $\mathbf{x}$, so that

$$(S_{ij})_{mn} = \tfrac{1}{2}(\delta_{im}\delta_{jn}-\delta_{in}\delta_{jm}), \tag{7}$$

while the field canonically conjugate to $q_n = A_n$ is $p_n = (4\pi c)^{-1}(\dot{A}_n/c) = -(4\pi c)^{-1}E_n$. The above formulas then yield

$$\mathbf{p} = (4\pi c)^{-1}\int d^3\mathbf{x}\, \{E_x\, \nabla A_x+E_y\, \nabla A_y+E_z\, \nabla A_z\}, \tag{8a}$$

$$\mathbf{L} = (4\pi c)^{-1}\int d^3\mathbf{x}\, \{E_x\mathbf{r}\times \nabla A_x+E_y\mathbf{r}\times \nabla A_y+E_z\mathbf{r}\times \nabla A_z\}, \tag{8b}$$

$$\mathbf{S} = (4\pi c)^{-1}\int d^3\mathbf{x}\, \mathbf{E}\times\mathbf{A}. \tag{8c}$$

The total angular momentum of the radiation field, given by

$$\mathbf{J} = \mathbf{L}+\mathbf{S}, \tag{9}$$

[4] See Belinfante (1939b, 1940), Rosenfeld (1940).

[5] We use the Gaussian system. We consider here only the transverse fields satisfying div $\mathbf{E} = 0$, div $\mathbf{A} = 0$.

is by integration by parts and by div $\mathbf{E} = 0$ then found to be equal to

$$\mathbf{J} = (4\pi c)^{-1} \int d^3\mathbf{x}\, \mathbf{r}\times(\mathbf{E}\times\mathbf{B}) \tag{10}$$

with $\mathbf{B} = \text{curl}\, \mathbf{A}$, if we may neglect the surface integrals

$$\mathbf{L}+\mathbf{S}-\mathbf{J} = (4\pi c)^{-1} \oint\!\!\oint (d\mathbf{S}\cdot\mathbf{E})\, [\mathbf{r}\times\mathbf{A}] \tag{11}$$

at the boundary (at infinity). For wave fields describing finite wave packets, as in Bohm's quantum electrodynamics (see Appendix H of Part II), this is alright.

Similarly, the "orbital momentum" (8a) is found by integration by parts to be equal to the "total momentum"

$$\mathbf{p} = (4\pi c)^{-1} \int d^3\mathbf{x}\, \mathbf{E}\times\mathbf{B}, \tag{12}$$

because the total spin momentum [eq. (12) minus eq. (8a)] is equal to the surface integral

$$-(4\pi c)^{-1} \oint\!\!\oint (d\mathbf{S}\cdot\mathbf{E})\, \mathbf{A}, \tag{13}$$

which like (11) vanishes for wave packets that do not extend to infinity.

Comparison of the expansion (II:H-12) and (II:H-13) of **A** and **B** shows that each term in **B** contains an extra factor ηk. It follows that the spin (8c) of a photon is related to its total momentum (12) by

$$\mathbf{S} = \mathbf{p}/\eta k. \tag{14}$$

As the momentum of a photon is[6]

$$\mathbf{p} = \hbar\mathbf{k}, \tag{15}$$

it follows that

$$\mathbf{S} = \eta\hat{p}\hbar \quad \text{or} \quad S_p = \eta\hbar, \tag{16}$$

where $p = \mathbf{p}/|\mathbf{p}|$.

We conclude that the photon is a spin-one particle, but that for a photon which has its momentum **p** directed along the $\pm z$-direction, the wavefunction contains components with $S_z = \pm\hbar$, but the component for which S_z would be zero is absent. Another way of saying the same is that, for a photon wavefunction corresponding to a definite direction of the photon spin **S**, the momentum **p** and the direction of propagation $\hat{p}$ of the wave are always parallel to $\pm\mathbf{S}$. According to Pauli (1940), this is a general property of all particles of rest-mass zero.

Considering the meaning of η mentioned above eqs. (4), we see that right-hand polarized photons point their spin forward, and left-hand polarized photons point it backward. By (4), linearly polarized photons are superpositions of spin forward and backward.

2.4. Expectation values for observables in polarized photon states in momentum space

The observables **p**, **L**, and **S** in eqs. (5a–c) all were expressed in the form

$$Q = \int d^3\mathbf{x} \sum_n [p_n(\mathbf{x})/i\hbar]\, Q_{\text{op}}\, q_n(\mathbf{x}), \tag{17}$$

where

$$\left.\begin{aligned} \mathbf{p}_{\text{op}} = (\hbar/i)\nabla, \quad \mathbf{L}_{\text{op}} = (\hbar/i)\mathbf{r}\times\nabla, \\ S_{x,\,\text{op}}q_n = (\hbar/i)\sum_m (S_{yz}-S_{zy})_{nm}q_m, \end{aligned}\right\} \tag{18}$$

[6] Compare eqs. (II:H-39) and (II:H-44) with (II:H-38).

as in electron wave mechanics. When q_n are the components of a vector **A**, as in the photon case, we see from (18) with (7) that

$$\sum_{n=x,y,z} E_n \mathbf{S}_{op} A_n = (\hbar/i)\mathbf{E}\times\mathbf{A}. \tag{19}$$

In Appendix H of Part II we discussed how to calculate expectation values of such observables of the photon field. The $q_n(\mathbf{x}) = A_n(\mathbf{x})$ and the $p_n(\mathbf{x}) = -(4\pi c)^{-1}E_n(\mathbf{x})$ are first expanded according to eqs. (II:H-12) and (II:H-14), in which the terms with **k** and with $-\mathbf{k}$ are taken together and expressed in terms of their real and imaginary parts by (II:H-17), so that **A**(**x**) and **E**(**x**) become sums over **k** in a halfspace:

$$\mathbf{A}(\mathbf{x}) = c(2\pi/\mathcal{V})^{\frac{1}{2}} \sum_{\mathbf{k}/2}\sum_{\eta} [\mathbf{e}^{\eta}_{\mathbf{k}}\,(q^{\eta}_{R\mathbf{k}} + iq^{\eta}_{I\mathbf{k}})\, e^{i\mathbf{k}\cdot\mathbf{x}} + \text{conj.}], \tag{20a}$$

$$\mathbf{E}(\mathbf{x}) = -(2\pi/\mathcal{V})^{\frac{1}{2}} \sum_{\mathbf{k}/2}\sum_{\eta} [\mathbf{e}^{\eta *}_{\mathbf{k}}(p^{\eta}_{R\mathbf{k}} - ip^{\eta}_{I\mathbf{k}})\, e^{-i\mathbf{k}\cdot\mathbf{x}} + \text{conj.}]. \tag{20b}$$

If we define the q_f (that is, the $q^{\eta}_{R\mathbf{k}}$ and the $q^{\eta}_{I\mathbf{k}}$) and similarly the p_f also in the other halfspace, by

$$\left.\begin{aligned} q^{\eta}_{R,-\mathbf{k}} &\equiv -q^{\eta}_{R,\mathbf{k}}, \quad q^{\eta}_{I,-\mathbf{k}} \equiv +q^{\eta}_{I,\mathbf{k}},\\ p^{\eta}_{R,-\mathbf{k}} &\equiv -p^{\eta}_{R,\mathbf{k}}, \quad p^{\eta}_{I,-\mathbf{k}} \equiv +p^{\eta}_{I,\mathbf{k}}, \end{aligned}\right\} \tag{21}$$

we may write (20a, b) as sums over all of **k**-space:

$$\{q_n(\mathbf{x})\} = \mathbf{A}(\mathbf{x}) = c(2\pi/\mathcal{V})^{\frac{1}{2}} \sum_{\mathbf{k}}\sum_{\eta} [\mathbf{e}^{\eta}_{\mathbf{k}}\,(q^{\eta}_{R\mathbf{k}} + iq^{\eta}_{I\mathbf{k}})\, e^{i\mathbf{k}\cdot\mathbf{x}}], \tag{22a}$$

$$\{p_n(\mathbf{x})/i\hbar\} = -\mathbf{E}(\mathbf{x})/4\pi i\hbar c = (i\hbar c)^{-1}\,(8\pi\mathcal{V})^{-\frac{1}{2}} \sum_{\mathbf{k}}\sum_{\eta} [\mathbf{e}^{\eta *}_{\mathbf{k}}(p^{\eta}_{R\mathbf{k}} - ip^{\eta}_{I\mathbf{k}})\, e^{-i\mathbf{k}\cdot\mathbf{x}}]. \tag{22b}$$

Quantization

We then interpret the p_f as

$$p_f = (\hbar/i)\,\partial/\partial q_f \tag{23}$$

[see (II:H-24)], by which Q [eq. (17)] becomes a differential operator $Q(q_f)_{\text{op}}$. Where products $p_f q_f$ occur, this operator should be hermitized by writing $\frac{1}{2}(p_f q_f + q_f p_f)$ instead. Expectation values are then calculated according to

$$\langle Q\rangle = \int \psi^{(F)*}(q_f)\, Q(q_f)_{\text{op}}\, \psi^{(F)}(q_f) \prod_f dq_f, \tag{24}$$

where the field state vector $\psi^{(F)}$ is expressed as a function of the q_f. These $\psi^{(F)}(q_f)$ contain the vacuum state vector $\psi^{(F)}_{\text{vac}}(q_f)$ as a factor [see (II:H-34) and (II:H-36)], where, by (II:H-32),

$$\psi^{(F)}_{\text{vac}}(q_f) = \prod_{\mathbf{k}/2}\prod_{\eta}\left\{(ck/\pi\hbar)^{\frac{1}{2}} \exp\left(-(ck/2\hbar)\,[(q^{\eta}_{R\mathbf{k}})^2 + (q^{\eta}_{I\mathbf{k}})^2]\right)\right\}. \tag{25}$$

Because of the Gaussian factor in (25), the integrals (24) always converge. However, if the vacuum value

$$\langle Q\rangle_{\text{vac}} = \int \psi^{(F)*}_{\text{vac}}(q_f)\, Q(q_f)_{\text{op}}\, \psi^{(F)}(q_f) \prod_f dq_f \tag{26}$$

for some observables Q would not vanish, then for finding $\langle Q\rangle$ for one or two photons one should subtract (26) from (24) calculated from a state vector $\psi^{(F)}(q_f)$ describing these photons in addition to the zero point radiation of the vacuum:

$$\langle Q\rangle_{n\ \mathrm{phot}} = \langle Q\rangle - \langle Q\rangle_{\mathrm{vac}}. \tag{27}$$

For the observables Q of interest to us here, we will find $\langle Q\rangle_{\mathrm{vac}} = 0$ after the hermitization below eq. (23).

Expectation values in configuration space and in momentum space

When (17) with (22a, b) and (23) is used for $Q(q_f)_{\mathrm{op}}$ in (24), the integrals in (24) over the q_f may be performed first, leaving the integration over $\mathbf{x}$ to be performed last. Thus, for $\langle Q\rangle$ we find an expression of the form

$$\langle Q\rangle = \int d^3\mathbf{x} \sum_n f_n^*(\mathbf{x}) Q_{\mathrm{op}} g_n(\mathbf{x}), \tag{28}$$

with Q_{op} given for instance by (18)–(19). This shows some similarity to $\langle Q\rangle = \int d^3\mathbf{x}\psi^\dagger(\mathbf{x})\times Q_{\mathrm{op}}\psi(\mathbf{x})$ in the case of electron wave mechanics, but there are some differences. In the first place, $f_n(\mathbf{x}) \neq g_n(\mathbf{x})$. A more fundamental difference is that while $\psi^\dagger\psi$ in nonrelativistic wave mechanics is interpreted as a particle density, such an interpretation of $\sum_n f_n^*(\mathbf{x})g_n(\mathbf{x})$ is impossible. This is related to the fact that photons cannot be sharply localized. If they could, we could define the photon density as the number of photons per unit volume in some arbitrarily small volumne. However, in a relativistic field described by **E**, **A**, and **B**, we cannot define such a density. If this density $(=j^0)$ would exist, it would have to form a four-vector j^μ together with $\mathbf{j} = \mathbf{J}/c$, where $\mathbf{J} = j^0\mathbf{v}$ would be the corresponding current density, satisfying with j^0 a continuity equation $\partial_\mu j^\mu = 0$. Moreover, j^0 then should be positive definite. Such density four-vector j^μ does not exist for the Maxwell field.[7]

It, therefore, is impossible to interpret the argument $\mathbf{x}$ of the functions f_n and g_n in (28) as the position of the photon. On the other hand, it *is* possible for a photon to have a sharply defined *momentum*. For this reason, it is desirable to perform a transformation from $\mathbf{x}$-space to momentum space. If we write $f_n^*(\mathbf{x})$ and $g_n(\mathbf{x})$ in (28) as

$$f_n^*(x) = \mathcal{V}^{-\frac{1}{2}} \sum_{\mathbf{k}} \tilde{f}_n^*(\mathbf{k})\, e^{-i\mathbf{k}\cdot\mathbf{x}}, \quad g_n(x) = \mathcal{V}^{-\frac{1}{2}} \sum_{\mathbf{k}'} \tilde{g}_n(\mathbf{k}')\, e^{i\mathbf{k}'\cdot\mathbf{x}}, \tag{29}$$

we obtain (omitting the tilde over f_n and g_n)

$$\langle Q\rangle = \sum_{\mathbf{k}} \sum_n f_n^*(\mathbf{k})\, \tilde{Q}_{\mathrm{op}} g_n(\mathbf{k}), \tag{30}$$

where $\tilde{Q}_{\mathrm{op}}$ operates on functions of $\mathbf{k}$ according to

$$\tilde{Q}_{\mathrm{op}} g_n(\mathbf{k}) = \sum_{\mathbf{k}'} g_n(\mathbf{k}') \int d^3\mathbf{x} e^{-i\mathbf{k}\cdot\mathbf{x}}\, Q_{\mathrm{op}} e^{i\mathbf{k}'\cdot\mathbf{x}}/\mathcal{V}. \tag{31}$$

[7] If **A** is generalized to a four-vector A^μ satisfying the Lorentz condition $\partial_\mu A^\mu = 0$, it is possible to form a four-vector $j^\mu = (4\pi)^{-1}F^{\mu\lambda}A_\lambda$, but then $\partial_\mu j^\mu = (8\pi)^{-1}F^{\mu\lambda}F_{\mu\lambda} - j^\lambda A_\lambda \neq 0$, and j^0 is not positive definite.

We may here change the sums over discrete **k**-values (making $e^{i\mathbf{k}\cdot\mathbf{x}}$ periodic between the boundaries of a volumne ($\mathcal{V} = L^3$) into integrals over continuous **k**-values, if at the same time we replace $\mathcal{V}$ by $(2\pi)^3$. In this manner, we find for $\tilde{Q}_{op}$ in (30), for the choices (18) of Q_{op},[8]

$$\tilde{\mathbf{p}}_{op} = \hbar\mathbf{k}; \quad \tilde{\mathbf{L}}_{op} = (\hbar/i)\mathbf{k}\times(\partial/\partial\mathbf{k}); \tag{32a}$$

$$\sum_{n=x,y,z} E_n(\mathbf{k})\tilde{\mathbf{S}}_{op}A_n(\mathbf{k}) = (\hbar/i)\,\mathbf{E}(\mathbf{k})\times\mathbf{A}(\mathbf{k}). \tag{32b}$$

We shall now first consider $\langle Q\rangle_{1\text{ phot}}$ for one single photon. We describe this photon, together with the zero-point radiation field, by the wave function (II:H-36):

$$\psi^{(F)}(q_f) = \psi^{(F)}_{\text{vac}}(q_f)\sum_{\mathbf{k}}\sum_{\eta} f_\eta(\mathbf{k})\,(ck/\hbar)^{\frac{1}{2}}(q^{\eta}_{R\mathbf{k}} - iq^{\eta}_{I\mathbf{k}}). \tag{33}$$

The calculation of $\langle Q\rangle_{1\text{ phot}}$ by (27) in this case is performed in Appendix A. We find [see eqs. (A-26) and (A-27)]

$$\langle Q\rangle_{1\text{ phot}} = \tfrac{1}{2}\sum_{\mathbf{k}}\{\mathbf{f}(\mathbf{k})^*\,\tilde{Q}(\mathbf{k})_{op}\,\mathbf{f}(\mathbf{k}) - \mathbf{f}(\mathbf{k})\,\tilde{Q}(-\mathbf{k})_{op}\,\mathbf{f}(\mathbf{k})^*\}, \tag{34}$$

where

$$\mathbf{f}(\mathbf{k}) = \sum_{\eta} f_\eta(\mathbf{k})\mathbf{e}^{\eta}_{\mathbf{k}} \tag{35}$$

is the Fourier transform of the wave-packet function $\sum_\eta \mathbf{g}^{\eta}_0(\mathbf{x})$ of (II:H-41) for $n = 0$. For $Q = \mathbf{p}$, by $\tilde{Q}(\mathbf{k})_{op} = \hbar\mathbf{k} = -\tilde{Q}(-\mathbf{k})_{op}$, this gives

$$\langle \mathbf{p}\rangle_{1\text{ phot}} = \sum_{\mathbf{k}}\sum_{n} f_n(\mathbf{k})^*\,\hbar\mathbf{k} f_n(\mathbf{k}), \tag{36a}$$

where $\sum_n$ is a sum over $n = x, y, z$. Therefore,

$$\langle \mathbf{p}\rangle_{1\text{ phot}} = \sum_{\mathbf{k}}\sum_{\eta}\sum_{\eta'} f_\eta(\mathbf{k})^* f_{\eta'}(\mathbf{k})\,(\mathbf{e}^{\eta *}_{\mathbf{k}}\cdot\mathbf{e}^{\eta'}_{\mathbf{k}})\hbar\mathbf{k} = \sum_{\mathbf{k}}\sum_{\eta} |f_\eta(\mathbf{k})|^2\,\hbar\mathbf{k}. \tag{36b}$$

For $Q = \mathbf{L}$, by $\tilde{Q}(\mathbf{k})_{op} = \mathbf{k}\times(\hbar/i)\partial/\partial\mathbf{k} = +\tilde{Q}(-\mathbf{k})_{op}$. (34) gives

$$\langle \mathbf{L}\rangle_{1\text{phot}} = \frac{\hbar}{2i}\sum_{\mathbf{k}}\sum_{n}\left\{f_n(\mathbf{k})^*\,\mathbf{k}\times\frac{\partial}{\partial\mathbf{k}}f_n(\mathbf{k}) - f_n(\mathbf{k})\,\mathbf{k}\times\frac{\partial}{\partial\mathbf{k}}\,f_n(\mathbf{k})^*\right\},$$

which, by integration by parts, gives

$$\langle \mathbf{L}\rangle_{1\text{ phot}} = \sum_{\mathbf{k}}\sum_{n} f_n(\mathbf{k})^*\left(\mathbf{k}\times\frac{\hbar}{i}\frac{\partial}{\partial\mathbf{k}}\right)f_n(\mathbf{k}) = \sum_{\mathbf{k}}\sum_{n} f_n(\mathbf{k})^*\tilde{\mathbf{L}}_{op}f_n(\mathbf{k}). \tag{37a}$$

For $Q = \mathbf{S}$, by $\sum_n f_n\tilde{\mathbf{S}}_{op}g_n = (\hbar/i)\mathbf{f}\times\mathbf{g}$, eq. (34) gives

$$\langle \mathbf{S}\rangle_{1\text{ phot}} = \frac{\hbar}{2i}\sum_{\mathbf{k}}\{\mathbf{f}(\mathbf{k})^*\times\mathbf{f}(\mathbf{k}) - \mathbf{f}(\mathbf{k})\times\mathbf{f}(\mathbf{k})^*\} = \frac{\hbar}{i}\sum_{\mathbf{k}}\mathbf{f}^*\times\mathbf{f}$$

$$= \sum_{\mathbf{k}}\sum_{n} f_n(\mathbf{k})^*\tilde{\mathbf{S}}_{op}f_n(\mathbf{k}), \tag{38a}$$

[8] The derivation of $\tilde{\mathbf{L}}_{op}$ involves an integration by parts, omitting again the boundary surface integral.

so, by (35),

$$\langle \mathbf{S} \rangle_{1\text{ phot}} = \frac{\hbar}{i} \sum_{\mathbf{k}} \sum_{\eta} \sum_{\eta'} f_\eta(\mathbf{k})^* f_{\eta'}(\mathbf{k})\, \mathbf{e}_{\mathbf{k}}^{\eta *} \times \mathbf{e}_{\mathbf{k}}^{\eta'},$$

which, by $\mathbf{e}_{\mathbf{k}}^{\eta *} \times \mathbf{e}_{\mathbf{k}}^{\eta'} = i\eta \mathbf{k}/k$, gives

$$\langle \mathbf{S} \rangle_{1\text{ phot}} = \sum_{\mathbf{k}} \sum_{\eta} |f_\eta(\mathbf{k})|^2\, \eta \hbar \mathbf{k}/k = \sum_{\mathbf{k}} \sum_{\eta} |f_\eta(\mathbf{k})^2|\, \eta \hbar \mathbf{e}_{\mathbf{k}}^0. \tag{38b}$$

Thus, for $Q = \mathbf{p}$ or $\mathbf{L}$ or $\mathbf{S}$, (34) simplifies to

$$\langle Q \rangle_{1\text{ phot}} = \int d^3\mathbf{k}\, \mathbf{f}(\mathbf{k})^* \cdot \tilde{Q}(\mathbf{k})_{\text{op}}\, \mathbf{f}(\mathbf{k}). \tag{39}$$

This simplification, however, is not valid for $Q_{\text{op}} = 1$ or $\mathbf{p}_{\text{op}}^2$ or $\mathbf{L}_{\text{op}}^2$ or $\mathbf{S}_{\text{op}}^2$, where we have to use the two-term expression (34) $\left(\text{possibly writing } \int d^3\mathbf{k} \text{ for } \sum_{\mathbf{k}}\right)$.

Interpreting **f(k)** *as a photon wavefunction*

From (36a), (37a), (38a) we see that in momentum space we may regard the vector $\mathbf{f}(\mathbf{k})$ of eq. (35) as a "photon wavefunction" from which for $Q = \mathbf{p}$ or $\mathbf{L}$ or $\mathbf{S}$ the expectation values (24) of the quantized field observables $Q(q_f)_{\text{op}}$ can be calculated by eq. (39) in much the same way as expectation values are calculated for electrons in configuration space from the spinor wavefunction $\psi(\mathbf{x})$. From (36a) and (38a) we also see that the polarization components $f_\eta(\mathbf{k})$ in the expansion (35) of $\mathbf{f}(\mathbf{k})$, in the expansion (33) of $\psi^{(F)}(q_f)$, and in the expansion (II:H-41) of the wave-packet functions $\mathbf{g}_n^\eta(\mathbf{x})$, may be regarded as probability amplitudes for photon states with polarization $\mathbf{e}_{\mathbf{k}}^\eta$ and momentum $\hbar\mathbf{k}$. They are normalized according to (II:H-38) by

$$\sum_{\mathbf{k}} \sum_{\eta} |f_\eta(\mathbf{k})|^2 = 1. \tag{40}$$

Warning

Let quantization by (23) and hermitization be indicated by the subscript "qu". Then, we obtained $Q(q_f)_{\text{op}}$ from (17) by

$$Q(q_f)_{\text{op}} = Q_{\text{qu}} = \left\{ \int d^3\mathbf{x} \sum_n [p_n(\mathbf{x})/i\hbar] Q_{\text{op}} q_n(\mathbf{x}) \right\}_{\text{qu}}. \tag{41}$$

Attention should be paid to the fact that our result, for $Q = \mathbf{p}$ or $\mathbf{S}$ or $\mathbf{L}$, that

$$\begin{aligned} \langle Q_{\text{qu}} \rangle &\equiv \int \psi^{(F)*}(q_f) \left\{ \int d^3\mathbf{x} \sum_n [p_n(\mathbf{x})/i\hbar] Q_{\text{op}} q_n(\mathbf{x}) \right\}_{\text{qu}} \psi^{(F)}(q_f) \prod_f dq_f \\ &= \int d^3\mathbf{k}\, \mathbf{f}(\mathbf{k})^*\, \tilde{Q}(\mathbf{k})_{\text{op}}\, \mathbf{f}(\mathbf{k}) \end{aligned} \tag{42}$$

does not mean at all that $\mathbf{f}(\mathbf{k})$ for every Q_{op} would be an eigenfunction of $\tilde{Q}(\mathbf{k})_{\text{op}}$ if $\psi^{(F)}(q_f)$ is an eigenfunction of Q_{qu} or vice versa, or that $\langle Q_{\text{qu}}^2 \rangle$ for every Q_{op} would be related to an expression (39) or even (34) with $\tilde{Q}_{op}$ replaced by $\tilde{Q}^2_{op}$. It is easy to give examples that this is not so, and below we shall give a few simple examples in order to dispel any ideas of the sort. The photon spin may serve as a good example.

Angular-momentum quantum numbers

The operators $\tilde{\mathbf{L}}_{op}$ and $\tilde{\mathbf{S}}_{op}$ in momentum space satisfy the customary angular-momentum commutation relations such as, for instance,

$$\tilde{S}_{x,\,op}\tilde{S}_{y,\,op}-\tilde{S}_{y,\,op}\tilde{S}_{x,\,op} = i\hbar\tilde{S}_{z,\,op}, \quad \text{etc.} \tag{43}$$

This guarantees that the operators $\tilde{S}_{z,\,op}$, $\mathbf{S}^2_{op}$, $\tilde{L}_{z,\,op}$, $\tilde{\mathbf{L}}^2_{op}$, $\tilde{J}_{z,\,op} \equiv \tilde{L}_{z,\,op}+\tilde{S}_{z,\,op}$, and $\tilde{\mathbf{J}}^2_{op} \equiv (\tilde{\mathbf{L}}_{op}+\tilde{\mathbf{S}}_{op})^2$ have the familiar eigenvalues, and that the usual Clebsch–Gordon coefficients may be used for constructing simultaneous eigenfunctions of $\tilde{\mathbf{L}}^2_{op}$, $\tilde{\mathbf{S}}^2_{op}$, $\tilde{\mathbf{J}}^2_{op}$, and $\tilde{J}_{z,\,op}$ out of products of simultaneous eigenfunctions of $\tilde{\mathbf{L}}^2_{op}$ and $\tilde{L}_{z,\,op}$, and of $\tilde{\mathbf{S}}^2_{op}$ and $\tilde{S}_{z,\,op}$. We thus may define angular momentum quantum numbers l, m_l, s, m_s, j, and m_j, for characterizing photon wavefunctions $\mathbf{f}$ in momentum space, where, for instance, the eigenvalues of $\tilde{\mathbf{J}}^2_{op}$ are $\hbar^2 j(j+1)$. (See Section 2.5 for some details on photon wavefunctions with given quantum numbers.)

However, as already indicated in the warning above, it would be wrong to believe that, for a state characterized by a vector wavefunction $\mathbf{f}$ that is an eigenfunction of $\tilde{\mathbf{S}}^2_{op}$ to the eigenvalue $s(s+1)\hbar^2$, the expectation value of the square of the spin angular momentum would be $s(s+1)\hbar^2$. As we shall see below, this is not so. In the first place, for $Q = \mathbf{S}^2_{op}$, eq. (42) is false, as not (39) but (34) holds for it. While in this case the right-hand side of (39) would give $s(s+1)\hbar^2$, the right-hand side of (34) would give zero. Moreover, inserting $Q_{op} = \mathbf{S}^2_{op}$ in the left-hand side of (42) does not give the expectation value of the square of (8c) (which would be of the fourth degree in the electromagnetic field variables), but gives $s(s+1)\hbar^2 = 2\hbar^2$ times the expectation value of the expression (17) for $Q_{op} = 1$ (which is of the second degree in the electromagnetic field). Finally, we shall show below that in a one-photon state the square of (8c) has the value $\hbar^2$.

The value of the photon spin and of its square

Since $\mathbf{f}(\mathbf{k})$ is in $\mathbf{k}$-space a vector field, it follows that

$$\tilde{S}_{x,\,op}\begin{pmatrix} f_x \\ f_y \\ f_z \end{pmatrix} = \frac{\hbar}{i}\begin{pmatrix} 0 \\ f_z \\ -f_y \end{pmatrix}, \quad \tilde{S}_{y,\,op}\begin{pmatrix} f_x \\ f_y \\ f_z \end{pmatrix} = \frac{\hbar}{i}\begin{pmatrix} -f_z \\ 0 \\ f_x \end{pmatrix}, \quad \tilde{S}_{z,\,op}\begin{pmatrix} f_x \\ f_y \\ f_x \end{pmatrix} = \frac{\hbar}{i}\begin{pmatrix} f_y \\ -f_x \\ 0 \end{pmatrix}. \tag{44}$$

Thence,

$$\begin{aligned} \tilde{\mathbf{S}}^2_{op}\mathbf{f} &= \tilde{S}_{x,\,op}(\tilde{S}_{x,\,op}\mathbf{f})+\tilde{S}_{y,\,op}(\tilde{S}_{y,\,op}\mathbf{f})+\tilde{S}_{z,\,op}(\tilde{S}_{z,\,op}\mathbf{f}) \\ &= -\hbar^2\left\{\begin{pmatrix} 0 \\ -f_y \\ -f_z \end{pmatrix}+\begin{pmatrix} -f_x \\ 0 \\ -f_z \end{pmatrix}+\begin{pmatrix} -f_x \\ -f_y \\ 0 \end{pmatrix}\right\} = 2\hbar^2\mathbf{f}. \end{aligned} \tag{45}$$

Thus any photon wavefunction $\mathbf{f}(\mathbf{k})$, by being a vector, is an eigenfunction of $\tilde{\mathbf{S}}^2_{op}$ belonging to the eigenvalue $2\hbar^2$. No photon wavefunction $\mathbf{f}(\mathbf{k})$, however, is a simultaneous eigenfunction of $\tilde{S}_{x,\text{ op}}$ *and* $\tilde{S}_{y,\text{ op}}$ *and* $\tilde{S}_{z,\text{ op}}$.

We shall now show that there exist simple radiation field wavefunctions $\psi^{(F)}(q_f)$ that are simultaneous eigenfunctions of $(S_x)_{\text{qu}}$ and $(S_y)_{\text{qu}}$ and $(S_z)_{\text{qu}}$, that is, of all three components of $\mathbf{S}(q_f)_{\text{op}}$. Moreover, they are eigenfunctions of $[\mathbf{S}(q_f)^2]_{\text{op}} = (\mathbf{S}^2)_{\text{qu}}$, to the eigenvalue $\hbar^2$ (not $2\hbar^2$). This will be an example of the various inequalities mentioned in the above *Warning*.

The details are given in Appendix C. Expressing Bohm's radiation field variables $q_{\mathbf{k},\eta} \equiv \sqrt{\tfrac{1}{2}}(q^{\eta}_{R\mathbf{k}} + iq^{\eta}_{I\mathbf{k}})$ (see Appendix H of Part II) as

$$q_{\mathbf{k},\eta} \equiv (\hbar/2ck)^{\frac{1}{2}}\,(u_{\mathbf{k}\eta} + iv_{\mathbf{k}\eta}), \tag{46}$$

we find in eq. (C-5)

$$\mathbf{S}(q_f)_{\text{op}} = i\hbar \sum_{\mathbf{k}/2} \sum_{\eta} \eta\,(\mathbf{k}/k)\,(u_{\mathbf{k}\eta}\partial/\partial v_{\mathbf{k}\eta} - v_{\mathbf{k}\eta}\partial/\partial u_{\mathbf{k}\eta}). \tag{47}$$

The wavefunction (33) for one photon of given $\mathbf{k}$ and η is

$$\psi^{(F)}_{\mathbf{k},\eta}(q_f) = \psi^{(F)}_{\text{vac}}(q_f)\cdot\{u_{\mathbf{k}\eta} - iv_{\mathbf{k}\eta}\}, \tag{48}$$

and it is easily found that

$$\mathbf{S}(q_f)_{\text{op}}\,\psi^{(F)}_{\mathbf{k},\eta}(q_f) = \eta\hbar(\mathbf{k}/k)\,\psi^{(F)}_{\mathbf{k},\eta}(q_f), \tag{49}$$

so that all three components of $\mathbf{S}(q_f)$ simultaneously have precise values. It follows that

$$\mathbf{S}(q_f)^2_{\text{op}}\,\psi^{(F)}_{\mathbf{k},\eta}(q_f) = \hbar^2\psi^{(F)}_{\mathbf{k},\eta}(q_f), \tag{50}$$

so that the quantum number $s = 1$ appearing in

$$\tilde{\mathbf{S}}^2_{\text{op}}\mathbf{f}(\mathbf{k}) = s(s+1)\hbar^2\mathbf{f}(\mathbf{k}) \tag{51}$$

does not give by $s(s+1)\hbar^2$ the value of the square of the spin angular momentum of the field.

For two-photon states, $\mathbf{S}(q_f)_{\text{op}}$ and its square are given, unaltered, by (47) and its square. In Section 2.6 we shall discuss $\tilde{\mathbf{S}}^2_{\text{op}}$ and s for the two-photon case.

Commutation relations for the photon spin

The fact that all components of $\mathbf{S}(q_f)_{\text{op}}$ can be diagonalized simultaneously [by wavefunctions like (48)] indicates that these components must be commutative. This is due to the transverse character of the fields $\mathbf{E}$ and $\mathbf{A}$ used in (8c). The proof of this commutativity is given in Appendix D.

2.5. Quantum numbers of the photon wavefunction

Because of their use in the literature, we shall here briefly review some of the classification of photon wavefunctions with the help of the quantum numbers j, l, s, and m mentioned below eq. (43).

Vector eigenfunctions of the $\tilde{S}_{z,\,\mathrm{op}}$ and $\tilde{\mathbf{S}}^2_{\mathrm{op}}$

While any $\mathbf{f}(\mathbf{k})$ by (45) is an eigenfunction of $\tilde{\mathbf{S}}^2_{\mathrm{op}}$, it is an eigenfunction of $\tilde{S}_{z,\,\mathrm{op}}$ only for $\mathbf{k}$ along the $+z$- or the $-z$-axis, and then only if the sum (35) contains one term only. More specifically, by

$$\mathbf{e}^{\eta}_{(\eta' k\hat{z})} = (\eta'\hat{x}+i\eta\hat{y})\sqrt{\tfrac{1}{2}} \quad (\text{for } k>0,\ \eta'=\pm1,\ \eta=\pm1) \tag{52}$$

[see below eq. (II:H-11)], we find

$$\tilde{S}_{z,\,\mathrm{op}}\{\mathbf{e}^{\eta}_{(\eta' k\hat{z})}f_\eta(\eta' k\hat{z})\} = \tilde{S}_{z,\,\mathrm{op}}\begin{Bmatrix}\eta' f\sqrt{\frac{1}{2}}\\ i\eta f\sqrt{\frac{1}{2}}\\ 0\end{Bmatrix} = \left(\frac{\hbar}{i}\right)\begin{Bmatrix}i\eta f\sqrt{\frac{1}{2}}\\ -\eta' f\sqrt{\frac{1}{2}}\\ 0\end{Bmatrix}$$
$$= \eta\eta'\hbar\{\mathbf{e}^{\eta}_{(\eta' k\hat{z})}f_\eta(\eta' k\hat{z})\}, \tag{53}$$

confirming that the spin has a component $\eta\hbar$ in the direction of propagation of the photon.

We could also ask for the vector eigenfunctions of $\tilde{S}_{z,\,\mathrm{op}}$, irrespective of whether these vectors are of the transverse form (35). Introducing also

$$\mathbf{e}^0_{\mathbf{k}} = \mathbf{k}/k, \quad \text{so} \quad \mathbf{e}^0_{(\eta' k\hat{z})} = \eta'\hat{z}, \tag{54}$$

we may consider any function $\mathbf{e}^{\mu}_{\hat{z}}F(\mathbf{k})$ an eigenfunction of $\tilde{S}_{z,\,\mathrm{op}}$ even if $\mathbf{k}$ is not parallel to $\hat{z}$

$$\tilde{S}_{z,\,\mathrm{op}}\{\mathbf{e}^{\mu}_{\hat{z}}F(\mathbf{k})\} = \mu\hbar\{e^{\mu}_{\hat{z}}F(\mathbf{k})\}. \tag{55}$$

It is convenient to introduce three vectors $\chi_\mu(\mu=-1, 0, 1)$ given by

$$\chi_+ = -\mathbf{e}^+_{\hat{z}} = \begin{pmatrix}-\sqrt{\frac{1}{2}}\\ -i\sqrt{\frac{1}{2}}\\ 0\end{pmatrix},\quad \chi_0 = +\mathbf{e}^0_{\hat{z}} = \begin{pmatrix}0\\0\\1\end{pmatrix},\quad \chi_- = +\mathbf{e}^-_{\hat{z}} = \begin{pmatrix}\sqrt{\frac{1}{2}}\\ -i\sqrt{\frac{1}{2}}\\ 0\end{pmatrix}. \tag{56}$$

Then not only

$$\tilde{S}_{z,\,\mathrm{op}}\chi_\mu = \mu\hbar\chi_\mu,\quad \chi^*_\mu\bullet\tilde{S}_{z,\,\mathrm{op}}\chi_{\mu'} = \hbar\begin{pmatrix}1&0&0\\0&0&0\\0&0&-1\end{pmatrix}, \tag{57a}$$

but also

$$\chi^*_\mu\bullet\tilde{S}_{x,\,\mathrm{op}}\chi_{\mu'} = \hbar\sqrt{\tfrac{1}{2}}\begin{pmatrix}0&1&0\\1&0&1\\0&1&0\end{pmatrix},\quad \chi^*_\mu\bullet\tilde{S}_{y,\,\mathrm{op}}\chi_{\mu'} = \hbar\sqrt{\tfrac{1}{2}}\begin{pmatrix}0&-i&0\\i&0&-i\\0&i&0\end{pmatrix}, \tag{57b}$$

so that the χ_μ give the customary representation of the spin angular momentum operators $\tilde{\mathbf{S}}_{\mathrm{op}}$ (like the spherical harmonics Y^μ_1 would for $\tilde{\mathbf{L}}_{\mathrm{op}}$).

Vector eigenfunctions of $\tilde{L}_{z,\,\mathrm{op}}$ and $\tilde{\mathbf{L}}^2_{\mathrm{op}}$

The simultaneous eigenfunctions of $\tilde{L}_{z,\,\mathrm{op}}$ and of $\tilde{\mathbf{L}}^2_{\mathrm{op}}$ are vectors that in $\mathbf{k}$-space are spherical harmonics. That is, if

$$k_x = k\sin\theta\cos\varphi,\quad k_y = k\sin\theta\sin\varphi,\quad k_z = k\cos\theta, \tag{58}$$

then, for any vector function $\mathbf{F}(k)$ of $k = |\mathbf{k}|$,

$$\tilde{L}_{z,\,\mathrm{op}}\{\mathbf{F}(k)\, Y_l^m(\theta, \varphi)\} = m\hbar\{\mathbf{F}(k)\, Y_l^m(\theta, \varphi)\},$$
$$\tilde{\mathbf{L}}^2_{\mathrm{op}}\{\mathbf{F}(k)\, Y_l^m(\theta, \varphi)\} = l(l+1)\,\hbar^2\{\mathbf{F}(k)\, Y_l^m(\theta, \varphi)\}. \tag{59}$$

Again, these vector functions in general will not be transverse fields, and therefore will not be good "photon wavefunctions" $\mathbf{f}(\mathbf{k})$.

Vector eigenfunctions of $\tilde{J}_{z,\,\mathrm{op}}$ and $\tilde{\mathbf{J}}^2_{\mathrm{op}}$

Simultaneous eigenfunctions of $\tilde{\mathbf{L}}^2_{\mathrm{op}}$, $\tilde{L}_{z,\,\mathrm{op}}$, $\tilde{\mathbf{S}}^2_{\mathrm{op}}$, and $\tilde{S}_{z,\,\mathrm{op}}$ are

$$F(k)\, \mathbf{Y}_l^{m\mu}(\theta, \varphi) \equiv F(k)\chi_\mu Y_l^m(\theta, \varphi). \tag{60}$$

These are also eigenfunctions of $\tilde{J}_{z,\,\mathrm{op}} = \tilde{L}_{z,\,\mathrm{op}} + \tilde{S}_{z,\,\mathrm{op}}$:

$$\tilde{J}_{z,\,\mathrm{op}}\mathbf{Y}_l^{m\mu}(\theta, \varphi) = (m+\mu)\hbar\mathbf{Y}_l^{m\mu}(\theta, \varphi). \tag{61}$$

Using Clebsch–Gordon coefficients $C^{jM}_{lms\mu}$ for "vector addition" of $\mathbf{L}$ and $\mathbf{S}$, we may make linear combinations of the $\mathbf{Y}_l^{m\mu}(\theta, \varphi)$ with a given value of $M = m+\mu$, which are simultaneous eigenfunctions of $\tilde{\mathbf{L}}^2_{\mathrm{op}}$, $\tilde{\mathbf{S}}^2_{\mathrm{op}}$, $\tilde{\mathbf{J}}^2_{\mathrm{op}}$, and $\tilde{J}_{z,\,\mathrm{op}}$. That is,[9]

$$\mathbf{Y}_{jlM}(\theta, \varphi) = \sum_{m+\mu=M} C^{jM}_{lm1\mu} Y_l^m(\theta, \varphi)\chi_\mu, \tag{62a}$$

$$\left.\begin{aligned} \tilde{\mathbf{J}}^2_{\mathrm{op}}\mathbf{Y}_{jlM} &= j(j+1)\hbar^2\mathbf{Y}_{jlM}, & \tilde{\mathbf{J}}_{z,\,\mathrm{op}}\mathbf{Y}_{jlM} &= M\hbar\mathbf{Y}_{jlM}, \\ \tilde{\mathbf{L}}^2_{\mathrm{op}}\mathbf{Y}_{jlM} &= l(l+1)\hbar^2\, \mathbf{Y}_{jlM}, & \tilde{\mathbf{S}}^2_{\mathrm{op}}\mathbf{Y}_{jlM} &= 2\hbar^2\, \mathbf{Y}_{jlM}, \end{aligned}\right\} \tag{62b}$$

The $C^{jM}_{lm1\mu}$ differ from zero only if

$$M = m+\mu, \tag{63a}$$
$$l = j \quad \text{or} \quad j \pm 1 \quad \text{for} \quad j \neq 0, \qquad \text{and} \quad l = 1 \quad \text{for} \quad j = 0. \tag{63b}$$

They are given in table 1, page 26, of Akhiezer and Berestetskii (1965), or by table 27.9.2 of Abramowitz and Stegun (1965). The Y_l^m are given by eq. (3.9′) on page 22 of Akhiezer and Berestetskii (1965).

Longitudinal and transverse vector spherical harmonics

The vector spherical harmonics $\mathbf{Y}_{jlM}(\mathbf{n})$ [with $\mathbf{n} \equiv \mathbf{e}^o_{\mathbf{k}} \equiv \mathbf{k}/k$] are not acceptable photon vector wavefunctions because they are not transverse fields. That is, contrary to $\mathbf{f}(\mathbf{k})$ given by (35), they are not perpendicular to $\mathbf{k}$ and to $\mathbf{n}$.

For given $j(\neq 0)$ and given M, there are according to (63b) usually three different $\mathbf{Y}_{jlM}(\mathbf{n})$. We shall consider linear combinations of them:[10]

$$\mathbf{f}_{jM} = \sum_{l=j-1}^{j+1} a_l \mathbf{Y}_{jlM}. \tag{64}$$

[9] See Akhiezer and Berestetskii (1965), eq. (4.3).

We will consider in particular three linear combinations denoted[10] by $Y_{jM}^{(\lambda)}$, with $\lambda = -1, 0$, or 1, where $Y_{jM}^{(-1)}$ is a longitudinal field ($\| \mathbf{n}$) given by[11]

$$Y_{jM}^{(-1)} \equiv \mathbf{n} Y_j^M(\mathbf{n}) = \sqrt{\frac{j}{2j+1}}\, \mathbf{Y}_{j,j-1,M}(\mathbf{n}) - \sqrt{\frac{j+1}{2j+1}}\, \mathbf{Y}_{j,j+1,M}(\mathbf{n}). \tag{65}$$

The two other combinations $\mathbf{f}_{jM}$ (64) which we shall consider are perpendicular to each other and to (65) and $\mathbf{k}$, and therefore represent transverse fields that may serve as vector photon wavefunctions (35). They are,[12] for $j \neq 0$,

$$Y_{jM}^{(0)} \equiv \mathbf{Y}_{jjM}(\mathbf{n}) \tag{66}$$

and

$$Y_{jM}^{(1)} \equiv -i\mathbf{n} \times Y_{jM}^{(0)}(\mathbf{n}) = \sqrt{\frac{j}{2j+1}}\, \mathbf{Y}_{j,j+1,M} + \sqrt{\frac{j+1}{2j+1}}\, \mathbf{Y}_{j,j-1,M}. \tag{67}$$

Using

$$k\frac{\partial}{\partial \mathbf{k}} Y_j^M(\mathbf{n}) = \sqrt{\frac{j(j+1)}{2j+1}} \left(\sqrt{j}\,\mathbf{Y}_{j,j+1,M} + \sqrt{j+1}\, \mathbf{Y}_{j,j-1,M}\right), \tag{68}$$

the transverse spherical harmonics (66) and (67) may also be written as[13]

$$Y_{jM}^{(0)} = \frac{\tilde{\mathbf{L}}_{\text{op}}}{\hbar\sqrt{j(j+1)}} Y_j^M(\mathbf{n}), \quad Y_{jM}^{(1)} = \frac{k}{\sqrt{j(j+1)}} \frac{\partial}{\partial \mathbf{k}} Y_j^M(\mathbf{n}). \tag{69}$$

Along unit vectors in the directions of the gradients (in $\mathbf{k}$-space) of k, θ, and φ, the components of the latter two $Y_{jM}^{(\lambda)}$ then are[14]

$$\left.\begin{aligned} (Y_{jM}^{(0)})_n &= (Y_{jM}^{(1)})_n = 0, \\ (Y_{jM}^{(1)})_\theta &= +i(Y_{jM}^{(0)})_\varphi = \frac{1}{\sqrt{j(j+1)}} \frac{\partial Y_j^M}{\partial \theta}, \\ (Y_{jM}^{(1)})_\varphi &= -i(Y_{jM}^{(0)})_\theta = \frac{1}{\sqrt{j(j+1)}} \frac{1}{\sin\theta} \frac{\partial Y_j^M}{\partial \varphi}. \end{aligned}\right\} \tag{70}$$

Equations (65) through (70) were taken from Akhiezer and Berestetskii without verification.

Parity of the one-photon states of given angular momentum

We define the inversion operation I_{op} of a vector field $\mathbf{f}$ by[15]

$$I_{\text{op}}\mathbf{f}(\mathbf{k}) = -\mathbf{f}(-\mathbf{k}). \tag{71}$$

By $Y_l^m(-\mathbf{n}) = (-1)^l Y_l^m(\mathbf{n})$ and by (62a) and (66)–(67), then, for $j \geqslant 1$, $\lambda \geqslant 0$,

$$I_{\text{op}} Y_{jM}^{(\lambda)} = (-1)^{j+1-\lambda} \mathbf{Y}_{jM}^{(\lambda)}, \tag{72}$$

[10] We use here the notation of Akhiezer and Berestetskii (1965), sec. 4.2.

[11] Akhiezer and Berestetskii (1965), eq. (4.9).

[12] Akhiezer and Berestetskii (1965), eqs. (4.11)–(4.13).

[13] Akhiezer and Berestetskii (1965), eqs. (4.14)–(4.16).

[14] Akhiezer and Berestetskii (1965), eq. (4.17).

[15] The minus sign here is introduced so that the interaction between the electromagnetic vector potential and the velocity of the charge may be invariant under inversion ($\mathbf{x} \to -\mathbf{x}$, so $d\mathbf{x}/dt \to -d\mathbf{x}/dt$) with a coupling constant e that is a scalar and not a pseudoscalar.

so that for each $j \geqslant 1$ both parities are possible. [For $j = 0$, there is only the longitudinal field $\mathbf{Y}_{00}^{(-1)} = \mathbf{n}Y_0^0(\mathbf{n})$.]

Photon vector wave functions (35) may depend yet on the value of $|\mathbf{k}|$. For given j, M, and parity $j+1-\lambda$ ($\lambda = 0$ or 1), then, the one-photon wavefunction $\mathbf{f}(\mathbf{k})$ is of the form

$$\mathbf{f}_{jM\lambda}(\mathbf{k}) = a(k)\ \mathbf{Y}_{jM}^{(\lambda)}(\mathbf{n}), \tag{73}$$

with $a(k)$ normalized by

$$\int_0^\infty |a(k)|^2 k^2\, dk = 1. \tag{74}$$

Orbital photon angular momentum

For $\lambda = 0$, by (66), the angular-momentum quantum number l is equal to j. For $\lambda = \pm 1$, the states (65) and (67) are superpositions of $l = j-1$ and $l = j+1$, so that these states have no definite orbital angular momentum.

Angular distribution

The angular distribution of the direction of motion of a photon in a state (73) is given by the angular dependence of $|\mathbf{Y}_{jM}^{(\lambda)}|^2$. By $\mathbf{Y}_{jM}^{(1)} = -i\mathbf{n}\times\mathbf{Y}_{jM}^{(0)}$, this angular dependence is the same for $\lambda = 1$ as for $\lambda = 0$. Therefore it depends merely upon the angular momentum of the photon state and not upon its parity.

2.6. Two-photon wavefunctions

We shall now calculate the momentum and angular momenta of two-photon states. For this purpose, we calculate the expectation values (24) of the same field observables $Q(q_f)_{\text{op}}$ obtained from (17)–(18) by (22)–(23) and hermitization, thus yielding eq. (A-5), but we calculate these expectation values now for a two-photon state $\psi^{(F)}(q_f)$ given by

$$\psi^{(F)}(q_f) = \psi_{\text{vac}}^{(F)}(q_f)\cdot\sum_{\mathbf{k}_1}\sum_{\mathbf{k}_2}\sum_{\eta_1}\sum_{\eta_2} f_{\eta_1\eta_2}(\mathbf{k}_1, \mathbf{k}_2)\,(ck_1/\hbar)^{\frac{1}{2}}\,(q_{R\mathbf{k}_1}^{\eta_1} - iq_{I\mathbf{k}_1}^{\eta_1})\cdot(ck_2/\hbar)^{\frac{1}{2}}\,(q_{R\mathbf{k}_2}^{\eta_2} - iq_{I\mathbf{k}_2}^{\eta_2}), \tag{75}$$

which is obtained from (II:H-34) in the same way as (II:H-36) and (33) were obtained from it, if we discount the possibility that both photons could be in the same state.

The calculation of $\langle Q\rangle$ is performed in Appendix B. We introduce the dyadic or tensor field

$$F_{n'n''}(\mathbf{k}', \mathbf{k}'') \equiv \sum_{\eta'}\sum_{\eta''}(e_{\mathbf{k}'}^{\eta'})_{n'}\,(e_{\mathbf{k}''}^{\eta''})_{n''}\,F_{\eta'\eta''}(\mathbf{k}', \mathbf{k}'') = F_{n''n'}(\mathbf{k}'', \mathbf{k}'), \tag{76}$$

where

$$F_{\eta'\eta''}(\mathbf{k}', \mathbf{k}'') \equiv \sqrt{\tfrac{1}{2}}\,[f_{\eta'\eta''}(\mathbf{k}', \mathbf{k}'')+f_{\eta''\eta'}(\mathbf{k}'', \mathbf{k}')] = F_{\eta''\eta'}(\mathbf{k}'', \mathbf{k}'). \tag{77}$$

Then we find for $Q = \mathbf{p}$ or $\mathbf{L}$ or $\mathbf{S}$ or $\mathbf{J}$ $(= \mathbf{L}+\mathbf{S})$ that

$$\langle Q\rangle = \int d^3\mathbf{k}\int d^3\mathbf{k}'\sum_n\sum_{n'}\{F_{nn'}^*(\mathbf{k}, \mathbf{k}')\,\tilde{Q}(\mathbf{k})_{\text{op}}F_{nn'}(\mathbf{k}, \mathbf{k}')-F_{nn'}(\mathbf{k}, \mathbf{k}')\,\tilde{Q}(-\mathbf{k})_{\text{op}}\,F_{nn'}^*(\mathbf{k}, \mathbf{k}')\}. \tag{78}$$

[See (B-43).] When $f_{\eta'\eta''}(\mathbf{k}', \mathbf{k}'')$ can be factorized into two orthogonal normalized functions

$$\begin{aligned} f_{\eta'\eta''}(\mathbf{k}', \mathbf{k}'') &= f_{\eta'}^{(\alpha)}(\mathbf{k}')f_{\eta''}^{(\beta)}(\mathbf{k}''), \\ f_{n'n''}(\mathbf{k}', \mathbf{k}'') &= f_{n'}^{(\alpha)}(\mathbf{k}')f_{n''}^{(\beta)}(\mathbf{k}''), \end{aligned} \tag{79a}$$

with

$$\int d^3\mathbf{k} \sum_{\eta} |f_{\eta}^{(\alpha)}(\mathbf{k})|^2 = \int d^3\mathbf{k} \sum_{\eta} |f_{\eta}^{(\beta)}(\mathbf{k})|^2 = 1 \tag{79b}$$

and

$$\int d^3\mathbf{k} \sum_{\eta} f_{\eta}^{(\alpha)}(\mathbf{k})^* f_{\eta}^{(\beta)}(\mathbf{k}) = \int d^3\mathbf{k}\, \mathbf{f}^{(\alpha)}(\mathbf{k})\cdot\mathbf{f}^{(\beta)}(\mathbf{k}) = 0, \tag{79c}$$

in that special case $\langle Q\rangle$ is a sum

$$\langle Q\rangle = \langle Q\rangle_{1\ \mathrm{phot}}^{(\alpha)} + \langle Q\rangle_{1\ \mathrm{phot}}^{(\beta)}. \tag{80}$$

In general, this is not so, as neither (79a) nor (79c) need be true.

Anyhow, eq. (78) permits us to regard (76) or (77) as the "two-photon wavefunction."

For even more similarity to the two-electron case, we introduce separate operators to act on the two sets of arguments ($\mathbf{k}_1$, n_1 and $\mathbf{k}_2$, n_2) of $F_{n_1n_2}(\mathbf{k}_1, \mathbf{k}_2)$, and define

$$\tilde{Q}_{\mathrm{op}}^{\mathrm{sum}} = \tilde{Q}(\mathbf{k}_1, n_1)_{\mathrm{op}} + \tilde{Q}(\mathbf{k}_2, n_2)_{\mathrm{op}}. \tag{81}$$

Then (78) becomes

$$\begin{aligned} \langle Q\rangle = \int d^3\mathbf{k}_1 \int d^3\mathbf{k}_2 \sum_{n_1}\sum_{n_2} \tfrac{1}{2}\{&F_{n_1n_2}^*(\mathbf{k}_1, \mathbf{k}_2)\, \tilde{Q}(\mathbf{k}_1, \mathbf{k}_2)_{\mathrm{op}}^{\mathrm{sum}}\, F_{n_1n_2}(\mathbf{k}_1, \mathbf{k}_2) \\ &- F_{n_1n_2}(\mathbf{k}_1, \mathbf{k}_2)\, \tilde{Q}(-\mathbf{k}_1, -\mathbf{k}_2)_{\mathrm{op}}^{\mathrm{sum}}\, F_{n_1n_2}(\mathbf{k}_1, \mathbf{k}_2)\}. \end{aligned} \tag{82}$$

For $Q = \mathbf{p}$ or $\mathbf{L}$ or $\mathbf{S}$ or $\mathbf{J}$, this might be simplified even further, to

$$\langle Q\rangle = \int d^3\mathbf{k}_1 \int d^3\mathbf{k}_2 \sum_{n_1}\sum_{n_2} F_{n_1n_2}^*(\mathbf{k}_1, \mathbf{k}_2)\tilde{Q}_{\mathrm{op}}^{\mathrm{sum}} F_{n_1n_2}(\mathbf{k}_1, \mathbf{k}_2). \tag{83}$$

[Compare the derivation of (39) from (34).]

Angular-momentum quantum numbers for two-photon wavefunctions in momentum space

We may apply (81) to the angular-momentum operators **L**, **S**, and **J**, and, as in two-electron wave mechanics, use the eigenvalues of the squares and the z-components of these for defining quantum numbers for characterization of particular two-photon wavefunctions. Thus a state F_j satisfying

$$(\tilde{\mathbf{J}}_{\mathrm{op}}^{\mathrm{sum}})^2\, F_{n_1n_2}^{jm}(\mathbf{k}_1, \mathbf{k}_2) = j(j+1)\hbar^2\, F_{n_1n_2}^{jm}(\mathbf{k}_1, \mathbf{k}_2) \tag{84}$$

and

$$\tilde{J}_{z,\,\mathrm{op}}^{\mathrm{sum}} F_{n_1n_2}^{jm}(\mathbf{k}_1, \mathbf{k}_2) = m\hbar F_{n_1n_2}^{jm}(\mathbf{k}_1, \mathbf{k}_2) \tag{85}$$

could be called a state with quantum numbers j and m. It is now true by (83) that

$$\begin{aligned} \langle J_z\rangle &\equiv \int \psi_{jm}^{(F)*}(q_f)\left\{\int d^3\mathbf{x} \sum_n [p_n(\mathbf{x})/i\hbar]\, J_{z,\,\mathrm{op}}\, q_n(\mathbf{x})\right\}_{\mathrm{qu}} \psi_{jm}^{(F)}(q_f) \prod_f dq_f \\ &= \iint d^3\mathbf{k}_1\, d^3\mathbf{k}_2 \sum_{n_1}\sum_{n_2} F_{n_1n_2}^{jm}(\mathbf{k}_1, \mathbf{k}_2)^*\, \tilde{J}_{z,\,\mathrm{op}}^{\mathrm{sum}}\, F_{n_1n_2}^{jm}(\mathbf{k}_1, \mathbf{k}_2), \end{aligned} \tag{86}$$

where $F^{jm}_{n_1n_2}(\mathbf{k}_1, \mathbf{k}_2)$ is obtained by (76)–(77) from the coefficients $f^{jm}_{\eta_1\eta_2}(\mathbf{k}_1, \mathbf{k}_2)$ occurring in the expansion (75) of $\psi^{(F)}_{jm}(q_f)$. It is important that this relation between the $\psi^{(F)}(q_f)$ and the $F(\mathbf{k}_1, \mathbf{k}_2)$ is linear.

A formula similar to (86) would also hold with J_z replaced by S_z or L_x. We have seen before that we cannot in general draw many conclusions from such an equality of expectation value, calculated once from $\psi^{(F)}(q_f)$ and once from a photon wavefunction in momentum space. However, for j and m some conclusion can be drawn from the theory of representations of groups.

The meaning of j

From the invariance of nature under rotations of the coordinate system, this theory concludes that solutions for $\psi^{(F)}(q_f)$ occur in sets of $(2j+1)$ solutions labeled by quantum numbers j and m, that under rotations transform linearly among each other in a way that forms a "representation" of the rotation group.[16] Since by (75)–(77) the transformations of the $\psi^{(F)}_{jm}$ induce corresponding transformations of the $F^{jm}_{n_1n_2}(\mathbf{k}_1, \mathbf{k}_2)$, it follows that the $F^{jm}_{n_1n_2}(\mathbf{k}_1, \mathbf{k}_2)$, too, form the basis of a representation of the rotation group, with the same meaning of the j and the m.

It follows that the same group properties hold for the transformations of $\psi^{(F)}_{jm}$ and of F^{jm}, and that therefore the same commutation relations hold for both transformations.[16] Since the total angular-momentum components of $\mathbf{J}_{\text{qu}}$ [for the free photon field given by quantizing (10) by (23)] are the operators for infinitesimal rotations of $\psi^{(F)}$ around the coordinate axes, it follows that the components of $\mathbf{J}_{\text{qu}}$ have the same commutation relations as the components of $\tilde{\mathbf{J}}_{\text{op}}$. Consequently, they also have the same eigenvalues, and $\mathbf{J}^2_{\text{qu}}$ will have the eigenvalue $j(j+1)\hbar^2$ for each set of $(2j+1)$ state vectors $\psi^{(F)}_{jm}$. It follows that $J^2_{\text{qu}}\psi^{(F)}_{jm} = j(j+1)\hbar^2\psi^{(F)}_{jm}$ whenever $\psi^{(F)}_{jm}$ is related by (75)–(77) to $F^{jm}_{n_1n_2}(\mathbf{k}_1, \mathbf{k}_2)$ for which

$$\tilde{\mathbf{J}}^2_{\text{op}}F^{jm}_{n_1n_2}(\mathbf{k}_1, \mathbf{k}_2) = j(j+1)\hbar^2 F^{jm}_{n_1n_2}(\mathbf{k}_1, \mathbf{k}_2). \tag{87}$$

Therefore a classification of the tensor wavefunctions $F_{n_1n_2}(\mathbf{k}_1, \mathbf{k}_2)$ in terms of quantum numbers j and m is important for answering questions about the value of the total field angular momentum $\mathbf{J}_{\text{qu}}$.

The reasoning here offered for the total angular momentum breaks down when one tries to apply it to partial angular momenta like $\mathbf{L}$ or $\mathbf{S}$, because they correspond to rotations applied to only *part* of the formalism, and the relations (75)–(77) in that case would be affected and would not guarantee equivalent transformations for the $\psi^{(F)}$ and the $F_{n_1n_2}$.

Transversality of $F_{n_1n_2}(\mathbf{k}_1, \mathbf{k}_2)$

While we may speak formally about the "partial-angular-momentum quantum numbers" l and s of the two-photon wavefunction F, we have to keep in mind that these numbers are not always "good" quantum numbers. For instance, it follows from (76) and $\mathbf{k}\cdot\mathbf{e}^{\eta}_{\mathbf{k}} = 0$ that

$$\sum_n k_n F_{nn'}(\mathbf{k}, \mathbf{k}') = 0 \quad (\text{"transversality" of } F) \tag{88}$$

[16] See, for instance, Wigner (1959).

and the eigenfunctions of angular-momentum operators do not always have this property, in which case physical two-photon states satisfying (88) must be superpositions of several such eigenfunctions. [In the one-photon case we have already met a similar case when we found that l was not a good quantum number for a state $\boldsymbol{Y}_{jM}^{(\lambda)}(\mathbf{k})$ with $\lambda = 1$.]

Parity of two-photon wavefunctions

We may define the parity of F by considering the effect of the inversion operator

$$I_{\text{op}}F_{n_1n_2}(\mathbf{k}_1, \mathbf{k}_2) = F_{n_1n_2}(-\mathbf{k}_1, -\mathbf{k}_2). \tag{89}$$

Wavefunctions for annihilation photons

A case of interest to us at the end of Chapter 5 is the case of a photon pair created by annihilation of a positon–negaton pair in the center-of-mass frame of this electron pair. In this case, the total photon momentum is zero, and $F_{n_1n_2}(\mathbf{k}_1, \mathbf{k}_2)$ differs from zero for $\mathbf{k}_2 = -\mathbf{k}_1$ only.

Thus $F_{n_1n_2}$ is a function of

$$\mathbf{k} = \tfrac{1}{2}(\mathbf{k}_1 - \mathbf{k}_2) \tag{90}$$

only, while

$$\mathbf{K} = \mathbf{k}_1 + \mathbf{k}_2 \tag{91}$$

is given to have the value zero.

In this case, putting

$$\mathbf{n} \equiv \mathbf{k}/k, \tag{92}$$

we may consider

$$F_{n_1n_2}(\mathbf{k}_1, \mathbf{k}_2) = F_{n_1n_2}(\mathbf{k}, -\mathbf{k}) \tag{93}$$

to be a function of k, $\mathbf{n}$, n_1, and n_2. Ignoring transversality, we could expand this for given k in terms of products of spherical harmonics $Y_l^m(\mathbf{n})$ that are eigenfunctions of $\tilde{\mathbf{L}}_{\text{op}}^2$ and of $\tilde{L}_{z,\,\text{op}}$ and of spin functions $(\chi_{s,\,\mu})_{n_1n_2}$ that are linear combinations of products $(\chi_{\mu_1})_{n_1}(\chi_{\mu_2})_{n_2}$ of the spin functions (56) chosen so that

$$(\tilde{S}_{z_1,\,\text{op}} + \tilde{S}_{z_2,\,\text{op}})\,\chi_{s,\,\mu} = \mu\hbar\chi_{s,\,\mu}, \tag{94}$$

and

$$(\tilde{\mathbf{S}}_{1,\,\text{op}} + \tilde{\mathbf{S}}_{2,\,\text{op}})^2\chi_{s,\,\mu} = s(s+1)\hbar^2\chi_{s,\,\mu}. \tag{95}$$

Among these $\chi_{s,\,\mu}$, three will be antisymmetric between n_1 and n_2 and will have $s = 1$; six will be symmetric between n_1 and n_2 and will have $s = 0$ or 2.

Consider the product $Y_l^m(\mathbf{n})\,(\chi_{s,\,\mu})_{n_1n_2}$. Its parity will be $(-1)^l$. From the symmetry (76) of F, we see that, for a two-photon wavefunction of the kind (93), inversion of $\mathbf{k} = \frac{1}{2}(\mathbf{k}_1 - \mathbf{k}_2)$ is equivalent to an interchange of n_1 and n_2. So, if $Y_l^m(\mathbf{n})\,(\chi_{s,\,\mu})_{n_1n_2}$ is to be part of (93), we want $(\chi_{s,\,\mu})_{n_1n_2}$ to satisfy

$$(\chi_{s,\,\mu})_{n_2n_1} = (-1)^l\,(\chi_{s,\,\mu})_{n_1n_2}. \tag{96}$$

Therefore, "even" annihilation-photon wavefunctions [with $(-1)^l = 1$ and $l =$ even] should be symmetric in n_1 and n_2 and correspond to "sum of spins" quantum numbers $s = 0$ and 2, while "odd" annihilation-photon wavefunctions with $(-1)^l = -1$ and $l =$ odd should be antisymmetric in n_1 and n_2 and should have "sum of spins" quantum number $s = 1$. In the latter case, the tensor $F_{n_1n_2}(\mathbf{k}_1, \mathbf{k}_2)$ is an antisymmetric tensor (always in the center-of-mass frame of the photon pair, in which $\mathbf{k}_1+\mathbf{k}_2 = 0$). An antisymmetric tensor $F_{n_1n_2}$ may be related to an axial vector $\mathbf{F}$ according to

$$F_{n_1n_2} = \sum_{n_3} \varepsilon_{n_1n_2n_3} F_{n_3}, \tag{97}$$

where ε_{lmn} is the antisymmetric Levi–Civita tensor with components 0 and ± 1. Transversality of $F_{n_1n_2}$ now by (88) and (90)–(92) requires

$$0 = \sum_{n_2}\sum_{n_3} \varepsilon_{n_1n_2n_3} F_{n_3} n_{n_2} = (\mathbf{n}\times\mathbf{F})_{n_1}. \tag{98}$$

Thus $\mathbf{F}$ must in $\mathbf{k}$-space be a radial field $\|\mathbf{n}\|\mathbf{k}$. Since, for given quantum numbers $j \neq 0$ and $M \equiv m_j$, according to Section 2.5 two of the three orthogonal vector fields $\mathbf{Y}_{jM}^{(\lambda)}(\mathbf{n})$ (namely the ones with $\lambda = 0$ or 1) are perpendicular to $\mathbf{k}$, $\mathbf{F}$ then must be the third field $\mathbf{Y}_{jM}^{(-1)}(\mathbf{n})$, given by eq. (65). We therefore conclude that any annihilation-photon wavefunction (93) of odd parity and with given quantum numbers $j \neq 0$ and $m_j = M$ must depend on angles and on tensor indices according to (97) with

$$\mathbf{F} \propto \mathbf{Y}_{jM}^{(-1)} = \mathbf{n} Y_j^M(\mathbf{n}). \tag{99}$$

Since (99) has parity $j+1$, it follows that

odd parity of $F_{n_1n_2}(\mathbf{k}, -\mathbf{k})$ is possible for even values of the quantum number j only. (100)

The simplest example of such a wavefunction (for $j = 0$) is given by

$$f_{n_1n_2}(\mathbf{k}_1, \mathbf{k}_2) = \int \delta(\mathbf{k}_1-\mathbf{k})\, \delta(\mathbf{k}_2+\mathbf{k}) \sum_{n_3} \varepsilon_{n_1n_2n_3}(k_1-k_2)_{n_3} f(k)\, d^3\mathbf{k} \tag{101}$$

with $F_{n_1n_2}(\mathbf{k}_1, \mathbf{k}_2)$ obtained by symmetrization (77). When only photon pairs with $\mathbf{k}_1-\mathbf{k}_2$ in the $\pm z$-direction are observed, the effective part of the wavefunction (101) has components only with either $n_1 = x$ and $n_2 = y$, or $n_1 = y$ and $n_2 = x$, and it is antisymmetric between x and y. This is the kind of wave function which we will meet in Section 5.

The case of wavefunctions $F_{n_1n_2}(\mathbf{k}, -\mathbf{k})$ of even parity is discussed in the second half of Section 7.3 of Akhiezer and Berestetskii (1965). We shall not discuss it here.

Abbreviated notation

In polarization correlation experiments on photon pairs, the photons and their polarization are observed only in certain definite directions, that is, for given values of $\mathbf{k}_1$ and $\mathbf{k}_2$. [Which $\mathbf{k}$ is $\mathbf{k}_1$ and which is $\mathbf{k}_2$ in $F_{n_1n_2}(\mathbf{k}_1, \mathbf{k}_2)$ is, of course, undetermined, as the photons have no individuality.] Since thus the dependence of the wavefunction on $\mathbf{k}$ remains unobserved for other directions, it is of no direct interest to us.

The polarizations observed are in the experiments which we shall discuss always linear rather than circular polarizations. A linearly polarized wave is a superposition (with a definite phase relation) between a left-hand and a right-hand circularly polarized wave. If the wave propagates in the z-direction, the wavefunction is a linear combination of the waves described by χ_+ and χ_- in (56). In particular, polarization in the x-direction is characterized by a vector wavefunction with a nonvanishing x-component and vanishing y-component.

Therefore, the polarization of the wave is described by the direction of the photon vector wavefunction $\mathbf{f}(\mathbf{k})$.

Similarly, for a two-photon state, what is of interest to us is for which pairs of values of the tensor indices n_1 and n_2 the tensor wavefunctions $f_{n_1n_2}(\mathbf{k}_1, \mathbf{k}_2)$ or its symmetrized form $F_{n_1n_2}(\mathbf{k}_1, \mathbf{k}_2)$ differs from zero for the $\mathbf{k}_1$ and $\mathbf{k}_2$ values given by the experimental arrangement.

We can use unit vectors $\hat{x}$ and $\hat{y}$ for indicating polarization directions of waves traveling in the $\pm z$-direction. For indicating that these unit vectors are used for characterizing waves, we replace the circumflex by a tilde. This gives the notation already introduced in Section 2.2, in which $\tilde{y}_1\tilde{x}_2$ stands for a wave $f_{n_1n_2}$ which has a nonzero component (for the given $\mathbf{k}_1$ and $\mathbf{k}_2$) only for $n_1 = y$, $n_2 = x$. The wave function (101), for $\mathbf{k}_1 = k\hat{z}$ and $\mathbf{k}_2 = -k\hat{z}$, would be characterized by $f\cdot(\tilde{x}_1\tilde{y}_2 - \tilde{y}_1\tilde{x}_2)$, where the factor f may be used for some convenient normalization.

2.7. Derivation of the final wavefunction for two photons emitted by a cascading atom

We consider the closed system consisting initially of an excited atom and, finally, of the atom together with the two-photon field emitted by it. Including all interactions, there must be conservation of total angular momentum $\mathbf{J}$ if we may trust quantum theory.[17]

The 0–1–0 case

In the 0–1–0 case, we start from a state $j = 0$ of the atom, with zero total angular momentum. There is only one initial state to consider (as $2j+1 = 1$ in this case). By the Schrödinger equation, this state evolves into one single final state. Since also the atom by itself ends up in a $j = 0$ state, we do not have to worry about several final photon field states each occurring in combination with a different atomic state (as would be the case, for instance, in a 0–1–1 transition of the atom). We conclude that the photon field at the end is in a simple $j = 0$ state by itself. That is, its state is invariant under rotations.

Perturbation treatment

We shall use time-dependent perturbation theory[18] for getting some idea about this final photon state. All we really want to know about it is its dependence upon the polarizations of the two photons.

We use an abbreviated notation. For the photons we use the notation explained in Section 2.2. As the atoms in the Clauser experiment for the 0–1–0 case are always in singlet

[17] In some hidden-variables theories it is assumed that the observed final state of a system depends upon hidden variables describing the microstate of the apparatus used for observation. Such an interaction of our system with the apparatus would make the system an open system, and there would be no guarantee that angular momentum would be conserved. Therefore the following reasoning would not necessarily be valid in such a hidden-variables theory.

[18] See Merzbacher (1970), chap. 18, secs. 7–8. For a better treatment, taking into account the depletion of the initial state as the transition probablity increases, see Heitler (1954), pp.182–186 of the 3rd edition (or pp. 110–114 of the 1st edition).

states (see Section 2.1), we need not worry about their spin. Their orbital states are first an *S*-state described by a single isotropic scalar wavefunction which we shall denote by *S*, then an intermediate *P*-state to be discussed below, and, finally, a single isotropic ground-state wavefunction which we shall denote by *s*. For the intermediate state with $l = 1$, we must consider three linearly independent possibilities corresponding to $m_l \equiv \mu = 1, 0, -1$. Under rotations they transform among each other like $Y_l^\mu(\theta, \varphi)$ for $l = 1$ or like e^μ, where

$$\mathrm{e}^\mu = (4\pi/3)^{\frac{1}{2}}\, rY_1^\mu \tag{102}$$

has the values[19]

$$\mathrm{e}^1 = \sqrt{\tfrac{1}{2}}\,(-x-iy), \quad \mathrm{e}^0 = z, \quad \mathrm{e}^{-1} = \sqrt{\tfrac{1}{2}}\,(x-iy). \tag{103}$$

From time-dependent perturbation theory we know that the perturbation Hamiltonian $\mathscr{H}^{(1)}$ causes transitions from the initial state *A* to a new state *B*, if the matrix element $H_{BA}^{(1)}$ between these two states differs from zero. In quantum electrodynamics, for the emission of a photon with polarization unit vector $\mathbf{e}_\mathrm{k}$, in a transition from an atomic state Ψ_A to an atomic state Ψ_B, one has in good first approximation[20]

$$H_{BA}^{(1)} = \varepsilon(hck/\mathcal{V})^{\frac{1}{2}} \int \Psi_B^* \Sigma \mathbf{r} \boldsymbol{\cdot} \mathbf{e}_\mathrm{k}^* \Psi_A, \tag{104}$$

where $\mathcal{V}$ is the volume in which photon waves are normalized to have an energy $\hbar ck$ and where the sum is over the radius vectors **r** of all electrons involved in the emission. Due to the perturbation (104) causing transitions $A \to B$, and subsequent perturbations $H_{CB}^{(1)}$ causing transitions $B \to C$, we obtain in second order states *C* with an amplitude proportional (for $\mathscr{E}_C = \mathscr{E}_B = \mathscr{E}_A$) to

$$\sum_B H_{CB}^{(1)} H_{BA}^{(1)}, \tag{105}$$

where we sum over all intermediate states. This is a sum $\sum_{\mu=-1}^{1}$, due to the degeneracy of the intermediate *P*-state. We may, however, equally well sum over any different complete orthonormal set of intermediate states obtained from the set of states Ψ_B by linear combinations. Equation (103) suggests to use here a set of linear combinations that transform like

$$x = \sqrt{\tfrac{1}{2}}\,(\mathrm{e}^{-1}-\mathrm{e}^1), \quad y = i\sqrt{\tfrac{1}{2}}\,(\mathrm{e}^{-1}+\mathrm{e}^1), \quad z = \mathrm{e}^0. \tag{106}$$

This is particularly convenient if we use linear polarizations for $\mathbf{e}_\mathrm{k}^*$ in (104) and for the description of the photon states by $\tilde{x}$, $\tilde{y}$, and $\tilde{z}$. (The use of the e^μ is more convenient if we use circular polarizations for describing the light emitted.)

In the first transition, Ψ_A is spherically symmetric, so that the integral in (104) vanishes unless $\int \Psi_B^* \Sigma \mathbf{r} \boldsymbol{\cdot} \mathbf{e}_\mathrm{k}^*$ contains terms even in *x*, in *y*, and in *z*. Since $\Sigma \mathbf{r}$ is odd in the coordinates along the axes along which $\mathbf{e}_\mathrm{k}^*$ has a nonvanishing component, it follows that light polarized in the *x*-direction can be emitted only if Ψ_B is odd in *x*, but even in *y* and *z*. Among the

[19] By (56), $\mathrm{e}^\mu = \sum_n (\chi_\mu)_n x_n$.

[20] Compare, for instance, Kramers (1957), sec. 91, eq. (8.178) with $\mathbf{c}_\lambda \equiv \sigma\, \mathbf{e}^\mathrm{k}$, $\sigma \equiv \nu/c$, $n_\lambda^0 + 1 = 1$, $\Omega = \mathcal{V}$, $\exp(-2\pi i \boldsymbol{\sigma} \cdot \mathbf{x}) \sim 1$, in which eq. (8.24) and the following equation of sec. 82 give $[\Sigma e_i \mathbf{v}_i]_{kk_0} \approx$ (in our notation) $2\pi i \nu \varepsilon \int \Psi_B^* \Sigma \mathbf{x} \Psi_A$.

combinations transforming like (106), this is the case for the first of the three combinations shown.

However, for this choice of Ψ_B odd in x, the matrix element

$$H_{CB}^{(1)} = e(hck/\mathcal{V})^{\frac{1}{2}} \int \Psi_C^* \Sigma \mathbf{r} \cdot \mathbf{e}_{\mathbf{k}'}^{\prime *} \Psi_B \tag{107}$$

for the emission of a second photon with polarization $\mathbf{e}'_{\mathbf{k}'}$ will now be different from zero only if also $\mathbf{e}'_{\mathbf{k}'}$ is directed along the x-axis, since $\Psi_C = s$ is spherically symmetric.

Similarly, if the first photon is to be polarized along $\mathbf{e}_{\mathbf{k}} = \hat{y}$, we must choose Ψ_B proportional to the second choice (y) in (106), and then we must have also $\mathbf{e}'_{\mathbf{k}'} = \hat{y}$ for keeping (107) different from zero. Similarly for $\mathbf{e}_{\mathbf{k}} = \mathbf{e}'_{\mathbf{k}'} = \hat{z}$.

Thus the polarization of the second photon in the 0–1–0 case will always be parallel to the polarization of the first photon. Thus the final photon wavefunction will contain terms with $\tilde{x}_1\tilde{x}_2$, with $\tilde{y}_1\tilde{y}_2$, and with $\tilde{z}_1\tilde{z}_2$, and, if this were all on which it would depend, for rotational invariance we might think that this photon wavefunction would be proportional to

$$\tilde{x}_1\tilde{x}_2 + \tilde{y}_1\tilde{y}_2 + \tilde{z}_1\tilde{z}_2 . \tag{108}$$

Transversality

We have, however, not discussed yet the dependence of the photon wavefunction upon the directions $\hat{k}_1 \equiv \mathbf{k}_1/k_1$ and $\hat{k}_2 \equiv \mathbf{k}_2/k_2$ of the two photons. It is clear that, for instance, for $\hat{k}_1 = \hat{z}$ there can be no polarization $\tilde{z}_1$. For the case of particular importance in the experiment, that $\hat{k}_1 = \hat{k}_a = \hat{z}$ and $\hat{k}_2 = \hat{k}_b = -\hat{z}$ or $\hat{k}_1 = \hat{k}_b = -\hat{z}$ and $\hat{k}_2 = \hat{k}_a = \hat{z}$ (where $\mathbf{k}_a$ and $\mathbf{k}_b$ are momenta of observed photons with definite wavelengths $\lambda_a = 2\pi/k_a = 5513$ Å and $\lambda_b = 2\pi/k_b = 4227$ Å), we would expect not (108) but a polarization dependence

$$\tilde{x}_1\tilde{x}_2 + \tilde{y}_1\tilde{y}_2 \tag{109}$$

of the two-photon wavefunction.

The $\tilde{x}$ in (108), therefore, stands for a wave emitted from a source that would like to emit light polarized in the $\hat{x}$-direction, as classically would be emitted by an electric dipole vibrating in the x-direction. This light is not radiated with equal intensity in all directions, but, in directions making an angle θ with the x-axis, the amplitude is proportional to $\sin\theta$, and the polarization direction is the component of $\hat{x}$ in the plane perpendicular to the propagation direction $\hat{k}$.[21] Thus, by replacing in (108) the polarization vector $\hat{x}$ with components $\hat{x}_n = \delta_{xn}$, by $\hat{x} - \hat{k}(\hat{k}\cdot\hat{x})$ with components $\delta_{xn} - \hat{k}_x\hat{k}_n$, we not only give the polarization the direction expected classically, but also make the amplitude of the light wave proportional to $\sin\theta$.

Thus (108) or $\sum_{l=1}^{3} (\tilde{x}_l)_1(\tilde{x}_l)_2$ by this procedure is changed into a transverse two-photon field proportional to

$$\begin{aligned}(\psi_{010}^{\text{pol}})_{n_1n_2} &= \sum_l (\delta_{n_1l} - \hat{k}_{1n_1}\hat{k}_{1l})(\delta_{n_2l} - \hat{k}_{2n_2}\hat{k}_{2l}) \\ &= \delta_{n_1n_2} - \hat{k}_{1n_1}\hat{k}_{1n_2} - \hat{k}_{2n_1}\hat{k}_{2n_2} + (\hat{k}_1\cdot\hat{k}_2)\hat{k}_{1n_1}\hat{k}_{2n_2} ,\end{aligned} \tag{110}$$

[21] See, for instance, Marion (1956), eqs. (8.22) and (8.62).

or

$$\begin{aligned}\tilde{\psi}_{010}^{\text{pol}} &\equiv \sum_{n_1}\sum_{n_2} (\psi_{010}^{\text{pol}})_{n_1 n_2} (\tilde{x}_{n_1})_1 (\tilde{x}_{n_2})_2 \\ &= (\tilde{\mathbf{x}}_1 \cdot \tilde{\mathbf{x}}_2) - (\hat{k}_1 \cdot \tilde{\mathbf{x}}_1)(\hat{k}_1 \cdot \tilde{\mathbf{x}}_2) - (\hat{k}_2 \cdot \tilde{\mathbf{x}}_1)(\hat{k}_2 \cdot \tilde{\mathbf{x}}_2) + (\hat{k}_1 \cdot \hat{k}_2)(\hat{k}_1 \cdot \tilde{\mathbf{x}}_1)(\hat{k}_2 \cdot \tilde{\mathbf{x}}_2) \\ &= [\tilde{\mathbf{x}}_1 - \hat{k}_1(\hat{k}_1 \cdot \tilde{\mathbf{x}}_1)] \cdot [\tilde{\mathbf{x}}_2 - \hat{k}_2(\hat{k}_2 \cdot \tilde{\mathbf{x}}_2)]. \end{aligned} \tag{111}$$

Note that the factors $\delta_{nl} - \hat{k}_n \hat{k}_l$ also may be written in the form

$$\delta_{nl} - \hat{k}_n \hat{k}_l = \delta_{nl} - (e_{\mathbf{k}}^0)_l^* (e_{\mathbf{k}}^0)_n = \sum_{\eta=\pm} (e_{\mathbf{k}}^{\eta})_l^* (e_{\mathbf{k}}^{\eta})_n . \tag{112}$$

Given photon momenta

The above expressions (110)–(111) still do not fully express the photon wavefunction. They do not express the fact that the field only contains $k_1 = k_a$, $k_2 = k_b$, and $k_1 = k_b$, $k_2 = k_a$ (i.e. definite frequencies). When applied to a setup which observes $\mathbf{k}_a$ only in one particular direction and $\mathbf{k}_b$ only in a particular different direction, the part of the wavefunction that is observed has even either $\mathbf{k}_1 = \mathbf{k}_a$, $\mathbf{k}_2 = \mathbf{k}_b$, or $\mathbf{k}_1 = \mathbf{k}_b$, $\mathbf{k}_2 = \mathbf{k}_a$. We may express these facts by multiplying (110) and (111) by a factor[22]

$$\Delta_{ab}(\mathbf{k}_1, \mathbf{k}_2) = C(\delta_{\mathbf{k}_1, \mathbf{k}_a}\delta_{\mathbf{k}_2, \mathbf{k}_b} + \delta_{\mathbf{k}_1, \mathbf{k}_b}\delta_{\mathbf{k}_2, \mathbf{k}_a}). \tag{113}$$

Then $F_{n_1 n_2}(\mathbf{k}_1, \mathbf{k}_2)$ in the 010 case is proportional to the product of (113) and (110):

$$F_{n_1 n_2}(\mathbf{k}_1, \mathbf{k}_2) = (\psi_{010}^{\text{pol}})_{n_1 n_2} \Delta_{ab}(\mathbf{k}_1, \mathbf{k}_2). \tag{114}$$

It is easily seen to satisfy the transversality condition (88). From $F_{n_1 n_2}$ we obtain $F_{\eta_1 \eta_2}$ according to (76) by

$$F_{\eta_1 \eta_2}(\mathbf{k}_1, \mathbf{k}_2) = \sum_{n_1}\sum_{n_2} F_{n_1 n_2}(\mathbf{k}_1, \mathbf{k}_2)(e_{\mathbf{k}_1}^{\eta_1})_{n_1}^* (e_{\mathbf{k}_2}^{\eta_2})_{n_2}^* . \tag{115}$$

For calculating this, it is convenient to use (112) for each of the factors in (110). By $\mathbf{e}_{\mathbf{k}_1}^{\eta_1 *} \cdot \mathbf{e}_{\mathbf{k}_1}^{\eta} = \delta_{\eta_1 \eta}$, this gives[22]

$$F_{\eta_1 \eta_2}(\mathbf{k}_1, \mathbf{k}_2) = C(\mathbf{e}_{\mathbf{k}_1}^{\eta_1 *} \cdot \mathbf{e}_{\mathbf{k}_2}^{\eta_2 *})(\delta_{\mathbf{k}_1, \mathbf{k}_a}\delta_{\mathbf{k}_2, \mathbf{k}_b} + \delta_{\mathbf{k}_1, \mathbf{k}_b}\delta_{\mathbf{k}_2, \mathbf{k}_a}). \tag{116}$$

Then, eq. (75), with $q_{\mathbf{k},\eta} = \sqrt{\tfrac{1}{}}(q_{R\mathbf{k}}^{\eta} + i q_{I\mathbf{k}}^{\eta})$ [see (II: H-17)] and with $f_{\eta_1 \eta_2} = \sqrt{\tfrac{1}{2}}\, F_{\eta_1 \eta_2}$, gives[22]

$$[\psi_{010}^{(F)}(q_f)]_{\mathbf{k}_a, \mathbf{k}_b} = \psi_{\text{vac}}^{(F)}(q_f)(2cC/\hbar)(2k_a k_b)^{\frac{1}{2}} \sum_{\eta}\sum_{\eta'} (\mathbf{e}_{\mathbf{k}_a}^{\eta *} \cdot \mathbf{e}_{\mathbf{k}_b}^{\eta' *}) q_{\mathbf{k}_a}^{\eta *} q_{\mathbf{k}_b}^{\eta' *} . \tag{117}$$

Opposite photon directions

In the special case $\hat{k}_a = \hat{z} = -\hat{k}_b$, eqs. (110) and (113) give

$$F_{n_1 n_2}(k_1 \hat{z}, -k_2 \hat{z}) = \Delta_{ab}(\mathbf{k}_1, \mathbf{k}_2)(\psi_{010}^{\text{pol}})_{n_1 n_2} = \begin{pmatrix} 1 & 0 & 0 \\ 0 & 1 & 0 \\ 0 & 0 & 0 \end{pmatrix} \Delta_{ab}(\mathbf{k}_1, \mathbf{k}_2), \tag{118}$$

[22] Here, C is a normalization factor depending upon $(\hat{k}_a \cdot \hat{k}_b)$. If the directions of $\mathbf{k}_a$ and $\mathbf{k}_b$ are not given, we have a *mixture* of states which each are superpositions for given $|\mathbf{k}_a|$, $|\mathbf{k}_b|$, and $\mathbf{K}_c = \mathbf{k}_a + \mathbf{k}_b$, of states (114) or (116)–(117) in the 0–1–0 case, or (132), (133) or (135) in the 1–1–0 case. See Appendix F for details.

where the matrix has rows and columns labeled by n_1 and n_2. This matrix represents the result (109).

This result may also be obtained from (116), which by

$$\mathbf{e}^{\eta_1 *}_{\hat{z}k} \cdot \mathbf{e}^{\eta_2 *}_{-\hat{z}k} = -\mathbf{e}^{\eta_1 *}_{\hat{z}k} \cdot \mathbf{e}^{\eta_2}_{\hat{z}k} = -\delta_{\eta_1 \eta_2} \tag{119}$$

becomes

$$[F_{\eta_1 \eta_2}(\mathbf{k}_1, \mathbf{k}_2)]_{\mathbf{k}_a, \mathbf{k}_b} = -\Delta_{ab}(\mathbf{k}_1, \mathbf{k}_2)\, \delta_{\eta_1 \eta_2}, \tag{120}$$

showing in the notation of Section 2.2 that circular polarizations occur in the combinations

$$-(\tilde{R}_1 \tilde{R}_2 + \tilde{L}_1 \tilde{L}_2)\, \Delta_{ab}(\mathbf{k}_1, \mathbf{k}_2). \tag{121}$$

By (4b), we have

$$-(\tilde{R}_1 \tilde{R}_2 + \tilde{L}_1 \tilde{L}_2) = \tilde{x}_1 \tilde{x}_2 + \tilde{y}_1 \tilde{y}_2, \quad (\tilde{R}_1 \tilde{R}_2 - \tilde{L}_1 \tilde{L}_2) = i(\tilde{x}_1 \tilde{y}_2 - \tilde{y}_1 \tilde{x}_2), \tag{122}$$

so that (121) is equivalent to (113) times (109), that is, it is equivalent to (118).

The 1–1–0 case

In the 1–1–0 case, we start from a state $j = 1$ of the atom, so that we must consider as possible initial states any linear combinations of three linearly independent states that could be labeled by $m_j = 1, 0, -1$, though we could equally well consider linear combinations of states labeled by $n = x, y, z$, where the latter three states are derived from the former ones like x, y, z in (106) are derived from the e^{μ}. Actually, the initial state will be a mixed state, with these linear combinations

$$\sum_{\mu} c_{\mu} \Psi_{A\mu} \tag{123}$$

occurring with equal probabilities for different values of the c_{μ}. More precisely, in a six-dimensional space with the real and imaginary parts of the three coefficients c_{μ} plotted as coordinates, the initial states lie evenly distributed over a unit sphere of radius 1 around the origin.

To each of the initial states $\Psi_{A\mu}$, an evolution of the state according to the Schrödinger equation will correspond, and, since in all cases we end up with the *atom* in the *same* final state Ψ_C, we would end up in three *different* final *photon* states. In general, the final photon state then will be a linear combination, with the same coefficient c_{μ} as in (123), of these three final photon states:

$$F_{n_1 n_2}(\mathbf{k}_1, \mathbf{k}_2) = \sum_{\mu} c_{\mu} F^{\mu}_{n_1 n_2}(\mathbf{k}_1, \mathbf{k}_2). \tag{124}$$

Perturbation treatment

Again, we shall look at the matrix elements for finding up to second order what final states evolve from an initial state $\Psi_{A\mu}$ which is in the case of Holt's experiments for the 1–1–0 case a $9P_1$ state. That is, we have $j = 1$, $m_j = \mu$, $l = 1$, and, since the 7^3S_1 state is the intermediate atomic state responsible for the emission of the wavelength $\lambda_a = 5677$ Å in the first step of the process, we are interested merely in the triplet contribution to the $9P_1$

state (with $s = 1$). We shall indicate by X, Y, Z the three linear combinations of the initial orbital factors $Y_1^m(\theta, \varphi)$ in the P-state for Ψ_A that transform like x, y, and z. Similar combinations of the three possible spin states with $m_j = 1, 0, -1$ we shall indicate by ξ, η, ξ. As the vector sum of **L** and **S** in this case is to be **J** with $j = 1$, the $\Psi_{A\mu}$ states in (123) that transform like x, y, z are those linear combinations of products of X, Y, Z with ξ, η, ζ that behave like the three components of a vector. It is well known that these combinations are

$$\Psi_{Ax} = \sqrt{\tfrac{1}{2}}\cdot(Y\zeta - Z\eta), \quad \Psi_{Ay} = \sqrt{\tfrac{1}{2}}\cdot(Z\xi - X\zeta), \quad \text{and} \quad \Psi_{Az} = \sqrt{\tfrac{1}{2}}\cdot(X\eta - Y\xi). \tag{125}$$

We now look for nonvanishing matrix elements (104) between these states for Ψ_A and the three possible states $\Psi_{B\mu}$, which are

$$\Psi_{Bx} = S\xi, \quad \Psi_{By} = S\eta, \quad \Psi_{Bz} = S\zeta, \tag{126}$$

where by S we indicate the orbital factor in this 7^3S_1 state, while ξ, η, ζ are the same triplet spin functions as before. As the spin does not change in photon emission [the matrix elements (104) and (107) vanish unless the spin factors in Ψ^* and in Ψ match with each other], we see for $\mu = x$ that the $Y\zeta$ part of Ψ_{Ax} can give transitions only to $\Psi_{Bz} = S\zeta$, and that the $-Z\eta$ part of Ψ_{Ax} gives transitions only to $\Psi_{By} = S\eta$, so that the Ψ_B state evolving from the Ψ_{Ax} state is a superposition of Ψ_{By} and Ψ_{Bz}, with a difference in sign due to the minus sign in Ψ_{Ax}, and with different polarizations of the photon emitted. The latter difference arises from the fact that for nonvanishing of (104) the oddness in y of the $Y\zeta$ term in Ψ_{Ax} must be matched by oddness in y in $\Sigma\mathbf{r}\cdot\mathbf{e}_k$ which is possible only for $\mathbf{e}_k = \hat{y}$, while oddness in z of the $Z\eta$ term requires similarly $\mathbf{e}_k = \hat{z}$. Thus, Ψ_{Ax} evolves in first order into a superposition proportional to

$$\Psi_{Bz}\tilde{y}_a - \Psi_{By}\tilde{z}_a. \tag{127}$$

From here, under emission of $\lambda_b = 4048$ Å, we should reach the 6^3P_0 state Ψ_C, which is a single state with $l = 1$, $s = 1$, $j = 0$. We shall denote by x, y, z the linear combinations (106) of the orbital $l = 1$ functions $Y_1^m(\theta, \varphi)$ for this P-state Ψ_C. Since Ψ_C is invariant under rotations, it then must be the normalized scalar product

$$\Psi_C = \sqrt{\tfrac{1}{3}}\cdot(x\xi + y\eta + z\zeta) \tag{128}$$

of the vectors $\{x, y, z\}$ and $\{\xi, \eta, \zeta\}$.

By conservation of spin in (107), it follows that the $\Psi_{Bx} = S\xi$ state can have nonvanishing matrix elements (107) at most with the term $x\xi$ in Ψ_C, Ψ_{By} can at most go into the $y\eta$ term, and Ψ_{Bz} can go into Ψ_C only if the term $z\zeta$ in Ψ_C gives a nonvanishing contribution to (107). However, the matrix element for $\Psi_{Bx} \to \Psi_C$ is odd in x (from the $x\xi$ factor) unless $\Sigma\mathbf{r}\cdot\mathbf{e}_k$ is odd in x, which is only the case if $\mathbf{e}_k = \hat{x}$. Similarly, $\Psi_{By} \to \Psi_C$ requires $\mathbf{e}_k = \hat{y}$ and gives rise to emission of the photon wave $\tilde{y}_b$, while a photon wave $\tilde{z}_b$ must arise by $\mathbf{e}_k = \hat{z}$, when $H_{CB}^{(1)}$ is to be nonzero for the part Ψ_{Bz} in Ψ_B.

Consequently, the wave (127), which arose from Ψ_{Ax}, will give rise in second order to

$$(\Psi_C\tilde{z}_b)\tilde{y}_a - (\Psi_C\tilde{y}_b)\tilde{z}_a = (\tilde{y}_a\tilde{z}_b - \tilde{z}_a\tilde{y}_b)\Psi_C. \tag{129}$$

Similarly, Ψ_{Ay} leads to $(\tilde{z}_a\tilde{x}_b - \tilde{z}_b\tilde{x}_a)\Psi_C$ and Ψ_{Az} to $(\tilde{x}_a\tilde{y}_b - \tilde{y}_a\tilde{x}_b)\Psi_C$. The three final photon waves by this reasoning would be proportional to

$$\tilde{y}_a\tilde{z}_b - \tilde{z}_a\tilde{y}_b, \quad \tilde{z}_a\tilde{x}_b - \tilde{x}_a\tilde{z}_b, \quad \text{and} \quad \tilde{x}_a\tilde{y}_b - \tilde{y}_a\tilde{x}_b. \tag{130}$$

Transversality

The result so far obtained for the 110 case is comparable to the result (108) in the 010 case. We may summarize these results (130) as $\sum_i \sum_j \varepsilon_{\mu ij}(\tilde{x}_i)_a (\tilde{x}_j)_b$. For the special case of interest to us, with $\hat{k}_a = \hat{z} = -\hat{k}_b$, we see at once that $\tilde{x}_a\tilde{y}_b - \tilde{y}_a\tilde{x}_b$ contains only allowable polarizations, and that the other two possibilities in (130) contain in each term a factor $\tilde{z}_a$ or $\tilde{z}_b$ that would be impossible. That is, only the Ψ_{Az} part in the initial state would give rise to photon pairs observable with the photon detectors on the $\pm z$-axes.

If we want to describe the entire photon field as it is emitted from the atomic source, in whatever directions these photons may travel, our next step is to replace the polarizations $\hat{x}_l$ of the $\tilde{x}_l$-waves in (130) or in $\sum_i \sum_j \varepsilon_{\mu ij}(\tilde{x}_i)_a (\tilde{x}_j)_b$ by polarizations $\hat{x}_l - \hat{k}(\hat{k}\cdot\hat{x}_l)$. We thus obtain the three photon polarization wave functions

$$(\psi_{110}^{\text{pol}})^{\mu}_{n_1 n_2} = \sum_i \sum_j \varepsilon_{\mu ij}(\delta_{in_1} - \hat{k}_{1i}\hat{k}_{1n_1})(\delta_{jn_2} - \hat{k}_{2j}\hat{k}_{2n_2})$$
$$= \varepsilon_{\mu n_1 n_2} - \sum_j \varepsilon_{\mu n_1 j}\hat{k}_{2j}\hat{k}_{2n_2} - \sum_j \varepsilon_{\mu j n_2}\hat{k}_{1j}\hat{k}_{1n_1} + \sum_i \sum_j \varepsilon_{\mu ij}\hat{k}_{1i}\hat{k}_{2j}\hat{k}_{1n_1}\hat{k}_{2n_2}, \tag{131}$$

which is comparable to eq. (110) in the 010 case, and which is not yet complete, as it is not normalized, and as we did not yet note that k_1 and k_2 have the specific values k_a and k_b. This information, however, is not to be indicated now by the symmetric factor $\Delta_{ab}(\mathbf{k}_1, \mathbf{k}_2)$ given by (113), because the $\sum_i \sum_j \varepsilon_{\mu ij}(\tilde{x}_i)_a (\tilde{x}_j)_b$ from which we started was antisymmetric in a and b. The wavefunctions $F^{\mu}_{n_1 n_2}(\mathbf{k}_1, \mathbf{k}_2)$, however, must be symmetric between $\{\mathbf{k}_1, n_1\}$ and $\{\mathbf{k}_2, n_2\}$.

Therefore, we note that (131) with our use of $\hat{k}_1$ and $\hat{k}_2$, instead of the $\mathbf{k}_a$ and $\mathbf{k}_b$ suggested by (130), is correct only if $\hat{k}_1 = \hat{k}_a$ and $\hat{k}_2 = \hat{k}_b$. So behind (131) we place merely a factor $\delta_{\mathbf{k}_1, \mathbf{k}_a}\delta_{\mathbf{k}_2, \mathbf{k}_b}$. Then the result thus obtained is finally symmetrized between $\{\mathbf{k}_1, n_1\}$ and $\{\mathbf{k}_2, n_2\}$.

In the 010 case this two-step procedure would have led to the same result as we obtained there by simply multiplying (110) by (113). In our present 110 case, however, the two-step procedure is the only way for getting the correct final result.

We thus obtain for the 110- case, for the part with chosen directions[22] of $\mathbf{k}_a$ and $\mathbf{k}_b$:

$$F^{\mu}_{n_1 n_2}(\mathbf{k}_1, \mathbf{k}_2) = C \sum_i \sum_j \varepsilon_{\mu ij}(\delta_{in_1} - \hat{k}_{1i}\hat{k}_{1n_1})(\delta_{jn_2} - \hat{k}_{2j}\hat{k}_{2n_2})[\delta_{\mathbf{k}_1, \mathbf{k}_a}\delta_{\mathbf{k}_2, \mathbf{k}_b} - \delta_{\mathbf{k}_1, \mathbf{k}_b}\delta_{\mathbf{k}_2, \mathbf{k}_a}]$$
$$= C\left[\varepsilon_{\mu n_1 n_2} - \sum_j \varepsilon_{\mu n_1 j}\hat{k}_{bj}\hat{k}_{bn_2} + \sum_j \varepsilon_{\mu n_2 j}\hat{k}_{aj}\hat{k}_{an_1} + \sum_i \sum_j \varepsilon_{\mu ij}\hat{k}_{ai}\hat{k}_{bj}\hat{k}_{an_1}\hat{k}_{bn_2}\right]\delta_{\mathbf{k}_1, \mathbf{k}_a}\delta_{\mathbf{k}_2, \mathbf{k}_b}$$
$$+ C\left[-\varepsilon_{\mu n_1 n_2} + \sum_j \varepsilon_{\mu n_1 j}\hat{k}_{aj}\hat{k}_{an_2} - \sum_j \varepsilon_{\mu n_2 j}\hat{k}_{bj}\hat{k}_{bn_1} - \sum_i \sum_j \varepsilon_{\mu ij}\hat{k}_{bi}\hat{k}_{aj}\hat{k}_{bn_1}\hat{k}_{an_2}\right]\delta_{\mathbf{k}_1, \mathbf{k}_b}\delta_{\mathbf{k}_2, \mathbf{k}_a}, \tag{132}$$

where C is a normalization factor. From the first form (132), using (112), we easily calculate $F^{\mu}_{\eta_1 \eta_2}(\mathbf{k}_1, \mathbf{k}_2)$ by eq. (115) and find[22]

$$F^{\mu}_{\eta_1 \eta_2}(\mathbf{k}_1, \mathbf{k}_2) = C \sum_i \sum_j \varepsilon_{\mu ij}(e^{\eta_1}_{\mathbf{k}_1})^*_i (e^{\eta_2}_{\mathbf{k}_2})^*_j [\delta_{\mathbf{k}_1, \mathbf{k}_a}\delta_{\mathbf{k}_2, \mathbf{k}_b} - \delta_{\mathbf{k}_1, \mathbf{k}_b}\delta_{\mathbf{k}_2, \mathbf{k}_a}]. \tag{133}$$

The transversality (88) of $F^{\mu}_{n_1n_2}$ follows most easily from the first form of (132). Under inversion of the signs of $\mathbf{k}_1$, $\mathbf{k}_2$, $\mathbf{k}_a$, and $\mathbf{k}_b$, eq. (132) is invariant, so that it represents an *even* photon state. This is as it should be, as the atom goes from a P_1 state ($l = 1$, parity $-$) to a P_0 state ($l = 1$, parity $-$), so it does not change parity.

The symmetry of $F^{\mu}_{n_1n_2}(\mathbf{k}_1, \mathbf{k}_2)$ between $\{\mathbf{k}_1, n_1\}$ and $\{\mathbf{k}_2, n_2\}$ was built in. Note, however, the antisymmetry of $F^{\mu}_{n_1n_2}(\mathbf{k}_1, \mathbf{k}_2)$ between $\mathbf{k}_a$ and $\mathbf{k}_b$ when these have given directions.[23]

The three states $F^{\mu}_{n_1n_2}$ under rotation transform like x, y, z. By combining them linearly like x, y, z are combined linearly in (103), we can construct a new set of $F^{\mu}_{n_1n_2}$, with $\mu = 1, 0, -1$ instead of $\mu = x, y, z$. These will transform under rotations like the Y^{μ}_{1}. In other words, our three wavefunctions form a representation with quantum number $j = 1$. This also means that the square of the total field angular momentum $\mathbf{J}$ has the value $j(j+1)\hbar^2 = 2\hbar^2$:

$$\mathbf{J}^2_{\text{qu}}\psi^{(F)\mu}_{110}(q_f) = 2\hbar^2\, \psi^{(F)\mu}_{110}(q_f), \tag{134}$$

where $\psi^{(F)\mu}_{110}(q_f)$ is obtained from $F^{\mu}_{n_1n_2}(\mathbf{k}_1, \mathbf{k}_2)$ by (75). As we derived (117) from (116), from (133) we find[22, 23]

$$[\psi^{(F)\mu}_{110}(q_f)]_{\mathbf{k}_a,\mathbf{k}_b} = \psi^{(F)}_{\text{vac}}(q_f)\,(2Cc/\hbar)\,(2k_ak_b)^{\frac{1}{2}} \sum_{\eta}\sum_{\eta'} (\mathbf{e}^{\eta *}_{\mathbf{k}_a} \times \mathbf{e}^{\eta' *}_{\mathbf{k}_b})_{\mu}\, q^{*}_{\mathbf{k}_a,\eta}\, q^{*}_{\mathbf{k}_b,\eta'}\,. \tag{135}$$

Opposite photon directions

We consider again the special case $\hat{k}_a = \hat{z} = -\hat{k}_b$. In that case, we find from the second form of (132), in the form of 3×3 matrices with rows and columns labeled by n_1 and n_2:

$$F^{x}_{n_1n_2}(k_a\hat{z}, -k_b\hat{z}) = \left[\begin{pmatrix}0&0&0\\0&0&1\\0&-1&0\end{pmatrix} - \begin{pmatrix}0&0&0\\0&0&1\\0&0&0\end{pmatrix} + \begin{pmatrix}0&0&0\\0&0&0\\0&1&0\end{pmatrix} + 0\right]\delta_{\mathbf{k}_1,\mathbf{k}_a}\delta_{\mathbf{k}_2,\mathbf{k}_b}$$
$$+\left[\begin{pmatrix}0&0&0\\0&0&-1\\0&1&0\end{pmatrix} + \begin{pmatrix}0&0&0\\0&0&1\\0&0&0\end{pmatrix} - \begin{pmatrix}0&0&0\\0&0&0\\0&1&0\end{pmatrix} + 0\right]\delta_{\mathbf{k}_1,\mathbf{k}_b}\delta_{\mathbf{k}_2,\mathbf{k}_a} = 0. \tag{136x}$$

Similarly, $F^{y}_{n_1n_2}(k_a\hat{z}, -k_b\hat{z}) = 0$, while

$$F^{z}_{n_1n_2}(k_a\hat{z}, -k_b\hat{z}) = \begin{pmatrix}0&1&0\\-1&0&0\\0&0&0\end{pmatrix}(\delta_{\mathbf{k}_1,\mathbf{k}_a}\delta_{\mathbf{k}_2,\mathbf{k}_b} - \delta_{\mathbf{k}_1,\mathbf{k}_b}\delta_{\mathbf{k}_2,\mathbf{k}_a}). \tag{136z}$$

This confirms our remarks below eq. (130), that only Ψ_{Az} is going to provide photon pairs along the $\pm z$-axis, and that for these polarizations occur in the combinations $n_1 = x$, $n_2 = y$, and vice versa with an opposite sign in the wavefunction. In our abbreviated notation, therefore, $(\psi^{\text{pol}}_{110})^z$ in that case is given by

$$\left.\begin{aligned}(\psi^{\text{pol}}_{110}) &= \tilde{x}_1\tilde{y}_2 - \tilde{x}_2\tilde{y}_1 \quad \text{for} \quad \mathbf{k}_1 = k_a\hat{z},\ \mathbf{k}_2 = -k_b\hat{z},\\ &= \tilde{x}_2\tilde{y}_1 - \tilde{x}_1\tilde{y}_2 \quad \text{for} \quad \mathbf{k}_1 = -k_b\hat{z},\ \mathbf{k}_2 = k_a\hat{z},\\ &= \tilde{x}_a\tilde{y}_b - \tilde{x}_b\tilde{y}_a \quad \text{in either case.}\end{aligned}\right\} \tag{137}$$

The polarizations are crossed; parallel polarizations do not occur in this case.

[23] Averaged over all $\hat{k}_a$ and $\hat{k}_b$ this would give zero in case $|k_a| = |k_b|$. See, however, Appendix F, why the final photon state is not such a superposition of waves with all directions of $\hat{k}_a$ and $\hat{k}_b$.

A quick derivations of the results obtained

A quick but somewhat sloppy derivation of the important results of Section 2.7 is found in Appendix E. It uses conservation of angular momentum and of parity.

2.8. The observable measured by an ideal linear polarization filter

We consider here light traveling in the $\pm z$-direction and perpendicularly incident upon polarizing plates or films. If in the way of Section 5.2 of Part II the wavefunction ψ of an incident photon is decomposed by

$$\psi_{\text{in}} = c_{x'}\tilde{x}' + c_{y'}\tilde{y}' \quad (\text{with } |c_{x'}|^2 + |c_{y'}|^2 = 1) \tag{138}$$

into orthogonal polarized photon wavefunctions of which $\tilde{x}'$ corresponds to the polarization direction $\hat{x}'$ of the polarizer, the polarizer will transmit most of $c_{x'}\tilde{x}'$ and little of $c_{y'}\tilde{y}'$. An *ideal* polarization filter is one behind which the incident wave (138) is changed into

$$\psi_{\text{out}} = c_{x'}\tilde{x}' + c_{y'}\tilde{0} \tag{139}$$

if $\tilde{0}$ is the wavefunction for absence of a photon. In the following quantum-theoretical considerations, we shall mainly consider *ideal* filters.

Justification that (139) is the effect of an ideal filter upon the incident wave (138) is provided by Malus's law, that is, by the fact that the intensity of light behind two filters with an angle θ between their polarization direction is $\cos^2\theta$ times the intensity behind the first filter:

$$\mathcal{J}_2 = \mathcal{J}_1 \cos^2 \theta. \tag{140}$$

This fact is explained by (139), if $\hat{x}'$ and $\hat{x}''$ are the two polarization directions, because two applications of (138)–(139) yield

$$\begin{aligned} \psi_0 = c_{x'}\tilde{x}' + c_{y'}\tilde{y}' &\rightarrow \psi_1 = c_{x'}\tilde{x}' + c_{y'}\tilde{0} = c_{x'}[\tilde{x}'' \cos\theta - \tilde{y}'' \sin\theta] + c_{y'}\tilde{0} \\ &\rightarrow \psi_2 = c_{x'}[\tilde{x}'' \cos\theta - \tilde{0} \sin\theta] + c_{y'}\tilde{0}, \end{aligned} \tag{141}$$

so that, for many separate incident photons,

$$\mathcal{J}_1/\mathcal{J}_0 = |c_{x'}|^2 \quad \text{and} \quad \mathcal{J}_2/\mathcal{J}_0 = |c_{x'} \cos\theta|^2 \tag{142}$$

are the probabilities per incident photon for traversing the first filter or both filters.

We see from (138)–(139) that the observable measured by a nondepolarizing polarization filter polarized in the x'-direction is characterized by an operator of which the eigenfunctions are $\tilde{x}'$ and $\tilde{y}'$. The observable itself is a quantity that can have two values: It has one value (say X') if the incident photon was polarized in the filter's polarization direction $\hat{x}'$; it has a different value (say Y') if the incident photon was polarized in the direction $\hat{y}'$ perpendicular to the filter's polarization direction. Thus (138) is an expansion of the initial wavefunction in terms of the eigenfunctions of the observable measured; $|c_{x'}|^2$ and $|c_{y'}|^2$ are the probabilities for finding either result. From (139) we see that the measurement is a *reproducible* one when the value X' is found, but is by absorption of the photon an *irreproducible* measurement when the observable has the value Y'.

Warning

Note that, prior to putting in a filter with direction $\hat{x}'$, the observable has neither the value X' nor the value Y'. For instance, suppose the incident light emerged from a filter with polarization direction $\hat{x}$, so that $\psi_{\text{in}} = \hat{x}$, and $c_{x'} = \cos(\angle \hat{x}, \hat{x}')$, $c_{y'} = \cos(\angle \hat{x}, \hat{y}')$. The light incident upon the second filter then is all polarized in the $\hat{x}$-direction. According to quantum theory it is meaningless to say that it is polarized either in the $\hat{x}'$- or in the $\hat{y}'$-direction (instead of either in the $\hat{x}''$- or the $\hat{y}''$-direction) before we have chosen a filter with polarization direction $\hat{x}'$ (rather than $\hat{x}''$) by which to analyze the light.

The importance of this warning is most easily seen if we consider again the derivation of Malus's law. Let α' and α'' be the angles of $\hat{x}'$ and of $\hat{x}''$ in the x, y-plane measured from $\hat{x}$ toward the y-axis. The light emerging from the first filter has for each photon a wave-function

$$\tilde{x}' = \tilde{x} \cos \alpha' + \tilde{y} \sin \alpha'. \tag{143}$$

It would now be incorrect, with the next filter's polarization direction along $\hat{x}''$, to say that (143) represents a state polarized either along $\hat{x}$ (with probability $\cos^2 \alpha'$), or along $\hat{y}$ (with probability $\sin^2 \alpha'$), because such reasoning would give for passing filter 2 a probability

$$(\cos^2 \alpha') \cos^2 \alpha'' + (\sin^2 \alpha') \sin^2 \alpha'', \tag{144}$$

where $\cos^2 \alpha''$ and $\sin^2 \alpha''$ are the probabilities for a photon with wave function $\tilde{x}$ or $\tilde{y}$ to pass through the second filter. The correct answer is obtained by decomposing (143) in terms of the eigenfunctions $\tilde{x}''$ and $\tilde{y}''$ of the second filter:

$$\begin{aligned} \tilde{x}' &= (\tilde{x}'' \cos \alpha'' - \tilde{y}'' \sin \alpha'') \cos \alpha' + (\tilde{x}'' \sin \alpha'' + \tilde{y}'' \cos \alpha'') \sin \alpha' \\ &= \tilde{x}''(\cos \alpha'' \cos \alpha' + \sin \alpha'' \sin \alpha') + \tilde{y}'' (\cos \alpha'' \sin \alpha' - \sin \alpha'' \cos \alpha'), \end{aligned} \tag{145}$$

and then interpreting the absolute square of the coefficient of $\tilde{x}''$ as the probability of passing the filter:

$$\begin{aligned} |\cos \alpha'' \cos \alpha' + \sin \alpha'' \sin \alpha'|^2 &= \cos^2 \alpha'' \cos^2 \alpha' + 2 \cos \alpha'' \sin \alpha'' \cos \alpha' \sin \alpha' + \sin^2 \alpha'' \sin^2 \alpha' \\ &= \cos^2 (\alpha'' - \alpha') = \cos^2 \theta. \end{aligned} \tag{146}$$

Comparison of Malus's law (146) with the erroneous result (144) shows that in the latter the "interference terms" are missing.

2.9. Polarization correlations between different photons passing through different polarization filters

We now return to the consideration of the pair of photons emitted by a single atom cascading to lower energy levels as in the 0–1–0 or the 1–1–0 case. For this pair of photons we calculated the photon wavefunction in Section 2.7.

The coincidence counting arrangement

Photon detectors are placed in the $+z$- and the $-z$-direction from the emitting atoms, with color filters so that the detector at $+z$ is sensitive for the first photon only, and the detector at $-z$ only for the second photon. We can measure the coincidence counting rate R_0 of the two detectors. Successive counts of photons 1 and 2 are counted as a "coincidence," if photon 2 is counted within a certain adjustable time length after photon 1 has been counted. The measurement may also be arranged in such a way that the amount of time lag of count 2 after count 1 is measured. By measuring the distribution of this time lag we can measure the life time of the intermediate atomic state between the two photon emissions. The occupancy of this level decays exponentially after the first photon has been emitted by the atom. This exponential decay is observed in the frequency of coincidence counts with increasing time lag. The observations, however, also show a background that does not decay exponentially and that is independent of the time lag. This background coincidence count should be subtracted from the observed coincidence count, as it is due to chance coincidences between photons from different atoms and other irrelevant phenomena.

Coincidence counts with one polarization filter in place

Say we place perpendicularly to the $+z$-axis a polarization filter with its polarization direction making an angle a with the x-direction. The coincidence rate drops from R_0 to $R_1(a)$. If we place the filter, instead, so as to intercept photon 2 on the $-z$-axis, and b is the angle of its polarization direction with the x-axis (toward the y-axis), the corresponding coincidence counting rate is called $R_2(b)$.

According to quantum theory,

$$R_1(a) = R_2(b) = \tfrac{1}{2}R_0, \tag{147}$$

as the light from the atom is unpolarized, and therefore has a 50% chance of passing through an ideal polarization filter of given orientation.

Two-filter coincidence counting rate

Let $R(a, b)$ be the coincidence counting rate with polarization filters in directions a and b intercepting both photons. Then, if the polarizations of both photons would have been uncorrelated, we might expect $R(a, b) = \frac{1}{4}R_0$. (One factor $\frac{1}{2}$ for each filter.)

This, however, is not found experimentally, and should not be expected to be found, because the polarizations of the two photons *are* correlated by the two-photon wavefunction calculated in Section 2.7. We shall normalize them here as in (138), so that for our experimental arrangement our previous results (109) and (137) become

$$\psi_{010} = \sqrt{\tfrac{1}{2}}\,(\tilde{x}_1\tilde{x}_2+\tilde{y}_1\tilde{y}_2), \quad \psi_{110} = \sqrt{\tfrac{1}{2}}\,(\tilde{x}_1\tilde{y}_2-\tilde{y}_1\tilde{x}_2). \tag{148}$$

Calculation of $R(a, b)$ according to quantum theory

With the two filters in place, introduce coordinate systems $\{x', y', z'\}$ and $\{x'', y'', z''\}$ with $\hat{z}' \,||\, \hat{z} \,||\, \hat{z}''$, with $\hat{x}'$ in the polarization direction (a) of the first filter, and with $\hat{x}''$ in the polarization direction (b) of the second filter. The coincidence counting arrangement then measures an observable of which the eigenfunctions are $\tilde{x}_1'\tilde{x}_2''$, $\tilde{x}_1'\tilde{y}_2''$, $\tilde{y}_1'\tilde{x}_2''$, and $\tilde{y}_1'\tilde{y}_2''$. Counts occur with a probability proportional to the absolute square of the coefficient of $\tilde{x}_1'\tilde{x}_2''$ in the expansion of ψ in terms of these four eigenfunctions.

Since ψ in Section 2.7 was expressed in terms of $\tilde{x}$ and $\tilde{y}$, we obtain the expansion needed by using

$$\left.\begin{aligned} \tilde{x}_1 &= \tilde{x}_1' \cos a - \tilde{y}_1' \sin a, & \tilde{y}_1 &= \tilde{x}_1' \sin a + \tilde{y}_1' \cos a, \\ \tilde{x}_2 &= \tilde{x}_2'' \cos b - \tilde{y}_2'' \sin b, & \tilde{y}_2 &= \tilde{x}_2'' \sin b + \tilde{y}_2'' \cos b. \end{aligned}\right\} \tag{149}$$

Inserting (149) in ψ_{010} given by (148), we obtain (in front of the two filters)

$$\psi_{010} = \sqrt{\tfrac{1}{2}}\cdot(\tilde{x}_1'\tilde{x}_2'' + \tilde{y}_1'\tilde{y}_2'')\cos(a-b) + \sqrt{\tfrac{1}{2}}\cdot(\tilde{x}_1'\tilde{y}_2'' - \tilde{y}_1'\tilde{x}_2'')\sin(a-b), \tag{150}$$

while in the 1–1–0 case of the mercury atom, (148) gives

$$\psi_{110} = \sqrt{\tfrac{1}{2}}\cdot(\tilde{x}_1'\tilde{y}_2'' - \tilde{y}_1'\tilde{x}_2'')\cos(a-b) - \sqrt{\tfrac{1}{2}}\cdot(\tilde{x}_1'\tilde{x}_2'' + \tilde{y}_1'\tilde{y}_2'')\sin(a-b). \tag{151}$$

As the coincidence rate, with many atoms emitting pairs of photons, was R_0 in absence of the filters, the coincidence rate with the filters in place will be

$$R(a, b) = w(a, b)R_0, \tag{152}$$

where $w(a, b)$ is the absolute square of the coefficient of $\tilde{x}_1'\tilde{x}_2''$, so that

$$R(a, b)/R_0 = \tfrac{1}{2}\cos^2(a-b) \text{ in the 0–1–0 case,} \tag{153a}$$
$$R(a, b)/R_0 = \tfrac{1}{2}\sin^2(a-b) \text{ in the 1–1–0 case.} \tag{153b}$$

2.10. The nonlocality paradox

We find from (153a) and from (153b) the nonlocality aspect of the Einstein–Podolsky–Rosen paradox, as $R(a, b) = 0$ for $a-b = 90°$ in the 0–1–0 case, or for $a-b = 0°$ in the 1–1–0 case. Thus, at these angles between the two polarization filters, coincidences are not permitted by quantum theory. It, then, is hard to understand how filter 2 at angle b knows it should not pass a photon if filter and photon could not know that photon 1 has met a filter at the corresponding angle a and that it has passed that filter. Hidden-variables theories of the second kind are inventions which are to explain in deterministic language the existence of correlations of this general type. In Chapter 3 we will give some simple examples of such inventions, and will find that they predict correlations which do not agree with the quantum-mechanical prediction. In Chapter 4 we will discover that this lack of agreement with quantum theory is unavoidable, and how it therefore is possible to test experimentally whether *any* such scheme of the second kind could be the true law of nature or not, without having to specify the particularities of that scheme.

CHAPTER 3

SOME EXAMPLES OF HIDDEN-VARIABLES THEORIES OF THE SECOND KIND FOR "EXPLAINING" POLARIZATION CORRELATIONS

THE leading idea in theories of the second kind for pairs of photons emitted by an atom is that they assume that the two photons will have some properties described by one or more parameters which for the two photons will have correlated values, and that these values will help the photon decide, upon reaching a polarization filter, whether or not it should pass through that filter.

3.1. Minimum requirements

In order to give such theories an air of possibility, we want to invent them in such a way that at least *in the following simple cases* they will predict the same as quantum theory predicts:

(A) Of unpolarized light reaching one ideal polarization filter, half should be transmitted by the filter and half should be stopped. [This will explain (147).]

(B) If a second filter is placed behind the first one, Malus's $\cos^2$ law should be valid.

However, we shall not, *a priori*, postulate that a $\cos^2$ law like (153a) would hold for correlations between the transmission or absorption of *different* photons in *different* beams. Instead, we shall apply each of these hidden-variables theories and find out what *do* they predict for these correlations. We then will find results *different* from (153a) in the 0–1–0 case, and *different* from (153b) in the 1–1–0 case.

3.2. Internal and external hidden variables

As mentioned above, some of the hidden variables of the second kind will be carried by the photons from their creation in the atom toward the polarization filters. It is, of course, thinkable that such parameters will be affected as the photon would pass through one of these filters, so that the polarized photons leaving a filter may have these parameters changed from what they were before reaching the filter.

The hidden variables carried by the photon as the photon is freely traveling through space are called the photon's *internal* hidden variables.

What happens to a photon upon reaching a filter depends not only upon the photon but also upon the filter. For one thing, it will depend upon the filter polarization direction. It might depend upon other microscope parameters not known to us at this time. These would be hidden variables internal to the filter but external to the photon.

The hypothesis of hidden-variables theories of the second kind is that, when a photon reaches a polarization filter, the angle of the filter's polarization direction together with all the internal and external hidden variables of the photon will determine locally (without asking questions about a second photon at a distant filter) whether or not the photon will traverse the filter.

3.3. Photon polarization as a possible hidden variable

None of the photons passing through a filter "at angle" a can pass through a second filter "at angle" $b = a+90°$ behind it. We thus could imagine that, among many other hidden variables carried by a photon, one is its polarization α, which we may define as the direction a of the last filter it passed before, or, if it did not pass such a filter, a direction perpendicular to the direction b of some filter that is necessarily going to stop that photon.

We then can easily satisfy the minimum requirements (A) and (B) by making the following additional assumption: consider all possible values of the set of hidden variables that determine whether or not a photon with polarization hidden variable α is going to pass a filter in the position b. Among these values, there shall be a probability

$$P(\alpha, b) = \cos^2 (b-\alpha) \tag{154}$$

for those values which determine that the photon shall pass the filter, and a probability $\sin^2 (b-\alpha)$ for those values of the hidden variables that will stop the photon in the filter.

By $\alpha = a$, this postulate will reproduce Malus's law for photons that passed through a preceding filter at angle a.

Like α became a for the photon as it passed through the preceding filter, the photon's polarization hidden variable will change from α to $\beta = b$ for those photons that pass the second filter, and for these photons there will be a probability $\cos^2 (c-\beta)$ that their hidden variables will have values which make such a photon pass a third filter with polarization direction c. All angles a, b, c, α, β, γ, are here in the x, y-plane from the x-axis toward the y-direction, for light propagating in the z-direction.

The polarization hidden variable here introduced is, of course, a quantity which does not exist in quantum theory. While in the above hidden-variables theory Malus's law was a statistical law about the probability that the hidden variables of a photon of polarization α and the hidden variables of a filter at polarization angle b would agree upon letting a photon pass, in quantum theory no such a thing as α even exists. Indeed, Malus's law followed from the decomposition of a *photon wavefunction* $(\tilde{a})$ along the eigenfunctions $(\tilde{b})$ and $(\widetilde{b+90°})$ of the second filter, where, for any angle φ, $(\tilde{\varphi})$ stands for $\tilde{x} \cos \varphi + \tilde{y} \sin \varphi$.

3.4. Correlations between two photons in hidden-variables theory

In a hidden-variables theory of the second kind, one refuses to believe that the question whether photon 1 passes through filter 1 would depend causally upon what filter is met by photon 2. As the components of ψ_{010} or of ψ_{110} along the eigenfunctions $\tilde{x}_1' = (\tilde{a})$ and $\tilde{y}_1' = \widetilde{(a+90°)}$ of Filter One did depend upon the wave functions of photon 2, hidden-variables theories of the second kind therefore in the case of spatially separated filters deny the validity of the quantum-theoretical prediction that the probabilities for passing or not passing filter 1 would be directly determined by the absolute squares of these amplitudes. Instead, these probabilities are assumed to be determined by statistics of the hidden-variables values, and the correlations in quantum theory derived from the two-photon wave-function ψ_{010} or ψ_{110} are replaced by correlations between the hidden variables of the two photons.

We shall here, as examples, consider only some of the most simple correlations of this kind that one could imagine.

In particular, consider the case that the polarizations α and β of the photons 1 and 2 would be correlated in such a way that

$$I(\alpha, \beta)\, d\alpha\, d\beta/(2\pi)^2 \tag{155a}$$

would be the probability that the two photons from one atom carry polarizations α and β, while all other hidden variables might be randomly distributed. The sum of these probabilities should be $= 1$, so that

$$\oint \oint I(\alpha, \beta)\, d\alpha\, d\beta/(2\pi)^2 = 1. \tag{155b}$$

Again, for $I(\alpha, \beta)$ we shall consider only the simplest possibilities. For the 0–1–0 case, which according to (153a) is characterized by a preference for parallel polarization of the two photons emitted, we will consider here only the following two extreme polarization correlations:

$$\text{(I)} \qquad I(\alpha, \beta) = 2\pi\, \delta(\alpha-\beta), \tag{156a}$$

in which case it is certain that $\beta = \alpha$, while α is randomly distributed; or, alternatively:

$$\text{(II)} \qquad I(\alpha, \beta) = 2\cos^2(\alpha-\beta), \tag{156b}$$

in which case there is a preference for $\alpha = \beta$, and α and β are never in orthogonal directions. Both (156a) and (156b) satisfy (155b).

The reader may want to try out additionally some *other* choices of $I(\alpha, \beta)$ for the following calculations.

For the 1–1–0 case, for which there is according to (153b) a preference for α and β perpendicular to each other, we shall in the following consider only the two possibilities

$$\text{(I)} \qquad I(\alpha, \beta) = 2\pi\, \delta(\alpha-\beta-\tfrac{1}{2}\pi), \tag{157a}$$

$$\text{(II)} \qquad I(\alpha, \beta) = 2\sin^2(\alpha-\beta). \tag{157b}$$

For the correlations which in hidden-variables theory replace the correlation (153b) of quantum theory for the 1–1–0 case, the assumptions (157) will lead to results which are similar to the results which according to (156) would replace (153a) for the 0–1–0 case, except for a 90° shift of these results as a function of $(a-b)$, interchanging $\sin^2(a-b)$ and $\cos^2(a-b)$ in the results.

3.5. Coincidence correlations

Consider a pair of photons from one atom, with polarization hidden variables α and β. Let a and b be the orientations of the polarization directions of the two filters.

Among all hidden-variables values there is by (154) a probability $\cos^2(\alpha-a)$ that photon 1 will pass filter 1, and there is a probability $\cos^2(\beta-b)$ that photon 2 will pass filter 2.

If the choices of hidden variables for which these two events occur are not correlated among each other, there is a probability

$$P_{\text{s.p.}} = \cos^2(\alpha-a)\cos^2(\beta-b) \tag{158}$$

that both photons will *simultaneously pass* the filters. If the counters are perfectly effective, (158) would be the probability of a coincidence count in this case of given α, β, a, and b. In that case, it would follow that

$$R(a,b) = R_0 \oint\oint (2\pi)^{-2}\, d\alpha\, d\beta\, I(\alpha,\beta)\cos^2(\alpha-a)\cos^2(\beta-b). \tag{159}$$

It is, however, thinkable that somehow the same hidden variables that make photon 1 pass filter 1 have a tendency of making photon 2 pass filter 2. If this were so, the coincidence rate $R(a,b)$ might be enhanced.

Or, it might be the other way around, so that hidden variables favorable to letting photon 1 pass are unfavorable to photon 2. This might depress $R(a,b)$.

We can easily find upper limits both for favorable and for unfavorable coincidence correlations of this kind. Obviously, the number of photons that pass both filters cannot be larger than what passes either filter. That is, if the probability $P_{\text{s.p.}}$ for simultaneously passing is not equal to the value (158) found in absence of coincidence correlations, we certainly must have

$$P_{\text{s.p.}} \leqslant P_{\max} = \text{Min}\,\{\cos^2(\alpha-a), \cos^2(\beta-b)\}. \tag{160}$$

On the other hand, there is also a minimum value for $P_{\text{s.p.}}$. Consider the cases where photon 1 passes but yet there is no coincidence. For given a, b, α, β, the probability for this is $\cos^2(\alpha-a)-P_{\text{s.p.}}$. Similarly, consider the cases where photon 2 passes. The probability for these is $\cos^2(\beta-b)$. Since the latter cases were not included in the former cases, the sum $[\cos^2(\alpha-a)-P_{\text{s.p.}}]+\cos^2(\beta-b)$ is the probability that *not both photons are stopped.* This probability cannot be more than 100%, so that

$$\cos^2(\alpha-a)+\cos^2(\beta-b)-P_{\text{s.p.}} < 1$$

or

$$P_{\text{s.p.}} > \cos^2(\alpha-a)+\cos^2(\beta-b)-1.$$

Since, also, $P_{\text{s.p.}}$ cannot be negative, we find

$$P_{\text{s.p.}} > P_{\min} = \text{Max}\,\{0, [\cos^2(\alpha-a)+\cos^2(\beta-b)-1]\}. \tag{161}$$

In general, then, in these simplest theories we have

$$R(a, b)/R_0 = (2\pi)^{-2} \oint\oint d\alpha\, d\beta\, I(\alpha, \beta) P_{s.p.}(\alpha, \beta, a, b), \tag{162}$$

and we can consider as especially simple examples for the 0–1–0 case the ones where $I(\alpha, \beta)$ has the value (156a) or (156b) [or whatever other function the reader wants to use for $I(\alpha, \beta)$ instead], while for $P_{s.p.}$ as a function of α, β, a, and b we might consider the three extreme cases, (158), and $P_{s.p.} = P_{max}$ (160), and $P_{s.p.} = P_{min}$ (161).

3.6. Results for $R(a, b)$, for the sample hidden-variables theories considered

For all cases considered above, the double integral (162) is easy to evaluate. For the maximum polarization correlation (156a), it reduces to a single integral over $\alpha = \beta$. We will here list the results of these single integrals for the three choices made above for $P_{s.p.}$, and the result of the double integral using (156b) with (158). We leave it to the reader to try other combinations.

We find, *for the 0–1–0 case:*

For (156a), $\alpha = \beta$:

With $P_{s.p.} = \cos^2(\alpha - a)\cos^2(\beta - b)$:

$$R(a, b)/R_0 = \tfrac{1}{8} + \tfrac{1}{4}\cos^2(a-b). \tag{163}$$

With $P_{s.p.} = P_{max}$:

$$R(a, b)/R_0 = \tfrac{1}{2} - |\sin(a-b)|/\pi. \tag{164}$$

With $P_{s.p.} = P_{min}$:

$$R(a, b)/R_0 = |\cos(a-b)|/\pi. \tag{165}$$

For (156b), $I(\alpha, \beta) = 2\cos^2(\alpha - \beta)$:

With $P_{s.p.} = \cos^2(\alpha - a)\cos^2(\beta - b)$:

$$R(a, b)/R_0 = \tfrac{3}{16} + \tfrac{1}{8}\cos^2(a-b). \tag{166}$$

The results (153a), (163), (164), and (165) are shown in Fig. 3.1. The curve for (166) would be flatter than for (163), varying between 3/16 and 5/16 instead of between 2/16 and 6/16.

For the 1–1–0 case, the replacement of (156a, b) by (157a, b) would shift the curves of Fig. 3.1 by 90° to the side, corresponding to an interchange of cos and sin in eqs. (163)–(166).

Discussion of the results

Looking at Fig. 3.1, we see that the quantum-theoretical prediction (153a) varies more pronouncedly with the angle $(a-b)$ than any of the hidden-variables predictions plotted. In the 0–1–0 case, the quantum-theoretical curve is at $22\frac{1}{2}°$ the highest, and at $67\frac{1}{2}°$ the lowest one.[24]

[24] In this connection, see eqs. (203) and (204a, b) in Section 4.7.

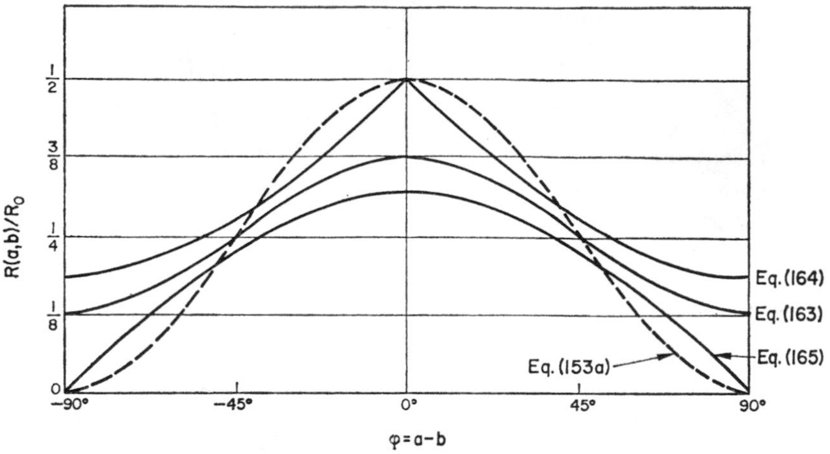

FIG. 3.1. Two-photon polarization correlation for the 0–1–0 case (parallel correlation). Solid lines = hidden-variables theories of the second kind. Broken line = quantum theory [eq. (153a)]

At 0°, only the $\alpha = \beta$, $P_{s.p.} = P_{max}$ hidden-variables curve reaches up to $R(a, b) = \frac{1}{2} R_0$ like the quantum-mechanical curve does, and at 90° only the $\alpha = \beta$, $P_{s.p.} = P_{min}$ curve reaches as low as 0, like the quantum-theoretical one.

The latter facts can easily be understood qualitatively. Even with complete correlation (156a) between the hidden-variable polarizations of the two photons, we still by hidden-variables theories of this type expect to find some coincidence counts at crossed filters ($a-b = 90°$), because the value of $\alpha = \beta$ is stochastically distributed, so there would be pairs of photons with α and β making slant angles with both filters, so both photons would have a chance of coming through, except perhaps in the case of the strongest anticoincidence correlation given by $P_{s.p.} = P_{min}$. Similarly, for the special case of parallel filters ($a = b$) in our 0–1–0 case, in quantum theory we would expand ψ according to (150) in terms of polarizations parallel and perpendicular to the filters and would find probabilities $1 \times 1 = 1$ and $0 \times 0 = 0$ for coincidences for these two cases, so on an average we find $R(a, b)/R_0 = \frac{1}{2}$. In hidden variables theory with full polarization correlation ($\alpha = \beta$), however, α and β can also become slant angles for which $\cos^2(\alpha - a) \cos^2(\beta - b) = [\cos^2(\alpha - a)]^2$ becomes the square of a fraction, and the average of squares of fractions tends to be *less* than $\frac{1}{2}$. Only maximum coincidence correlation $P_{s.p.} = P_{max}$ can in this case restore the quantum-theoretical value $\frac{1}{2}$.

CHAPTER 4

THE EXPERIMENTS OF CLAUSER AND OF HOLT

4.1. The experiments of Kocher and Commins

The first persons to try observing some of the angular correlation given in Fig. 3.1 were Kocher and Commins (1967). The observations were made in 1966 in Berkeley. For technical details, see their *Phys. Rev. Letter*. An atomic beam of calcium atoms from a 700° C oven is met in the center of the apparatus by a beam of ultraviolet light from a hydrogen arc (see Fig. 4.1). This beam of light, before meeting the atomic beam, passes through a filter F, which allows a spectral band 300 Å wide and centered at 2275 Å to excite some of the calcium atoms, mainly to the 6^1P_1 state. The atoms reach from there the ground state within about 10^{-8} s = 10 nanosecond, before they have moved much down the beam. While some atoms jump straight down to the 4^1S_0 ground state, other atoms cascade, in particular via the 4^1P_1 level (see Fig. 4.2). About 10% of the latter atoms reach the 4^1P_1 level via the 6^1S_0 level, and these are the ones which cause the coincidence counts between the two photomultipliers PM_1 and PM_2 in Fig. 4.1, shielded by filters f_1 and f_2 against wavelengths much different

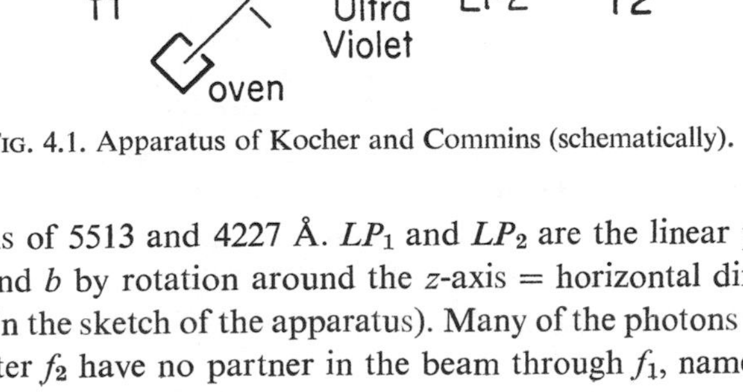

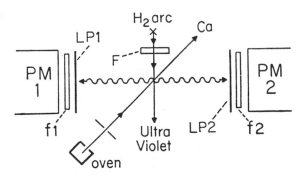

FIG. 4.1. Apparatus of Kocher and Commins (schematically).

from the wavelengths of 5513 and 4227 Å. LP_1 and LP_2 are the linear polarization filters turned at angles a and b by rotation around the z-axis = horizontal direction of the two photons (wavy lines in the sketch of the apparatus). Many of the photons in the spectral line observed through filter f_2 have no partner in the beam through f_1, namely for those cases where the 4^1P_1 level was reached from the 6^1P_1 level via the 5^1S_0 or the 4^1D_2 level, so that there is no photon of wavelength $\lambda_1 = 5513$ Å. Because of the excess of these incoherent λ_2 photons, some of them will give chance coincidence counts with λ_1 photons that are not their partners, thus causing an undesirable background.

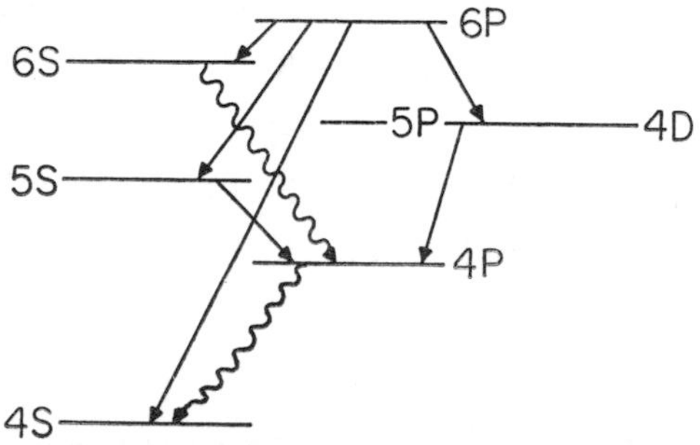

FIG. 4.2. Singlet spectrum of Ca (not scaled).
The wavy arrows are the transitions used (allowed by f_1 and f_2).

In the Kocher–Commins experiments, the coincidence mechanism measured within about ± 6 ns ($= \pm 6 \times 10^{-9}$ s) the time delay between the photons in PM_1 and PM_2. This inaccuracy was larger than the decay time ($\approx 4\frac{1}{2}$ ns) of the $4P$ state. Thus their measurements did not show the steep rise of the probability of finding photon 2 at the instant photon 1 was counted, with the typical exponential decay of this probability afterwards. This made the separation of coincidences from background somewhat rough and the measured coincidence rate $R(a, b)$ somewhat inaccurate.

Instead of measuring the entire curve of Fig. 3.1, they measured R as a function of $\varphi = a-b$ only at $\varphi = 0°$ and at $\varphi = 90°$. Within the crude "coincidence peak" of *measured number of coincidences* versus *time delay between the two photons* (in their observations between -7 and $+11$ ns, where the apparent "negative delays" were due to the inaccuracy of the measurement of the time delay), they measured, including background, some 837 counts at $\varphi = 0°$ in $10\frac{1}{2}$ h observation time, versus 492 counts at 90°. The background can be estimated from their data at larger positive or negative delay times, and was about 387 counts. This left 450 coherent coincidence counts at $\varphi = 0°$ versus 105 at 90°, with a rather large possible error. Since also the polarization filters used were not ideal ones, their experiments were not in reliable contradiction to the quantum-theoretical prediction $R(a, b) = 0$ at $\varphi = 90°$.

Mere improvement of the Kocher–Commins experiment is not sufficient for a reliable experimental decision whether for two-photon polarization correlations the quantum-theoretical prediction (153a) or whether some hidden-variables prediction is the correct one. Even at $\varphi = 90°$, experiments will never give $R(a, b) = 0$, because we cannot restrict the direction of the photon exactly to the z-direction. For obtaining a finite counting rate at $\varphi = 0°$, the solid angle of directions of photons traveling toward the photomultipliers should not be infinitesimal, and some photons will make nonzero angles θ with the z-axis. For them, the polarization filters LP_1 and LP_2 will not be ideal filters, even if they would be so for $\theta = 0$.

Moreover, by lack of a comparison with R_0, it is impossible by this method to distinguish the quantum-mechanical result from a hidden-variables result of the type of eq. (165) (for $P_{s.p.} = P_{min}$), which, too, gives $R(90°) = 0$ with $R(0°) > 0$. (See Fig. 3.1.)

4.2. Our lack of understanding of hidden-variables theories of the second kind

From the concluding remarks of Section 4.1, it is clear that for a reliable experimental distinction of quantum theory from hidden-variables theories of the second kind, two things are necessary:

(1) One should measure $R(\varphi) \equiv R(a-b) = R(a, b)$ not just at $\varphi = 0°$ and $\varphi = 90°$, but also at some inbetween values.

(2) One should study the effects of nonideal polarizers and of finite angles of deviation of the photons from the z-axis.

As to point (1), the question arises at how many angles φ are measurements needed for arriving at a conclusive result. The difficulty here lies in the fact that the hidden-variables theories considered in Chapter 3 were just some arbitrary sample theories, and by no means the only possible ones. The reader may easily construct some more fancy theories. Then, if we find a result that disagrees with all theories invented, how do we know that nobody could invent a different hidden-variables theory yet which would explain the observed facts better than quantum theory could?

We should keep in mind that the range of theories possible is very wide because all kinds of assumptions made naturally in quantum theory need not necessarily hold for hidden-variables theories of the second kind.

Take, for instance, conservation of angular momentum. We use it in Appendix E for calculating the quantum-theoretical two-photon polarization wavefunctions. Thus we obtained for photons traveling in the $\pm$ z-directions $\psi_{010} = \sqrt{\frac{1}{2}} \cdot (\tilde{x}_1\tilde{x}_2 + \tilde{y}_1\tilde{y}_2)$, which does *not* contain the component $\tilde{y}_1\tilde{x}_2$ which for crossed filters with $a = 90°$ and $b = 0°$ is according to eq. (149) the filter eigenfunction $\tilde{x}_1'\tilde{x}_2''$ corresponding to simultaneous passage of both photons through the filters. Therefore a combination of conservation of angular momentum (and of parity, see Appendix E) with the idea that probabilities would be given by squares of coefficients in expansions in terms of eigenfunctions, leads to the conclusion that in the 0–1–0 case crossed filters do not allow coincidences.

However, most hidden-variables theories considered in Chapter 3 predicted nonzero probabilities at $\varphi = a-b = 90°$ (see Fig. 3.1), so that either probabilities are no longer squares of amplitudes of eigenfunctions, or somehow $\tilde{y}_1\tilde{x}_2$ should have sneaked into ψ_{010}. In the latter case, either parity would not be conserved [as $\tilde{y}_1\tilde{x}_2$ for $j = 0$ occurs only in the state given by eq. (E-1), which has odd parity], or angular momentum would not be conserved (as $\tilde{y}_1\tilde{x}_2$ does not occur in the $j = 0$ state of even parity). Some people might claim that this breakdown of well-known conservation laws would suffice for disproving the possibility of these hidden-variables theories.

Against this attitude, objection must be made. In the first place, the fact that the rules of quantum theory for calculating probabilities from squares of coefficients may not be valid in hidden-variables theories of the second kind is readily admitted. According to hidden-variables theory, the rules of quantum theory hold only as a crude first approximation for systems of atomic size, and need not hold when applied to coincidences behind filters separated by macroscopic distances. Moreover, these rules are valid *only statistically*. As far as quantum theory has validity, then, one might expect that angular momentum would be conserved *on the average*. Individual photons, however, would be steered by hidden variables. Angular momentum quantum numbers, which are *properties of the wavefunctions*,[25] may or may not be rigorously and uniquely definable for single photon pairs in hidden-variables theory. Its conservation may or may not hold in each individual photon emission. Something that is valid in quantum theory need not always be valid in hidden-variables theory. If it is found not to hold, this cannot be an *a priori* objection against the theory. It can be held against the theory only where *experiments* show the validity of quantum-theoretical conclusions that conflict with predictions of hidden-variables theory.

In the latter case, however, one would ask whether by a *modification* of the hidden-variables theory used one could not avoid the conflict.

For this reason, in our attempts to decide experimentally whether any kind of hidden-variables theory would be right or cannot be right, we should be wary of trusting this or that particular brand of hidden-variables theory. If we want to disprove hidden-variables theory, we should disprove experimentally some consequence of hidden-variables theory that is entirely *independent* of any specific assumptions about the form of such a theory. The experiment should disprove some prediction of hidden-variables theory that is based merely on the *existence* of a set of hidden variables that predict the future behavior of particles as they sooner or later might meet specific pieces of apparatus by which we make measurements or observations for which the outcome quantum-theoretically could be predicted merely statistically.

In the next section we shall discuss how this can be done.

4.3. The idea of Clauser and of Horne and Shimony

Inspired by a brief paper by J. S. Bell (1964), John F. Clauser, while at Columbia University, had an idea how the Kocher and Commins experiment could be made conclusive by performing it at angles different from 0° and 90°, and combining measurements made at more than one angle. [Clauser (1969)]. Simultaneously, Abner Shimony and his student Michael Horne at Boston University had the same idea.

They suggested repeating the experiments of Kocher and Commins, not just for parallel and crossed polarization filters, but at wisely chosen different angles. The coincidence counting rate $R(a, b)$ was to be measured for all four combinations of two choices a_1 and a_2 for the angle a of the one filter, and two choices b_1 and b_2 for the angle b of the other filter. Moreover, coincidence counts were to be made with the one and with the other filter removed. The angles should be chosen in a particular way to be discussed below. The following combination of the measured coincidence rates is then calculated:

$$\Delta \equiv R(a_1, b_1)+R(a_2, b_1)+R(a_2, b_2)+R(a_1, b_2)-R(a_1, b_2)-R_1(a_2)-R_2(b_1), \quad (167)$$

[25] Compare Bohm and Aharonov (1960). For example, the sign in the photon wavefunctions $\sqrt{\frac{1}{2}}(\tilde{x}_1\tilde{x}_2 \pm\tilde{y}_1\tilde{y}_2)$ for photons traveling in the $\pm z$-direction would determine whether this photon pair would be in a state with spin quantum number 0 or 2.

where $R_1(a)$ is the coincidence rate with polarization filter 2 removed and filter 1 with its polarization direction at angle a, while $R_2(b)$ is the coincidence rate without filter 1 and with filter 2 at angle b. They showed then that for a "perfectly efficient" experimental setup (with 100% effective counters and coincidence electronics) this quantity Δ for any hidden-variables theory of the second kind must be negative. [Clauser *et al.* (1969); see Sections 4.7 and 4.9 below.] On the other hand, by an appropriate choice of the angles one can make this quantity Δ positive according to quantum theory. (See Section 4.5.) The angles then are chosen so as to maximize the quantum-theoretical value of Δ. A mere verification of the sign of Δ as calculated from the measured coincidence rates then can verify whether the hidden-variables theory is false, or whether quantum theory is false.

There are, of course, some complications. For obtaining a sufficient counting rate, the detectors will register photons traveling in directions within a finite solid angle. For those starting out at a slight angle with the $\pm z$-axis, and possibly by lenses made normally incident upon the polarization filters, the polarization correlations will not be identical with those calculated above for photon pairs along the $\pm z$-axes. Moreover, even for photons on the axis itself, the filters are not ideal because of absorption and because of imperfect polarization of the light transmitted.

Therefore it is necessary to verify in the first place whether this absence of ideal conditions cannot change the sign of the quantum-theoretical predictions, or, rather, how wide deviaons from the z-axis for filters with a given amount of absorption and of imperfect polarization will keep the sign of the quantum-theoretical prediction for Δ positive at the angles a and b chosen for the experiment. These calculations were made by Michael Horne for his 1970 Ph.D. thesis. The calculations showed that the experiment is feasible: With existing polarization filters, wide enough a solid angle is allowed to give a reasonable counting rate. [See Clauser *et al.* (1969), Horne (1970), Shimony (1970).]

In the second place, it is necessary to verify that lack of perfect efficiency of the detector mechanism cannot change the sign of the hidden-variables prediction for Δ from negative to positive. We will discuss this point in Section 4.8.

4.4. The experimental results

John Clauser went to Berkeley hoping to use the Kocher and Commins apparatus for doing the proposed experiments for the 0–1–0 case on calcium. The work was done together with a student of Professor Commins, Stuart Freedman. For better results, they rebuilt the apparatus, e.g. replacing the original polaroid films LP_1 and LP_2 by wide-spaced stacked glass plates.

Richard Holt at Harvard University, for his Ph.D. thesis work, started similar experiments using mercury, for studying the 1–1–0 case.

Freedman and Clauser, for obtaining sufficient intensity, needed large aspheric lenses in the photon beams from source to photomultipliers. Strains in the glass at first caused depolarization effects. After the lenses had been annealed, serious measurements could be started. Measurements were made of $R(a, b)$, for all combinations of a and b values which were multiples of $22\frac{1}{2}°$, so that $16 \times 16 = 256$ pairs of angles formed a complete cycle. Between any two of these 256 runs, each of 100 s duration, R_0 would be measured during 100 s for calibration. In $R(a, b)/R_0$ for any particular a and b, the value of R_0 used was the average between the R_0 values measured before and after the run determining $R(a, b)$.

While we will see below that for ideal circumstances the quantum-theoretical prediction for Δ/R_0 at

$$a_1 - b_1 = b_1 - a_2 = a_2 - b_2 = 22\tfrac{1}{2}^\circ \quad \text{(for the 0–1–0 case)} \tag{168}$$

is 0.207, a calculation of the type performed by Horne (1970) for the nonideal circumstances actually prevailing in Clauser's experiments gave a quantum-theoretical prediction $\Delta R_0 = 0.11$. Measurements so far available are in agreement with this prediction which, by being nonnegative, is in disagreement with the predictions of any hidden-variables theory of the second kind under ideal circumstances. The statistical error at the time this is written was still too large to make the results conclusive. A longer running time of the experiment should cure this trouble.

Holt uses 99.8%-pure ^{198}Hg. Mercury isotopes of odd atomic weight were mostly removed in order to avoid coupling of the angular momentum j with a nuclear angular momentum $I \neq 0$. Holt's experiments at the time this is written have not yet yielded any results. In some regards, his experiment is simpler, as the efficient excitation mechanism allows the mercury light source to be nearly a point source. Thus Holt can use small lenses and calcite polarizers.

Another advantage of Holt's mercury over Clauser's calcium is that the occupancy of the 6^3P_0 metastable state of Hg is practically zero, while with the calcium the problem arises that the second photon may be absorbed by vapor in the instrument and may excite a different calcium atom which will re-emit the 4227 Å photon later. This decreases the polarization correlation between the 5513 Å and the 4227 Å photons, and causes an apparent lengthening of the decay time of the intermediate 4^1P_1 state, in particular when the calcium vapor pressure increases.

At the time the above was written, Holt was still waiting for delivery of more satisfactory polarizing filters ordered for his experiment.

New phone calls (October, 1971) reveal that in the meantime Freedman and Clauser increased their accuracy and statistics, so that their results now are quite conclusive in favor of quantum theory, with the assumptions concerning detector efficiency discussed in Section 4.8. The results are to be published in the near future in the *Phys. Rev. Letters.*[25a] Clauser is now devising a new experiment which is to eliminate *any* need for assumptions about the efficiency of the counters, like our assumption eq. (205).

Holt still had not made his measurements, but expected to make them soon.[25b]

Thus we soon will have a definite experimental answer to the question whether for explaining two-photon polarization correlations quantum theory is insufficient and a hidden-variables theory of the second kind is needed, or whether the data are explainable by quantum theory and contradict any hidden-variables theory of the second kind irrespective of its particular form, and not only for the 0–1–0 case by Freedman and Clauser's measurements of Δ/R_0 at the angles of eq. (168), but by Holt's measurements at

$$a_1 - b_1 = b_1 - a_2 = a_2 - b_2 = 67\tfrac{1}{2}^\circ \quad \text{(for the 1–1–0 case)} \tag{169}$$

also for his photons from mercury.[25b]

[25a] S. J. Freedman and J. F. Clauser, *Phys. Rev. Letters* **28**, 938 (1972).

[25b] Results obtained by Holt until April 1973 at $67\frac{1}{2}^\circ$ and at $22\frac{1}{2}^\circ$ apparently contradict quantum theory. Others plan to continue measurements with his apparatus.

4.5. Calculation of the extreme values of Δ/R_0 according to quantum theory under ideal circumstances

Since the theoretical values for $R(a, b)$ are functions of $(a-b)$ only, we shall introduce the angles

$$\varphi_1 = a_1 - b_1, \quad \varphi_2 = b_1 - a_2, \quad \varphi_3 = a_2 - b_2, \tag{170}$$

so, $a_1 - b_2 = \varphi_1 + \varphi_2 + \varphi_3$. Since the theoretical values of $R_1(a)$ and of $R_2(b)$ for all theories are $= \frac{1}{2}R_0 =$ constant, we can write the theoretical value of Δ as

$$\Delta \equiv R(\varphi_1) + R(\varphi_2) + R(\varphi_3) - R(\varphi_1 + \varphi_2 + \varphi_3) - R_0. \tag{171}$$

For the 0–1–0 case, quantum theory predicts $R(\varphi)/R_0 = \frac{1}{2}\cos^2\varphi$ [see eq. 153a)]. We shall rewrite this here as

$$R(\varphi) = \frac{1+\cos 2\varphi}{4} R_0, \tag{172}$$

and we shall introduce the double angles

$$A = 2\varphi_1, \quad B = 2\varphi_2, \quad C = 2\varphi_3. \tag{173}$$

Then (171) becomes

$$f \equiv \Delta/R_0 = \tfrac{1}{4}\,[\cos A + \cos B + \cos C - \cos(A+B+C) - 2]. \tag{174}$$

For finding the maxima and minima of this expression, we differentiate with respect to each of the variables A, B, and C. We obtain the three conditions

$$\sin A = \sin B = \sin C = \sin(A+B+C). \tag{175}$$

The first two of these equations give

either $\qquad A = B = C \quad$ (Case I),

or $\qquad A = B = 180° - C \quad$ or a permutation thereof (Case II).

[Note that $A = 180° - B = 180° - C$ would be a permutation of Case II with A, B, C changed into B, C, A.]

We will not consider the permutations separately. For any solution we obtain for A, B, C, we can always regard permutations of them other solutions of the above equations (175).

In Case II we have $B + C = 180°$, so, $\sin A = \sin(A+B+C) = -\sin A$ gives $\sin A = 0$. Thus we find two solutions of (175) for Case II, viz.:

(1) $\quad A = B = 0°, \quad C = 180°, \quad \varphi_1 = \varphi_2 = 0, \quad \varphi_3 = 90° \quad (f = 0),$ (176a)

(2) $\quad A = B = 180°, \quad C = 0°, \quad \varphi_1 = \varphi_2 = 90°, \quad \varphi_3 = 0° \quad (f = -1),$ (176b)

and, of course, permutations of this.

In Case I we have

$$\varphi_1 = \varphi_2 = \varphi_3 \; (= \varphi), \tag{177a}$$

and (171) and (174) become

$$f = [3R(\varphi)-R(3\varphi)]/R_0-1$$
$$= [3 \cos (2\varphi)-\cos (6\varphi)-2]/4. \tag{177b}$$

In this case we find from (175) that, at the extrema of f, $\sin A = \sin 3A$, so that

either $$3A-A = 2A = \text{multiple of } 360°, \tag{178a}$$
or $$3A+A = 4A = \text{odd multiple of } 180°. \tag{178b}$$

Thence we find for A (modulo 360°) and for φ (modulo 180°) and for f [eq. (177b)],

from (178a):

either $$A = B = C = 0°, \quad \varphi = A/2 = 0° \quad (f = 0), \tag{179a}$$
or $$A = B = C = 180°, \quad \varphi = A/2 = 90° \quad (f = -1), \tag{179b}$$

or, from (178b):

either $$A = B = C = \pm 45°, \quad \varphi = \pm 22\tfrac{1}{2}°, \quad [f = \tfrac{1}{2}(\sqrt{2}-1)], \tag{179c}$$
or $$A = B = C = \pm 135°, \quad \varphi = \pm 67\tfrac{1}{2}°, \quad [f = -\tfrac{1}{2}(\sqrt{2}+1)]. \tag{179d}$$

The polarization angles, and therefore φ, are meaningful modulo 180°, so, the above (176a, b) and (179a–d) are all the solutions there are.

The Case I is illustrated by the broken curve in Fig. 4.3. The maxima and minima given by eqs. (179a–d) are easily recognized on that curve.

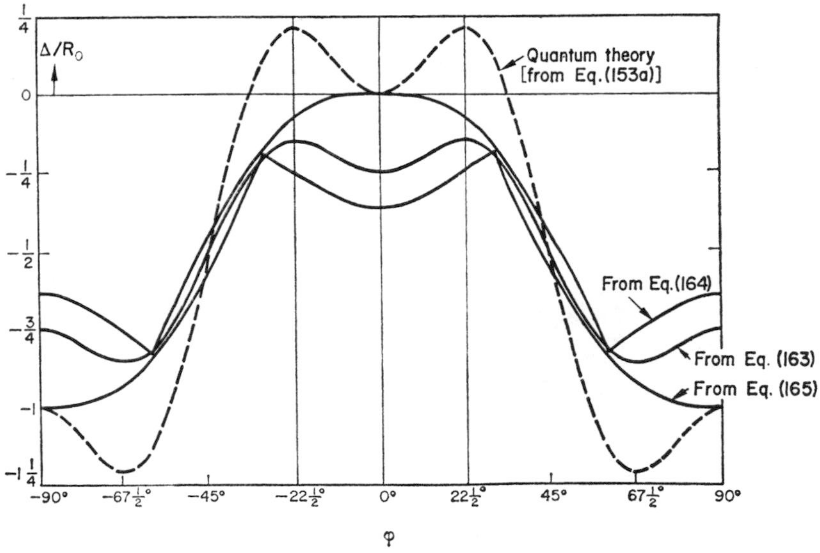

FIG. 4.3. Δ/R_0 as a function of φ. $\Delta \equiv R(a_1, b_1)+R(a_2, b_1)+R(a_2, b_2)-R(a_1, b_2)-R_1(a_2)-R_2(b_1)$ in units R_0 for $a_1-b_1 = b_1-a_2 = a_2-b_2 = \varphi$, $a_1-b_2 = 3\varphi$ (that is, $\Delta = 3R(\varphi)-R(3\varphi)-2R_1$), plotted here for parallel correlation in the 0–1–0 case.

We are interested in the absolute maxima, which are given by (179c), and which lie at

$$\varphi = \pm 22\tfrac{1}{2}^\circ. \tag{180}$$

Here, $f = 0.2071$, so that

$$\Delta = 3R\,(22\tfrac{1}{2}^\circ) - R(67\tfrac{1}{2}^\circ) - R_1 - R_2 = 0.207R_0. \tag{181}$$

The question to be settled experimentally, therefore, is whether, *in the 0–1–0 case*, Δ as defined by eqs. (167), (170), and (177a) for $\varphi = 22\frac{1}{2}^\circ$ has a value $\approx 0.2R_0$, or whether it is negative. (As mentioned already in Section 4.4, in practice we should expect $\Delta \simeq 0.1R_0$ at $\varphi = 22\frac{1}{2}^\circ$, due to imperfections of the experiment.)

In the 1–1–0 case, the prediction $R(\varphi) = \frac{1}{2}R_0\cos^2\varphi$ is replaced by $R(\varphi) = \frac{1}{2}R_0\sin^2\varphi$ [see eq. (153b)]. Therefore, in this case, (171) becomes

$$\Delta/R_0 = \tfrac{1}{2}[\sin^2\varphi_1 + \sin^2\varphi_2 + \sin^2\varphi_3 - \sin^2(\varphi_1+\varphi_2+\varphi_3) - 2]. \tag{182}$$

This differs from the formula for the 0–1–0 case (with cos instead of sin) by a 90° shift of the angles φ_1, φ_2, and φ_3. That is, (182) has at the angles φ_i the value which the formula for the 0–1–0 case has at values that are 90° less. [It is true that $\sin(\varphi_1+\varphi_2+\varphi_3)$ has a value that is *the opposite* of the value of $\cos(\varphi_1+\varphi_2+\varphi_3)$ at 90° less value of *each* of the three φ_i, but luckily these quantities appear squared in (182) and in the corresponding equation for the 0–1–0 case.]

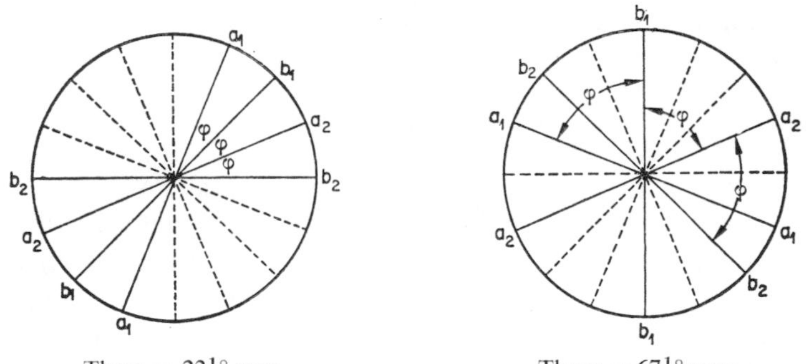

The $\varphi = 22\frac{1}{2}^\circ$ case. The $\varphi = 67\frac{1}{2}^\circ$ case.

FIG. 4.4. Relative polarization directions used for maximal quantum-theoretical Δ/R_0 for the 0–1–0 case and the 1–1–0 case.

We conclude that the curves for Δ in Fig. 4.3 will for the 1–1–0 case look similar, but will be 90° shifted. The maximum of Δ in this case will lie at

$$\varphi = \pm 67\tfrac{1}{2}^\circ. \tag{183}$$

This is the case studied by Richard Holt at Harvard University, using mercury atoms as the source of his photon pairs.

Figure 4.4 shows the relative positions of the polarization directions a_1, a_2, b_1, and b_2 in each of the two cases (168) and (169), that is, (177a) with (180) or with (183).

4.6. Predictions of hidden-variables theories for the experiments of Clauser and of Holt

Inserting eqs. (163) through (166) and (170) in eq. (171), we obtain predictions for Δ/R_0 by some hidden-variables theories of the second kind. The curves derived from eqs. (163) through (165) are shown in Fig. 4.3. The one obtained from (166) has the general shape of the ones derived from eqs. (153a) and (163) but is flatter: it starts at $\varphi = \pm 90°$ from $\Delta/R_0 = -5/8$, dips at $\varphi = \pm 67\frac{1}{2}°$ to $\Delta/R_0 = -0.6768$, goes through $\Delta/R_0 = -\frac{1}{2}$ at $\varphi = \pm 45°$, reaches a top of $\Delta/R_0 = -0.3232$ at $\varphi = \pm 22\frac{1}{2}°$, and dips to $\Delta/R_0 = -3/8$ at $\varphi = 0°$.

For the 1–1–0 case, all curves of Fig. 4.3 are shifted sideways by 90°.

We see that for all of these hidden-variables predictions of what Δ/R_0 should be in a perfectly efficient experiment, we have not only $\Delta/R_0 \leq 0$ as predicted by Clauser *et al.* (1969), but we also have $\Delta/R_0 \geq -1$. We shall give in the following section a simple general derivation of both these inequalities for any hidden-variables theory of the second kind.

4.7. A simple general proof of $-1 \leq \Delta/R_0 \leq 0$ for perfectly efficient experiments, according to any hidden-variables theory of the second kind

We shall here prove $-1 \leq \Delta/R_0 \leq 0$ by a method of our own making.[25c] [In Section 4.9 we shall show how Clauser *et al.* obtained this result by a generalization of "Bell's inequality". See eq. (221).]

For simplicity we shall reason below as if the color filters f_1 and f_2 (see Fig. 4.1) were placed between the emitting atom and the polarization filters LP_1 and LP_2, as in fact they are in the new apparatus of Clauser.

Hidden-variables space

In the following reasoning we allow the filters to be nonideal. We will consider here all external as well as internal hidden variables that might possibly influence the directions of emission of the photons, their passage through the color filters in directions that would make them hit the photomultipliers, and the passage or nonpassage of each photon through any of the polarization filters if the latter would be in place. We may imagine these hidden variables to be plotted as coordinates in a many-dimensional "hidden-variables space." We need consider in this space only that region H_0 that corresponds to the values of the hidden variables that guarantee that the direction of the photons is toward the photomultipliers and that they are not absorbed by the color filters.

The prevailing statistical distribution of hidden-variable values determines for each part H_i of hidden-variables space a corresponding probability P_i that the hidden variables have values inside H_i. We normalize the P_i so that $P_0 = 1$ for H_0. For perfectly efficient detectors and counters, then, in absence of polarization filters the coincidence counting rate R_0 is equal to $P_0 = 1$ times the rate A at which photon pairs emitted by the atom emerge from the

[25c] F. J. Belinfante, "Experiments to disprove that nature would be deterministic" (mimeographed lecture notes, February, 1970). Independently, the same method has been suggested by E. P. Wigner, *American Journal of Physics* **38**, 1005 (1970).

color filters and head toward the photomultipliers:

$$R_0 = AP_0 = A. \tag{184}$$

If, however, photomultiplier 1 has an efficiency α_0 in absence of polarizaiton filter 1, and photomultiplier 2 has an efficiency β_0 in absence of polarization filter 2, and the coincidence counting mechanism has in this case an efficiency ε_{00}, we have

$$R_0 = (P_0) \cdot A\alpha_0\beta_0\varepsilon_{00}. \tag{185}$$

Four local questions

A characteristic of any hidden-variables theory of the second kind is that supposedly the values of the hidden variables tell us for each of the two photons separately what will happen to it when it meets a polarization filter in a particular orientation. That is, each point in part H_0 of hidden-variables space provides answers "yes" or "no" simultaneously to each of the following four questions separately.

Question "i": If photon 1 meets a polarization filter LP_1 in orientation a_1, will it pass through the filter and emerge from its back toward the detector PM_1?
Question "j": If photon 1 meets LP_1 in orientation a_2, will it pass through toward PM_1?
Question "k": If photon 2 meets LP_2 in orientation b_1, will it pass through toward PM_2?
Question "l": If photon 2 meets LP_2 in orientation b_2, will it pass through toward PM_2?

Note that these are all *local* questions. For instance, we ask whether LP_2 will allow photon 2 to pass, without telling what photon 1 is doing at LP_1. Therefore, neither quantum theory nor any hidden-variables theory of the first kind can provide unambiguous answers to *any* of these four questions. Quantum theory could at most answer such questions as: With filters LP_1 and LP_2 in positions a_1 *and* b_1, what is the probability that both photons (or that only photon 1 or only photon 2) would pass its filter? Similarly, when the two-photon wavefunction is not factorizable into two one-photon wavefunctions, hidden-variables theories of the *first* kind should be able to answer by "yes" or "no" only questions on what would happen to the *pair* of photons at the *pair* of filters. [Compare Section 1.6 of Part I] Since hidden-variables theories of the second kind are theories that supposedly *can* answer such *local* questions, they are often called "*local*" *hidden-variables theories.*

Probabilities of answers to local questions

Each of the above four questions has its own answers "yes" (+) or "no" (−). Each point in H_0 answers all four of the questions. Therefore, we can subdivide H_0 into sixteen (= 2^4) parts H^{ij}_{kl} each corresponding to a particular set of answers to the four questions. Each of the four indices i, j, k, l takes either of the two values + and −. For instance, H^{+-}_{-+} is the locus of hidden-variables points for which filter 1 transmits photon 1 when in orientation a_1 but not in orientation a_2, while filter 2 transmits photon 2 when its angle is b_2, but stops it when its angle is b_1.

Correspondingly, among all hidden variables in H_0, the probability of those inside H^{ij}_{kl} is called P^{ij}_{kl}. We then must have, for the sum of all sixteen P^{ij}_{kl},

$$\sum_i \sum_j \sum_k \sum_l P^{ij}_{kl} = P_0 = 1. \tag{186}$$

The various coincidence rates

Suppose we have filter 1 in position a_1, and filter 2 in position b_1. What is the coincidence rate $R(a_1, b_1)$?

If all counters and detectors were perfectly effective, this rate would be the product of the rate A with which photon pairs are emerging toward the polarization filters, and the probability that among these photons (described by part H_0 of hidden-variables space) the questions asked above have the right answers. The right answers here are "yes" to questions i and k, but the answers to questions j and l in this case may be "*no*" as well as "yes", so we add the corresponding probabilities:

$$R(a_1, b_1) = A[(P^{++}_{++})+(P^{+-}_{++})+(P^{++}_{+-})+(P^{+-}_{+-})]. \tag{187}$$

Actually, the photon detectors PM_1 and PM_2 may have efficiencies less than 1. Let α_n be the efficiency of PM_1 with filter LP_1 in position a_n (and α_0 its efficiency in absence of LP_1). Let β_n be the efficiency of PM_2 with filter LP_2 in position b_n (and β_0 its efficiency in absence of LP_2). Let ε_{mn} be the efficiency of the coincidence counting mechanism with LP_1 in position a_m (or absent for $m = 0$), and with LP_2 in position b_n. Then,

$$R(a_1, b_1) = A\,[P^{++}_{++} + P^{+-}_{++} + P^{++}_{+-} + P^{+-}_{+-}]\,\alpha_1\beta_1\varepsilon_{11}. \tag{188}$$

Similarly (with $i = \pm, j = k = +, l = \pm$):

$$R(a_2, b_1) = A[P^{++}_{++} + P^{-+}_{++} + P^{++}_{+-} + P^{-+}_{+-}]\alpha_2\beta_1\varepsilon_{21}, \tag{189}$$

and (with $i = \pm, j = +, k = \pm, l = +$):

$$R(a_2, b_2) = A\,[P^{++}_{++} + P^{++}_{-+} + P^{-+}_{++} + P^{-+}_{-+}]\,\alpha_2\beta_2\varepsilon_{22}. \tag{190}$$

In Δ, the above three terms are added. We then must subtract

$$R(a_1, b_2) = A\,[P^{++}_{++} + P^{++}_{-+} + P^{+-}_{++} + P^{+-}_{-+}]\,\alpha_1\beta_2\varepsilon_{12} \tag{191}$$

and (with $i = \pm, j = +, k = \pm, l = \pm$)

$$R_1(a_2) = A[P^{++}_{++} + P^{++}_{+-} + P^{++}_{-+} + P^{++}_{--} + P^{-+}_{++} + P^{-+}_{+-} + P^{-+}_{-+} + P^{-+}_{--}]\,\alpha_2\beta_0\varepsilon_{20} \tag{192}$$

and (with $k = +$, other indices $= \pm$)

$$R_2(b_1) = A\,[P^{++}_{++} + P^{+-}_{++} + P^{-+}_{++} + P^{--}_{++} + P^{++}_{+-} + P^{+-}_{+-} + P^{-+}_{+-} + P^{--}_{+-}]\,\alpha_0\beta_1\varepsilon_{01}. \tag{193}$$

Note that these expressions remain correct when the filters are not ideal or when the photons make angles with the z-axis. (This would merely affect the *values* of the individual P^{ij}_{kl}.)

Matrix notation

Below we shall use matrix notation. We start by replacing the four-index P^{ij}_{kl} by a two-index P_{rs}, where r and s run from 1 to 4. The value pairs $++$, $+-$, $-+$, and $--$ for the index pair $\{i, j\}$ will be indicated by $r = 1, 2, 3, 4$. Similarly, $s = 1, 2, 3, 4$ denote $\{k, l\} = ++$ or $+-$ or $-+$ or $--$.

Then the eight expressions (186) through (193) all are linear combinations $\sum_r\sum_s c_{rs}P_{rs}$. We shall use the matrix c_{rs} alone, as an abbreviated notation for this sum.

Thus, the c_{rs}-matrix $\begin{pmatrix}1&1&1&1\\1&1&1&1\\1&1&1&1\\1&1&1&1\end{pmatrix}$ represents the sum (186), or, formally:

$$\begin{pmatrix}1&1&1&1\\1&1&1&1\\1&1&1&1\\1&1&1&1\end{pmatrix} = P_0 = 1 = R_0/A\alpha_0\beta_0\varepsilon_{00}. \tag{186'}$$

[See (185).] Similarly, (188) becomes

$$R(a_1, b_1) = A\alpha_1\beta_1\varepsilon_{11}\begin{pmatrix}1&1&0&0\\1&1&0&0\\0&0&0&0\\0&0&0&0\end{pmatrix}, \tag{188'}$$

and so on.

The "perfectly efficient" case

When detectors, counters, and electronics are perfectly efficient, we may put $\alpha_m = \beta_n = \varepsilon_{mn} = 1$, and Δ becomes a much simpler expression. In matrix notation, then, the sum of (188) through (190) is

$$A\left\{\begin{pmatrix}1&1&0&0\\1&1&0&0\\0&0&0&0\\0&0&0&0\end{pmatrix}+\begin{pmatrix}1&1&0&0\\0&0&0&0\\1&1&0&0\\0&0&0&0\end{pmatrix}+\begin{pmatrix}1&0&1&0\\0&0&0&0\\1&0&1&0\\0&0&0&0\end{pmatrix}\right\} = A\begin{pmatrix}3&2&1&0\\1&1&0&0\\2&1&1&0\\0&0&0&0\end{pmatrix}, \tag{194}$$

while the sum of (191) through (193) is in this case

$$A\left\{\begin{pmatrix}1&0&1&0\\1&0&1&0\\0&0&0&0\\0&0&0&0\end{pmatrix}+\begin{pmatrix}1&1&1&1\\0&0&0&0\\1&1&1&1\\0&0&0&0\end{pmatrix}+\begin{pmatrix}1&1&0&0\\1&1&0&0\\1&1&0&0\\1&1&0&0\end{pmatrix}\right\} = A\begin{pmatrix}3&2&2&1\\2&1&1&0\\2&2&1&1\\1&1&0&0\end{pmatrix}. \tag{195}$$

Subtracting (195) from (194) we obtain, by (184) in this "efficient" case,

$$\Delta = A\begin{pmatrix} 0 & 0 & -1 & -1 \\ -1 & 0 & -1 & 0 \\ 0 & -1 & 0 & -1 \\ -1 & -1 & 0 & 0 \end{pmatrix} = R_0\begin{pmatrix} 0 & 0 & -1 & -1 \\ -1 & 0 & -1 & 0 \\ 0 & -1 & 0 & -1 \\ -1 & -1 & 0 & 0 \end{pmatrix}. \tag{196}$$

Now remembering that the matrix c_{rs} here stands as an abbreviation for $\sum c_{rs}P_{rs}$, we find that actually

$$\begin{aligned} \Delta/R_0 &= -[P_{13}+P_{14}+P_{21}+P_{23}+P_{32}+P_{34}+P_{41}+P_{42}] \\ &= -[P^{++}_{-+}+P^{++}_{--}+P^{+-}_{++}+P^{+-}_{-+}+P^{-+}_{+-}+P^{-+}_{--}+P^{--}_{++}+P^{--}_{+-}]. \end{aligned} \tag{197}$$

Since $\sum_i\sum_j\sum_k\sum_l P^{ij}_{kl} = \sum_r\sum_s P_{rs} = 1$ by (186), and since each of the P^{ij}_{kl} must be nonnegative, we see immediately from (197) that

$$-1 \leq \Delta/R_0 \leq 0, \tag{198}$$

which was the promised result.

Freedman's inequality

Above, we have proved for the "perfectly efficient" case two inequalities, which for the choice (177a) of the three angles (170) may be written as

$$-R_0 \leq \Delta = 3R(\varphi)-R(3\varphi)-R_1-R_2 \leq 0. \tag{199}$$

We write down the one inequality for $\varphi = 22\frac{1}{2}°$ and the other one for $\varphi = 67\frac{1}{2}°$, or vice versa. We note that, for $\varphi = 67\frac{1}{2}°$,

$$R(3\varphi) = R(202\tfrac{1}{2}°) = R(22\tfrac{1}{2}°) \tag{200}$$

because the polarization directions are defined modulo 180°. Thence,

$$-R_0 \leq 3R(22\tfrac{1}{2}°)-R(67\tfrac{1}{2}°)-R_1-R_2 \leq 0, \tag{201}$$

$$0 \geq 3R(67\tfrac{1}{2}°)-R(22\tfrac{1}{2}°)-R_1-R_2 \geq -R_0. \tag{202}$$

Subtracting and dividing by 4, one finds[26]

$$-\tfrac{1}{4}R_0 \leq R(22\tfrac{1}{2}°)-R(67\tfrac{1}{2}°) \leq \tfrac{1}{4}R_0,$$

that is,

$$|R(22\tfrac{1}{2}°)-R(67\tfrac{1}{2}°)| \leq \tfrac{1}{4}R_0 \tag{203}$$

for the coincidence rates in perfectly efficient experiments predicted by hidden-variables theories of the second kind. The inequality (203) has the advantage that $R_1(a)$ and $R_2(b)$ cancel out merely assuming that they do not depend upon a or b. An advantage of (202)

[26] Private communication from S. Freedman. Clauser, by similarly combining eq. (199) for *any* φ and for $\varphi' = 3\varphi$, further generalizes eq. (203) to $|3R(\varphi)-4R(3\varphi)+R(9\varphi)| \leq R_0$ for *any* φ.

over (201) pointed out by Freedman[26] is that experimentally $R(67\frac{1}{2}°)$ can be measured more accurately than $R(22\frac{1}{2}°)$.

Quantum theory violates the Freedman inequality (203), as by (153a, b) it predicts

in the 0–1–0 case, $$R(22\tfrac{1}{2}°)-R(67\tfrac{1}{2}°) = \left(\tfrac{1}{2}\sqrt{\tfrac{1}{2}}\right)R_0 = 0.354R_0 > \tfrac{1}{4}R_0, \tag{204a}$$

in the 1–1–0 case, $$R(22\tfrac{1}{2}°)-R(67\tfrac{1}{2}°) = -\left(\tfrac{1}{2}\sqrt{\tfrac{1}{2}}\right)R_0 = -0.354R_0 < -\tfrac{1}{4}R_0. \tag{204b}$$

(See Fig. 3.1, p. 284.)

Inapplicability of the above derivation to hidden-variables theories of the first kind

The above derivations are not applicable to hidden-variables theories of the first kind because such theories do no answer local questions, but answer by "yes" or "no" the questions that quantum theory answers only statistically. In the notation of Section 1.6 of Part I, the above theory of the second kind answers for Photon One two questions i and j ($M = 2$) and has two possible answers to each question ($R = 2$). Similarly, for Photon Two, in this notation, $N = S = 2$, and hidden-variables space is subdivided into $(M^R)(N^S) = 16$ different parts. If we would try to treat a theory of the first kind in a similar way, the questions would not be i, j, k, l separately, but would be the four pairs (i, k), (i, l), (j, k) and (j, l), so, there would be $MN = 4$ questions, and each question would have $RS = 4$ answers (yes–yes or yes–no or no–yes or no–no), and hidden-variables space of the first kind would be subdivided into $(MN)^{RS} = 256$ parts. What would result is a 16×16 matrix of coefficients c_{rs} of probabilities P_{rs}, so Δ would not be as simple as (196). Thus a theory of the first kind *need* not satisfy the inequality (198), and it *could* not, as a theory of the first kind would predict the quantum-theoretical result, which contradicts (198) around $\varphi_1 = \varphi_2 = \varphi_3 = 22\frac{1}{2}°$ or $67\frac{1}{2}°$, where Δ/R_0 according to (179c–d) has the value $\frac{1}{2}(\sqrt{2}-1)$ or $-\frac{1}{2}(\sqrt{2}+1)$ in the 0–1–0 case, or vice versa in the 1–1–0 case.

4.8. The case of imperfect efficiency

The question arises, how much of the above proof of (198) remains valid, if the α, β, and ε are not all equal to 1? If all products $(\alpha_m \beta_n \varepsilon_{mn})$ were equal among each other $(=\alpha\beta\varepsilon)$, they would cancel out in the quotient Δ/R_0, and (198) would remain valid. Let us therefore consider the reasons why the α_m and the β_n and the ε_{mn} *might* depend upon which of the values 1 or 2 or 0 the indices m and n take.

A dependence of α or β upon whether its index is 1 or 2 would mean a difference of sensitivity of the photon detector for different polarizations of the incident light. If, however, a run of observations includes a rotation of Fig. 4.4 in sixteen $22\frac{1}{2}°$ turns, such polarization dependence should cancel out in the average.

A dependence of α or β upon whether the index is 0 or whether it is 1 or 2, would mean mainly a dependence upon whether the incident light is unpolarized or polarized. (We assume here that the photon leaving the polarization filter has practically the same energy as the photon incident upon the polarization filter.)

It would seem that the assumption that

$$\alpha_0 \approx \alpha_1 \approx \alpha_2 (= \alpha), \quad \beta_0 \approx \beta_1 \approx \beta_2 (= \beta) \tag{205}$$

is a good approximation.

It seems unlikely that ε_{11}, ε_{12}, ε_{21}, and ε_{22} would differ among each other. If m and n mean each 1 or 2, but *not zero,* it is in principle thinkable that ε_{00} and ε_{0m} and ε_{m0} would differ from each other and from ε_{mn}, because in absence of a polarization filter the photon would be slightly faster in reaching the photomultiplier, thus affecting the delay between the two counts by PM_1 and PM_2. When, however, one puts in the numerical values for this difference, it is seen to be of the order of 10^{-11} s, which is negligible compared to the decay times of the order of 10^{-9} s of the intermediate atomic energy level. Therefore, in good approximation, also

$$\varepsilon_{00} \approx \varepsilon_{01} = \varepsilon_{02} \approx \varepsilon_{10} = \varepsilon_{20} \approx \varepsilon_{11} = \varepsilon_{12} = \varepsilon_{21} = \varepsilon_{22} (= \varepsilon). \tag{206}$$

Therefore, for imperfect efficiency, the values of $R(a, b)$ and of $R_1(a)$ and of $R_2(b)$ and of R_0 and of Δ all are decreased by approximately the same factor $\alpha\beta\varepsilon$, and the inequalities (198) remain valid in good approximation.

If somehow (206) would be valid but (205) would be replaced by

$$\alpha_0 \geqslant \alpha_1, \quad \alpha_0 \geqslant \alpha_2, \quad \beta_0 \geqslant \beta_1, \quad \beta_0 \geqslant \beta_2, \tag{207}$$

it is easily shown by explicit calculation of Δ not neglecting the $\alpha_m\beta_m\varepsilon$ terms that $\Delta \leqslant 0$ remains valid.[27]

4.9. Bell's inequality and the derivation of $\Delta \leqslant 0$ by Clauser and by Horne and Shimony

Bell's inequality

For showing that hidden-variables theories of the second kind will not agree with quantum theory, Bell (1964) considered simultaneous measurements of the spins of two electrons which together were in a singlet spin state, measuring by a Stern–Gerlach experiment the component of σ_1 in the direction of a unit vector **a**, and, similarly, the component of σ_2 in the direction of a unit vector **b**. While each of these spin components in units $\hbar/2$ can have the values ± 1 only, the expectation value of their product according to quantum theory is in the singlet state

$$\langle(\sigma_1 \cdot \mathbf{a})(\sigma_2 \cdot \mathbf{b})\rangle = -\mathbf{a} \cdot \mathbf{b}. \tag{208}$$

He then asked himself whether the same result could possibly be derived from a "local" hidden-variables theory.[28] Let λ represent the hidden variables, and $\varrho(\lambda)\, d\lambda$ the probability

[27] One finds for Δ a linear combination of the $P^{ij}_{kl} = P_{rs}$ with coefficients of which each one is negative. For instance, $P^{++}_{++} = P_{11}$ would have the coefficient

$$\varepsilon[(\alpha_1-\alpha_0)\beta_1+\alpha_2(\beta_1-\beta_0)+(\alpha_2-\alpha_1)\beta_2] = \varepsilon[\alpha_1(\beta_1-\beta_0)+(\alpha_2-\alpha_0)\beta_1+(\alpha_2-\alpha_1)(\beta_2-\beta_0)],$$

of which the first expression is clearly nonpositive by (207) if $\alpha_2 \leqslant \alpha_1$, while the second expression is clearly nonpositive by (207) if $\alpha_2 \geqslant \alpha_1$.

[28] This "Bell theory of spins" is, of course, entirely different from the one discussed in Section 2.14 of Part II. There we discussed a specific theory. Here, we discuss a rather general (unspecific) theory. It is not of the first kind, because its assumptions (210)–(211) lead by (212) and (215) to a contradiction between the hidden-variables prediction (210) and the quantum-mechanical prediction (208).

that their values be within $d\lambda$. Let

$$A(\mathbf{a}, \lambda) = \pm 1 \quad \text{and} \quad B(\mathbf{b}, \lambda) = \pm 1 \tag{209}$$

be the values of $(\sigma_1 \cdot \mathbf{a})$ and of $(\sigma_2 \cdot \mathbf{b})$ predicted for hidden variables λ. Then, by hidden-variables theory,

$$\langle(\sigma_1 \cdot \mathbf{a})\,(\sigma_2 \cdot \mathbf{b})\rangle_{\text{h.v.}} = P(\mathbf{a}, \mathbf{b}), \tag{210}$$

where

$$P(\mathbf{a}, \mathbf{b}) \equiv \int d\lambda\, \varrho(\lambda)\, A(\mathbf{a}, \lambda)\, B(\mathbf{b}, \lambda). \tag{211}$$

The question therefore was whether the absolute difference between (208) and (211) could be made arbitrarily small; that is, if we assume

$$|\,P(\mathbf{a}, \mathbf{b}) + \mathbf{a} \cdot \mathbf{b}\,| \leqslant \varepsilon, \tag{212}$$

can the positive number ε be made arbitrarily small? For answering this question, Bell first proved the inequality[29]

$$|\,P(\mathbf{a}, \mathbf{b}) - P(\mathbf{a}, \mathbf{c})\,| \leqslant 1 + P(\mathbf{b}, \mathbf{c}) + \varepsilon, \tag{213}$$

and thence, using (212) another few times,

$$4\varepsilon \geqslant |\,\mathbf{a} \cdot \mathbf{c} - \mathbf{a} \cdot \mathbf{b}\,| + \mathbf{b} \cdot \mathbf{c} - 1. \tag{214}$$

Taking as a special case $\mathbf{a} \perp \mathbf{c}$, and $\mathbf{b}$ bisecting the angle between $\mathbf{a}$ and $\mathbf{c}$, we find from (214)

$$4\varepsilon > \sqrt{2} - 1, \tag{215}$$

which shows that ε can*not* be made arbitrarily small, so that hidden-variables theories of the second kind cannot reproduce the result (208) of quantum theory.

The work of Clauser and of Horne and Shimony

The formulas used by Bell in discussing a singlet state of two electrons, suggested Clauser (1969) and Horne and Shimony [Clauser *et al.* (1969)] to treat the two-photon case similarly. In their application, $A(a, \lambda) = +1$ or -1 meant transmission or stopping of photon 1 by polarization filter 1 with a polarization direction a, while $B(b, \lambda)$ meant a similar thing for photon 2 and filter 2 at angle b. Defining again the correlation function $P(a, b)$ by (211), they proved an inequality more general than (213), namely,[30]

$$|\,P(a_1, b_1) - P(a_1, b_2)\,| \leqslant 2 - P(a_2, b_1) - P(a_2, b_2), \tag{216}$$

provided that $P(a_2, b_1) > 0$. If $w[A(a)_{\pm}, B(b)_{\pm}]$ is the probability that $A(a) = \pm 1$ and $B(b) = \pm 1$, one has

$$P(a, b) = w[A(a)_+, B(b)_+] - w[A(a)_+, B(b)_-] - w[A(a)_-, B(b)_+] + w[A(a)_-, B(b)_-], \tag{217}$$

[29] For a proof of this formula, see its derivation from (225) and (212) at the end of this section.
[30] For a proof of (216), see the derivation below, of eq. (225) from (222)–(223).

and, for perfect efficiency of the coincidence counting,

$$\left.\begin{aligned} R(a, b)/R_0 &= w[A(a)_+, B(b)_+], \\ R_1(a)/R_0 &= w[A(a)_+, B(b)_+]+w[A(a)_+, B(b)_-], \\ R_2(b)/R_0 &= w[A(a)_+, B(b)_+]+w[A(a)_-, B(b)_+]. \end{aligned}\right\} \tag{218}$$

Thence,

$$P(a, b) = [4R(a, b)-2R_1(a)-2R_2(b)]/R_0+1. \tag{219}$$

Inserting this in (216), one finds

$$\begin{aligned} &|4R(a_1, b_1)-4R(a_1,b_2)-2R_2(b_1)+2R_2(b_2)| \\ &\leqslant -4R(a_2, b_1)-4R(a_2, b_2)+4R_1(a_2)+2R_2(b_1)+2R_2(b_2). \end{aligned} \tag{220}$$

Then, the same inequality will also hold without absolute sign on the left. Bringing for that case all to the left, and dividing by 4, Clauser *et al.* obtained

$$\Delta \leqslant 0 \tag{221}$$

with Δ given by eq. (167).

Further generalization by Bell

Bell (1970) has generalized the inequality (216) even further. Suppose the measuring results (209) depend not only on the hidden variables λ, but also on some additional external hidden variables over which all measurements should be averaged. We indicate these averages by a bar over the quantity averaged. Then,

$$|\bar{A}(a, \lambda)| \leqslant 1, \quad |\bar{B}(b, \lambda)| \leqslant 1. \tag{222}$$

Let us now define

$$P(a, b) \equiv \int d\lambda\, \varrho(\lambda)\, \bar{A}(a, \lambda)\, \bar{B}(b, \lambda). \tag{223}$$

(We assume that the averages over external hidden variables at the two spatially separated apparatus for measuring A and B can be taken independently.) Then,

$$\begin{aligned} P(a_1, b_1)-P(a_1, b_2) &= \int d\lambda\, \varrho(\lambda)\, [\bar{A}(a_1, \lambda)\, \bar{B}(b_1, \lambda)-\bar{A}(a_1, \lambda)\, \bar{B}(b_2, \lambda)] \\ &= \int d\lambda\, \varrho(\lambda)\{\bar{A}(a_1, \lambda)\, \bar{B}(b_1, \lambda)\, [1\pm\bar{A}(a_2, \lambda)\, \bar{B}(b_2, \lambda)]\} \\ &\quad - \int d\lambda\, \varrho(\lambda)\{\bar{A}(a_1, \lambda)\, \bar{B}(b_2, \lambda)\, [1\pm\bar{A}(a_2, \lambda)\, \bar{B}(b_1, \lambda)]\}, \end{aligned} \tag{224}$$

so, by $|P-Q| \leqslant |P|+|Q|$ and by (222),

$$\begin{aligned} |P(a_1, b_1)-P(a_1, b_2)| &\leqslant \int d\lambda\, \varrho(\lambda)\, [1\pm\bar{A}(a_2, \lambda)\, \bar{B}(b_2, \lambda)] \\ &\quad +\int d\lambda\, \varrho(\lambda)\, [1\pm\bar{A}(a_2, \lambda)\, \bar{B}(b_1, \lambda)] \\ &= 2\pm[P(a_2, b_2)+P(a_2, b_1)], \end{aligned} \tag{225}$$

so,

$$|P(a_1, b_1)-P(a_1, b_2)| \leqslant 2-|P(a_2, b_2)+P(a_2, b_1)|. \tag{226}$$

This is more general than (216), because the condition (222) for its validity is more general than the condition (209) for validity of (216). On the other hand, the result (216) of Clauser *et al.* follows immediately from (225) as a special case.

Taking the $+$sign in (225), putting $a_1 = a$, $a_2 = b_1 = b$, and $b_2 = c$, we obtain from (225) as a special case

$$|P(a, b)-P(a, c)| \leq 2+P(b, c)+P(b, b). \tag{227}$$

For Bell's singlet two-electron case, with ε chosen in agreement with eq. (212), this eq. (212) for $\mathbf{a} = \mathbf{b}$ with $\mathbf{b} \cdot \mathbf{b} = 1$ gives $|P(b, b)+1| \leq \varepsilon$, or $P(b, b) < \varepsilon-1$. In that case, (213) is a consequence of (227).

CHAPTER 5

THE EXPERIMENTS OF BLEULER AND BRADT, WU AND SHAKNOV, OF LANGHOFF, AND OF KASDAY, ULLMANN, AND WU

IN THIS chapter we consider experiments in which polarization correlations are observed for pairs of photons traveling in opposite directions, where this time the source of the photon pair is an annihilating positronium atom[31] at rest. Since no effective polarization filters are available for photons as hard as annihilation radiation, the polarization is measured indirectly through the effectiveness of Compton scattering at chosen angles. Due to the indirect character of this polarization measurement, the results of these measurements are not conclusive for proving or disproving hidden-variables theories of the second kind quite in general, but, *if we assume that quantum theory describes Compton scattering correctly*, they do show that it is not possible to explain the two-photon correlation function $P(a, b)$ either by any hidden-variables theory of the second kind, or by a different "solution of the nonlocality paradox" suggested by Furry (1936) and discussed by Bohm and Aharonov (1957). [See Kasday (1970).]

As the assumption mentioned makes the outcome of these experiments a less convincing proof of the invalidity of hidden-variables theories of the second kind, we shall here only briefly discuss these experiments.

5.1. The annihilation two-photon wavefunction

Experiments of this kind were suggested by Wheeler (1946). The annihilation of positronium might be described crudely (neglecting the Coulomb binding) as a transition of a positive-energy electron at rest into a "hole" in the negative-energy states at rest.
Thus, the initial state is characterized by the Dirac equation

$$E_0\psi_0 \approx mc^2\beta\psi_0 = mc^2\psi_0. \tag{228a}$$

and the final state by

$$E_f\psi_f \approx mc^2\beta\psi_f = -mc^2\psi_f. \tag{228b}$$

We thus see that in the transition the value of β jumps from $+1$ to -1. Since β in the Dirac

[31] The "positronium atom" was by Wheeler (1946) more aptly called the "bielectron."

theory is the operator of spatial reflection, we conclude that there is a change of parity of the electronic wavefunction in the annihilation process. Therefore, the final state of the photon field must have odd parity.

The initial state of the positronium is assumedly the ground level S-state,[32] so that the total angular momentum is $j = 1$ or 0 depending upon whether the spins of the positon and the negaton line up or not. The final state has all of this angular momentum, so that the odd final photon state has $j = 0$ if it arises from annihilation of positronium in the singlet state, and has $j = 1$ when it arises from annihilation of positronium in the triplet state.

However, according to Section 2.6 [see (100)], for pairs of annihilation photons odd states with $j = 1$ do not exist. We conclude that positronium in the triplet state can be annihilated only into *three* photons, and that two-photon annihilation of positronium always starts from the singlet state.

Therefore the photon wavefunction arising in two-photon annihilation has $j = 0$. This wavefunction is given by eq. (97) with $\mathbf{F}$ from (99) with $j = 0$. In our abbreviated notation, this wavefunction is given by eq. (E-1) in Appendix E. In the notation of Section 2.6 it is given by

$$F_{n_1 n_2}(\mathbf{k}_1, \mathbf{k}_2) \propto \sum_n \varepsilon_{n n_1 n_2} k_n \quad (\text{with } \mathbf{k} = \mathbf{k}_1 = -\mathbf{k}_2). \tag{229}$$

For photon pairs observed in the $\pm z$-directions, (E-1) or (229) is proportional to

$$\sqrt{\tfrac{1}{2}}\,(\tilde{x}_1\tilde{y}_2 - \tilde{y}_1\tilde{x}_2), \tag{230}$$

so that we should find a perpendicular correlation of the photon polarizations in this odd $j = 0$ state, just like in Holt's experiments one finds in an even $j = 1$ state. (In both cases, the photon spin quantum number of the two-photon state is $s = 1$.)

5.2. The experimental setup

Some properly chosen radioactive substance is used as the source of positons. It is surrounded by material to slow down and stop the positons and to emit the annihilation radiation. The entire capsule is placed in the center of a narrow channel bored through a mass of lead. We choose this channel as the z-direction. Photon pairs that are emitted approximately in the $\pm z$-directions emerge from both ends of this channel and hit scintillation counters which serve two purposes. In the first place they serve as targets in which Compton scattering takes place. In the second place, they are part of the coincidence arrangement. The scattered photons are counted when they hit two other scintillation counters, and when there is coincidence within 5×10^{-9} s between the first two counters, and within 3×10^{-8} s between the first pair of counters and each of the second pair. Moreover, those counts are rejected in which the pulse in either of the second pair of counters does not have approximately the correct height. This discriminates against multiple-scattered photons which would have lesser energy [Langhoff (1960); Kasday (1970)].

[32] If the positronium is not in an S-state, there is insufficient probability that the electron pair is close enough together for annihilating each other.

The planes through the z-axis and each of the second pair of counters make an adjustable azimuthal angle φ with each other. In each of these planes, the scattering angle θ of the photon may be adjustable or may have been given a fixed value.

5.3. Quantum-theoretical dependence of the number of coincidence counts upon the angles φ and θ

The theory was worked out by Snyder *et al.* (1948). [See also Pryce and Ward (1947).] We refer to their paper for details. For given directions φ_1, θ_1 and φ_2, θ_2 of the two observed scattered beams, by the Compton formula the energy of scattered photon number n will be

$$h\nu_n = [2-\cos\theta_n]^{-1}mc^2. \tag{231}$$

The coincidence rate in this case, for the photon wavefunction (230) before scattering, is from the Klein–Nishina Compton scattering formula found to be proportional to

$$d\Omega_1\, d\Omega_2[F(\theta_1)\, F(\theta_2) - G(\theta_1)\, G(\theta_2) \cos 2(\varphi_1 - \varphi_2)], \tag{232}$$

where $d\Omega_1$ and $d\Omega_2$ are the solid angles in which scattering is observed, and where

$$F(\theta) \equiv [2+(1-\cos\theta)^3]/(2-\cos\theta)^3, \tag{233a}$$

$$G(\theta) \equiv \sin^2\theta/(2-\cos\theta)^2. \tag{233b}$$

For given $\theta = \theta_1 = \theta_2$, the coincidence rate at $\varphi = 90°$ is then larger than the coincidence rate at $\varphi = 0°$ by a factor

$$\varrho = \frac{F^2(\theta)+G^2(\theta)}{F^2(\theta)-G^2(\theta)}. \tag{234}$$

This ϱ, as a function of θ, has a maximum $\varrho \approx 2.85$ at $\theta \approx 82°$. (At $\theta = 90°$, $\varrho \approx 2.60$.)

5.4. Experimental results

Experiments have been performed by Bleuler and Bradt (1948), Hanna (1948), Vlasow and Dzehelepov (1949), Wu and Shaknov (1950), Bertolini, Bettoni, and Lazzarini (1955), and by Langhoff (1960). Recently, the experiment has been repeated by Kasday *et al.* (1970). [See also Kasday (1970, 1971).] Not all of these experiments were equally accurate. Within possible errors, agreement with quantum theory was always found.

5.5. Furry's solution of the nonlocality paradox

As a solution for the "nonlocality aspect of the Einstein–Podolsky–Rosen paradox" (see Chapter 1), Furry (1936) suggested an "Assumption A" which amounts to a breakdown of quantum theory of many-particle systems when the particles would become macroscopically separated. This idea was further discussed by Bohm and Aharonov (1957).

According to this idea, the two-photon wavefunction (230), which by (149) and by (4) may be written in a variety of ways, like

$$\psi_{\text{phot}} = \sqrt{\tfrac{1}{2}}(\tilde{x}_1\tilde{y}_2 - \tilde{y}_1\tilde{x}_2) = \sqrt{\tfrac{1}{2}}(\tilde{x}'_1\tilde{y}'_2 - \tilde{y}'_1\tilde{x}'_2) = -i\sqrt{\tfrac{1}{2}}(\tilde{R}_1\tilde{R}_2 - \tilde{L}_1\tilde{L}_2), \tag{235}$$

would upon macroscopic separation of the photons change into a mixed state made up of factorizable two-photon wavefunctions that might be terms in one of many possible ways of writing the original ψ_{phot} as a sum of products. That is, ψ_{phot} might change into a mixture of states like $\tilde{x}_1\tilde{y}_2$ or $\tilde{y}_1\tilde{x}_2$ or $\tilde{x}'_1\tilde{y}'_2$ or $\tilde{R}_1\tilde{R}_2$ or $\tilde{L}_1\tilde{L}_2$, etc., or maybe only into expressions like the first three, or maybe only into expressions like the latter two. (This somewhat vague hypothesis that a pure state ψ_{phot} would change into a mixed state certainly is a deviation from conventional quantum theory.) Then there still would be correlations between the polarizations of the two photon beams, but it would be of a different nature, and therefore lead to a different value of the ratio ϱ (see Section 5.3) for given θ and for $\varphi = 90°$. Bohm and Aharonov calculated for an ideal experiment that the maximum of ϱ according to Furry's ideas would be down from the quantum-theoretical value 2.85 to 1.00 if ψ_{phot} would change into $\tilde{R}_1\tilde{R}_2$ or $\tilde{L}_1\tilde{L}_2$, or to some value < 2 if ψ_{phot} would change into $\tilde{x}'_1\tilde{y}'_2$ in a randomly oriented coordinate system with $\hat{z}' \parallel \hat{z}$.

For the nonideal circumstances valid in the experiments of Wu and Shaknov (1950), Bohm and Aharonov (1957) found, for the average of ϱ over the solid angles used, a quantum-theoretical value of 2.0, and still found 1.0 when they applied Furry's idea using circular polarizations, while they found a value of approximately 1.5 when they applied Furry's ideas using linear polarizations ($\psi_{\text{phot}} \to \tilde{x}'_1\tilde{y}'_2$). The experimental result of Wu and Shaknov was 2.04 ± 0.08, showing that Furry's idea was in disagreement with experimental results.

The correlation function P(a, b) for annihilation radiation

When $A = \pm 1$ and $B = \pm 1$ for the Clauser and Holt experiments have the meaning explained above eq. (216), and when the quantum-theoretical value of $P(a, b)$ by analogy to (211) is defined as the expectation value $\langle AB \rangle_{\text{qu.th.}}$ [compare (210) with (208)], eq. (217) holds for $P(a, b)_{\text{qu.th.}}$ with the quantum-theoretical values of the probabilities w, which for ψ_{phot} given by eq. (151) are given by

$$\left.\begin{aligned} w[A_+, B_+] &= w[A_-, B_-] = \tfrac{1}{2}\sin^2(a-b),\\ w[A_+, B_-] &= w[A_-, B_+] = \tfrac{1}{2}\cos^2(a-b),\\ P(a, b)_{\text{qu.th.}} &= -\cos[2(a-b)]. \end{aligned}\right\} \tag{236}$$

This quantum theoretical $P(a, b)$, of course, for the angles a_1, b_1, a_2, b_2 with differences given by (169), will not satisfy Bell's generalized inequality (226). (This is the disagreement upon which Clauser *et al.* based their experiments.)

Kasday (1970) shows that any mixture with rotational and reflective symmetry, of products of two linearly, circularly, or elliptically polarized single-photon states of the kind into which ψ_{phot} according to Furry might change, will always lead to a $P(a, b)$ of the form

$$P(a, b)_{\text{Furry}} = -C\cos[2(a-b)] \quad \text{with} \quad |C| \leqslant \tfrac{1}{2}. \tag{237}$$

With this symmetry, we also have for the Furry probabilities $w[A_\pm, B_\pm]$ for any given directions a and b

$$w[A_+, B_+] = w[A_-, B_-], \quad w[A_+, B_-] = w[A_-, B_+], \tag{238}$$

where, again, A_+ (that is, $A = +1$) means polarization of the first photon parallel to a, and $A_-(A = -1)$ means its polarization perpendicular to a. (Similarly for $B_\pm$, photon 2, and b.) Then, eqs. (217), (238), and $\Sigma w = 1$ together with the Klein–Nishina formula give for the coincidence counting rate of the Compton-scattered photons, by reasoning similar to that used by Snyder *et al.* (1948),

$$d\Omega_1\, d\Omega_2[F(\theta_1)\, F(\theta_2)+G(\theta_1)\, G(\theta_2)\, P(\varphi_1, \varphi_2)], \tag{239}$$

of which (232) is the special quantum-theoretical case for which (236) is valid. If, however, the value (237) of $P(a, b)$ would be valid according to Furry's idea, and yet in each scatterer the Klein–Nishina formula would be valid, (239) in this case would predict

$$d\Omega_1\, d\Omega_2\, [F(\theta_1)\, F(\theta_2)-G(\theta_1)\, G(\theta_2)\, C \cos 2(\varphi_2-\varphi_1)] \tag{240}$$

with $|C| \leqslant \frac{1}{2}$.

Kasday (1970) discusses the factor $K(\varphi_1, \varphi_2)$ by which the coincidence counting rate at given θ_1 and θ_2 depends on $\varphi = \varphi_2-\varphi_1$. If we may believe quantum mechanics for the scattering, this factor is, by (239),

$$K(\varphi) = 1+\beta P(\varphi_1, \varphi_2), \tag{241a}$$

where the coefficient

$$\beta = G(\theta_1)\, G(\theta_2)/F(\theta_1)\, F(\theta_2) \tag{241b}$$

is given by eqs. (233a, b). Values of K for $\varphi = 0°, 30°, 45°, 60°$, and $90°$ observed by Kasday, Ullman, and Wu all agreed with an azimuthal dependence

$$K(\varphi) = 1-B \cos (2\varphi), \tag{242}$$

which agrees with (241a, b) for

$$P(a, b) = -C \cos [2(a-b)] \tag{243}$$

with

$$B = C\beta. \tag{244}$$

In (243), according to quantum theory, $C = 1$ by (236), while Furry's hypothesis gives $|C| \leqslant \frac{1}{2}$. The values of B obtained by (242) from observed values of K gave by (244) and (241b) values for C in agreement with $C = 1$ and in total disagreement with $|C| \leqslant \frac{1}{2}$. (Actually, these values were averages over certain intervals of θ_1 and θ_2.)

Most earlier observers (see Section 5.4) have measured K at $\varphi = 0°$ and $90°$ only, and then expressed their results in terms of the ratio

$$\varrho = \frac{K(90°)}{K(0°)} = \frac{1+B}{1-B} = \frac{F(\theta_1)\, F(\theta_2)+CG(\theta_1)\, G(\theta_2)}{F(\theta_1)\, F(\theta_2)\, -CG(\theta_1)\, G(\theta_2)}, \tag{245}$$

which for $C = 1$ gives the quantum-mechanical result (234) if $\theta_1 = \theta_2$.

5.6. Proofs that experiments with Compton scattering of annihilation radiation cannot conclusively exclude *all* hidden-variables theories of the second kind

Bell has shown [see Kasday (1970)] that, if one does not accept for granted that quantum theory correctly describes Compton scattering, one can construct an *ad hoc* hidden-variables theory of the second kind which yields for the experiments here discussed the quantum-theoretical predictions (232). While Bell's *ad hoc* theory would be impossible for Compton scattering of *soft* photon pairs, Kasday (1970) has proposed a more general *ad hoc* theory of the second kind for which the limitation to hard photons does not exist. See Appendix G for details on these *ad hoc* theories of Bell and of Kasday.

In these not-disprovable hidden-variables theories of Bell and Kasday, however, arbitrarily *ad hoc* postulates are made for the scattering probabilities. If we assume that the latter, for given incident polarized photon wavefunctions, are given by quantum-mechanical scattering theory, these arbitrary scattering probabilities are not possible, and the experiments of Kasday, Ullman, and Wu become more comparable to the experiments of Clauser and of Holt, with assumedly fully understood Compton scattering replacing fully understood polarization filters. The difference remaining is that people wanting to replace quantum theory by hidden-variables theory of the second kind dislike using quantum theory for understanding Compton scattering, while in Section 4.7 we used no quantum theory for explaining how the polarization filters worked in the experiments of Clauser and of Holt.

5.7. The kind of hidden-variables theory that is excluded by the experiments using Compton scattering of annihilation radiation

Consider a hidden-variables theory that does predict a polarization correlation $P(a, b)$ between polarizations in directions a and b of the two photons of two-photon annihilation radiation. (That is, assume that, in principle, ideal linear polarizers would exist by which this radiation could be analyzed and to which analysis this hidden-variables theory would be applicable.) Then, Bell's generalized inequality (226) would be valid for $P(a, b)$.

For now applying the Klein–Nishina formula for calculating the coincidence counting rate with the photon detectors at θ_1 and θ_2 in two scattering planes in directions φ_1 and φ_2 (see Section 5.2), we do as if the photons incident upon the scatterers are in a mixed state with the polarizations of their wavefunctions correlated with correlation function $P(\varphi_1, \varphi_2)$. Then for perfect counter efficiency the coincidence rate would be proportional to (239).

The use of φ_1 and φ_2 as a and b in $P(a, b)$ means resolving for each scatterer the incident light wave into polarizations parallel and perpendicular to the scattering plane. This is necessary for avoiding interference terms between different polarizations, like the one in (146).[33]

[33] Instead of using a mixed state of incident waves polarized either parallel or perpendicular to the scattering planes (φ_1 and φ_2), one could in principle use any mixture of arbitrarily polarized light waves incident upon scatterers 1 and 2, as long as the probability distribution over simultaneously incident polarizations at both scatterers is chosen in such a way that it would yield the correlation function $P(a, b)$ if the scatterers would be replaced by ideal polarization filters with polarization directions a and b. In the cross-sections for scattering into the pair of planes φ_1 and φ_2, then, all interference terms, of incident waves polarized parallel to these planes with those polarized perpendicularly to them, should cancel out. [Compare the purely quantum-mechanical case considered by Snyder *et al.* (1948).] Therefore we may as well leave out such interference cross-terms from the beginning in the way indicated in the text.

Examples of hidden-variables theories for which $P(a, b)$ is well defined were those leading to eqs. (163)–(166) for the case of parallel polarization correlation, and the same with sin and cos interchanged for the case of perpendicular polarization correlation as in the case considered here. In each case, $P(a, b)$ can be obtained from $R(a, b)$ by eq. (219) with $R_1(a) = R_2(b) = \frac{1}{2} R_\theta$, so that

$$P(a, b) = 4R(a, b)/R_0 - 1. \tag{246}$$

This gives in the four cases mentioned above

$$P(a, b) = \sin^2 (a-b) - \tfrac{1}{2} = -\tfrac{1}{2} \cos [2(a-b)], \tag{163'}$$
$$P(a, b) = 1 - 4|\cos (a-b)|/\pi, \tag{164'}$$
$$P(a, b) = 4|\sin (a-b)|/\pi - 1, \tag{165'}$$
$$P(a, b) = -\tfrac{1}{4} + \tfrac{1}{2} \sin^2 (a-b) = -\tfrac{1}{4} \cos [2(a-b)], \tag{166'}$$

as contrasted to the result

$$P(a, b) = -C \cos [2(a-b)] \tag{247}$$

confirmed by the measurements of Kasday *et al.*[34] Consider, therefore, a hidden-variables theory like in (163′) or (166′), for which (247) would hold. Then Bell's generalized inequality (226) gives by (170)

$$|C|\{|\cos [2(\varphi_1+\varphi_2+\varphi_3)] - \cos (2\varphi_1)| + |\cos(2\varphi_2)+\cos(2\varphi_3)|\} \leqslant 2. \tag{248}$$

Picking $\varphi_1 = \varphi_2 = \varphi_3 = (22\frac{1}{2}^\circ$ or $67\frac{1}{2}^\circ)$, we find from this

$$|C| \leqslant \sqrt{\tfrac{1}{2}} \tag{249}$$

for this type of a theory.

We may discuss such theories like we discussed Furry's theories with $|C| \leqslant \frac{1}{2}$ below eq. (237). As before, we can relate C to an experimental ϱ-value [compare (245) for the Furry case], or express it in terms of a θ-dependent coefficient B like below eq. (242). While C and therefore B in Furry's theory had at most half the value they have in quantum theory, in the type of theory described above they have at most $\sqrt{\frac{1}{2}} \times$ the quantum-theoretical value. As the theoretical value of ϱ in quantum theory might go as high as $2.85 = \dfrac{1+0.48}{1-0.48}$, for which $B = 0.48$, in Furry's theory we have at most $B = 0.24$ and, by (242), $\varrho = \dfrac{1+0.24}{1-0.24} = 1.63$, and in the hidden-variables theories of the type described above we have at most $B = 0.24\sqrt{2} = 0.34$ and $\varrho = \dfrac{1+0.34}{1-0.34} = 2.03$, for optimal scattering directions and infinitesimal solid angles.

In practice, we must average over finite solid angles. Kasday (1970) gives results for various regions of the angles θ_1 and θ_2 on which B depends, and compares the maximal values for B allowed for this type of a hidden-variables theory according to Bell's inequality (248),

[34] See the sentence containing eq. (243). Note that (247) excludes (164′)–(165′).

with the experimental values obtained by Kasday *et al.* He concludes that the experimental results exclude this type of hidden variables theory which we described in the first three paragraphs of this section together with the condition (247). They exclude also all hidden-variables theories which do *not* satisfy the condition (247), for reasons explained in the sentence containing (242). But all these exclusions of hidden-variables theories by these experiments are qualified by the assumption that we know quantum theory to hold for Compton scattering. This makes the disproof of hidden-variables theories of the second kind by these experiments much less convincing than their disproof by the experiments of Clauser and of Holt (see the conclusion of Section 5.6), though it must be granted that in the latter case the equality of the efficiencies (205)–(206) in Section 4.8 is an assumption (but it is a plausible one).

CHAPTER 6

CONCLUSIONS

IN THIS three-part book we have considered the reasons why some people are not satisfied by various aspects of quantum theory and how they have tried to construct theories that to their taste would be more satisfactory. The purpose of this all was either to complement quantum theory by amendments which would give the theory a cryptodeterministic character, or (in the case discussed in this Part III) to make the theory look like a local causal theory for elementary particles (rather than for wave fields).

Mathematicians can disprove any theory by adopting axioms that contradict the theory. We have seen how several authors have taken this approach and have constructed so-called theories of the zeroth kind. These theories were useful as warnings what axioms should be avoided if one would try to construct a theory of the first kind.

The theories of the first kind considered were of two types. There was Bohm's 1951 "cryptodeterministic nonrelativistic wave mechanics" in which the electron positions were the hidden variables. We did not succeed in making a satisfactory relativistic theory out of this.

The other type used the polychotomic algorithm. While this theory was also formulated in a nonrelativistic way, as the division of spacetime into time and space was tacitly assumed to be given and invariant, this might be ascribed to the fact that there is a preferred frame in which the apparatus is at rest by which one makes the measurements of which the algorithm predicts the outcome. While we made no attempt, for instance in the Bohm–Bub phenomenological generalization of the Schrödinger equation during a measurement, to formulate the theory in a relativistically covariant way, it might be imaginable that it could be done.

"Gleason troubles" could be avoided by adopting Tutsch's rule, which led back to the Wiener and Siegel 1953 generalization of the polychotomic algorithm. This theory, however, led to serious paradoxes as discussed in Appendix K of Part II.

The theory led to deviations from quantum theory where measurements followed each other rapidly. Experiments by Papaliolios trying to find these nonquantum effects failed to find them. More experimentation is needed to be sure that these effects really do not exist.

Theories of the second kind lead in a less sophisticated way to deviations from quantum theory. In fact, these theories were constructed with the *purpose* of contradicting quantum theory, where quantum theory does not explain the behavior of composite systems in a way that makes the individual elementary particles or quanta behave individually locally causally.

Experiments so far finished, however, have given results in agreement with unaltered quantum theory. More conclusive experiments, like those by Clauser and by Holt, should decide which theory is wrong—quantum theory, or the entire category of hidden-variables theories of the second kind considered in Section 4.7. Results obtained by Freedman and Clauser seem to show that the latter are wrong.[25a] Holt's results, however, would make quantum theory appear to be wrong.[25b]

Unless Holt's experimental results would be confirmed and the results of Freedman and Clauser would be shown to be faulty, we would have to drop theories of the second kind for their disagreement with experimental data, while on pages 214–217 we found theories of the first kind to lead paradoxes.

It must be granted that, for those trying to explain all happenings in nature on the basis of cryptodeterministic local laws, the arguments brought forward in favor of hidden-variables theories have great attractiveness. It would be unfortunate for them if nature is found to refuse cooperation. It then should be concluded honestly that the attempts toward reintroducing some kind of determinism into physics would have failed.

We then can take one of several possible approaches. We can retreat to positivism and refuse to consider questions meaningful to which the answers cannot be verified directly by observation. Quantum theory in the past has been able to deal with many of the questions that remain. We all know that relativistic quantum field theory is still in bad shape, which indicates that the theory in its present form has no finality. There are, however, indications that corrections yet to be made upon the theory will take the theory even farther away from a local cryptodeterministic theory.

Another possibility is that we resign to the fact that nature in principle acts in whatever way it likes to act, and that the so-called laws of nature observed in the past are merely statistical averages over a large number of free-will acts of nature. This point of view can easily be extended in a way to make it attractive to those inclined to ascribe much of what is happening to the existence of some principle called God.

For those more mathematically inclined, there may be attractiveness in the Everett type of quantum theory,[35] in which we live in one branch of a universe that is continuously splitting up into infinitely many branches of different possibilities. This, of course, leaves the question unanswered why we live in the particular branch we live in. Thus it does not really answer more questions than ordinary quantum theory. (It merely avoids "reductions of wave-functions" that are practical from a physical point of view but that are illogical from a purely mathematical point of view.) Nor does it answer more than the religious answer that "what happens is in the hands of God."

The reader may draw a different conclusion, yet.[36]

Whatever point of view we personally feel forced to take under these circumstances, it will mean resignation to the fact that today's science does not have all the answers. And this conclusion is satisfactory, if not trivial.

[35] See Everett (1957), Wheeler (1957), Shimony (1963), Pearle (1967), DeWitt (1970), Graham (1970), DeWitt and Graham (1971), and Belinfante, "Measurements in objective quantum theory" (1973, to be published).

[36] If he is an open-minded physicist and not some kind of an axiomatic orthodox mathematician who is not interested in experimental results, it does not seem likely that this different conclusion would be an undisturbed belief that hidden-variables theory has solved the problem.

And, if some sharp minds will keep trying to come up with a better kind of hidden-variables theory, I wish them all possible success in demonstrating *experimentally* that their theory is better than any previous one, as such a theory would find its right of existence primarily in predicting deviations from the old theory which by observation would be found to have factual reality.

APPENDICES TO PART III

APPENDIX A

THE EXPECTATION VALUE OF AN OBSERVABLE Q FOR ONE PHOTON OF GIVEN POLARIZATION AND MOMENTUM

IN THE state given by eq. (33) we shall here evaluate the value (27) of $\langle Q\rangle_{1\,\text{phot}}$ in the form (30) of a sum over a momentum and polarization distribution.

We insert (22a, b) and (23) in (17) and find for Q the quantum operator

$$Q(q_f)_{\text{op}} = -(2\mathcal{V})^{-1}\int \sum_{\mathbf{k}}\sum_{\eta}\sum_{\mathbf{k}'}\sum_{\eta'} \mathbf{e}_{\mathbf{k}}^{\eta *} e^{-i\mathbf{k}\cdot\mathbf{x}} \left\{\frac{1}{2}\left(\frac{\partial}{\partial q_R^{\eta}} - i\frac{\partial}{\partial q_{I\mathbf{k}}^{\eta}}\right)(q_{R\mathbf{k}'}^{\eta'} + i q_{I\mathbf{k}'}^{\eta'}) + \frac{1}{2}(q_{R\mathbf{k}'}^{\eta'} + i q_{I\mathbf{k}'}^{\eta'})\left(\frac{\partial}{\partial q_{R\mathbf{k}}^{\eta}} - i\frac{\partial}{\partial q_{I\mathbf{k}}^{\eta}}\right)\right\}\cdot Q_{\text{op}} e^{i\mathbf{k}'\cdot\mathbf{x}} \mathbf{e}_{\mathbf{k}'}^{\eta'}\, d^3\mathbf{x}, \tag{A-1}$$

which should be inserted in eq. (24) for $\langle Q\rangle$. Using (33) [or, for $\langle Q\rangle_{\text{vac}}$, using (25)] for $\psi^{(F)}(q_f)$, we can perform all integrations over the q_f. This leaves the integration in (A-1) over the $\mathbf{x}$, as in eq. (28).

Let λ be an abbreviation for $\mathbf{k}$ and η together, and, for given λ, let

$$\left.\begin{aligned} u &= (ck/\hbar)^{\frac{1}{2}} q_{R\mathbf{k}}^{\eta}, \quad v = (ck/\hbar)^{\frac{1}{2}} q_{I\mathbf{k}}^{\eta}, \\ r^2 &= u^2+v^2, \quad \theta = \arctan(v/u), \quad u \pm iv = re^{\pm i\theta}, \end{aligned}\right\} \tag{A-2}$$

so that

$$(ck/\pi\hbar)\, dq_{R\mathbf{k}}^{\eta}\, dq_{I\mathbf{k}}^{\eta} = du\, dv/\pi = r\, dr\, d\theta/\pi. \tag{A-3}$$

Then, by

$$\frac{\partial}{\partial u} \pm i\frac{\partial}{\partial v} = e^{\pm i\theta}\left[\frac{\partial}{\partial r} \pm \frac{i}{r}\frac{\partial}{\partial \theta}\right], \tag{A-4}$$

(A-1) becomes

$$Q(q_f)_{\text{op}} = -\frac{1}{2}\sum_{\lambda}\sum_{\lambda'}\int \frac{d^3\mathbf{x}}{\mathcal{V}} e^{-i\mathbf{k}\cdot\mathbf{x}}\mathbf{e}^{*}\cdot\left\{\frac{1}{2}e^{-i\theta}\left[\frac{\partial}{\partial r} - \frac{i}{r}\frac{\partial}{\partial\theta}\right](r'e^{i\theta'}) + \frac{1}{2}(r'e^{i\theta'})e^{-i\theta}\left[\frac{\partial}{\partial r} - \frac{i}{r}\frac{\partial}{\partial\theta}\right]\right\} Q_{\text{op}} e^{i\mathbf{k}'\cdot\mathbf{x}}\mathbf{e}', \tag{A-5}$$

where in $\sum_\lambda = \sum_{\mathbf{k}}\sum_\eta$ the u and v of "the other halfspace of $\mathbf{k}$" are related to those of the preferred first halfspace by

$$u_{-\mathbf{k}} \equiv -u_{\mathbf{k}}, \quad v_{-\mathbf{k}} \equiv +v_{\mathbf{k}}, \tag{A-6}$$

in accordance with (21). In general, the relation $\mathbf{k}' = -\mathbf{k}''$, $\eta' = \eta''$ between two waves we shall indicate by saying that "$\lambda' = -\lambda''$," so that

$$\begin{aligned} &\mathbf{k}' = -\mathbf{k}'', \quad \eta' = \eta'', \quad u' = -u'', \quad v' = v'', \\ &r' = r'', \quad \theta' = \pi - \theta'', \quad \text{if} \quad \lambda' = -\lambda''. \end{aligned} \tag{A-7}$$

Equation (25) now becomes

$$\psi^{(F)}_{\text{vac}} = \prod_f \left\{ (ck_f/\pi\hbar)^{\frac{1}{2}} \exp\left(-\tfrac{1}{2}r_f^2\right) \right\}, \tag{A-8}$$

where $\prod_f = \prod_{\mathbf{k}/2}\prod_\eta$, and (33) becomes

$$\psi^{(F)}(q_f) = \psi^{(F)}_{\text{vac}} \sum_a f_a r_a \exp(-i\theta_a), \tag{A-9}$$

where the index a labels another set of values of λ. $\left(\sum_a = \sum_{\mathbf{k}_a}\sum_{\eta_a}.\right)$

Inserting (A-3), (A-8), and (A-9) in (24), we find

$$\langle Q \rangle = \left\{ \prod_f \int 2r_f\, dr_f\, (d\theta_f/2\pi) \cdot \exp\left(-\tfrac{1}{2}r_f^2\right) \right\} \sum_a f_a^* r_a \exp(i\theta_a) \ldots$$

$$\sum_b f_b r_b \exp(-i\theta_b) \cdot \left\{ \prod_f \exp\left(-\tfrac{1}{2}r_f^2\right) \right\}, \tag{A-10}$$

where ... stands for $Q(q_f)_{\text{op}}$ given by (A-5), in which we will write λ_j for λ and λ_k for λ', and $\sum_j\sum_k$ for $\sum_\lambda\sum_{\lambda'}$.

The factors $\int_0^\infty 2r_f\, dr_f \exp(-r_f^2) \oint d\theta_f/2\pi$ arising from the vacuum factors (A-8) in $\int \psi^{(F)*} \ldots \psi^{(F)} \prod_f dq_f$ yield factors 1 in (A-10) with (A-5) unless q_f happens to coincide with q_j or q_k or q_a or q_b. The latter may partially or all be identical among each other, so, for obtaining factors $\neq 1$ we carry explicitly only as many factors in $\prod_f$ of (A-10) as correspond to "essentially different" field coordinates among the q_j, q_k, q_a, and q_b, where we call two of these q_λ "not essentially different" if their λ are either equal or opposite and equal to each other. [Compare (A-7).]

Notice that (A-10) with (A-5) contains

$$\left\{ \prod_f \oint d\theta_f \right\} \exp i(\theta_a - \theta_j + \theta_k - \theta_b). \tag{A-11}$$

This fact is not altered by a possible differentiation of $\exp(i\theta_k)$ with respect to θ_j when the indices j and k in $\sum_j\sum_k$ stand for the same λ_f. If any of the θ_a, θ_j, θ_k, and θ_b is essentially different from all the other three of them, the corresponding $\oint d\theta \exp(i\theta)$ contained in (A-11) gives a factor zero. Therefore, for a nonvanishing contribution, the λ_a, λ_j, λ_k, and λ_b must be either all essentially equal, or pairwise essentially equal. However, even in those cases (A-11) would give zero unless the essential equalities make $\theta_a - \theta_j + \theta_k - \theta_b \equiv$ constant. [For instance, if λ_j and λ_k are essentially equal but essentially different from λ_a, we want $\lambda_j = \lambda_k$ and not $\lambda_j = -\lambda_k$; but, if λ_a and λ_k are essentially equal but are essentially different from λ_b, we want $\lambda_a = -\lambda_k$ (so $\theta_a = \pi - \theta_k$ and $\theta_a + \theta_k =$ constant), and not $\lambda_a = \lambda_k$.]

This leaves six cases in which (A-11) might possibly not vanish:

$$\left.\begin{array}{l}
\text{Case 1: } \lambda_j = \lambda_k = \lambda_a = \lambda_b. \left[\text{In } \prod_f, \text{ then merely consider } f|=|j, \text{ where } f|=|j \text{ is an abbreviation for } \lambda_f|=|\lambda_j, \text{ that is, } \lambda_f = (\lambda_j \text{ or } -\lambda_j). \text{ Note that the sign in } \lambda_f \text{ is determined because } \mathbf{k}_f \text{ must lie in the preferred halfspace.}\right] \\
\text{Case 2. } \lambda_j = \lambda_k|\neq|\lambda_a = \lambda_b, \text{ where } \lambda_k|\neq|\lambda_a \text{ means } \lambda_k \neq \lambda_a \neq -\lambda_k. \left[\text{In } \prod_f, \text{ then consider merely } f| = |j \text{ and } f| = |b.\right] \\
\text{Case 3. } \lambda_j = \lambda_k = -\lambda_a = -\lambda_b. \left[\text{In } \prod_f \text{ then merely consider } f| = |j.\right] \\
\text{Case 4. } \lambda_k = -\lambda_a| \neq |\lambda_j = -\lambda_b. \left[\text{In } \prod_f \text{ then consider merely } f| = |j \text{ and } f| = |k.\right] \\
\text{Case 5. } \lambda_j = -\lambda_b = -\lambda_k = \lambda_a. \left[\text{In } \prod_f \text{ then consider merely } f| = |j.\right] \\
\text{Case 6. } \lambda_j = \lambda_a|\neq|\lambda_k = \lambda_b. \left[\text{In } \prod_f \text{ then consider merely } f| = |j \text{ and } |f| = k.\right]
\end{array}\right\} \quad \text{(A-12)}$$

In each case, we should consider on what factors the operator

$$\Omega_j \equiv \frac{\partial}{\partial r_j} - \frac{i}{r_j}\frac{\partial}{\partial \theta_j} \quad \text{(A-13)}$$

in (A-5) is acting. There is always a factor $\exp(-\frac{1}{2}r_j^2)$ hiding in the $\prod_f$ at the end of (A-10). However, in the above six cases some of the a, b, k will be identical with j.

Keeping provisionally $Q(q_f)_{\text{op}}$ in the form (A-5) containing $\sum_j \sum_k \int \exp(-i\mathbf{k}_j \cdot \mathbf{x}) Q_{\text{op}} \exp(i\mathbf{k}_k \cdot \mathbf{x})\, \mathbf{e}_k\, d^3x/\mathcal{V}$ rather than an expression containing $\tilde{Q}_{\text{op}}$, there are no differentiations by Q_{op} of the various r and θ over which we must integrate, nor any differentiations of their subscripts $\boldsymbol{k}$ or $\boldsymbol{x}$. In all six cases, the terms contributing to $\langle Q \rangle$ have in common the factors

$$-\frac{1}{2}\int \frac{d^3\mathbf{x}}{\mathcal{V}} e^{-i\mathbf{k}_j \cdot \mathbf{x}} \mathbf{e}_j^* f_a^* \cdot \{Q_{\text{op}} e^{i\mathbf{k}_k \cdot \mathbf{x}} \mathbf{e}_k\} f_b. \quad \text{(A-14)}$$

Also, there are in each case the following factors which are never differentiated by Ω_j:

$$\left\{\prod_f 2r_f\, dr_f (d\theta_f/2\pi) \exp(-\tfrac{1}{2}r_f^2)\right\} r_a e^{i\theta_a - i\theta_j}, \quad \text{(A-15)}$$

See here (A-12) in each case for the values of the index f to be considered. Also, $\theta_a-\theta_j$ should be expressed in terms of the θ_f; that is, in cases 1 through 6 its respective values are

$$\theta_a-\theta_j \quad = \quad 0, \quad (\theta_b-\theta_j), \quad (\pi-2\theta_j), \quad (\pi-\theta_k-\theta_j), \quad 0, \quad \text{and} \quad 0. \tag{A-16}$$

Finally, there are the factors which are sensitive to the differentiations by Ω_j. We shall use here the notation

$$\text{Sym}\,[\Omega]\{[B]\cdot C\} \equiv \tfrac{1}{2}[\Omega]\{BC\}+\tfrac{1}{2}B[\Omega]\{C\}. \tag{A-17}$$

Then, the factors not yet listed in (A-14) and (A-15) are:

$$\left.\begin{aligned}
&\text{In Case 1: Sym}\,[\Omega_j]\{[r_j e^{i\theta_j}]\cdot r_j e^{-i\theta_j}\exp(-\tfrac{1}{2}r_j^2)\} = (r_j-r_j^3)\exp(-\tfrac{1}{2}r_j^2).\\
&\text{In Case 2: Sym}\,[\Omega_j]\{[r_j e^{i\theta_j}]\cdot r_b e^{-i\theta_b}\exp(-\tfrac{1}{2}r_j^2-\tfrac{1}{2}r_b^2)\}\\
&\qquad = (1-r_j^2)e^{i(\theta_j-\theta_b)}r_b\exp(-\tfrac{1}{2}r_j^2-\tfrac{1}{2}r_b^2).\\
&\text{In Case 3: Sym}\,[\Omega_j]\{[r_j e^{i\theta_j}]\cdot r_j e^{i(\theta_j-\pi)}\exp(-\tfrac{1}{2}r_j^2)\} = (3r_j-r_j^3)e^{2i\theta_j-i\pi}\exp(-\tfrac{1}{2}r_j^2).\\
&\text{In Case 4: Sym}\,[\Omega_j]\{[r_k e^{i\theta_k}]\cdot r_j e^{i(\theta_j-\pi)}\exp(-\tfrac{1}{2}r_j^2-\tfrac{1}{2}r_k^2)\}\\
&\qquad = (2-r_j^2)e^{i(\theta_j+\theta_k-\pi)}r_k\exp(-\tfrac{1}{2}r_j^2-\tfrac{1}{2}r_k^2).\\
&\text{In Case 5: Sym}\,[\Omega_j]\{[r_j e^{i(\pi-\theta_j)}]\cdot r_j e^{i(\theta_j-\pi)}\exp(-\tfrac{1}{2}r_j^2)\} = (2r_j-r_j^3)\exp(-\tfrac{1}{2}r_j^2).\\
&\text{In Case 6: Sym}\,[\Omega_j]\{[r_k e^{i\theta_k}]\cdot r_k e^{-i\theta_k}\exp(-\tfrac{1}{2}r_j^2-\tfrac{1}{2}r_k^2)\}\\
&\qquad = -r_j r_k^2\exp(-\tfrac{1}{2}r_j^2-\tfrac{1}{2}r_k^2).
\end{aligned}\right\} \tag{A-18}$$

Multiplying these expressions (A-18) by (A-15) with (A-16) and with $\prod_f$ from (A-12), and performing the integrations over the r_f and θ_f, we find products of integrals

$$I_m \equiv \int_0^\infty 2\,r^m\,dr\,e^{-r^2}\int_0^{2\pi} d\theta/2\pi, \quad (I_1 = I_3 = 1, \quad I_5 = 2), \tag{A-19}$$

namely:

$$\left.\begin{aligned}
&\text{In Case 1: the integral over } r_j \text{ gives} && I_3-I_5 = -1,\\
&\text{In Case 2: the integrals over } r_j \text{ and } r_b \text{ give} && (I_1-I_3)I_3 = 0,\\
&\text{In Case 3: the integral over } r_j \text{ gives} && 3I_3-I_5 = 1,\\
&\text{In Case 4: the integrals over } r_j \text{ and } r_k \text{ give} && (2I_1-I_3)I_3 = 1,\\
&\text{In Case 5: the integral over } r_j \text{ gives} && 2I_3-I_5 = 0,\\
&\text{In Case 6: the integrals over } r_j \text{ and } r_k \text{ give} && -I_3I_3 = -1.
\end{aligned}\right\} \tag{A-20}$$

In each case, (A-14) is to be multiplied by the numerical result found in (A-20), and the results summed over a, j, k, b subject to the restrictions (A-12). It is convenient to add and subtract the expression (A-14) for Case 5; so, we obtain the sum of (A-14) for Cases 3, 4, 5 minus the sum of (A-14) for Cases 1, 5, 6. Cases 1, 5, 6 together form Case I characterized by $\lambda_j = \lambda_a$ and $\lambda_k = \lambda_b$, while Cases 3, 4, 5 together form Case II with $\lambda_j = -\lambda_b$ and $\lambda_k = -\lambda_a$. Thus

$$\langle Q\rangle = \tfrac{1}{2}\Big(\sum_{\text{I}}-\sum_{\text{II}}\Big)\int \frac{d^3\mathbf{x}}{\mathcal{V}}\, e^{-i\mathbf{k}_j\cdot\mathbf{x}}\,\mathbf{e}_j^* f_a^*\cdot\{Q_{\text{op}}\,e^{i\mathbf{k}_k\cdot\mathbf{x}}\,\mathbf{e}_k\}\,f_b. \tag{A-21}$$

With Q_{op} given by one of the operators (18), we may make now the transformation from (28)–(29) to (30), with $\tilde{Q}_{op}$ given by (32). In the result, $\sum_j \sum_k$ becomes a sum $\sum_{\mathbf{k}} \sum_{\eta} \sum_{\eta'}$ with $\eta = \eta_j$, $\eta' = \eta_k$, and $\mathbf{k} = \mathbf{k}_j = \mathbf{k}_k$, but, for $Q \equiv \mathbf{L}$, the differentiation $\partial/\partial\mathbf{k}$ contained in $\mathbf{L}_{op}$ will differentiate the coefficient of $e^{i\mathbf{k}_k \cdot \mathbf{x}}$ in (A-21) *before* $\mathbf{k}_j$ is equated to $\mathbf{k}_k$. In $\sum_{\mathrm{I}}$ with $\lambda_k = \lambda_b$, this coefficient is $\varepsilon_k f_b^* = \mathbf{e}_k f_k^*$, while, in $\sum_{\mathrm{II}}$ with $\lambda_k = -\lambda_a$, this coefficient is $\mathbf{e}_k f_a^* = \mathbf{e}_k f_{-k}^*$. Thus we find

$$\langle Q \rangle = \tfrac{1}{2} \sum_{\mathbf{k}} \sum_{\eta} \sum_{\eta'} \{\mathbf{e}_{\mathbf{k}}^{\eta *} f_{\eta}^*(\mathbf{k}) \cdot \tilde{Q}_{op}[\mathbf{e}_{\mathbf{k}}^{\eta'} f_{\eta'}(\mathbf{k})] - \mathbf{e}_{\mathbf{k}}^{\eta *} f_{\eta}(-\mathbf{k}) \cdot \tilde{Q}_{op}[\mathbf{e}_{\mathbf{k}}^{\eta'} f_{\eta'}^*(-\mathbf{k})]\}. \tag{A-22}$$

From this, we should possibly subtract $\langle Q \rangle_{vac}$, which is obtained by omitting from (A-10) the factors with indices a and b. Because of the factor $\oint d\theta \ldots \exp\{i(\theta_k - \theta_j)\}$, we need consider only the case $\theta_k \equiv \theta_j$, that is, $\lambda_k = \lambda_j$. Instead of (A-18), we now have

$$\mathrm{Sym}\left[\frac{\partial}{\partial r_j} - \frac{i}{r_j}\frac{\partial}{\partial \theta_j}\right] \{[r_j e^{i\theta_j}] \cdot \exp(-\tfrac{1}{2} r_j^2)\} = (1 - r_j^2)\, e^{i\theta_j} \exp(-\tfrac{1}{2} r_j^2), \tag{A-23}$$

while (A-15) is replaced by

$$2r_j\, dr_j\, (d\theta_j/2\pi) \exp(-\tfrac{1}{2} r_j^2)\, e^{-i\theta_j}. \tag{A-24}$$

Integration of the product of (A-23) and (A-24) gives a factor $I_1 - I_3 = 0$. Thence,

$$\langle Q \rangle_{vac} = 0. \tag{A-25}$$

Thus, $\langle Q \rangle_{1\,\mathrm{phot}}$ is given by (A-22).

In the last term of (A-22), we may write $-\mathbf{k}$ for the dummy $\mathbf{k}$, and we may interchange the dummies η and η'. Also using $\mathbf{e}_{-\mathbf{k}}^{\eta} = -e_{\mathbf{k}}^{\eta *}$, and putting

$$\mathbf{f}(\mathbf{k}) \equiv \sum_{\eta} f_{\eta}(\mathbf{k})\, \mathbf{e}_{\mathbf{k}}^{\eta}, \tag{A-26}$$

and indicating as an argument of $\tilde{Q}$ the value of the argument upon which $\tilde{Q}_{op}$ operates, we then obtain

$$\langle Q \rangle_{1\,\mathrm{phot}} = \tfrac{1}{2} \sum_{\mathbf{k}} \{\mathbf{f}(\mathbf{k})^* \cdot \tilde{Q}(\mathbf{k})_{op}\, \mathbf{f}(\mathbf{k}) - \mathbf{f}(\mathbf{k}) \cdot \tilde{Q}(-\mathbf{k})_{op}\, \mathbf{f}(\mathbf{k})^*\}. \tag{A-27}$$

Here, $\sum_{\mathbf{k}}$ may be replaced by $\int d^3\mathbf{k}$, as below eq. (31).

The normalization condition (39) gives

$$\begin{aligned}\sum_{\mathbf{k}} |\mathbf{f}(\mathbf{k})|^2 &= \sum_{\mathbf{k}} \sum_{\eta} \sum_{\eta'} f_{\eta}(\mathbf{k})^* f_{\eta'}(\mathbf{k})\, (\mathbf{e}_{\mathbf{k}}^{\eta *} \cdot \mathbf{e}_{\mathbf{k}}^{\eta'}) \\ &= \sum_{\mathbf{k}} \sum_{\eta} |f_{\eta}(\mathbf{k})|^2 = 1.\end{aligned} \tag{A-28}$$

APPENDIX B

THE EXPECTATION VALUE OF A FIELD OBSERVABLE Q IN A TWO-PHOTON STATE

We calculate here the expectation value $\langle Q\rangle$ of the observable given by eq. (A-1) or (A-5) in a state (75) which may be written in the notation of (A-9) as

$$\psi^{(F)}(q_f) = \psi^{(F)}_{\text{vac}} \sum_a \sum_b f_{ab} r_a r_b \exp(-i\theta_a - i\theta_b). \tag{B-1}$$

In case of two uncorrelated photons in states $f_\eta^{(\alpha)}(\mathbf{k})$ and $f_\eta^{(\beta)}(\mathbf{k})$, we could put

$$f^{(\alpha,\beta)}_{\eta_1,\eta_2}(\mathbf{k}_1, \mathbf{k}_2) \equiv f^{(\alpha,\beta)}_{ab} = f^{(\alpha)}_a f^{(\beta)}_b \equiv f^{(\alpha)}_{\eta_1}(\mathbf{k}_1) f^{(\beta)}_{\eta_2}(\mathbf{k}_2), \tag{B-2}$$

where we need not symmetrize, as the photon *states* of the radiation field are distinguishable (even if photons are not). It is seen from (B-1) that the symmetric part $\frac{1}{2}(f_{ab}+f_{ba})$ of f_{ab} determines $\psi^{(F)}$. It is convenient to define

$$F_{ab} = \sqrt{\tfrac{1}{2}}\,(f_{ab}+f_{ba}). \tag{B-3}$$

In this state of the radiation field, we calculate $\langle Q\rangle$ by

$$\langle Q\rangle = \left\{\prod_f \int 2r_f\, dr_f\, (d\theta_f/2\pi) \exp(-\tfrac{1}{2}r_f^2)\right\}$$
$$\sum_a \sum_b \sum_c \sum_d \left[f^*_{ab} r_a r_b \exp(i\theta_a + i\theta_b) \ldots f_{cd} r_c r_d \exp(-i\theta_c - i\theta_d) \left\{\prod_f \exp(-\tfrac{1}{2}r_f^2)\right\}\right], \tag{B-4}$$

where ... stands for (A-5). Following the reasoning of Appendix A, we find this time that the factors

$$\left\{\prod_f \oint d\theta_f\right\} \exp i(\theta_a+\theta_b+\theta_k+\theta_j-\theta_c-\theta_d) \tag{B-5}$$

contained in (B-4) with (A-5) will be zero when any θ in (B-4) is independent. For (B-5) $\neq 0$, a necessary condition is that each θ be paired off with another θ, so the two terms in (B-5) will combine to 0 or to $\pm\pi$. The six θ can be paired off in 15 different ways. For the calculation of the differentiations in (A-5) and integrations in (B-4) it is necessary to know which r and which θ are independent of each other. If A, B, and C stand for the three pairs into which the θ's are combined, we therefore must distinguish the eleven different possibilities,

$$\left.\begin{array}{llll} A=B=C, & A=B=-C, & A=-B=C, & -A=B=C, \\ A=B\,|\neq|\,C, & A=-B\,|\neq|\,C, & A=C\,|\neq|\,B, & A=-C\,|\neq|\,B, \\ B=C\,|\neq|\,A, & B=-C\,|\neq|\,A, & A\,|\neq|\,B\,|\neq|\,C\,|\neq|\,A. & \end{array}\right\} \tag{B-6}$$

Here, $A|\neq|B$ means that A differs from B as well as from $-B$.

We thus would distinguish $11\times 15 = 165$ different cases, but, when we write them out we see that they are not all different from each other. Of the $4\times 15 = 60$ cases in which all $\lambda(\equiv r, \theta$ pairs, as in Appendix A) are equal or opposite to each other, we find only 10 to be different from each other. Among the $6\times 15 = 90$ cases in which two pairs are equal or opposite to each other and the third pair consists of variables independent of the first two pairs, we find only 45 cases to be different among each other. Of the $1\times 15 = 15$ cases where all three pairs are independent, all 15 cases differ among each other. Therefore, in our calculation we must distinguish $10+45+15 = 70$ cases, as in Appendix A we distinguished $3+3 = 6$ cases.

That is, the integrals or sums over a, b, k, j, c, d which yield the value of $\langle Q\rangle$ must be split up according to these possible cases. We find

$$\langle Q\rangle = -\tfrac{1}{2}\sum_{a,b,k,j,c,d}\int \frac{d^3\mathbf{x}}{\mathcal{V}}\, e^{-i\mathbf{k}_j\cdot\mathbf{x}}\mathbf{e}_j^* f_{ab}^* \cdot\{Q_{\text{op}}\, e^{i\mathbf{k}_k\cdot\mathbf{x}}\,\mathbf{e}_k\} f_{cd}\, C_{abkjcd}, \tag{B-7}$$

where

$$C_{abkjcd} = \left\{\prod_f \int_0^\infty 2r_f\,dr_f \oint (d\theta_f/2\pi)\exp(-\tfrac{1}{2}r_f^2)\right\} r_a r_b\, e^{i(\theta_a+\theta_b-\theta_j)}\, S_{kjcd}, \tag{B-8}$$

with, in the notation of (A-13) and (A-17),

$$S_{kjcd} = \text{Sym}\,[\Omega_j]\,\{[r_k e^{i\theta_k}]\cdot r_c r_d\, e^{-i(\theta_c+\theta_d)} \prod_f \exp(-\tfrac{1}{2}r_f^2)\}. \tag{B-9}$$

In 10 cases, then, $\prod_f$ has only one factor; in 45 cases there are two factors; and, in 15 cases, there are three factors. In the latter cases, there are each time three sets of each two among the λ_a, λ_b, λ_k, λ_j, λ_c, and λ_d that are paired off and that are either equal (if the one of the pair is from among λ_a, λ_b, λ_k, and the other from among λ_j, λ_c, λ_d) or are opposite (if both are from among either λ_a, λ_b, λ_k, or λ_j, λ_c, λ_d). The index f labels these three sets. The label f of the set containing λ_j will be called F.

Let $A_{f\mu} = 1$ if λ_μ belongs to set f, and $A_{f\mu} = 0$ otherwise. Then,

$$\sum_f A_{f\mu} = 1, \quad \sum_\mu\sum_f A_{f\mu} = 6. \tag{B-10}$$

The same notations we also use for the other $45+10$ cases. There the number of sets is two or one instead of three.

Of the λ_μ belonging to set f, we pick one to represent the independent variables (r_f and θ_f) for the set. All λ_μ in a set then are equal or opposite to that λ_f. We put $\lambda_\mu = \varepsilon_\mu\lambda_f$ (where $A_{f\mu} = 1$). We may also write

$$\lambda_\mu = \sum_f A_{f\mu}\varepsilon_\mu\lambda_f. \tag{B-11}$$

The number of λ_μ contained in set f is always even. We call it $N_f = 2M_f$. Then

$$\sum_\mu A_{f\mu} = 2M_f. \tag{B-12}$$

We note that, for $A_{f\mu} = 1$, we have $r_\mu = r_f$, but we have $\theta_\mu = \theta_f$ only if $\varepsilon_\mu = 1$, while for

$\varepsilon_\mu = -1$ we have $\theta_\mu = \pi - \theta_f$. [See (A-7).] Therefore

$$e^{i\theta_\mu} = \varepsilon_\mu e^{i\varepsilon_\mu \theta_f}. \tag{B-13}$$

Therefore the integrand in (B-5) is

$$\exp i \sum_\mu \eta_\mu \theta_\mu = \varepsilon_a \varepsilon_b \varepsilon_k \varepsilon_j \varepsilon_c \varepsilon_d \exp\left(i \sum_f \theta_f \sum_\mu \eta_\mu \varepsilon_\mu A_{f\mu}\right), \tag{B-14}$$

where

$$\eta_a = \eta_b = \eta_k = 1 \quad \text{and} \quad \eta_j = \eta_c = \eta_d = -1. \tag{B-15}$$

For nonvanishing results, (B-14) must be constant. Therefore, for each f,

$$\sum_\mu \eta_\mu \varepsilon_\mu A_{f\mu} = 0. \tag{B-16}$$

The sign relation inside each pair in the original pairing off of λ was such that $\sum_{\mu \text{ in pair}} \eta_\mu \varepsilon_\mu = 0$ for each pair. Therefore also, summing over all pairs,

$$\sum_\mu \eta_\mu \varepsilon_\mu = 0 = \varepsilon_a + \varepsilon_b + \varepsilon_k - \varepsilon_j - \varepsilon_c - \varepsilon_d. \tag{B-17}$$

This is also obtainable from (B-16) by summing over μ and using (B-10).

We can now evaluate (B-9) as a product over f, where Ω_j is kept in the factor with $f = F$. We use (B-13) and

$$r_\mu = \prod_f r_f^{A_{f\mu}}. \tag{B-18}$$

The factors not containing λ_j are

$$r_f^{(A_{fk}+A_{fc}+A_{fd})} \exp\left(-\tfrac{1}{2} r_f^2\right) \varepsilon_k^{A_{fk}} \varepsilon_c^{A_{fc}} \varepsilon_d^{A_{fd}} \exp i\theta_f (A_{fk}\varepsilon_k - A_{fc}\varepsilon_c - A_{fd}\varepsilon_d). \tag{B-19}$$

The remaining factor is

$$\varepsilon_k^{A_{Fk}} \varepsilon_c^{A_{Fc}} \varepsilon_d^{A_{Fd}} \operatorname{Sym}[\Omega_j] \left\{ \left[r_F^{A_{Fk}} e^{iA_{Fk}\varepsilon_k \theta_F} \right] \cdot r_F^{A_{Fc}+A_{Fd}} \exp\left[-i\theta_F (A_{Fc}\varepsilon_c + A_{Fd}\varepsilon_d) - \tfrac{1}{2} r_F^2 \right] \right\}, \tag{B-20}$$

where

$$\Omega_j = \left[\frac{\partial}{\partial r_F} - \frac{i\varepsilon_j}{r_F} \frac{\partial}{\partial \theta_F} \right]. \tag{B-21}$$

We thus find for this factor

$$\varepsilon_k^{A_{Fk}} \varepsilon_c^{A_{Fc}} \varepsilon_d^{A_{Fd}} r_F^{(A_{Fk}+A_{Fc}+A_{Fd})} \exp\left(-\tfrac{1}{2} r_F^2\right) \exp i\theta_F (A_{Fk}\varepsilon_k - A_{Fc}\varepsilon_c - A_{Fd}\varepsilon_d).$$
$$\cdot \left[\frac{1}{r_F} \left(A_{Fc} + A_{Fd} + \frac{1}{2} A_{Fk} \right) - r_F + \frac{\varepsilon_j}{r_F} \left(\frac{1}{2} A_{Fk}\varepsilon_k - A_{Fc}\varepsilon_c - A_{Fd}\varepsilon_d \right) \right]. \tag{B-22}$$

We multiply this by the other factors (B-19), and obtain

$$S_{kjcd} = \varepsilon_k^{\sum_f A_{fk}} \varepsilon_c^{\sum_f A_{fc}} \varepsilon_d^{\sum_f A_{fd}} \left\{ \prod_f r_f^{(A_{fk}+A_{fc}+A_{fd})} \right\} \exp\left(-\sum_f \tfrac{1}{2} r_f^2 \right)$$
$$\cdot \exp\left[i \sum_f \theta_f (A_{fk}\varepsilon_k - A_{fc}\varepsilon_c - A_{fd}\varepsilon_d) \right] \cdot \left[\frac{K}{r_F} - r_F \right]. \tag{B-23}$$

where

$$K = \tfrac{1}{2} A_{Fk}(1+\varepsilon_j\varepsilon_k)+A_{Fc}(1-\varepsilon_j\varepsilon_c)+A_{Fd}(1-\varepsilon_j\varepsilon_d). \tag{B-24}$$

We insert (B-23) in (B-8), and use (B-10):

$$C_{abkjcd} = \varepsilon_a\varepsilon_b\varepsilon_j\varepsilon_k\varepsilon_c\varepsilon_d\left\{\prod_f \int_0^\infty 2r_f\,dr_f \oint (d\theta_f/2\pi) r_f^{A_{fa}+A_{fb}+A_{fk}+A_{fc}+A_{fd}} \exp(-r_f^2)\right.$$
$$\left.\cdot\exp i\theta_f(A_{fa}\varepsilon_a+A_{fb}\varepsilon_b-A_{fj}\varepsilon_j+A_{fk}\varepsilon_k-A_{fc}\varepsilon_c-A_{fd}\varepsilon_d)\right\}\cdot\left(\frac{K}{r_F}-r_F\right). \tag{B-25}$$

The latter exponent vanishes by (B-16). From (B-17) we conclude that the number of $\varepsilon_\mu = -1$ among the ε_b, ε_a, ε_k is equal to this number among the ε_j, ε_c, ε_d. So the total number of negative signs among the ε_μ is even, and $\varepsilon_a\varepsilon_b\varepsilon_j\varepsilon_k\varepsilon_c\varepsilon_d = 1$. Thus we get, by (B-12),

$$C_{abkjcd} = \left\{\prod_f \int_0^\infty 2r_f\,dr_f \oint (d\theta_f/2\pi) \exp(-r_f^2) r_f^{2M_f-A_{fj}}\right\}\cdot\left(\frac{K}{r_F}-r_F\right). \tag{B-26}$$

Thus the factors $f \neq F$ (with $A_{fj} = 0$) are (with $r_f^2 = x_f$)

$$\int_0^\infty dx_f \exp(-x_f)\, x_f^{M_f} = M_f!, \tag{B-27}$$

while the factor $f = F$ is

$$\int_0^\infty dx_f \exp(-x_f)\,(Kx_F^{M_f-1} - x_F^{M_F}) = \left(\frac{K}{M_F}-1\right)\cdot M_F!. \tag{B-28}$$

Thence,

$$C_{abkjcd} = \left(\frac{K}{M_F}-1\right)\left(\prod_f M_f!\right). \tag{B-29}$$

With the help of this formula, we can easily evaluate C_{abkjcd} numerically for all 70 cases to be considered. We find in this way C_{abkjcd} equal to 4 in one case, 2 in 11 cases, 1 in 12 cases, 0 in 22 cases, -1 in 12 cases, -2 in 11 cases, and -4 in 1 case. These results are listed in Table B.2. They are comparable to the results (A-20) of the one-photon case.

As mentioned before, we started out from 15 cases of pairing off, before we distinguished between the 11 possibilities (B-16). These 15 primary cases are comparable to three primary cases in the one-photon case, viz. cases I and II occurring in eq. (A-21), and a Case III which comprises cases 1, 2, 3 of (A-12).

In Appendix A, by adding and subtracting Case 5, we could write the sum of (A-20) times all cases 1 through 6, as a difference (A-21) between two of the original pairing-off cases. In the two-photon case, it turns out that the sum in (B-7) by adding and subtracting of equal terms can be written as the difference between two sums each over six original (pairing-off) cases. To see this, one needs not only the explicit values of the C_{abkjcd} for all 70 cases (see Table B.2), but also the splitting up (B-6) for the $6+6 = 12$ original cases concerned (see Table B.1).

TABLE B.1. BREAKING UP THE ORIGINAL PAIRING-OFF CASES INTO 70 SPECIAL CASES

Original Case No.	*a b k*	*j c d*	Subcases contained in original case											$C =$
I	*A B C*	*C E F*	1	2	3	4	5	6	7	8	9	10	11	0
II	*A B C*	*E C F*	1	12	13	4	14	15	16	17	9	18	19	1
III	*A B C*	*E F C*	2	12	13	3	20	21	22	23	10	18	24	1
IV	*A C B*	*C E F*	25	26	3	4	27	6	28	29	30	31	32	−1
V	*A C B*	*E C F*	25	33	13	4	34	15	35	36	30	37	38	1
VI	*A C B*	*E F C*	26	33	13	3	39	21	40	41	31	37	42	1
VII	*A C D*	*A E F*	25	26	1	2	27	5	43	44	45	46	47	−1
VIII	*A C E*	*A C E*	48	25	2	13	49	22	50	35	51	45	52	−1
IX	*A C E*	*A E C*	48	26	1	13	53	16	54	40	51	46	55	−1
X	*A C D*	*E A F*	25	33	1	12	34	14	43	56	57	58	59	1
XI	*A C E*	*C A E*	48	25	12	3	49	23	60	28	61	57	62	−1
XII	*A C E*	*E A C*	48	33	1	3	63	7	54	41	61	58	64	0
XIII	*A C D*	*E F A*	26	33	2	12	39	20	44	56	65	66	67	1
XIV	*A C E*	*C E A*	48	26	12	4	53	17	60	29	68	65	69	−1
XV	*A C E*	*E C A*	48	33	2	4	63	8	50	36	68	66	70	0
	Pairing off in original case		All pairs equal or opposite				Two pairs equal or opposite. The third one different						All pairs different	

The first columns in this table show how in the original Cases I through XV the variables λ_a, λ_b, λ_k, λ_j, λ_c and λ_d are paired off. A is paired off with A (by $A = A$) or with B (by $A = -B$); C is paired off with C or D ($C = C$ or $C = -D$); E is paired off with E or F ($E = E$ or $E = -F$). The next 11 columns show how each original case by (B-6) is split up into 11 cases for the purpose of calculating (B-9) and (B-8). The numbers 1 through 70 refer to the possible cases. The meaning of these cases is listed in Table 2 together with the value of C_{abkjcd} for each case. The last column in this Table 1 copies from Table 2 the value of C_{abkjcd} for the subcase with all pairs different. This C may be taken to characterize the original case as a whole. Then, *minus* the sum $\left(\sum_1 - \sum_2 + \sum_3 - \sum_4\right)$ occurring in eq. (B-30) turns out to be equal to a sum over all 15 original cases each multiplied by the factor C from this last column. This same $\left(-\sum_1 + \sum_2 - \sum_3 + \sum_4\right)$ also is equal to a sum over the 70 different subcases each multiplied by its individual factor C from Table 2.

TABLE B.2. THE 70 CASES AND THEIR VALUES OF C_{abkjcd} (= C)

Case No.	a	b	k	j	c	d	$C =$
1	1	2	1	1	1	2	0
2	1	2	1	1	2	1	0
3	1	2	2	2	1	2	0
4	1	2	2	2	2	1	0
5	1	2	1	1	3	4	−1
6	1	2	2	2	3	4	−1
7	1	2	3	3	1	2	0
8	1	2	3	3	2	1	0
9	3	4	1	1	1	2	1
10	3	4	1	1	2	1	1
11	3	4	5	5	7	8	0
12	1	2	1	2	1	1	2
13	1	2	2	1	2	2	2
14	1	2	1	3	1	4	2
15	1	2	2	3	2	4	2
16	1	2	3	1	3	2	0
17	1	2	3	2	3	1	0
18	3	4	1	2	1	1	2
19	3	4	5	7	5	8	1
20	1	2	1	3	4	1	2
21	1	2	2	3	4	2	2
22	1	2	3	1	2	3	0
23	1	2	3	2	1	3	0
24	3	4	5	7	8	5	1
25	1	1	2	1	1	2	−2
26	1	1	2	1	2	1	−2
27	1	1	2	1	3	4	−2
28	1	3	2	3	1	2	−2
29	1	3	2	3	2	1	−2
30	3	1	4	1	1	2	0
31	3	1	4	1	2	1	0
32	3	5	4	5	7	8	−1
33	1	1	2	2	1	1	4
34	1	1	2	3	1	4	2
35	1	3	2	1	3	2	0
36	1	3	2	2	3	1	1
37	3	1	4	2	1	1	2
38	3	5	4	7	5	8	1
39	1	1	2	3	4	1	2
40	1	3	2	1	2	3	0
41	1	3	2	2	1	3	1
42	3	5	4	7	8	5	1
43	1	3	4	1	1	2	0
44	1	3	4	1	2	1	0
45	3	1	2	3	1	2	−2
46	3	1	2	3	2	1	−2
47	3	5	6	3	7	8	−1
48	1	1	1	1	1	1	−4
49	1	1	3	1	1	3	−2
50	1	3	1	1	3	1	−1
51	3	1	1	3	1	1	−2
52	3	5	7	3	5	7	−1
53	1	1	3	1	3	1	−2
54	1	3	1	1	1	3	−1
55	3	5	7	3	7	5	−1
56	1	3	4	2	1	1	2
57	3	1	2	1	3	2	0
58	3	1	2	2	3	1	1
59	3	5	6	7	3	8	1
60	1	3	1	3	1	1	−2
61	3	1	1	1	3	1	−1
62	3	5	7	5	3	7	−1
63	1	1	3	3	1	1	0
64	3	5	7	7	3	5	0
65	3	1	2	1	2	3	0
66	3	1	2	2	1	3	1
67	3	5	6	7	8	3	1
68	3	1	1	1	1	3	−1
69	3	5	7	5	7	3	−1
70	3	5	7	7	5	3	0

For each of the 70 different cases, this table shows in the last column the value of C_{abkjcd} as calculated by equation (B-29) with (B-24). Here, A_{Fk}, A_{Fc}, and A_{Fd} are equal to 1 if k, c, or d is in the same set with j. The value of ε_μ is 1 if in the column headed by μ an odd number appears for the case considered, while $\varepsilon_\mu = -1$ if an even number appears. Numbers $\geqslant 3$ indicate that μ belongs to an independent pair ($N_f = 2$, $M_f = 1$). Numbers $\leqslant 2$ indicate $N_f = 4$ or 6 ($M_f = 2$ or 3); the number of numbers $\leqslant 2$ on the line for the case considered is N_f. Equal numbers in two columns μ and μ' indicate that $\lambda_\mu = \lambda_{\mu'}$. When the one number is 1 and the other is 2, then $\lambda_\mu = -\lambda_{\mu'}$. Similarly, $\lambda_\mu = -\lambda_{\mu'}$, if columns μ and μ' carry 3 and 4, or 5 and 6, or 7 and 8. For example, case 38 has $\lambda_a = -\lambda_k \mid\neq\mid \lambda_b = +\lambda_c \mid\neq\mid \lambda_j = -\lambda_d \mid\neq\mid \lambda_a$, while case 1 has $\lambda_a = -\lambda_b = \lambda_k = \lambda_j = \lambda_c = -\lambda_d$ with $N_F = 6$, $M_F = 3$, and case 5 has $\lambda_a = -\lambda_b = \lambda_k = \lambda_j \mid\neq\mid \lambda_c = -\lambda_d$ with $N_F = 4$, $M_F = 2$, while case 7 has $\lambda_a = -\lambda_b = \lambda_c = -\lambda_d \mid\neq\mid \lambda_k = \lambda_j$ with $N_F = 2$, $M_F = 1$ (as in the latter case j is contained in the independent pair of k and j).

The result thus obtained is

$$\langle Q\rangle = +\frac{1}{2}\left(\sum_1 - \sum_2 + \sum_3 - \sum_4\right)\int \frac{d^3\mathbf{x}}{\mathcal{V}}\, e^{-i\mathbf{k}_j\cdot\mathbf{x}}\mathbf{e}_j^* f_{ab}^* \cdot \{Q_{\text{op}} e^{i\mathbf{k}_k\cdot\mathbf{x}}\mathbf{e}_k\} f_{cd}, \tag{B-30}$$

where $\sum_1$, $\sum_2$, $\sum_3$, and $\sum_4$ all are sums over $a, \boldsymbol{b}, k, j, c, d$ subjected to certain restrictions corresponding to various of the original cases of pairing of the λ_μ.

$\sum_1$ corresponds to the *four* original cases, IX, VIII, XIV, and XI, viz.,

$$\left.\begin{array}{llllll} a=j, & b=d, & c=k; & a=j, & b=c, & d=k; \\ b=j, & a=d, & c=k; & b=j, & a=c, & d=k. \end{array}\right\} \tag{B-31a}$$

$\sum_2$ corresponds to four other original case, viz. cases VI, V, XIII, and X; that is:

$$\left.\begin{array}{llllll} a=-k, & b=d, & c=-j; & a=-k, & b=c, & d=-j; \\ b=-k, & a=d, & c=-j; & b=-k, & a=c, & d=-j. \end{array}\right\} \tag{B-31b}$$

Here, $a=-k$ is shorthand for $\lambda_a = -\lambda_k$, with all the consequences (A-7), so, $\mathbf{k}_a = -\mathbf{k}_k$ $\eta_a = \eta_k$.

$\sum_3$ corresponds to the two original cases VII and IV, viz.

$$a=j, \quad b=-k, \quad c=-d; \qquad b=j, \quad a=-k, \quad c=-d, \tag{B-31c}$$

while $\sum_4$ corresponds to two other original cases, III and II, for which

$$a=-b, \quad d=k, \quad c=-j; \qquad a=-b, \quad c=k, \quad d=-j. \tag{B-31d}$$

The combination $\sum_1 - \sum_2 + \sum_3 - \sum_4$ is then found from Table B.1 to contain (after some cancellations) each of the 70 special cases just as many times as given by C_{abkjcd} in Table B.2.

As in Appendix A, we can now interpret this result by transforming to momentum space. We write $\mathbf{k}, \eta$ for the index j, and $\mathbf{k}, \eta'$ for the index k, as again the factors $\int e^{i(\mathbf{k}_k - \mathbf{k}_j)\cdot\mathbf{x}}\, d^3\mathbf{x}/\mathcal{V}$ make $\mathbf{k}_k = \mathbf{k}_j$ (possibly after performing a differentiation, as in the case $Q = \mathbf{L}$). The remaining summation index will be represented by $\mathbf{k}''$ and η''. Writing $f^{\eta\eta''}(\mathbf{k}, \mathbf{k}'')$ for f_{jl}, and F according to (B-3) for the symmetrized function, we obtain

$$\begin{aligned}\langle Q\rangle = \sum_{\mathbf{k}}\sum_{\eta}\sum_{\eta'}\sum_{\mathbf{k}''}\sum_{\eta''} \{ & \mathbf{e}_{\mathbf{k}}^{\eta *} F_{\eta\eta''}^{*}(\mathbf{k}, \mathbf{k}'')\boldsymbol{\cdot}[\tilde{Q}_{\text{op}}\mathbf{e}_{\mathbf{k}}^{\eta'} F_{\eta'\eta''}(\mathbf{k}, \mathbf{k}'')] \\ & - \mathbf{e}_{\mathbf{k}}^{\eta *} F_{\eta\eta''}(-\mathbf{k}, \mathbf{k}'')\boldsymbol{\cdot}[\tilde{Q}_{\text{op}}\mathbf{e}_{\mathbf{k}}^{\eta'} F_{\eta'\eta''}^{*}(-\mathbf{k}, \mathbf{k}'')] \\ & + \tfrac{1}{2}\,\mathbf{e}_{\mathbf{k}}^{\eta *} F_{\eta''\eta''}(-\mathbf{k}'', \mathbf{k}'')\boldsymbol{\cdot}[\tilde{Q}_{\text{op}}\mathbf{e}_{\mathbf{k}}^{\eta'} F_{\eta'\eta}^{*}(-\mathbf{k}, \mathbf{k})] \\ & - \tfrac{1}{2}\,\mathbf{e}_{\mathbf{k}}^{\eta *} F_{\eta''\eta''}^{*}(-\mathbf{k}'', \mathbf{k}'')\boldsymbol{\cdot}[\tilde{Q}_{\text{op}}\mathbf{e}_{\mathbf{k}}^{\eta'} F_{\eta'\eta}(\mathbf{k}, -\mathbf{k})]\}.\end{aligned} \tag{B-32}$$

In all cases, $\tilde{Q}_{\text{op}}$ operates here only upon the first of the two arguments $\{\mathbf{k}, \eta\}$, and not on a second one.

For some observables Q of interest to us, the last two terms $\sum_3 - \sum_4$ in (B-32) each are zero:

$$\sum_{\mathbf{k}}\sum_{\eta}\sum_{\eta'} \mathbf{e}_{\mathbf{k}}^{\eta *}\boldsymbol{\cdot}\tilde{Q}_{\text{op}}\mathbf{e}_{\mathbf{k}}^{\eta'} F_{\eta'\eta}(\mathbf{k}, -\mathbf{k}) = 0. \tag{B-33}$$

For $Q = \mathbf{p}$, with $Q_{\text{op}} = \hbar\mathbf{k}$, this sum, by $\mathbf{e}_{\mathbf{k}}^{\eta *} = -\mathbf{e}_{-\mathbf{k}}^{\eta}$, equals

$$\sum_{\mathbf{k}}\sum_{\eta}\sum_{\eta'}\sum_{n}(-e_{-\mathbf{k}}^{\eta})_n\hbar\mathbf{k}(e_{\mathbf{k}}^{\eta'})_n F_{\eta'\eta}(\mathbf{k}, -\mathbf{k}). \tag{B-34}$$

Changing the name of the dummy $\mathbf{k}$ into $-\mathbf{k}$, and interchanging the names η and η' of two other dummies, we find for this

$$\sum_{\mathbf{k}}\sum_{\eta}\sum_{\eta'}\sum_{n} -(e_{\mathbf{k}}^{\eta'})_n(e_{-\mathbf{k}}^{\eta})_n(-\hbar\mathbf{k})F_{\eta\eta'}(-\mathbf{k}, \mathbf{k}), \tag{B-35}$$

but, by the symmetry $F_{\eta\eta'}(\mathbf{k}, \mathbf{k}') = F_{\eta'\eta}(\mathbf{k}', \mathbf{k})$, (B-35) is opposite and equal to (B-34), and therefore must vanish. This proves (B-33) for $Q = \mathbf{p}$.

For $Q = \mathbf{L}$, with $\tilde{Q}_{\text{op}} = (\hbar/i)\mathbf{k}\times\partial/\partial\mathbf{k}$, we must remember that $\sum_{\mathbf{k}}$ stands for an integral $\int d^3\mathbf{k}$. Then, (B-33) becomes

$$(\hbar/i)\int d^3\mathbf{k}\sum_{\eta}\sum_{\eta'}\sum_{n}(-e_{-\mathbf{k}}^{\eta})_n\mathbf{k}\times(\partial/\partial\mathbf{k})(e_{\mathbf{k}}^{\eta'})_n F_{\eta'\eta}(\mathbf{k}, -\mathbf{k}_c), \tag{B-36}$$

where the subscript c indicates an argument $\mathbf{k}$ that is not to be differentiated by $\partial/\partial\mathbf{k}$. By integration by parts, (B-36) becomes

$$(\hbar/i)\int d^3\mathbf{k}\sum_{\eta}\sum_{\eta'}\sum_{n}(e_{\mathbf{k}}^{\eta'})_n\,\mathbf{k}\times(\partial/\partial\mathbf{k})(e_{-\mathbf{k}}^{\eta})_n F_{\eta'\eta}(\mathbf{k}_c, -\mathbf{k}).$$

Change $\mathbf{k}\to-\mathbf{k}$ and $\eta\leftrightarrow\eta'$ of the dummies gives

$$(\hbar/i)\int d^3\mathbf{k}\sum_{\eta}\sum_{\eta'}\sum_{n}(e_{-\mathbf{k}}^{\eta})_n\mathbf{k}\times(\partial/\partial\mathbf{k})(e_{\mathbf{k}}^{\eta'})_n F_{\eta\eta'}(-\mathbf{k}_c, \mathbf{k}). \tag{B-37}$$

By the symmetry of $F_{\eta\eta'}(\mathbf{k}, \mathbf{k}')$, (B-37) is opposite and equal to (B-36), so, it must vanish. This proves (B-33) for $Q = \mathbf{L}$.

For $Q = \mathbf{S}$, we use

$$\sum_{n}(-e_{-\mathbf{k}}^{\eta})_n\mathbf{S}_{\text{op}}(e_{\mathbf{k}}^{\eta'})_n F_{\eta'\eta}(\mathbf{k}, -\mathbf{k}) = (\hbar/i)(-\mathbf{e}_{-\mathbf{k}}^{\eta}\times e_{\mathbf{k}}^{\eta'})F_{\eta'\eta}(\mathbf{k}, -\mathbf{k}). \tag{B-38}$$

Thence, by the same change of dummies as before,

$$\sum_{\mathbf{k}}\sum_{\eta}\sum_{\eta'}\sum_{n}(-e_{-\mathbf{k}}^{\eta})_n\mathbf{S}_{\text{op}}(e_{\mathbf{k}}^{\eta'})_n F_{\eta'\eta}(\mathbf{k}, -\mathbf{k})$$

$$= \quad -(\hbar/i)\sum_{\mathbf{k}}\sum_{\eta}\sum_{\eta'}\mathbf{e}_{-\mathbf{k}}^{\eta}\times\mathbf{e}_{\mathbf{k}}^{\eta'}F_{\eta'\eta}(\mathbf{k}, -\mathbf{k}) \tag{B-39}$$

becomes

$$-(\hbar/i)\sum_{\mathbf{k}}\sum_{\eta}\sum_{\eta'}\mathbf{e}_{\mathbf{k}}^{\eta'}\times\mathbf{e}_{-\mathbf{k}}^{\eta}F_{\eta\eta'}(-\mathbf{k}, \mathbf{k}). \tag{B-40}$$

By the symmetry of $F_{\eta\eta'}(\mathbf{k}, \mathbf{k}')$ and the antisymmetry of the cross product, (B-40) is opposite and equal to (B-39), and therefore must vanish. This proves (B-33) also for $Q = \mathbf{S}$.

Thus (B-32) simplifies to

$$\langle Q\rangle = \int d^3\mathbf{k}\int d^3\mathbf{k}''\sum_{\eta}\sum_{\eta'}\sum_{\eta''}\sum_{n}\{(e_{\mathbf{k}}^{\eta *})_n F_{\eta\eta''}^{*}(\mathbf{k}, \mathbf{k}'')\tilde{Q}(\mathbf{k})_{\text{op}}(e_{\mathbf{k}}^{\eta'})_n F_{\eta'\eta''}(\mathbf{k}, \mathbf{k}'')$$
$$-(e_{\mathbf{k}}^{\eta *})_n F_{\eta\eta''}(-\mathbf{k}, \mathbf{k}'')\tilde{Q}(\mathbf{k})_{\text{op}}(e_{\mathbf{k}}^{\eta'})_n F_{\eta'\eta''}^{*}(-\mathbf{k}, \mathbf{k}'')\}. \tag{B-41}$$

where we indicated the argument on which $\tilde{Q}_{\text{op}}$ operates. In the second term we change again the name of the dummy $\mathbf{k}$ into $-\mathbf{k}$, and use $\mathbf{e}^{\eta *}_{-\mathbf{k}} = -\mathbf{e}^{\eta}_{\mathbf{k}}$. We also introduce the dyadic field (tensor field)

$$F_{n'n''}(\mathbf{k}', \mathbf{k}'') \equiv \sum_{\eta'}\sum_{\eta''} (e^{\eta'}_{\mathbf{k}'})_{n'} (e^{\eta''}_{\mathbf{k}''})_{n''} F_{\eta'\eta''}(\mathbf{k}', \mathbf{k}''). \tag{B-42}$$

Then, writing $\sum_{\eta''}$ as $\sum_{\eta''}\sum_{\eta'''}\sum_{n''} (e^{\eta''*}_{\mathbf{k}''})_{n''} (e^{\eta'''}_{\mathbf{k}''})_{n''} = \sum_{\eta''}\sum_{\eta'''} \delta_{\eta''\eta'''}$, we obtain

$$\langle Q\rangle = \int d^3\mathbf{k} \int d^3\mathbf{k}'' \sum_{n}\sum_{n''} \{F^*_{nn''}(\mathbf{k},\mathbf{k}'')\tilde{Q}(\mathbf{k})_{\text{op}} F_{nn''}(\mathbf{k}, \mathbf{k}'') - F_{nn''}(\mathbf{k}, \mathbf{k}'')\tilde{Q}(-\mathbf{k})_{\text{op}} F^*_{nn''}(\mathbf{k}, \mathbf{k}'')\}. \tag{B-43}$$

When $f_{\eta'\eta''}(\mathbf{k}', \mathbf{k}'')$ is a product like (B-2), $F^{(\alpha)(\beta)}_{n'n''}(\mathbf{k}', \mathbf{k}'')$ by (B-3) and (A-26) becomes

$$F^{(\alpha)(\beta)}_{n'n''}(\mathbf{k}', \mathbf{k}'') = \sqrt{\tfrac{1}{2}}\,[f^{(\alpha)}_{n'}(\mathbf{k}')f^{(\beta)}_{n''}(\mathbf{k}'') + f^{(\beta)}_{n'}(\mathbf{k}')f^{(\alpha)}_{n''}(\mathbf{k}'')]. \tag{B-44}$$

When this is inserted in (B-43), we obtain by (A-27) and by the normalization condition (A-28)

$$\begin{aligned}\langle Q\rangle &= \langle Q\rangle^{(\alpha)}_{1\,\text{phot}} + \langle Q\rangle^{(\beta)}_{1\,\text{phot}} + \\ &+ \tfrac{1}{2}\Big\{\sum_{\mathbf{k}}\sum_{n} f^{(\alpha)}_{n}(\mathbf{k})^* \tilde{Q}(\mathbf{k})_{\text{op}} f^{(\beta)}_{n}(\mathbf{k}) \sum_{\mathbf{k}''} \mathbf{f}^{(\beta)}(\mathbf{k}'')^* \cdot \mathbf{f}^{(\alpha)}(\mathbf{k}'') + (\alpha \leftrightarrow \beta)\Big\} \\ &- \tfrac{1}{2}\Big\{\sum_{\mathbf{k}}\sum_{n} f^{(\alpha)}_{n}(\mathbf{k}) \tilde{Q}(-\mathbf{k})_{\text{op}} f^{(\beta)}_{n}(\mathbf{k})^* \sum_{\mathbf{k}''} \mathbf{f}^{(\beta)}(\mathbf{k}'') \cdot \mathbf{f}^{(\alpha)}(\mathbf{k}'') + (\alpha \leftrightarrow \beta)\Big\}.\end{aligned} \tag{B-45}$$

The last two terms vanish when $\mathbf{f}^{(\alpha)}$ and $\mathbf{f}^{(\beta)}$ are orthogonal solutions.

When the two photons are correlated, $f_{\eta'\eta''}(\mathbf{k}', \mathbf{k}'')$ is not a simple product like (B-2), but could be a superposition of such products in a pure state.

For obtaining some more similarity with electron wave mechanics, we may yet slightly reformulate our results, introducing the "two-photon sum operator"

$$\tilde{Q}^{\text{sum}}_{\text{op}} = \tilde{Q}(\mathbf{k}_1)_{\text{op}} + \tilde{Q}(\mathbf{k}_2)_{\text{op}}. \tag{B-46}$$

When $\tilde{Q}^{\text{sum}}_{\text{op}}$ operates on $F_{n_1n_2}(\mathbf{k}_1, \mathbf{k}_2)$, $\tilde{Q}(\mathbf{k}_1)_{\text{op}}$ operates on the first arguments n_1 and $\mathbf{k}_1$, and $\tilde{Q}(\mathbf{k}_2)_{\text{op}}$ on the second arguments n_2, $\mathbf{k}_2$. Or, when it operates on $\sum_{\eta_1,\eta_2} \mathbf{e}^{\eta_1}_{\mathbf{k}_1} F_{\eta_1\eta_2}(\mathbf{k}_1, \mathbf{k}_2)\, e^{\eta_2}_{\mathbf{k}_2}$, $\tilde{Q}(\mathbf{k}_i)_{\text{op}}$ operates on $\mathbf{e}^{\eta_i}_{\mathbf{k}_i}$ and on $\mathbf{k}_i$. Because of the symmetry of $F_{n_1n_2}(\mathbf{k}_1, \mathbf{k}_2)$, it makes no difference in (B-43) whether $\tilde{Q}_{\text{op}}$ operates on the first or on the second argument in F. Therefore, replacing $\tilde{Q}_{\text{op}}$ by $\tilde{Q}^{\text{sum}}_{\text{op}}$ would double $\langle Q\rangle$, and a factor $\frac{1}{2}$ is needed for undoing this:

$$\begin{aligned}\langle Q\rangle = \int d^3\mathbf{k} \int d^3\mathbf{k}' \sum_{n}\sum_{n'} \tfrac{1}{2}\{&F^*_{nn'}(\mathbf{k}, \mathbf{k}')\, \tilde{Q}(\mathbf{k}, \mathbf{k}')^{\text{sum}}_{\text{op}} F_{nn'}(\mathbf{k}, \mathbf{k}') \\ &- F_{nn'}(\mathbf{k}, \mathbf{k}')\, \tilde{Q}(-\mathbf{k}, -\mathbf{k}')^{\text{sum}}_{\text{op}} F^*_{nn'}(\mathbf{k}, \mathbf{k}')\}.\end{aligned} \tag{B-47}$$

Finally, as we did in Eqs. (36)–(38) in simplifying (34) to (39), we may for $Q = \mathbf{p}$ or $\mathbf{L}$ or $\mathbf{S}$ show that the two terms in (B-47) are equal to each other, so that $\iint \frac{1}{2}\{F^*\tilde{Q}F - F\tilde{Q}(-)F^*\}$ then simply becomes $\iint F^*\tilde{Q}F$.

APPENDIX C

THE PHOTON SPIN OBSERVABLE

We shall calculate here $\mathbf{S}(q_f)_{op}$. It is obtained by choosing $Q_{op} = \mathbf{S}_{op}$ in (A-5). We then use

$$\mathbf{e}^* \cdot Q_{op}\mathbf{e}' = \sum_{n=x,y,z} e_n^* \mathbf{S}_{op} e_n' = (\hbar/i)\mathbf{e}^* \times \mathbf{e}'. \tag{C-1}$$

Since

$$\sum_{\lambda}\sum_{\lambda'} \int \frac{d^3\mathbf{x}}{\mathcal{V}} e^{-i\mathbf{k}\cdot\mathbf{x}} \ldots e^{i\mathbf{k}'\cdot\mathbf{x}} = \sum_{\mathbf{k}}\sum_{\eta}\sum_{\eta'} \cdots \tag{C-2}$$

and

$$\sum_{\eta}\sum_{\eta'} \mathbf{e}_{\mathbf{k}}^{*\eta} \times \mathbf{e}_{\mathbf{k}}^{\eta'} = \sum_{\eta} i\eta \mathbf{e}_{\kappa}^{0} \equiv \sum_{\eta} i\eta \mathbf{k}/k, \tag{C-3}$$

we obtain

$$\begin{aligned} \mathbf{S}(q_f)_{op} &= -\frac{1}{2}\sum_{\mathbf{k}}\sum_{\eta}\left\{\frac{1}{2} e^{-i\theta}\left[\frac{\partial}{\partial r} - \frac{i}{r}\frac{\partial}{\partial \theta}\right](re^{i\theta}) \right. \\ &\qquad \left. + \frac{1}{2}(re^{i\theta})e^{-i\theta}\left[\frac{\partial}{\partial r} - \frac{i}{r}\frac{\partial}{\partial\theta}\right]\right\}\eta\hbar\mathbf{k}/k \\ &= -\frac{1}{2}\sum_{\mathbf{k}}\sum_{\eta}\{1 + r(\partial/\partial r) - i(\partial/\partial\theta\}\eta\hbar\mathbf{k}/k, \end{aligned} \tag{C-4}$$

where r and θ are the field variables introduced in eqs. (A-2).

The expression (C-4) is much simplified by combining terms belonging in $\sum_{\mathbf{k}}$ to opposite values of $\mathbf{k}$. By (A-7) they have the same r, but their θ are related by $\theta_{-\mathbf{k},\eta} = \pi - \theta_{\mathbf{k},\eta}$, so, $\partial/\partial\theta_{-\mathbf{k},\eta} = -\partial/\partial\theta_{\mathbf{k},\eta}$. We thus find

$$\begin{aligned} \mathbf{S}(q_f)_{op} &= -\tfrac{1}{2}\sum_{\mathbf{k}/2}\sum_{\eta}\eta\hbar(\mathbf{k}/k)\{(1 + r\partial/\partial r - i\partial/\partial\theta) - (1 + r\partial/\partial r + i\partial/\partial\theta)\} \\ &= i\hbar\sum_{\mathbf{k}/2}\sum_{\eta}\eta(\mathbf{k}/k)\partial/\partial\theta_{\mathbf{k},\eta}. \end{aligned} \tag{C-5}$$

We need not seek long for eigenfunctions of $\mathbf{S}(q_f)_{op}$. It suffices to consider the one-photon wavefunction (33) or (A-9) with precise values of $\mathbf{k}$ and η [only one term in $\sum_a$ in (A-9)]. Since $\psi_{vac}^{(F)}$ depends on the r_f only [and is an eigenfunction of (C-5) to the eigenvalue zero], we find

$$\mathbf{S}(q_f)_{op}\,\psi_{\mathbf{k},\eta}^{(F)}(q_f) = S(q_f)_{op}\,\psi_{vac}^{(F)} r_{\kappa,\eta} \exp(-i\theta_{\kappa,\eta}) = \eta\hbar(\mathbf{k}/k)\,\psi_{\kappa,\eta}^{(F)}(q_f), \tag{C-6}$$

so that $\psi_{\mathbf{k},\eta}^{(F)}(q_f)$ is a simultaneous eigenfunction of all components of $\mathbf{S}(q_f)_{op}$, to the eigenvalues = components of $\eta\hbar\mathbf{k}/k$, as would be expected classically on account of eq. (16).

Now the quantized field operator for the square of the total spin angular momentum in the electromagnetic field is

$$\mathbf{S}(q_f)^2_{\text{op}} = -\hbar^2\Big[\sum_{\mathbf{k}/2}\sum_{\eta}(\eta\mathbf{k}/k)\partial/\partial\theta_{\mathbf{k},\eta}\Big]^2. \tag{C-7}$$

For the field $\psi^{(F)}_{\mathbf{k},\eta}(q_f)$, this gives

$$\mathbf{S}(q_f)^2_{\text{op}}\,\psi^{(F)}_{\mathbf{k},\eta}(q_f) = [\eta\hbar\mathbf{k}/k]^2\,\psi^{(F)}_{\mathbf{k},\eta}(q_f) = \hbar^2\,\psi^{(F)}_{\mathbf{k},\eta}(q_f). \tag{C-8}$$

Thus the square of the spin angular momentum of the field is in this case $\hbar^2$, even though the eigenvalue of $\tilde{\mathbf{S}}(\mathbf{k})^2_{\text{op}}$ must be $2\hbar^2$.

Consider now also the two-photon field (B-1) for the special case that f_{ab} is nonvanishing only for specific values of a and b. For brevity, we assume here $a \neq b$, as (B-1) is not correct for $a = b$. [For two photons in the same $\mathbf{k}, \eta$ state, one should not use a wavefunction (II: H-36) of (II: H-35) derived from (II: H-34) with $n = 1$, but one should start, instead, from wavefunctions (II: H-34) containing the Hermite polynomial $H_2(q_f\sqrt{ck/\hbar})$.] If $a = -b$ we have

$$\psi^{(F)}_{\mathbf{k},\eta,-\mathbf{k},\eta}(q_f) = \psi^{(F)}_{\text{vac}}r^2_{\mathbf{k},\eta}\exp(-i\theta_{\mathbf{k},\eta}-i\pi+i\theta_{\mathbf{k},\eta}) \tag{C-9}$$

independent of $\theta_{k,\eta}$, and we find

$$\mathbf{S}(q_f)_{\text{op}}\,\psi^{(F)}_{\mathbf{k},\eta,-\mathbf{k},\eta}(q_f) = 0. \tag{C-10}$$

If $a \neq -b$ (if $\mathbf{k} \neq -\mathbf{k}'$ or $\eta \neq \eta'$), we have

$$\psi^{(F)}_{ab}(q_f) = \psi^{(F)}_{\text{vac}}r_a r_b\exp(-i\theta_a - i\theta_b) \tag{C-11}$$

and

$$\mathbf{S}(q_f)_{\text{op}}\,\psi^{(F)}_{ab} = \hbar[\eta_a\mathbf{k}_a/k_a+\eta_b\mathbf{k}_b/k_b]\psi^{(F)}_{ab}. \tag{C-12}$$

Again, the result is "classical", as $\psi^{(F)}_{ab}$ is a simultaneous eigenfunction of all components of $\mathbf{S}(q_f)_{\text{op}}$. The square of the spin angular momentum in this two-photon field therefore is given by

$$\mathbf{S}(q_f)^2_{\text{op}}\psi^{(F)}_{ab} = \hbar^2[\eta_a\mathbf{k}_a/k_a+\eta_b\mathbf{k}_b/k_b]^2\psi^{(F)}_{ab}, \tag{C-13}$$

and is not related to the spin quantum number s for the two-photon wavefunction $F_{n_1n_2}(\mathbf{k}_1, \mathbf{k}_2)$.

APPENDIX D

COMMUTATIVITY OF PHOTON SPIN COMPONENTS

The commutativity of S_x, S_y, and S_z for the quantized photon spin obtained from eq. (8c) by quantizing the transverse **A** and **E** fields by (23) is seen immediately from the representation (C-5) of $\mathbf{S}(q_f)_{\text{op}}$, as all differentiations $\partial/\partial\theta_{\mathbf{k},\eta}$ commute with each other.

We can also demonstrate this commutativity directly from (8c), using $\mathbf{E}^{\perp} = -4\pi c\,\mathbf{P}^{\perp}$, where $\mathbf{P}^{\perp}$, the transverse field canonically conjugated to $\mathbf{A}^{\perp}$, satisfies the commutation relations[37]

$$A_m^{\perp}(\mathbf{x})\,P_n^{\perp}(\mathbf{x}')-P_n^{\perp}(\mathbf{x}')\,A_m^{\perp}(\mathbf{x}) = i\hbar\,\delta_{mn}^{\perp}(\mathbf{x}-\mathbf{x}'). \tag{D-1}$$

Here, the transverse delta function $\delta_{mn}^{\perp}(\mathbf{x}-\mathbf{x}')$ is given by[38]

$$\delta_{mn}^{\perp}(\mathbf{r}) = \delta_{mn}\,\delta_3(\mathbf{r})+\nabla_m\nabla_n(1/4\pi r), \tag{D-2}$$

so that

$$\int d^3\mathbf{x}' \sum_n \delta_{mn}^{\perp}(\mathbf{x}-\mathbf{x}')\,F_n(\mathbf{x}') = F_m^{\perp}(\mathbf{x}) \tag{D-3}$$

with

$$\text{curl}\,\mathbf{F}^{\perp} = \text{curl}\,\mathbf{F}, \quad \text{div}\,\mathbf{F}^{\perp} = 0. \tag{D-4}$$

Now, with $\int = \int d^3\mathbf{x}$ and $\mathbf{A}' = \mathbf{A}(\mathbf{x}')$, from

$$\mathbf{S} = \int \mathbf{A}^{\perp}\times\mathbf{P}^{\perp}, \quad S_j = \sum_{m,n} \varepsilon_{jmn}\int A_m^{\perp}P_n^{\perp}, \tag{D-5}$$

we obtain

$$\begin{aligned} S_xS_y-S_yS_x = i\hbar \sum_{j,k,m,n} \varepsilon_{xjk}\varepsilon_{ymn} \int\int' \Big\{ & A_m^{\perp\prime}\Big[\delta_{nj}\,\delta(\mathbf{r})+\nabla_n'\nabla_j'\frac{1}{4\pi r}\Big]P_k^{\perp} \\ & -A_j^{\perp}\Big[\delta_{km}\,\delta(\mathbf{r})+\nabla_k\nabla_m\frac{1}{4\pi r}\Big]P_n^{\perp\prime}\Big\}, \end{aligned} \tag{D-6}$$

so, by $\sum_k\sum_m \varepsilon_{xjk}\varepsilon_{ymn}\delta_{km} = \delta_{xn}\,\delta_{yj}$, and integrating by parts,

$$\begin{aligned} S_xS_y-S_yS_x &= i\hbar\int(A_x^{\perp}P_y^{\perp}-A_y^{\perp}P_x^{\perp}) \\ &\quad + i\hbar\sum_{j,k,m,n}\varepsilon_{xjk}\varepsilon_{ymn}\Big\{\int P_k^{\perp}\int'\frac{\nabla_j'\nabla_n'A_m^{\perp\prime}}{4\pi r}-\int' P_n^{\perp}\int\frac{\nabla_m\nabla_k A_j^{\perp}}{4\pi r}\Big\} \\ &= i\hbar\Big\{S_z-\sum_{j,k}\int P_k^{\perp}\int'\frac{\varepsilon_{xjk}\nabla_j'\,\text{curl}_y'\,\mathbf{A}_{\perp}'}{4\pi r}+\sum_{m,n}\int P_n^{\perp}\int'\frac{\varepsilon_{ymn}\nabla_m'\,\text{curl}_x'\mathbf{A}_{\perp}'}{4\pi r}\Big\} \\ &= i\hbar\Big\{S_z-\int P_z^{\perp}\int'\frac{\nabla_y'\,\text{curl}_y'\mathbf{A}_{\perp}'}{4\pi r}+\int P_y^{\perp}\int'\frac{\nabla_z'\,\text{curl}_y'\mathbf{A}_{\perp}'}{4\pi r}+\int P_x^{\perp}\int'\frac{\nabla_z'\,\text{curl}_x'\mathbf{A}_{\perp}}{4\pi r} \\ &\quad -\int P_z^{\perp}\int'\frac{\nabla_x'\,\text{curl}_x'\mathbf{A}_{\perp}'}{4\pi r}\Big\} = i\hbar\Big\{S_z+\sum_n\int P_n^{\perp}\nabla_z\int'\frac{\text{curl}_n'\mathbf{A}_{\perp}'}{4\pi r}\Big\}. \end{aligned} \tag{D-7}$$

Let $\mathbf{A}^{\perp} = \text{curl}\,\mathbf{C}^{\perp}$. Then, $\mathbf{C}^{\perp} = \displaystyle\int'\frac{\text{curl}'\,\mathbf{A}_{\perp}'}{4\pi r}$, so

$$S_xS_y-S_yS_x = i\hbar\Big\{S_z+\sum_n\int P_n^{\perp}\nabla_z C_n^{\perp}\Big\}. \tag{D-8}$$

[37] See F. J. Belinfante (1946, eq. (47); 1961a).
[38] See F. J. Belinfante (1946, 1951b).

Since $\nabla_z C_x^\perp = \nabla_x C_z^\perp + \text{curl}_y\, \mathbf{C} = \nabla_x C_z^\perp + A_y^\perp$ and $\nabla_z C_y^\perp = \nabla_y C_z^\perp - A_x^\perp$, we obtain from (D-8) and (D-5)

$$S_x S_y - S_y S_x = i\hbar \left\{ S_z + \int \left(P_x^\perp A_y^\perp - P_y^\perp A_x^\perp + \sum_n P_n^\perp \nabla_n C_z^\perp \right) \right\}$$

$$= i\hbar \int \mathbf{P}^\perp \cdot \nabla C_z^\perp = -i\hbar \int C_z^\perp \,\text{div}\, P^\perp = 0. \qquad \text{(D-9)}$$

Thus the photon spin angular momentum, like the photon's linear momentum, can be treated classically, in contrast to the operators $\tilde{\mathbf{S}}_{\text{op}}$ in momentum space used for defining the spin quantum numbers of the photon wavefunction.

APPENDIX E

DERIVATION OF TWO-PHOTON POLARIZATION WAVEFUNCTIONS FROM CONSERVATION OF ANGULAR MOMENTUM AND OF RELATIVE PARITY

The following is a heuristic, quick, but sloppy derivation of the two-photon polarization wavefunctions of importance to us.

For the 0–1–0 case of the cascading atom, this reasoning is suggested by the start of Section 2-7, where we used our knowledge from conservation of angular momentum that the two photons should be in a $j = 0$ state, for concluding that the photon field should be invariant under rotations. Therefore, if ψ_{010} is to be bilinear in $\tilde{x}_1, \tilde{y}_1, \tilde{z}_1$ and $\tilde{x}_2, \tilde{y}_2, \tilde{z}_2$, we might conclude that it should be of the form (108). Thence, for $\hat{k}_1 = \hat{z}$ and $\hat{k}_2 = -\hat{z}$, we obtain (109) by omission of the impossible $\tilde{z}_1$ and $\tilde{z}_2$, asking no questions what happens to the postulated invariance of (108).

This reasoning leads quickly to the desired result, but it does not yield the dependence of ψ_{010} upon $\hat{k}_1$ and $\hat{k}_2$ [given in (110)]. Moreover, the argumentation is faulty, because in Chapter 5 we find for pairs of annihilation photons that they are in a state $j = 0$, too, while in that case the polarizations of the photons are perpendicularly correlated, like in the 1–1–0 case, for which the photon field has $j = 1$.

A similar quick but sloppy heuristic derivation may be given for our result (137) in the 1–1–0 case. Since the atom goes from $j = 1$ to $j = 0$, the missing angular momentum $j = 1$ must have gone to the photon field. So there must be three photon wavefunctions, transforming like a vector, and bilinear in $\tilde{x}_1, \tilde{y}_1, \tilde{z}_1$ and $\tilde{x}_2, \tilde{y}_2, \tilde{z}_2$. By inspection one sees that this triplet of wavefunctions is given by (130). Again, for $\hat{k}_1 = \hat{z}$ and $\hat{k}_2 = -\hat{z}$, only the third one of the three makes sense, and therefore is the one observed in the Holt experiment.

The question now arises, how is it possible that the same $\tilde{x}_1\tilde{y}_2 - \tilde{y}_1\tilde{x}_2$ is observed (with its perpendicular polarization correlation) for pairs of annihilation electrons, where one

starts from positronium in its singlet S-state ($j = 0$), so that the photon field has $j = 0$ ir this case?

The answer is that the wavefunction here is proportional to

$$\hat{k}_x(\tilde{y}_1\tilde{z}_2-\tilde{z}_1\tilde{y}_2)+\hat{k}_y(\tilde{z}_1\tilde{x}_2-\tilde{x}_1\tilde{z}_2)+\hat{k}_z(\tilde{x}_1\tilde{y}_2-\tilde{y}_1\tilde{x}_2) = \hat{\mathbf{k}}\cdot(\tilde{\mathbf{x}}_1\times\tilde{\mathbf{x}}_2), \qquad \text{(E-1}$$

which is a single scalar wavefunction and therefore has $j = 0$ indeed. Here, $\mathbf{k} = \mathbf{k}_1-\mathbf{k}_2$

The error we made in the 0–1–0 case was overlooking that also (E-1) is a possible $j =$ state for a two-photon field.

The difference between (E-1) and (108) is that (E-1) is linear in $\hat{\mathbf{k}}$, while (108) is not [an the more correct (110) is *even* in $\hat{\mathbf{k}}$]. This brings us to consider the parity of the photo wavefunction. But here a word of caution must be inserted.

"Absolute" parity and "relative" parity

In Section 2.5 we defined the parity of one-photon wavefunctions in terms of the eigen values of the inversion operator I_{op} given for vector wavefunctions by eq. (71). Correspon ingly, the parity of a two-photon tensor wavefunction $F_{n_1n_2}(\mathbf{k}_1, \mathbf{k}_2)$ may be defined by tl inversion operation

$$I_{\text{op}}F_{n_1n_2}(\mathbf{k}_1, \mathbf{k}_2) = +F_{n_1n_2}(-\mathbf{k}_1, -\mathbf{k}_2). \qquad \text{(E-}$$

Here we want to distinguish between "absolute" parity and "relative" parity.

For defining *absolute* parity of a wavefunction F, we use (E-2) literally as it stands; th is, I_{op} inverts in this case the arguments $\mathbf{k}_1$ and $\mathbf{k}_2$ of which $F_{n_1n_2}$ is a function, but it do not affect any parameters which may label this function.

In eqs. (113)–(114) with (110), and in eq. (132), we have found wavefunctions $F_{n_1n_2}(\mathbf{k}_1, \mathbf{k}$ that were labeled by parameters $\mathbf{k}_a$ and $\mathbf{k}_b$ which indicated the intermediate recoil mome tum $-\hbar\mathbf{k}_a$ and the final recoil momentum $-\hbar\mathbf{K}_c \equiv -\hbar(\mathbf{k}_a+\mathbf{k}_b)$ of the atom which emitt the two photons. In defining the *relative* parity of these photon fields (relative to the em ting atom), we assume that the inversion operator I_{op} will invert not only the vectors and $\mathbf{k}_2$, but also the vectors $\mathbf{k}_a$ and $\mathbf{k}_b$. Since a change of $\mathbf{k}_a$ and $\mathbf{k}_b$ means a change of t wavefunction considered, we may say that *absolute* parity compares the signs of one sin wavefunction in two "opposite" points $\mathbf{k}_1, \mathbf{k}_2$ and $-\mathbf{k}_1, -\mathbf{k}_2$ in momentum space, a that *relative* parity compares the signs of two corresponding but *different* wavefunctic in corresponding "opposite" points in momentum space.

"Conservation of parity" then tells us that the *relative* parity of the photon field emit is odd or even depending upon whether or not the emitting atomic system changes par It is then easily seen that, under simultaneous inversion of all four arguments $\mathbf{k}_1, \mathbf{k}_2, \mathbf{k}_a$, the wavefunctions $F_{n_1n_2}(\mathbf{k}_1, \mathbf{k}_2)$ given by (114) with (110) and given by (132) remain changed, so that these states have *even* relative parity. This is as it should be, because cascading atom twice changes l by ± 1, and thus flips parity twice, returning to its origi parity.

As explained in Chapter 5, in the annihilation of positronium there is a single chang parity. It follows that the two-annihilation-photons field should have *odd* relative par It is easily seen that (E-1), being odd in $(\mathbf{k}_1-\mathbf{k}_2)$, has odd relative as well as odd absol

parity. The full two-photon wavefunction is obtained from (E-1) as we obtained (132) from (131), by adding factors which tell us the complete dependence of the wavefunction upon the photon momenta. In the center-of-mass frame of the initial positronium, we must have $\mathbf{K} \equiv \mathbf{k}_1+\mathbf{k}_2 = 0$, and $|\mathbf{k}_1| = |\mathbf{k}_2| = mc/\hbar$, so, also $\mathbf{k} \equiv \frac{1}{2}(\mathbf{k}_1-\mathbf{k}_2)$ has magnitude $|\mathbf{k}| = k = mc/\hbar$. In a different inertial frame, let $\hbar\mathbf{K}_c$ be the initial positronium momentum. Then, with the use of the notation of eq. (229) for the factor (E-1) in the wavefunction, the more complete form of this wavefunction is

$$F_{n_1 n_2}(\mathbf{k}_1, \mathbf{k}_2) = C\delta_{\mathbf{K}, \mathbf{K}_c} \delta(k-k_d) \sum_{\mu=1}^{3} k_\mu \varepsilon_{\mu n_1 n_2}, \tag{E-3}$$

where $k_d \equiv mc/\hbar$. [The first δ here stands again for the three-dimensional delta function of $(\mathbf{K}-\mathbf{K}_c)$.] This has odd relative parity for all $\mathbf{K}_c$; it has odd absolute parity only in the center-of-mass frame of the positronium ($\mathbf{K}_c = 0$).

We see from the above that relative-parity conservation, together with angular-momentum conservation, determines in quantum theory which one of the three two-photon states (109), (137), or (E-1) [or, more elaborately, (114), (132), or (E-3)] results from each of the three different sources of a two-photon field here considered.

APPENDIX F

INDETERMINACY OF THE MOMENTA IN TWO-PHOTON STATES

In Section 2.7 we obtained in eqs. (132)–(133) the two-photon wavefunctions with $j = 1$ for the photon pair emitted by a cascading atom, as a function of the momenta $\mathbf{k}_a$ and $\mathbf{k}_b$ of these photons. Correspondingly, (114) and (116) were these wavefunctions for $j = 0$.

The two photons of a cascading atom are emitted (in good first approximation) independently, so that the directions $\hat{k}_a$ and $\hat{k}_b$ of their emission are uncorrelated and arbitrary. Only the magnitudes k_a and k_b of these photon momenta are given, as we know the colors emitted in the two transitions. If one would neglect the recoil of the emitting atom, one would perhaps tend to assume that the actual wavefunctions would be obtained by superposing the wavefunctions (132) or (133) or (114) or (116) for arbitrary directions of $\hat{k}_a$ and $\hat{k}_b$ with equal amplitudes for all directions, as done by integrating $\hat{k}_a$ and $\hat{k}_b$ over solid angles $d\Omega_a$ and $d\Omega_b$, or by multiplying by $\delta(k_a-k_a^o)\,\delta(k_b-k_b^o)/(4\pi k_a^o k_b^o)^2$ and integrating over $d^3\mathbf{k}_a d^3\mathbf{k}_b$. Here, $\hbar c k_a^o$ and $\hbar c k_b^o$ are the energy quanta emitted by the atom.

If we write $\Delta_{ab}^{+}(\mathbf{k}_1, \mathbf{k}_2)$ for the factor $\Delta_{ab}(\mathbf{k}_1, \mathbf{k}_2)$ in (114) and (116) given by (113), and $\Delta_{ab}^{-}(\mathbf{k}_1, \mathbf{k}_2)$ for the similar factor with a minus sign in the middle occurring for $j = 1$ in (132) and (133), this averaging of the wavefunction over directions of $\hat{k}_a$ and $\hat{k}_b$ would replace the $\Delta^{\pm}$ factors in these wavefunctions by

$$\overline{\Delta_{ab}^{\pm}}(\mathbf{k}_1, \mathbf{k}_2) = C'[\delta(k_1-k_a^o)\,(k_2-k_b^o) \pm \delta(k_2-k_a^o)\,\delta(k_1-k_b^o)], \tag{F-1}$$

where

$$C' = C/(4\pi k_a^o k_b^o)^2. \tag{F-2}$$

Thus we see that in the $j = 1$ case this would vanish in the rare case that k_a^o and k_b^o would happen to be equal. [See footnote 23, p. 274.] In the case of the cascading atom, however, it is *incorrect* to superpose the wavefunctions (114), (116), (132), or (133) with not only equal amplitudes, but also equal phases, as there is *no reason* for coherence of solutions with different directions of $\mathbf{k}_a$ and of $\mathbf{k}_b$. After emission of the first photon, there cannot exist coherent interference between states with different $\mathbf{k}_a$ because these photon states come multiplied by mutually orthogonal wavefunctions of the emitting atom with different recoil momenta $-\hbar\mathbf{k}_a$. After the second photon has been emitted, coherent interference is possible only between states in which the atomic total recoil momentum $-\hbar\mathbf{K}_c \equiv \equiv -\hbar(\mathbf{k}_a+\mathbf{k}_b)$ is the same.

Momenta of superposable two-photon states

For given $\mathbf{K}_c$ we find the allowable photon momenta $\hbar\mathbf{k}_a$ and $\hbar\mathbf{k}_b$ as follows. From the origin O of $\mathbf{k}$-space, draw the radius vector $\overrightarrow{OP_o} = \frac{1}{2}\mathbf{K}_c$. Around O, draw the spheres $\mathcal{S}_a$ ($k_a = k_a^o$) and $\mathcal{S}_b(k_b = k_b^o)$, and their mirror images $\mathcal{S}_a'$ and $\mathcal{S}_b'$ with respect to the point P_o. Let the circle $\mathcal{C}_a$ (in a plane $\perp \mathbf{K}_c$) be the intersection of $\mathcal{S}_a$ and $\mathcal{S}_b'$, and let $\mathcal{C}_b$ be the intersection of $\mathcal{S}_a'$ and $\mathcal{S}_b$. Let P_a be any point on $\mathcal{C}_a$, and let P_b on $\mathcal{C}_b$ be its mirror image with respect to P_o. Then, $\mathbf{k}_a = \overrightarrow{OP_a}$ and $\mathbf{k}_b = \overrightarrow{OP_b}$ satisfy not only $k_a = k_a^o$ and $k_b = k_b^o$, but also $\mathbf{k}_a+\mathbf{k}_b = \mathbf{K}_c$. Therefore only state vectors for which $\mathbf{k}_a$ and $\mathbf{k}_b$ are given by pairs of points P_a, P_b as just constructed can coherently interfere, and the final two-photon state resulting from a cascading atom will be a *mixture* of states in each of which the atom has a definite recoil momentum $-\hbar\mathbf{K}_c$, and in which the photon wavefunction is a superposition of states (114)–(117) or (132)–(135) with $\mathbf{k}_d$ [defined by $\mathbf{k}_d \equiv \frac{1}{2}(\mathbf{k}_a-\mathbf{k}_b)$] equal to one of the vectors $\overrightarrow{P_oP_a}$ taken from the cone of vectors that point from P_o toward the circle $\mathcal{C}_a$.

For each choice of $\mathbf{K}_c$ and $\mathbf{k}_d$, the factor $\Delta_{ab}^{\pm}(\mathbf{k}_1, \mathbf{k}_2)$ in the photon wavefunctions (114)–(117) or (132)–(135) may be written, in terms of

$$\mathbf{K} \equiv \mathbf{k}_1+\mathbf{k}_2 \quad \text{and} \quad \mathbf{k} \equiv \tfrac{1}{2}(\mathbf{k}_1-\mathbf{k}_2), \tag{F-3}$$

in the form

$$\Delta_{ab}^{\pm}(\mathbf{k}_1, \mathbf{k}_2) = C\delta_{\mathbf{K}, \mathbf{K}_c}[\delta_{\mathbf{k}, \mathbf{k}_d} \pm \delta_{\mathbf{k}, -\mathbf{k}_d}]. \tag{F-4}$$

Each pure component in the final mixed two-photon state may now be a wavefunction like (114) or (132) which contains as a factor, instead of (F-4), a superposition of expressions (F-4) all with the same (nonzero[39]) vector $\mathbf{K}_c$, and with different values of $\mathbf{k}_d$ lying around the cone reaching from P_o to $\mathcal{C}_a$. [If φ is the angle of arc measured around $\mathcal{C}_a$ and used for labeling its points P_a and thus the various allowed vectors $\mathbf{k}_d$, this superposition is an average calculated with some weight function $f(\varphi)$ that must be a periodic function of φ, and therefore may be expanded in a series $\sum c_m e^{im\varphi}$.]

[39] Note that $|k_a^o - k_b^o| < |\mathbf{K}_c| < k_a^o + k_b^o$.

Lack of absolute parity

We see immediately that, because of the three-dimensional delta function $\delta_{\mathbf{K},\mathbf{K}_c}$ occurring as a factor in these two-photon wavefunctions, these wavefunctions for $\mathbf{K}_c \neq 0$ have neither even, nor odd *absolute*[40] parity. If we start from some point $\mathbf{k}_1$, $\mathbf{k}_2$ for which the wavefunction is nonzero, inversion of $\mathbf{k}_1$ and $\mathbf{k}_2$ will lead to a point with opposite vector $\mathbf{K}$, and it is impossible for $\mathbf{K}$ and $-\mathbf{K}$ to be both equal to $\mathbf{K}_c$, so, the wavefunction is changed from nonzero to zero. (Remember that $\mathbf{K}_c$ had to have a definite value for each pure component in the mixed state, so, we cannot make superpositions of wavefunctions with $\mathbf{K}_c$ and with $-\mathbf{K}_c$.) This lack of either even or odd *absolute* parity, of course, does not contradict the fact discussed in Appendix E that these states have even *relative* parity.

Two-photon states with $j = 1$

The remark made by Akhiezer and Berestetskii that two-photon wavefunctions with $j = 1$ *would not exist*[41] holds only for wavefunctions of either even or odd *absolute* parity that are functions of $\mathbf{k} = \frac{1}{2}(\mathbf{k}_1 - \mathbf{k}_2)$ and of n_1 and n_2 as in eq. (E-3) with $\mathbf{K}_c = 0$, and that do not depend upon some nonvanishing $\mathbf{K}_c$, like a (possibly weighted) average of a function of $\mathbf{k}_d$ over an angle φ around a cone having $\mathbf{K}_c$ as its axis. Therefore, it does not apply to the photon wavefunctions appearing in the mixed two-photon states created by cascading atoms, and, in particular, it does not contradict the occurrence of a $j = 1$ two-photon field in the 1–1–0 case.

The situation is quite different in the case of the two-photon decay of positronium, where the two photons are emitted simultaneously, and where no atom is left behind to take up any recoil. This led in eq. (E-3) to $\mathbf{K}_c = 0$ in the center-of-mass frame of the positronium, and there is no mixture of states each with a different definite and nonzero $\mathbf{K}_c$. This allows the state here to have absolute parity.

APPENDIX G

UNDISPROVABLE *AD HOC* THEORIES OF THE SECOND KIND OF BELL AND OF KASDAY FOR COMPTON SCATTERING OF PAIRS OF ANNIHILATION PHOTONS

Bell [see Kasday (1970)] considers an *ad hoc* hidden-variables theory of the second kind in which the hidden-variable polarization angle λ of a photon somehow would predict that the first photon would be Compton-scattered into the solid angle $d\Omega_1$ in the direction

[40] See Appendix E.

[41] Akhiezer and Berestetskii (1965), sec. 7.3, p. 59, paragraph above their table 4. See also Yang (1950).

(θ_1, φ_1) with a probability[42]

$$P_1(\theta_1, \varphi_1, \lambda)\, d\Omega_1 = [F(\theta_1) - \sqrt{2}\, G(\theta_1) \cos 2(\varphi_1 - \lambda)]\, d\Omega_1, \tag{G-1a}$$

and that the second photon would scatter with probability

$$P_2(\theta_2, \varphi_2, \lambda)\, d\Omega_2 = [F(\theta_2) + \sqrt{2}\, G(\theta_2) \cos 2(\varphi_2 - \lambda)]\, d\Omega_2. \tag{G-1b}$$

If λ is uniformly distributed between 0 and 2π, the probability for simultaneous scattering within $d\Omega_1$ and $d\Omega_2$ is then given by

$$d\Omega_1\, d\Omega_2 \oint P_1(\theta_1, \varphi_1, \lambda)\, P_2(\theta_2, \varphi_2, \lambda)\, d\lambda/2\pi, \tag{G-2}$$

which is easily seen to be epual to Pryce and Ward's formula, eq. (232).

This model makes sense only if (G-1a) and (G-1b) both are positive for any values of θ_1, φ_1, θ_2, φ_2, and λ. That is, we need

$$F(\theta) > G(\theta)\sqrt{2}. \tag{G-3}$$

With F and G from (233a, b), it is easily verified that this is true for any value of θ.

The existence of the above model shows that the *experiments with Compton scattering of annihilation radiation, when giving results in agreement with quantum theory, do not conclusively exclude hidden-variables theories of the second king.*

However, the derivation of (232) uses eq. (231), which is valid for initial photons of energy mc^2 only. For softer photon pairs, the Compton scattering formulas give a result different from (232) for which a similar model would be impossible because it would lead to probabilities that are not always positive.

Kasday (1970) generalizes Bell's proof as follows (without need for assuming hard photons). Let $F(\hat{k}_1, \hat{k}_2)$ be the experimentally observed probability distribution for simultaneous scattering in the directions $\hat{k}_1$ and $\hat{k}_2$, and let the hidden variable for each photon be a unit vector $\hat{\lambda}$ which tells the photon to be scattered in the direction $\hat{k} = \hat{\lambda}$. Let $F(\hat{\lambda}_1, \hat{\lambda}_2)$ be the original probability distribution of the hidden variables. This *ad hoc* hidden-variables theory of the second kind would "explain" the observations.

The reason for possibility of such a theory is that we ask it to "explain" only observations of commutative observables $(\hat{k}_1, \hat{k}_2)$ for which all probability distributions can simultaneously be known and be ascribed to the hidden variables. In the Clauser-Holt experiments, where the observables measured were the ones discussed in Section 2.8, with eigenfunctions like $\tilde{x}' = \tilde{x} \cos \alpha' + \tilde{y} \sin \alpha'$ [see (143)], these observables had for different α' no simultaneous eigenfunctions, and thus do not commute. Therefore for them no simultaneous mixed-state probability distribution can exist that could be pirated as the hidden-variables distribution "explaining" measurements for all angles α' of the polarization filters [Kasday (1970)].

[42] This probability would be an integral over *other* hidden variables which would tell *exactly* how the photon would scatter.

REFERENCES

ABRAMOWITZ, M., and I. A. STEGUN (1965) *Handbook of Mathematical Functions*, Dover Publ., New York.

AKHIEZER, A. I., and V. B. BERESTETSKII (1965) *Quantum Electrodynamics*, 2nd edition, translated by G. M. Volkoff, Interscience Publishers, New York.

BALLENTINE, L. E. (1970) *Rev. Mod. Phys.* **42,** 358.

BELINFANTE, F. J. (1939a) *Physica* **6,** 849.

BELINFANTE, F. J. (1939b) *Physica* **6,** 887.

BELINFANTE, F. J. (1940) *Physica* **7,** 449.

BELINFANTE, F. J. (1946) *Physica* **12,** 1.

BELINFANTE, F. J. (1949) *Phys. Rev.* **75,** 1633 (abstract V10) and **76,** 461 (abstract H6).

BELINFANTE, F. J. (1951a, b). *Phys. Rev.* **84,** 541, 546.

BELINFANTE, F. J. (1953) *Physica* **19,** 849.

BELINFANTE, F. J. (1954) *Phys. Rev.* **93,** 935 (abstract T5).

BELINFANTE, F. J. (1972) *Bulletin Am. Phys. Soc.* **17,** 42, paper BH7 of the San Francisco meeting, January 1972.

BELL, J. S. (1964) *Physics (NY)* **1,** 195.

BELL, J. S. (1966) *Rev. Mod. Phys.* **38,** 447.

BELL, J. S. (1970) Introduction to the hidden-variable question, Proceedings of the International School of Physics "Enrico Fermi", Course 49 "Foundations of Quantum Mechanics" 29th June–11th July 1970, at Varenna sul Lago di Como. Published by Academic Press (New York–London) in *Scuola Internazionale di Fisica "Enrico Fermi", Rendiconti* **49,** 171 (1971).

BERTOLINI, G., and M. BETTONI, and E. LAZZARINI (1955). *Il Nuovo Cimento* **2,** 661.

BLEULER, E., and H. L. BRADT (1948) *Phys. Rev.* **73,** 1398.

BOHM, D. (1951) *Quantum Theory*, Prentice-Hall, New York.

BOHM, D. (1952a) *Phys. Rev.* **85,** 166.

BOHM, D. (1952b) *Phys. Rev.* **85,** 180.

BOHM, D. (1953a) *Phys. Rev.* **89,** 458.

BOHM, D. (1953b) *Progress Theor. Phys.* **9,** 273.

BOHM, D., and Y. AHARONOV (1957) *Phys. Rev.* **108,** 1070.

BOHM, D., and Y. AHARONOV (1960) *Il Nuovo Cimento* **17,** 964.

BOHM, D., and J. BUB (1966a) *Rev. Mod. Phys.* **38,** 453.

BOHM, D., and J. BUB (1966b) *Rev. Mod. Phys.* **38,** 470.

BOHM, D., and J. P. VIGIER (1954) *Phys. Rev.* **96,** 208.

BOHR, N. (1935) *Phys. Rev.* **48,** 696.

BUB, J. (1969) *Int. J. Theor. Phys.* **2,** 101.

CLAUSER, J. F. (1969) *Bulletin Am. Phys. Soc.* **14,** 578, paper EK1 of the Washington meeting, April 1969.

CLAUSER, J. F., M. A. HORNE, A. SHIMONY, and R. A. HOLT (1969) *Phys. Rev. Letters* **23,** 880.

DARWIN, C. G. (1928) *Proc. Roy. Soc. London* A **120,** 621.

DE BROGLIE, L. (1926) *C. r. Acad. Sci. Paris* **183,** 447.

DE BROGLIE, L. (1927a) *C. r. Acad. Sci. Paris* **184,** 273.

DE BROGLIE, L. (1927b) *J. Physique Radium* (6) **8,** 225.

DE BROGLIE, L. (1951) *C. r. Acad. Sci. Paris* **233,** 641.

DE BROGLIE, L. (1952a) *C. r. Acad. Sci. Paris* **234,** 265; **235,** 1345, 1453.

DE BROGLIE, L. (1952b) *C. r. Acad. Sci. Paris* **235,** 557.

DE BROGLIE, L. (1953) *La Physique quantique, restera-t-elle indéterministe?* Suivi d'une contribution de M. Jean-Pierre Vigier, Gauthier-Villars, Paris.

DeWitt, B. S. (1970) *Physics Today* **23** (9), 30.
DeWitt, B. S., and R. N. Graham (1971) *Resource Letter on the Interpretation of Quantum Mechanics* (mimeographed).
Einstein, A. (1949) Reply to criticisms, in *Albert Einstein, Philosopher Scientist* (ed. P. A. Schilpp), Library of Living Philosophers, **7**, 665. (Republished by Harper, by Open Court, and probably other publishing companies.)
Einstein, A., B. Podolsky, and N. Rosen (1935) *Phys. Rev.* **47,** 777.
Everett, H. (1957) *Rev. Mod. Phys.* **29,** 454.
Feynman, R. P. (1949) *Phys. Rev.* **76,** 749.
Fock, V. (1929) *Z. Physik* **57,** 261.
Fock, V. (1932) *Phys. Rev.* **75,** 622; **76,** 852.
Freedman, S. J., and J. F. Clauser (1972) *Phys. Rev. Letters* **28,** 938.
Furry, W. H. (1936) *Phys. Rev.* **49,** 393, 476.
Furry, W. H., and J. R. Oppenheimer (1934) *Phys. Rev.* **45,** 245.
Gleason, A. M. (1957) *J. Math. Mech.* **6,** 885.
Graham, N. (1970) The Everett interpretation of quantum mechanics. Ph.D. thesis, University of North Carolina, Chapel Hill.
Gudder, S. P. (1968) *Rev. Mod. Phys.* **40,** 229.
Gudder, S. P. (1970) *J. Math. Phys.* **11,** 431.
Hanna, R. C. (1948) *Nature* **162,** 332.
Heitler, W. (1954) *The Quantum Theory of Radiation*, 3rd edn., Clarendon Press, Oxford.
Horne, M. A. (1970) Ph.D. thesis, Boston University.
Jackson, J. D. (1962) *Classical Electrodynamics*, John Wiley, New York.
Jauch, J. M. (1968) *Foundations of Quantum Theory*, Addison-Wesley, Reading, Mass.
Jauch, J. M., and C. Piron (1963) *Helv. phys. Acta* **36,** 827.
Jauch, J. M., and C. Piron (1968) *Rev. Mod. Phys.* **40,** 228.
Kasday, L. R. (1970) Experimental test of quantum predictions for widely separated photons, Proceedings of the International School of Physics "Enrico Fermi," Course 49 "Foundations of Quantum Mechanics" 29th June–11th July 1970 at Varenna sul Lago di Como. Published by Academic Press (New York–London) in *Scuola Internazionale di Fisica "Enrico Fermi", Rendiconti* **49,** 195 (1971).
Kasday, L. R. (1971) The polarization correlation of two photons emitted in positron annihilation, and the Einstein–Podolsky–Rosen argument, Ph.D. thesis, Columbia University, New York.
Kasday, L. R., J. Ullman, and C. S. Wu (1970) *Bulletin Am. Phys. Soc.* **15,** 586.
Kochen, S., and E. P. Specker (1967) *J. Math. Mech.* **17,** 59.
Kocher, C. A., and E. D. Commins (1967). *Phys. Rev. Letters* **18,** 575.
Kramers, H. A. (1957) *Quantum Mechanics* (translated by D. ter Haar from the 1937 original), North Holland Publ., Amsterdam.
Landau, L. D., and E. M. Lifshitz (1960) *Electrodynamics of Continuous Media*, Pergamon Press, Oxford, and Addison-Wesley Publ., Reading, Mass.
Langhoff, H. (1960) *Z. Physik* **160,** 186.
Madelung, E. (1926) *Z. Physik* **40,** 322.
Marion, J. B. (1965) *Classical Electromagnetic Radiation*, Academic Press, New York.
Merzbacher, E. (1970) *Quantum Mechanics*, 2nd edn., John Wiley, New York.
Messiah, A. (1961) *Quantum Mechanics*, vol. I (translated by G. M. Temmer), North Holland Publ., Amsterdam, and John Wiley, New York.
Misra, B. (1966) *Il Nuovo Cimento* **47,** 841.
Paley, R. E. A. C., and N. Wiener (1934) *Fourier Transforms in the Complex Domain*, Vol. 19 of the Am. Math. Soc. Colloquium Publ., chap. 9, Am. Math. Soc., New York.
Papaliolios, C. (1967) *Phys. Rev. Letters* **18,** 622.
Pauli, W. (1933) *Die allgemeine Prinzipien der Wellenmechanik*, chap. 2, in H. Geiger and Karl Scheel, *Handbuch der Physik*, 2nd edn., vol. **24/1,** J. Springer, Berlin.
Pauli, W. (1940) *Phys. Rev.* **58,** 716.
Pearle, P. (1967) *Am. J. Phys.* **35,** 742.
Pearle, P. (1968a) Reply to Dr. Bloch, *Am. J. Phys.* **36,** 463.
Pearle, P. (1968b) Reply to Dr. Sachs, *Am. J. Phys.* **36,** 464.
Piron, C. (1963) Thesis, Université de Lausanne.
Pryce, M. H. L., and J. C. Ward (1947) *Nature* **160,** 435.
Rosenfeld, L. (1940) *Sur le tenseur d'impulsion-énergie*, Académie royale de Belgique, Classe des Sciences, Mémoires, tome 18, fascicule 6 (no. 1536).

SCHWEBER, S. S. (1961) *An Introduction to Relativistic Quantum Field Theory*, Harper & Row, New York.
SCHWINGER, J. (1949) *Phys. Rev.* **75,** 651.
SHIMONY, A. (1963) *Am. J. Phys.* **31,** 755.
SHIMONY, A. (1970) Experimental test of local hidden-variable theories, Procedings of the International School of Physics "Enrico Fermi," Course 49 "Foundations of Quantum Mechanics", 29th June–11th July 1970 at Varenna sul Lago di Como. Published by Academic Press (New York–London) in *Scuola Internazionale di Fisica "Enrico Fermi", Rendiconti* **49,** 182 (1971).
SIEGEL, A. (1962) *Synthese* **14,** 171.
SIEGEL, A. (1966) *The Differential-Space Theory of Quantum Systems*, chap. 5 of N. Wiener *et al.* (1966).
SIEGEL, A., and N. WIENER (1956) *Phys. Rev.* **101,** 429.
SNYDER, H. S., S. PASTERNACK, and J. HORNBOSTEL (1948) *Phys. Rev.* **73,** 440.
SOLVAY, INSTITUT (1928) *Électrons et photons, rapports et discussions de l'Institut International Physique Solvay*, Gauthier-Villars, Paris.
TURNER, J. E. (1968) *J. Math. Phys.* **9,** 1411.
TUTSCH, J. H. (1968) *Rev. Mod. Phys.* **40,** 231.
TUTSCH, J. H. (1969) *Phys. Rev.* **183,** 1116.
TUTSCH, J. H. (1971a) *J. Math. Phys.* **12,** 1711.
TUTSCH, J. H. (1971b) Preprints on applications of the method of Tutsch (1971a).
VIGIER, J. P. (1952) *C. r. Acad. Sci. Paris* **235,** 1107.
VIGIER, J. P. (1953) See DE BROGLIE, L. (1953).
VLASOW, N. A., and B. S. DZEHELEPOV (1949) *Dokl. Akad. Nauk SSSR* **69** (6), 777.
VON NEUMANN, J. (1932) *Mathematische Grundlagen der Quantenmechanik*, J. Springer, Berlin.
VON NEUMANN, J. (1955) *The Mathematical Foundations of Quantum Mechanics*, Princeton Univ. Press, Princeton, NJ.
WHEELER, J. A. (1946) *Ann. NY Acad. Sci.* **48,** 219.
WHEELER, J. A. (1957) *Rev. Mod. Phys.* **29,** 463.
WICK, G. S. (1950) *Phys. Rev.* **80,** 268.
WIENER, N., *et al.* (1966) *Differential Space, Quantum Systems, and Prediction* (ed. B. Rankin), The MIT Press, Cambridge, Mass.
WIENER, N., and A. SIEGEL (1953) *Phys. Rev.* **91,** 1551.
WIENER, N., and A. SIEGEL (1955) *Il Nuovo Cimento*, Suppl. **2** (4), 982.
WIGHTMAN, A. S., and S. S. SCHWEBER (1955) *Phys. Rev.* **98,** 812.
WIGNER, E. P. (1959) *Group Theory* (translated by J. J. Griffin), Academic Press, New York.
WIGNER, E. P. (1970) *Am. J. Phys.* **38,** 1005.
WU, C. S., and J. SHAKNOV (1950) *Phys. Rev.* **77,** 136 (L).
YANG, C. N. (1950) *Phys. Rev.* **77,** 242.

AUTHOR INDEX

SUBJECT INDEX

OTHER TITLES IN THE SERIES IN NATURAL PHILOSOPHY